HELMINTHOLOGY

HELMINTHOLOGY

EDITORS
N. Chowdhury
I. Tada

Springer-Verlag

Narosa Publishing House

Editors
N. Chowdhury
Professor and Head
Department of Parasitology
College of Veterinary Science
Ludhiana, India

I. Tada
Chairman
Department of Parasitology
Faculty of Medicine, Kyushu University
Higashi-ku, Fukuoka, Japan

Softcover reprint of the hardcover 1st edition 1994

Cover Illustration: Induced neurocysticercosis in a rhesus monkey (*Macaca mulatta*)

ISBN-13: 978-3-642-78840-6 e-ISBN-13: 978-3-642-78838-3
DOI: 10.1007/978-3-642-78838-3

To

Professor Lord Soulsby

Foreword

Study of parasitology, like any other branch of biological science, has in recent years been increasingly revealing and rewarding with enrichment and embellishment by basic sciences, specially with application of spectacular advances in molecular biology and biotechnology. Such a fruitful fusion of more than one discipline has now come to characterise more than ever before our approach to the subject. This volume of helminthology with contributions from a galaxy of distinguished scientists in specific areas, bears an eloquent testimony to the gratifying yield that accrues from cross fertilisation of multiple disciplines and able support from basic sciences.

Matazoan parasites, helminths, living in more than one host in different stages of development, present an intricate spectrum of host-parasite relationship, which has evolved through diverse frame and flow of ecologic circumstances. The situation is further complicated, as elaborated here, by genotypic and phenotypic variations, which profoundly influence the dynamic interaction between hosts and parasites and determine consequently survival and propagation of the latter. Indeed, one of the important messages upheld in this book is that genetic endowment of parasites plays a pivotal role in their immunogenicity, pathogenicity, response to variable environmental composition, including drug response and also their transmission dynamics and epidemiology.

Evolution of parasitic helminths with concomitant physiologic and morphologic alterations have been traced from free-living stage to development of host-specificity of different degrees and parasitic speciation.

The idea sought to be established that biochemical variations in helminths reflect their adaptive modifications in response to environmental changes is of considerable interest and importance.

Eosinophilic mobilisation, circulatory or infiltrative, has long been recognised almost as a hall-mark of helminthic infections. The immunologic significance of such assemblage and its evaluation as a measure of protection and scavenging have opened up new vista which may lead to better understanding of host defence against helminthic infections.

It has been aptly emphasised that helminthic antigen is a complex mosaic of variable complexion and composition, changing character at different points of time and space. The immune responses evoked, antibody-mediated as well as cell-mediated, their modality and modulation, call for a closer examination to ascertain the scope of their manipulation for necessary augmentation or abrogation.

It is in the fitness of things that the problem of helminthic zoonosis and its impact on human and animal health have received their due share of consideration. It assumes special significance when presence of reservoir hosts complicates the epidemiologic framework of helminthic infections and makes the task of containment difficult. Separate attention has rightly gone to the zoonotic potential of helminths in domestic animals.

Geographical distinctiveness being well known in helminthic infections, it is only proper that the problems peculiar to the developing countries, specially Indian subcontinent have received selective concern.

The book aims to provide comprehensive coverage of the topics included, without going for an exhaustive discussion on any for obvious reasons. Because, such an endeavour may run the risk of being counterproductive to reach a wide range of clientage. Nevertheless, essentials of recent advances in different areas have been recorded with proper appraisal of emphasis and importance, they deserve. Judicious use of newly evolved tools has been recommended with a critical awareness of their limitations. This will undoubtedly ensure their meaningful utilisation to understand on a clearer basis hitherto unsolved problems and subtle nuances of host-parasite relationship in helminthic infections of man and animal.

A rich repository of valuable information, this volume will meet the needs of teachers, students, researchers, and public health workers of medical and veterinary sciences.

A.B. CHOWDHURY
Emeritus Professor and Ex-Director
School of Tropical Medicine
Calcutta, India

Preface

Outstanding educationists interested in the development of Veterinary Science place Helminthology, a major branch of Parasitology, among para-clinical subjects in teaching curricula. However, teaching in helminthology, by and large, is based on books dealing with the problems of specific geographic region(s). During the last three decades, helminthology itself has expanded greatly. Consequently, some interest among the students in the newly evolved areas has to be evoked at an early stage for creative research in future. The present volume is primarily an outcome of these thoughts which have also been cherished by late Emeritus Professor B.P. Pande, an eminent helminthologist of the Indian subcontinent.

Helminths are known since time immemorial. Human helminth infections e.g. filariasis, onchocercosis and schistosomiasis were known to people for centuries now. These diseases spreading over three continents, not only impede national development but also cause tremendous human suffering by inflicting physical deformities, mental disabilities, blindness and even death. In spite of a lot of thrust on research, the prospects of development of vaccines against these diseases are still not in sight! Significantly, in highly industrialised countries, a large segment of population also harbours *Ascaris, Trichuris* and hook worm, if not others. With regard to diseases in domestic stocks—both in tropical and temperate countries—fascioliasis, parasitic gastroenteritis, stephanofilariasis, echinococcosis, bovine and porcine cysticercosis including neurocysticercosis, cause incalculable economic losses every year; some even pose threat to human life. Surprisingly, to date no satisfactory treatment is available for hydatidosis and neurocysticercosis.

In many countries, research in applied sciences is exacting and needs broad based education in fundamental sciences. This has been rather inadequate as far as veterinary and medical helminthology are concerned. The main concept behind this book is to bring the experts in veterinary, medical and fundamental sciences together for a concerted effort to understand these diverse problems in their proper perspective to create an awareness about the true intricacies and complexities of several aspects of parasitism and to assist teachers, researchers, practitioners and also stock-owners (and perhaps planners!).

In a world of spectacular advancement in science and rapidly expanding scientific knowledge and knowhow, it is indeed difficult to cover every area of a particular field in detail. However, it has been our utmost endeavour to incorporate as much important information in this book as the constraints of space, time and framework could allow. The present volume contains fifteen reviews written by highly distinguished scientists. The topics have been selected so as to provide a thorough understanding of the subject starting from 'evolution' upto the latest knowledge on it, which is essentially a prerequisite to arm the young researchers who will ultimately hold the realm.

N. Chowdhury
I. Tada

Acknowledgements

We would like to record our deepest sense of gratitude to all the distinguished contributors of this volume, and especially to those who have produced their chapters at a very short notice.

A few chapters of the book have been reviewed by eminent scientists including Professors Derek Wakelin and Daniel R. Brooks (Chapter 1); Professor H.L. Shah (Chapter 3); Professor S.P. Ahuja (Chapter 7); Professor Graham F. Mitchell (Chapter 9 and 10); and Professor V. Kumar (Chapter 14). We are indebted to these reviewers.

We are equally obliged to Professors J.F.A. Sprent, W.P. Rogers, Graham F. Mitchell, Y. Nawa, J.D. Smyth, W. Mulligan, G.A. Schad, R.K. Pritchard, Charles E. Tanner, Paul H. DeRycke, C.A. Himonas, M. Nooruddin, V.S. Alwar, C.S. Hayat and G.S. Nelson among many others from abroad and India who have extended their support, suggestion(s) and criticism during different phases of this project and in difficult times. Special mention may, however, be made of Drs. S.K. Dey, S.S. Saini and R.K. Biswas, PAU, without whose sincere cooperation this project would not have been completed so smoothly.

The permission granted by the Surveyor General of India, Dehra Dun, for inclusion of Fig. 1 in Chapter 3 is thankfully acknowledged.

Finally, we wish to thank the publisher, Narosa Publishing House, and Springer-Verlag for co-publishing this volume.

N. Chowdhury
I. Tada

Contents

HELMINTHOLOGY

1. Evolution, Parasitism and Host Specificity in Helminths

N. Chowdhury, M.L. Sood* and Richard T. O'Grady**
Department of Parasitology, College of Veterinary Science, Ludhiana, India
*Department of Zoology, College of Basic Science and Humanities, Punjab Agricultural University, Ludhiana, India
**Department of Invertebrate Zoology, Smithsonian Institution, National Museum of Natural History, Washington DC, USA

INTRODUCTION

Evolution results in adaptation to the environment, and is a change in the genetic composition of the population. It results from an interaction between genealogical and ecological processes. This is the basis of unified theory, which is an expansion of Darwinian theory. Macroevolutionary predictions of the unified theory require the basis of a combination of systematics, developmental biology and ecology, and parasitologists should be in a strong position to offer empirical evolutions of these new ideas (see [18]). Parasites are good model systems for studying general evolutionary principles [19, 96].

Adaptation and functionality are important concepts in any discussion on the role of natural selection. Evolution of parasitism involves a series of adaptations, to increase chances of infection, modify virulence of the parasite and evade the resistance of the host to the parasite. The adaptation of host to its own environment also has an important bearing on the evolution of parasitism [104]. In the development of ideas concerning parasitic evolution, one important aspect concerns the reasons why and how protozoa and helminths became separated from bacteria and viruses and other infectious agents [142].

In many cases parasites have evolved in parallel with their hosts. Direct evolutionary correlation has been observed in various host groups and their parasites. A parasite will only be able to survive a drastic evolutionary change in a host by adapting in some way or the other to that change. Thus, the relationships between the parasites of different hosts often given valuable clues to the relationships of the hosts themselves. During the evolution of host-parasite relationship, many modifications, both morphological and physiological, are required. Many characters are lost and others are gained and among several factors, the type of the host parasitized, the site of predilection and the micro-environment in the host play important roles. Biochemical modifications play an even more important role in determining both the survival of the parasite within the host, as well as the survival and development of the free-living stages outside the body of the host.

The host-parasite relationship might be regarded as an ultimate stable and balanced state on which selection might continue to act. As the host evolves, its own isolation may lead to the isolation of some of its parasites. Every parasite community necessarily has a history of accumulation of species onto the host—the development of the community occurs over a long period during which interactions between host and parasites constantly mould the community [97].

Recent consideration of the evolution of host-parasite relationship is focussed on immunological phenomena. In the course of evolution, parasitic forms may have come to share a number of common antigens with the host, the degree of antigenic disparity between the host and the parasite determining the success or otherwise of the parasitic organism [121]. Thus, parasites may share common macromolecules. More recently, DNA probes are being used to unravel evolutionary ralationship among parasites. Macromolecular sequence comparisons generally provide the most reliable basis for determining phylogenetic relationships. Sequence data can be interpreted quantitatively and also provide a continually expanding database for subsequent reference [55].

HOST-PARASITE COEVOLUTION

Price [95] rightly remarked, "parasites are specialists in a host environment with close co-evolutionary ties". Coevolution embodies two components, togetherness (Co-) and history (evolution). Coevolution encompasses both microevolutionary (short time scale) and macroevolutionary (long time scale) associations. Host specificity (by the parasite) and 'defence' mechanisms (by the host) play important roles in microevolutionary studies. It is important to integrate microevolutionary and macroevolutionary patterns and processes in order to generate a more robust and unified evolutionary biology (see [17]). Earlier, Brooks [16] made a complete analysis of the topic under "Analysis of host-parasite coevolution". It has been observed that principles and methods of host-parasite coevolution apply equally well to any ecological association, and this requires expertise in phylogenetic analysis, ecology and developmental biology.

Roza [108] while discussing the theories of coevolution and sequential evolution, has suggested that selective pressure is reciprocal, but asymmetrical between host and parasite populations, while cladogenetic relationships are unilateral. Recently, May & Anderson [75] in their essay on "Parasite-host coevolution", have developed three themes—the evolution of virulence, the population dynamics and population genetics of host-parasite association and invasions or 'emergence' of new parasites. Consideration of coevolutionary trajectories of host-parasite associations involves trade-offs not only between transmission and virulence, but also among costs and benefits of host resistance. Many coevolutionary trajectories are possible, depending on the details of the parasite's life history.

Coevolutionary patterns in parasite-host associations indicate that both classes of organisms show a phylogenetic retention of evolved traits that is sufficient to allow reconstruction of their historical relationships. Phylogenetic systematic techniques help to determine the host and parasite lineages. The parallel portions of host and parasite phylogenies are 'congruent' and indicate historical association. Incongruent portions represent episodes in which a parasite species involved in one host species and then colonized in another (host switching). Even though parasite and host phylogenies are not completely congruent, they still remain consistent with each other. However, we need more precise method of assessing the "fit" between host and parasite phylogenies.

Examples of coevolutionary association have been observed in different groups of helminths. Among nematodes, coevolution has been hypothesized between *Enterobius* and their primate hosts [48, 86] and between ascaridoids and their crocodilian hosts [20]. Cameron [24] first suggested that coevolution in Oxyurida paralleled that of the primate hosts, but at a slower rate so that a given parasite species corresponds to a genus of host. The finding has been confirmed subsequently by a number of workers (see [3]). However, lower level incongruence occurs, such as that of the phylogenies of the great apes and their parasites. Beveridge [13] investigated the coevolutionary relationships and origin of the helminths of Australian marsupials and found that the parasite fauna had a number of origins. In many cases, the evolution has been parallel to that of the hosts at the generic or family level, but there were also major shifts of helminths to distantly related groups. A similar pattern is evident at the species level. At all three levels, therefore, coevolution between host and parasite is mixed with much apparently random host-switching. Interestingly, a few nematode and cestode groups and representative trematode spp. were probably present in original colonizing marsupials. Other helminths have been acquired from monotremes, reptiles and birds present on the continent. Additional acquisitions believed to have been occurred from southeast Asian rodents and bats.

Davis [38] studied coevolution of *Schistosoma* spp. and the snails and concluded that *Schistosoma* and the relevant snail faunas evolved in Gondwanaland prior to the break up of the Pangaea; their coevolution involved reciprocal selective pressures affecting the course of evolution of both parasite and snail.

EVOLUTION AMONG HELMINTH GROUPS

Parasites have evolved from free-living ancestors. The isolation of hosts in the evolutionary process as mentioned earlier, influences the isolation of parasites. This would ultimately lead to the speciation of the parasite, and the development of phylogenetic specificity.

Digenea

Among trematodes, the evolution of digeneans is more complex than that of monogeneans because of the presence of an invertebrate host- a mollusc in the former group. It seems quite obvious that the digenetic trematodes were originally parasites of molluscs in which they became sexually mature, subsequently deferring maturity until they left the mollusc [25]. A study of the germ cycles in the digenea supports this theory. Most workers agree that the digenea and the Aspidogastrea (= Aspidobothrea) are sister groups (Figs. 1 and 2; see also [18]). Thus, the origin of digenea would have involved: (i) the

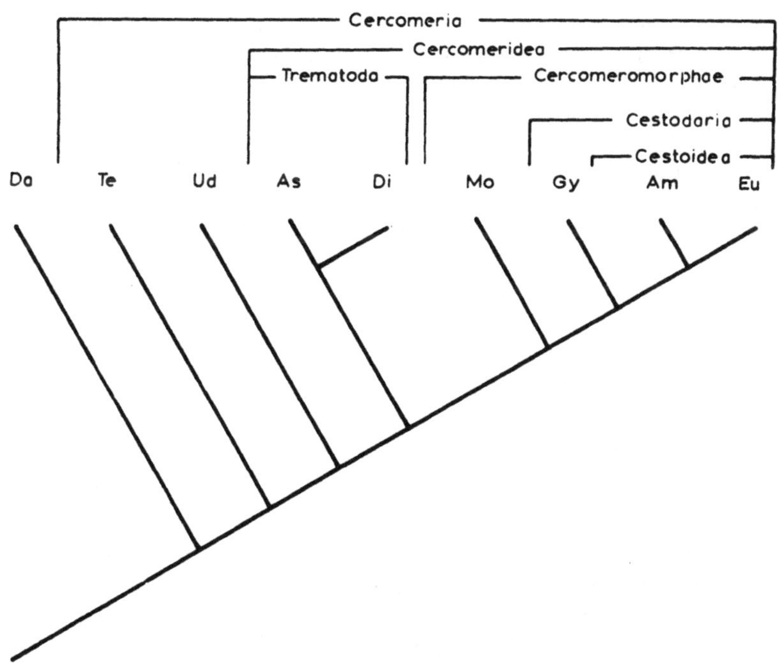

Fig. 1. A cladogram of the phylogenetic relationships among the parasitic platyhelminths, as hypothesized by Brooks, O'Grady and Glen (1985). Dalyelloidea (free-living outgroup) (DA); Temnocephalidea (Te); Udonellidea (Ud); Aspidiogastrea (As) (also referred to Aspidocotylea and Aspidobothrea); Digenea (Di); Monogenea (Mo); Gyrocotylidea (Gy); Amphilinidea (Am); Cestoda (Eu) (also referred to as Eucestoda).

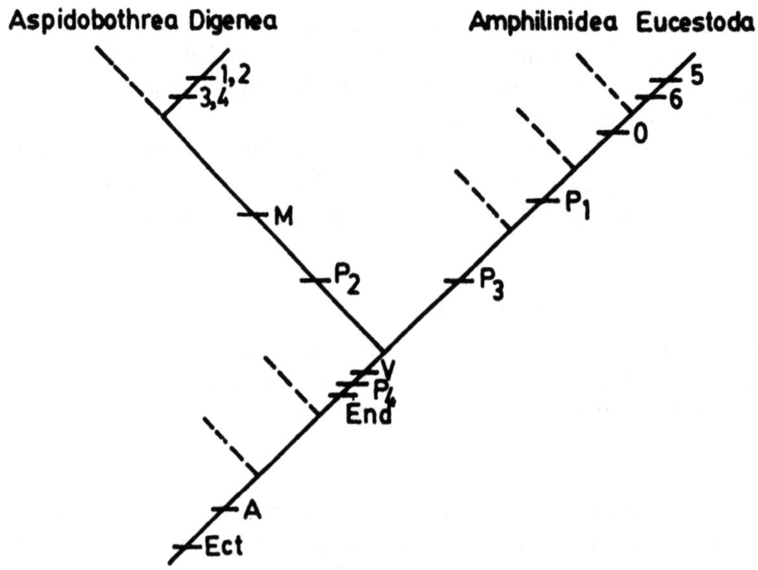

Fig. 2. Summary of phylogenetic correlates of ontogenic and ecological diversification of life cycle patterns of cercomerian platyhelminths in vertebrate and invertebrate hosts. Digeaneans: 1 = miracidium, sporocyst, redia, cercaria; 2 = asexual proliferation of radiae, cercariae; 3 = accelerated development signified by sexual maturation in second intermediate host (transversotrematids, schistosomatoids) or in molluscan host (e.g. some gymnophallids and microphallids); 4 = paedomorphic development of gut in miracidia; same in rediae resulting in "daughter sporocysts"; 5 = hexacanth larva; 6 = decelerated development signified by absence of cerebral development in hexacanth larva; loss of apical sucker in many lineages. P = placoderms; O = ostracoderms (to indicate the ancestors of all nonchondrichthyan gnathostomous vertebrates); M = molluscan host acquired in exchange for arthropod host (derived 2-host life cycle); V = vertebrate host acquired (primitive 2-host endoparasitic life cycle); End = endoparasitic adult; Ect = ectoparasitic adult; A = arthropod host acquired (primitive 1-host ectoparasitic life cycle). Note association between vertebrates and cercomerians apparently early in vertebrate evolution, with extensive radiation in placoderms, due either to host-switching or sympatric speciation, and subsequently close phylogenetic association between evolutionary divergence of chondrichthyans and ostracoderms and divergence of aspidobothreans and digeneans, of gyrocotylideans and cestoideans. Dotted lines indicate other taxa (Abstracted from Brooks, 1989).

addition of obligate development in a vertebrate host (most aspidogastrids have obligate development in only a mollusc, but some species are facultative parasites of vertebrates) and (ii) the evolution of the miracidium, sporocyst, redia and cercaria.

Rhode [99] suggested that pretrematodes were similar to the present day Aspidogastrea that have many archaic features, such as absence of alteration of generations and hosts. It has been postulated, that the protodigenean was a rhabdocoele ecotoparasite on snails. The digeneans and oncophoreans (monogeneans and cestodes) had separate origins from these free-living rhabdocoele-like ancestors.

In a study of phylogeny of digenean life cycle patterns, the most primitive digenean life cycle is a two-host system involving a snail and a fish that ingest the cercaria [92]. From the two-host cycle, the three-host and four-host systems evolved. Recently, Brooks [18] in an extensive discussion on major groups of cercomerian life cycles, pointed out that there is an evidence of a shift in intermediate host type for the trematodes from arthropods to molluscs, coupled with host-switching or sympatric speciation which supports theories that vertebrate hosts were added to the life cycles of trematodes by ingestion of infected arthropods. With regard to phylogenetic distribution of definitive host, there is an indication of association between vertebrates and cercomerians which began shortly after the first vertebrates evolved, early in the evolution of placoderms (Fig. 2). Brooks [18] believed that evolutionary diversification in trematodes (and eucestodes), has been highly conservative and phylogenetically coherent due to ecologically adaptive radiation most likely triggered by "developmental revolutions" rather than "ecological invasion". There is now a general agreement that, phylogenetically the mollusc was the first host, and the vertebrate acquired later. However, there is disagreement about the order in which the second and third hosts (incorporated as a means of increasing efficiency of transfer between primary mollusc and secondary vertebrate) were incorporated (see discussion Llewellyn [69, 70]).

Stunkard [129] believed that schistosomes first parasitised reptilian ancestors sometime in the early Mesozoic and possibly even in Paleozoic period since birds and mammals originated from an entirely different Mesozoic reptilian groups. Recently, Combes et al [36] have discussed the origin and evolution of Schistosomatidae in more detail. Among trematodes, schistosomes are considered to be a very old group which exceptionally selected the circulatory system as a habitat in warm-blooded vertebrates. As a result of host-switching, flukes in general and schistosomes in particular have been exposed to a wide range of "niches" having diverse physico-chemical properties and wide range of nutritional levels (see specificity).

Cestodes

The origin of tapeworms is rather speculative. However, a cestode is not degenerate. Probably, the protocestode was an early turbellarian. This turbellarian-like-ancestor was probably a freshwater form which became parasitic and matured in crustacea [25]. Fundamental similarities between cestodes and free-living platyhelminths suggest that cestodes evolved from some free-living or parasitic platyhelminths, rather than directly from free-living or parasitic protozoa. Most theories assume that the original parasitic precestode was the adult worm (see [44]).The protocestode developed originally in crustacea and later in mites and insects. Two opposing theories have been advanced about the evolution of cestode life cycles: (i) that the ancestral cestodes acquired invertebrete hosts first and vertebrate hosts later, (ii) it was vertebrate hosts that were acquired before invertebrate hosts (see references and discussion Llewellyn [69]; [70]. It is now evident that original single-host life cycle became two-host cycle by the ingestion of eggs by invertebrates and the third host was incorporated between the secondary (first intermediate) host and the primary vertebrete host.

According to Freeman [44] (see [71]), the precestode stage had an adult that was single-suckered and free-living; the onchosphere came from an anoperculate egg and developed parenterally in an invertebrate; the plerocercoid-preadult was tailed and free-swimming. The protocestode stage had an adult that was single-suckered and in the vertebrate gut; the onchosphere came from an anoperculate egg and the plerocercoid metamorphosed (?) parenterally in an invertebrate (Fig. 3). The most commonly accepted theory relating to cestode evolution suggests that the precestode ancestor developed to the adult in invertebrates, conceivably as *Archigetes* does today [44].

Cladistic analysis of adult and larval characters drawn from both gross morphologic and ultrastructural observations, places the cestodes as a highly derived group within the Cercomeromorpae, with amphilinids and then gyrocotylids as the sister groups [21, 42] (Fig. 1).

Freeman [44] in his excellent essay on cestode evolution emphasized that, in addition to other facts, there is a need for data on the origin, development and metabolism of cestodes in the vertebrate gut. Spasski [125] suggested that the evolutionary sequence of Anoplocephala-like tapeworms among terrestrial vertebrates is somewhat different from that accepted by most other authors. Anoplocephalids are known as most primitive among all cyclophyllidean cestodes and are confined to perissodactyles and paenungulates. The genus *Moniezia* is more advanced than other Anoplocephalan tapeworms having a reticulate uterus.

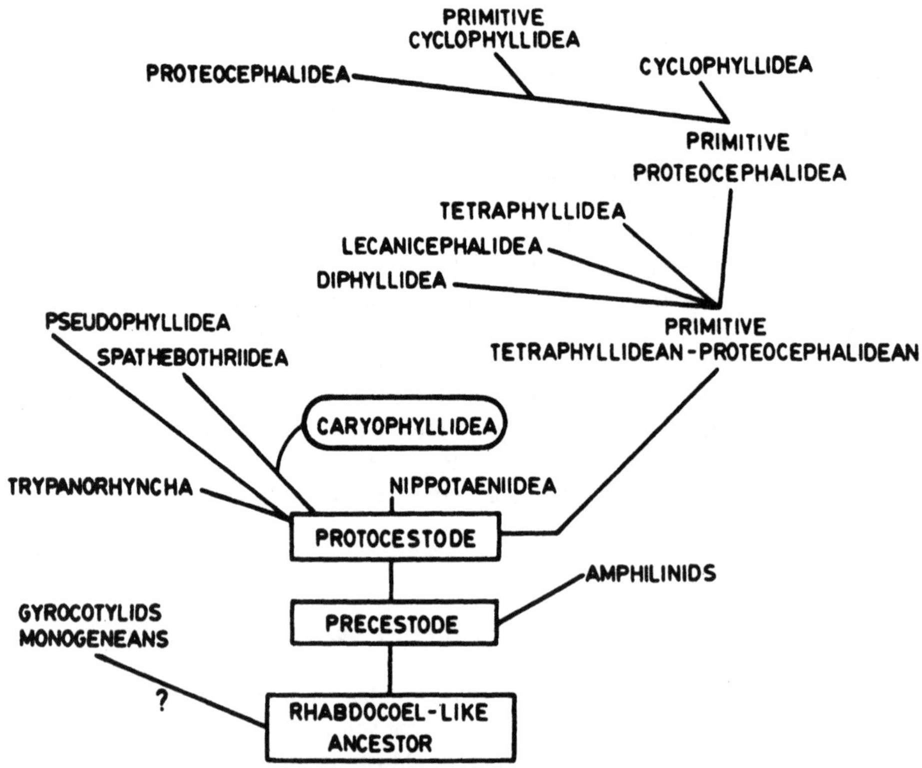

Fig. 3. Evolutionary scheme of cestodes according to Freeman (1973). The precestode stage had an adult that was single-suckered and free-living; the onchosphere came from an anoperculate egg and developed parenterally in an invertebrate host; the plerocercoid pre-adult was tailed and free-swimming. The protocestode stage had an adult that was single-suckered and in the vertebrate gut; the onchosphere came from an operculate egg; and the plerocercoid had a metamorphosis(?) parenterally in an invertebrate. Nothing should be implied from the angle or length of connecting lines. (After Mackiewicz, 1981; adapted and simplified from Freeman, 1973).

Mackiewicz [71] in his excellent treatment on Caryophyllids ragarded this group "as a dead end branch of a polyzoic line" (Fig. 3) though he believed that different cestode groups perhaps radiated from them due to "cryptic segmentation" discernible by the distribution of calcareous bodies. Earlier, we have reported that calcareous corpuscles in cestodes are not inert structures but subserve several vital functions [30, 33]. Extensive studies carried out later from different vertebrate hosts, revealed that in *M. expansa* a few corpuscles occurred only in the scolex and are absent in the rest of the body, a feature reported in *Hunterella nodulosa* and *Monobothrium hunteri* unlike *Glaridacris catstomi* and *G. laruei* [71] and other cyclophyllidean cestodes [33; 114]. If the distribution of corpuscles is a vestige of pattern in an ancestral *segmented* worm and there are separate genes regulating distribution of corpuscles and segmentation in different groups of cestodes then some caryophyllids are true vestige of an *adult* character of Polyzoic cestodes. Some cyclophyllidean cestodes e.g. *M. expansa* must have retained the primitive character and is an indication of "masked" gene related to the distribution of corpuscles. As predicted earlier by us (see [114]), the distribution of corpuscles in Anoplocephalan tapeworms, caryophyllids and Monozoic forms may now provide excellent clues not only to the mystery of evolution of this group but also a new dimension on the classification of cestodes.

Acanthocephala

Acanthocephalans undoubtedly representing an ancient group of parasitic worms, have become adapted to the bodies of freshwater fishes (rare in marine fishes), amphibians, freshwater turtles (absent in terrestrial reptiles), birds and mammals. In all probability, the acanthocephala originated as parasites of freshwater fish, with a single invertebrate host—an arthropod. Although sometimes classified with roundworms, they are not even remotely related to the nematoda. Their relationships with cestoda (with apparent similarities like the hooked embryo, absence of gut and so on) is quite distant, probably having no living relations. Cameron [25] has rightly called these pseudocoelomates "a small group with elusive affinities and puzzling origin".

Nicholas [88] while reviewing the biology of the acanthocephala rightly commented that little fresh information on the phylogeny has come to light.

Nematodes

Among nematodes, it is not possible to say, how parasitism began. However, it is certain that it occurred a number of times, giving rise to several distinct and unrelated groups of parasitic nematodes (see specificity).

Morphological traits reveal that parasitic nematodes are descendants of several independent parasitic events that have given rise to lineages differing greatly in numbers of species. Anderson [6] contributed substantially on the origin and evolution of zoo-parasitic nematodes. It has been proposed that nematode parasites of fishes were derived from terrestrial vertebrates. Adamson [1] produced a cladogram for the higher phylogeny of nematodes (Fig. 4).

The host distribution of nematodes suggests that these worms became abundant in vertebrates only after the hosts had taken to terrestrial life, and that the complexity of the parasites evolved with their hosts. Adamson [2] presented a hypothetical phylogeny of life history patterns in zoo-parasitic nematodes. It has been suggested that the course of evolution of life histories has been strongly influenced by conditions under which parasitism arose. The larvae of nematodes often enter the lungs of the final hosts. It has been suggested that this site preference represents an ancestral character, in which only lungs with their abundant O_2 supply were the location of infection. Anderson [7] argued that some lung parasitisms are the results of secondary monoxeny having evolved from heteroxeny i.e., the tissue migration phase in the final host is the phylogenetic remnant of development in an intermediate host.

Inglis [56]; (see Fig. 5) and Cameron [25] have summarized the nematode evolution as follows:

(i) all parasitic nematodes appear to have arisen from terrestrial free-living groups, the phasmids from the Rhabditida and aphasmids from the Dorylaimida.

(ii) speciation in parasitic nematodes is considered to be wholly allopatric and not due to any evolutionary change in the host.

(iii) trans-specific evolution is shown to be largely independent of the evolution of the hosts and to be due to the parasites speciating and evolving so as to utilize all available localities and niches within the host body.

(iv) the significance of the third-stage larva in the phasmidian parasites is stressed as being the stage at which the parasites transfer from one environment to another.

(v) the reduction of the length of time spent in the intermediate host in the life cycle of many nematode groups is considered to be a result of selection pressure acting to reduce "host-discordance" i.e., the many differences in physiology etc., between the two, or more, animals in an indirect life cycle.

(vi) oxyurids represent a second line of parasites. Members of *Oxyuris* in invertebrates may represent the stem from which the vertebrates acquired these parasites. Presence of *Enterobius* in both new

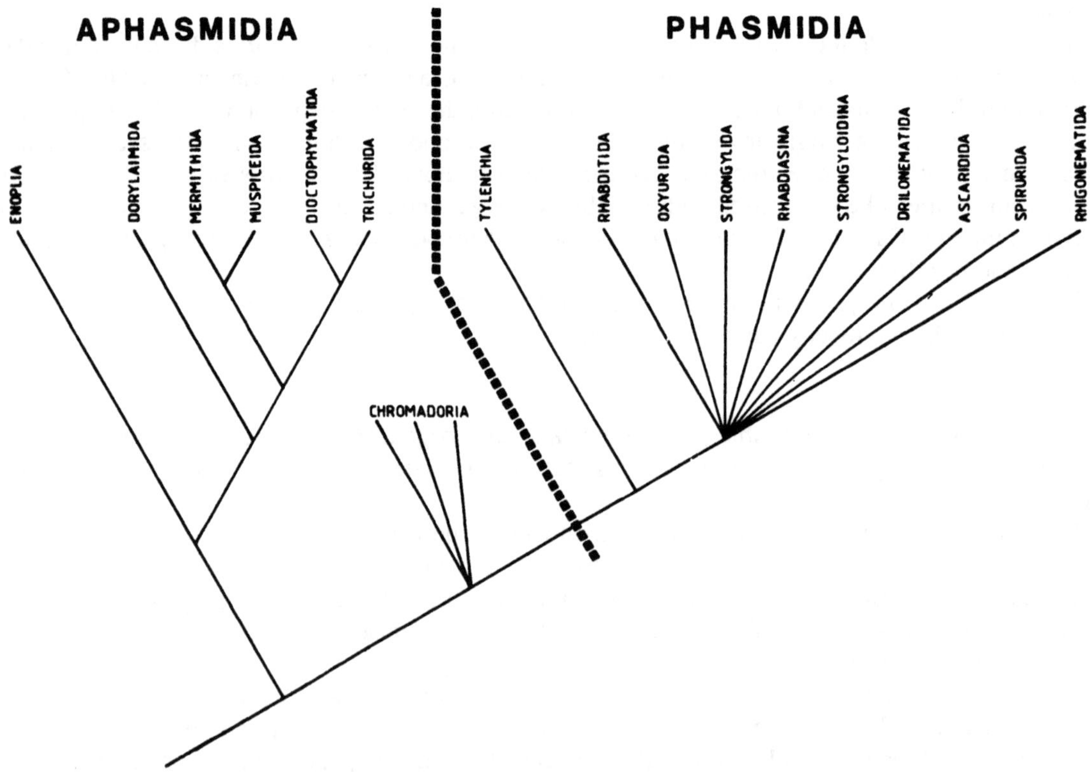

Fig. 4. Cladogram of the higher-level relationships among the Nematodes (Modified from Adamson, 1987). (From Noble, E.R. et al; Parasitology, 6th Edn., 1989, Courtesy: Lea & Febiger, Philadelphia).

and old world primates testifies to the common origin of both groups (see review by Adamson [3] on evolutionary biology of Oxyurida).

(vii) among bursate nematodes, complexity of mouth capsule tends to accompany complexity of burstal rays. This is attributed to evolutionary age, because such species occur in the more archaic groups of vertebrate hosts. The simplest but not necessarily the most ancient bursate nematodes are the Trichostrongyles. The bursate nematodes with true buccal capsules probably arose from *Amidostomum*-like ancestors.

(viii) spirurids and Filarids have become so specialized in their life cycles and habitats that their evolution remained completely obscure (see also specificity).

PARASITISM

In 1967, Justus Mueller argued, "no one has yet been able to define a parasite and thus classical connotation of the word "parasite" remained unexplained" [76]. The shepherd, Jehan de Brie believed to have first set his eyes on a liver fluke in 1379. Surprisingly, though parasites were known to science several hundreds of years ago, the term "parasitism" remained obscure until 1947 when Haldane recognised it as a "potential evolutionary force". Unfortunately, such a beautiful definition also remained "unhonoured" until mid-seventies. Ordinarily, "parasitism" is known as a social relationship but it also implies an ecological relationship. To evolutionary biologists "parasitism" is a theoretical concept. More recently, "parasitism" has been redefined by Graham Bell & Austin Bert (McGill University, Canada)

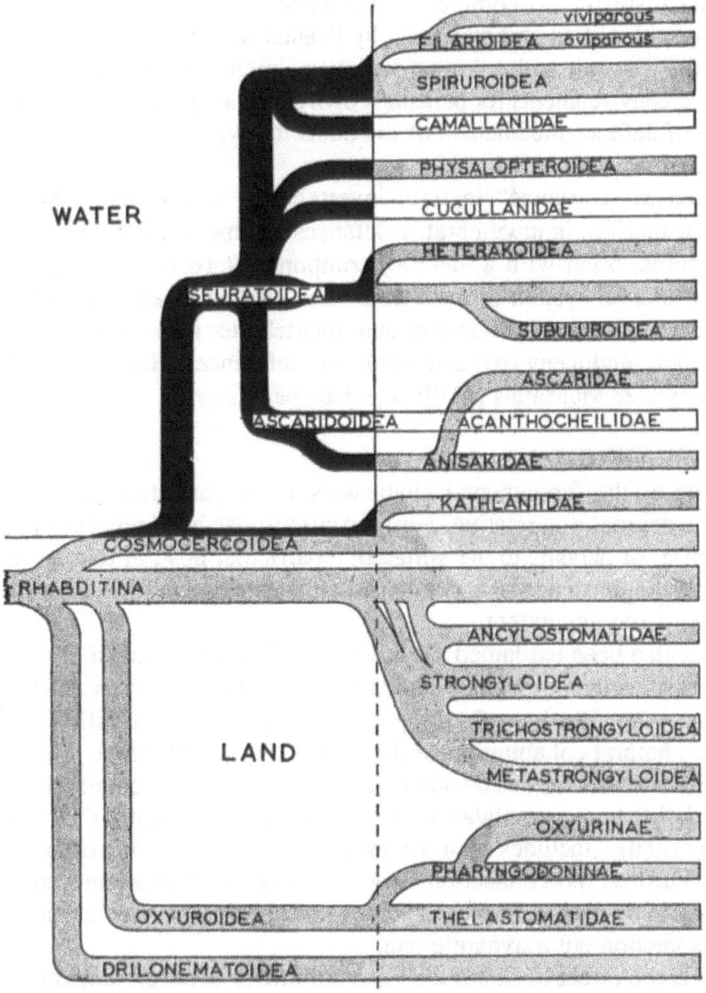

Fig. 5. Dendrogram (generalized evolutionary tree, not constructed with any particular methodology) illustrating the relationships among the major groups of phasmidian parasites, as postulated by Inglis. (From Inglis, W.G.: Patterns of evolution in parasitic nematodes. In: *Evolution of Parasites* (Editor: A.E.R. Taylor, 1965) (Courtesy of Blackwell Scientific Publications: Oxford).

echoing 'Red Queen' hypothesis as "recombination, favoured by antagonistic coevolution between hosts and parasites". All non-essential DNA may be regarded as parasitic, including the accessory or supernumery 'B' chromosomes which themselves have deleterious effects on host fitness (see [59, 134]). Parasites are not only associated with the evolution of disease virulence, reproductive strategies, pathology and so on, but now they are also known to play major roles in the areas of evolutionary biology, social behaviour, sexual reproduction, ornamentation, bright colouration and speciation of the hosts (see [59, 74, 144]).

Today, the major challenge before parasitologists is to understand as to how parasites succeed in establishing relationships with their hosts at a molecular level and also to understand the genetic versality of parasites and their inbuilt mechanism to evolve too rapidly and re-establish in new niches by virtue of adaptive specialization. Indeed, convergent evolution is believed to have led the emergence of certain themes in parasitism due to common selection pressures. Recently, four broad areas of such specialization have been identified and discussed by Bianco & Maizels [14] and are: (i) mechanisms by which parasites invade, transmit and colonize variety of niches within the hosts; (ii) optimization of conditions of host's microenvironments for parasite's own survival by inducing physical and/or biochemical changes; (iii) evasion of defence mechanism of the hosts in a variety of ways, and (iv) specialization to enhance dissemination (Table 1).

Recently, "evolution of resistance" in both invertebrates and vertebrates has been discussed by Harvell [49] and Mitchison [85]. In invertebrates, defences against parasites are characterised by inducible and amplifiable responses, often with a memory component (see Fig. 4; Harvell [49]). Although the reduced specificity of internal system of invertebrates is viewed as an *evidence* of their primitiveness, the differences in defence systems of vertebrates and invertebrates may be more related to their respective selection pressures than to phylogeny (see discussion and references: Harvell [49]; Klein [60]; McLennan & Brooks [78] and Smyth & McManus [119]; see also *specificity*).

PARASITE-HOST SPECIFICITY

Host specificity is one of the fundamental characteristics of parasitism. It is seldom absolute, more often, but to a variable degree, it is relative. Host specificity has been regarded as the adaptation of one species (in a broad sense) of parasite to the *milieu* of a particular host, or to a more or less limited group of hosts. It depends on the existence of a stable (host-parasite) system; in abnormal hosts, the parasite will be unable to establish to any extent.

Host specificity has also been explained as the ability of different parasites to infect or to develop to maturity in different hosts, which obviously indicates fundamental differences in their physiology. Host range reflects two interacting situations; firstly, physico-chemical compatibility on the part of the parasite and host and secondly, natural cohabitation of the two in the same biotope [114].

Parasite-host specificity can also be considered as the result of co-accomodation between parasite and its host. Brooks [15] defined co-accomodation as "the mutual adaptation of a given parasite species and its host(s) through time...(it)...includes such parameters as pathogenicity, host-specificity and synchrony of life cycle stages". Further, the connection between co-accomodation and co-speciation of parasites and hosts together constitute what is known as "co-evolution" as discussed earlier. Therefore, specificity is not a constant phenomenon but a dynamic one.

Indeed, specificity is the effect of biological compatibilities between parasites and hosts. Apart from ecological, physiological and anatomical considerations, a very fundamental aspect of host specificity is the evasion of the host's immune response. Parasites employ various mechanisms to achieve this, e.g., mounting effective suppressive immune responses, providing antigenic disguise or modulating surface antigens and so on (see Table 1). The topic has been reviewed by Rickard & Williams [102]; Parkhouse [91]; McLaren [77]; Rickard [101]; Flissar et al [43]; Simpson [113]; Behnke [11]; Bianco & Maizels [14]; Capron & Dessaint [26]; Smyth & McManus [119]; Terry [131] (See Chapter 10).

Phylogenetic Specificity

Many parasites are host specific and complete their life cycles within one species of host. Many parasites, however, develop in more than one kind of hosts in their life cycles. Therefore, conditions which determine the degree of host specificity often differ in several stages of life cycle, and the metabolic requirements of a parasite generally vary with its developmental changes. Even for a given species within a particular host the molecular basis of host-parasite specificity largely remained unexplored, in spite of the fact that this is instrumental in controlling parasitic infections [122].

Sprent [126] suggested that evolution of host-parasite system has taken a long time and that stable

Table 1. Some of the principal adaptations to parasitism among helminths

Helminths	Feature	Mechanism
	Host invasion	
Trichinella spiralis and filarial larvae in arthropods	Penetration of individual cells	Binding to cell surface receptors, perturbation of cell membrane/ cytoskeleton
Schistosomes, metacestodes, hookworms, *Toxocara* spp., filarial nematodes	Penetration of tissues	Secretion of proteinases
Trichuris spp., adult cestodes	Colonization of gut	Ingestion (with or without attachment to mucosa)
	Host modification	
Trichinella spiralis	Molecular interactions with cellular compartments	Re-differentiation of muscle cell
Helminths in general	Secretion of pharmacological mediators	Superoxide dismutase, Acetylcholinesterase, Glutathione-s-transferase, Protease inhibitors
	Host defence evasion	
Schistosomes	Host antigen acquisition	Surface adsorbed blood proteins and glycolipids
Toxocara canis	Host mimicry	Blood group carbohydrates
Toxocara canis	Surface turnover	Release of complexed antigens
Taenia crassiceps?, *Toxocara canis*? filarial nematodes	Immunosuppression	Secretion of suppressor factor
Schistosomes	Immune dysfunction	Promotion of non-functional responses
Toxocara canis	Surface defence	Surface protease (? cleavage of complement or immunoglobulin)
	Specialization for dissemination	
Toxocara spp.	Vertical transmission of infective larvae	Migration across placenta (? host hormonal cues)
Filarial nematodes	Periodicity of blood microfilariae (mf)	Movement of mf between general and pulmonary circulation. Stimuli part defined, mechanism unknown
Filarial nematodes	Spatial distribution of skin mf	Tropism to sites of vector biting on body. Mechanisms unknown
Schistosomes	Tissue egg excretion	Proteinase secretion and expression of antigens mediating immune-dependent transport across tissue.

Adapted mainly from Bianco & Maizels [14].

systems e.g., monospecific systems, are not necessarily old ones. "Narrow specificity" is generally supposed to reflect an old relationship but it does not reflect the phylogeny of both host and parasite (see further). At a very basic level, an adequate structural complemantarity must exist between a parasite and a host. Sufficient commonality in the conditions provided by different hosts species can allow a common parasitism.

In many cases there also exists a relationship between the age of a host-parasite system and the specificity of a parasite; the relationship, however, not being always very precise one. Palaentological studies have also contradicted the phylogenetic host-parasite relationship as an indicator of narrow specificity (see discussion [57]).

Apart from phylogenetic relationship, feeding habit or behaviour of the host in a broad sense, of course, is a very important factor in determining specificity. For example, *Fasciola hepatica* and *F. gigantica* occur in grazing herbivorous animals including sheep, goat, cattle, buffalo, camel, rabbit, rarely in humans and pig and exceptionally in dog and cat [63] because of diverse food habits. Similarly, *Gigantocotyle explanatum* a common bile duct amphistome of ruminants in India, has also been recorded only once in a dog. *Killigrewa* spp., which normally parasitise pigeons and doves, also sometime occur in humans (see Appendix; Chapter 3).

It must be remembered that during evolution of parasitism, repeated speciation did occur along with niche diversification of the parasite. This ultimately lead to an increasingly broad co-accommodation or narrow host range confined to taxonomically related hosts (which means increased host specificity). Another dynamic nature of specificity is that once host-parasite systems have come into existence they may remain stable, or may speciate together, or show only parasite speciation which again may result in narrow specificity. The host may speciate but not the parasite, which then may be able to infect the newly evolved host species. The parasite may retain its original host but under the influence of constraint spread to new host(s) or abandon the old one for ever. An example of this situation often cited is the introduction of *F. hepatica* in Australia from Europe where it adapted to a new host, *Lymnaea tomentosa*; the original host, *L. truncatula* being absent in the continent. Within Indian subcontinent, *L. truncatula* acts as intermediate host of *F. hepatica* in hill-tracts of Pakistan; in the higher altitude of Kashmir valley in Indian region, on the other hand, both *F. hepatica* and *F. gigantica* coexist. But the two snails, *L. truncatula* and *L. auricularia rufescens* (the latter being the intermediate host of *F. gigantica* in the plains of Indian region) are conspicuous by their absence there. Instead, *L. auricularia sensu stricto* is widely prevalent in the valley and is believed to act as intermediate host for both the flukes.

Thus, it is obvious that phylogenetic specificity is likely to alter with changing habitat and under ecological constraint, within both short or long periods of time. More precisely, phylogenetic specificity is dependent on ecological or physiological factor(s) or a combination of both [57].

Physiologic Specificity

The factors and processes which can affect host specificity and occur during each of the three major stages in the host parasite association have been well documented [104,106]. Ecological factors(s) initially determine the host range of a parasite. The presence or absence of physico-chemical factors then play the decisive roles in regulating the degree of specificity by determining whether or not development of a parasite can resume. Reducing conditions, for example undissociated H_2CO_3 and/or CO_2 help exsheathment of *Haemonchus contortus* larvae. Though these factors are believed to play a significant role in limiting the host range to ruminants, in other parasites, for example, *Trichostrongylus axei* and *T. colubriformis* or *Ascaris* spp., precise requirements limiting the host range or influencing the development in particular type of host(s) are not yet known. The triggering of development in different species of helminths is known to be dependent upon complex relationship between (CO_2), (O_2), Eh, (H^+) and temperature [29, 93, 105, 106]; see reviews by Mercer [81]; Hurd [54]; McLennan & Brooks [78].

The host's digestive enzymes sometimes directly induce hatching or excystment, for example, metacercariae of *F. gigantica, F. hepatica, Opisthorchis* spp. and metacestodes of *Taenia* spp. (see Chapter 3). Host specificity in such instances is often low, because such enzymes are widely distributed.

After infection of the host, the invading juvenile faces a crucial stage within the host which largely determines host specificity. During this critical phase, the juvenile needs adequate nourishment not only to withstand the damaging physical and chemical factors, such as enzymes, excitants, surface active agents, immune complexes and so on (see [119]; Figs. 6 and 7) which are present in the micro-environments in the host but also for its establishment and growth.

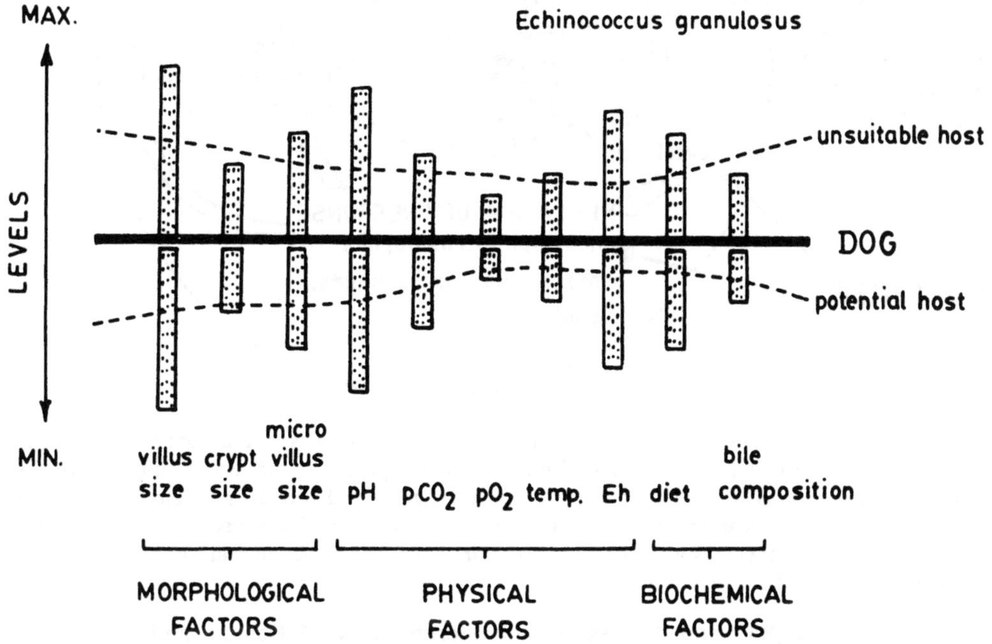

Fig. 6. Some parameters of the intestinal environment which play a part in the determination of the host specificity of a cestode; based on *Echinococcus granulosus*. The levels of the various factors are only notional, as the limits of tolerance have not been defined precisely. A host with an intestine with parameters ranges outside those shown cannot act as a suitable host (After Smyth, 1969) (From Physiology of cestodes. Courtesy: Longman Group, U.K.).

It is now well recognised that the primary importance of bile which is the key factor in determining the host specificity in addition to establishment and maturation of parasites, is due to its salts [118]. Haslewood [50] found that, except in bovidae, bile salt types have distinct evolutionary relationships in Eutherian mammals and the conjugates in bile, in fact, determine host specificity (see discussion [32]).

The major role played by the "gut factors" in parasite host specificity was imperfectly known until it was demonstrated that larval *Echinococus granulosus* could develop in mice after intraperitoneal route, mice normally being a refractory host [112]. The altered physico-chemical factors associated with emptying of the stomach and intestine are also believed to influence host specificity. Juvenile stages sometime may pass through the gut too rapidly to be acted upon by the essential factors or be retained and be adversely affected by deleterious factors.

The influence of age and sex of the host on the distribution and tissue specificity of helminths both in the intermediate and definitive hosts was first discussed by Ulmer [133] (see also recent reviews

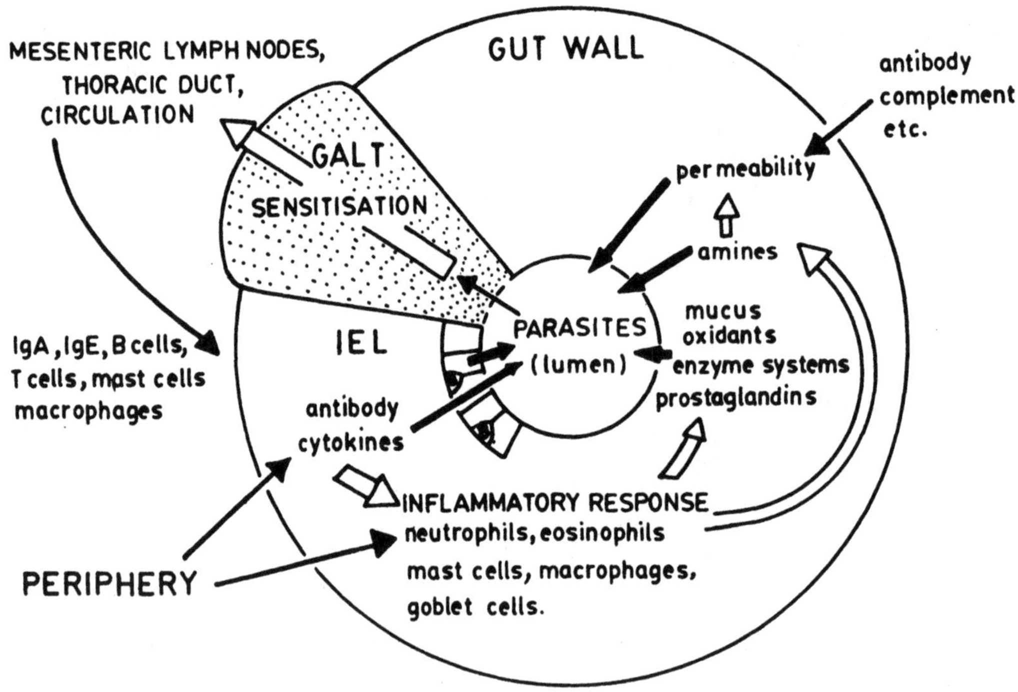

Fig. 7. Summary of the host response to mucosal parasitic infections. Sensitisation occurs initially in gut-associated lymphoid tissue (GALT); various cells of mucosal or peripheral origin are recruited largely by thymus-dependent mechanisms; the inflammatory response so generated produces an environment hostile to the continued survival of the parasites. Heavy solid arrows denote possible sources of effector molecules; IEL, intraepithelial lymphocytes, some of which are granulated, and probably natural killer cells (After Befus & Bienenstock, 1982).

by Hurd [54]; Minchella [83]; and discussion and references by McLennan and Brooks [78] on this topic).

The period of growth and reproduction is an important event in the biology of a parasite but the relationship between specificity and reproduction is not clear. Experimental data, however, suggest that the bile and pancreatic juices stimulate oogenesis in the cestode, *Hymenolepis microstoma* [31, 32]. Though not well studied, vitamin A, D_3 and E and trace element like zinc play a role in oogenesis (and spermatogenesis) like other physical, chemical and immunological factors [28, 33, 34, 39, 51, 140]; see reviews by Clegg and Smyth [35]; Gibbs [47]; Wakelin [141]; Jackson [58]; Hurd [54]; see also von Brand [135,136]; Smyth & Halton [117]; Smyth & Macmanus [119]; Table 2). Interestingly, analogy with insects, presence of ecdysteroids in eggs of *S. mansoni* and *F. hepatica* suggests a regulating role in embryogenesis. In adult nematode, these hormones are found concentrated around reproductive organs (see Hurd [54]; Mercer [81]).

Fasciolopsis buski is an intestinal fluke of pigs and humans and is also found in dog and rabbit, but does not grow in monkeys, ruminant or rodents, except guinea pig (partially susceptible) [73]. Similarly, *E. granulosus* which normally develops in dogs, does not grow in cats and only rarely in foxes whereas all these hosts are equally susceptible to *E. multilocularis*. Earlier, such specificity was thought to be due to intense physico-chemical differences in the gut or in broad sense, the host's diet. The turning point came when it was experimentally demonstrated that the immunological responses manifested by the abnormal or unsuitable host are key factors in determining parasite-host specificity. Thus, *Hymenolepis*

Table 2. Parameters of anti-helminth immune responses known to be influenced by host genotype

Functional	
	Parasite reproduction
	Survival of primary infection
	Survival of secondary infection
	Threshold levels of infections for immune response
	Response to vaccination
	Susceptibility to parasite-induced immunodepression
Cellular	
	Antigen recognition by lymphocytes
	Eosinophilia
	Granulomatous hypersensitivity
	Intestinal mastocytosis
	Macrophage activation
Serological	
	Antibody isotype
	Antibody specificity
	Complement components
	Hypergammaglobulinaemia
	Rate of antibody production

Adapted from Wakelin [140].

diminuta does not stimulate protective immunity in rat which is its normal host. However, in mice, it grows normally but in some strains it starts shedding proglottids tenth day post-infection (Fig. 8), and does not grow further until transplanted in a new host [52]. This phenomenon was first explained as a balance between growth and apolysis. Later work confirmed it to be the result of a second line of host defense [51].

Besides *H. diminuta* classical, immune expulsion phenomenon has been experimentally demonstrated in *H. citelli* and several other nematodes; for example, *Strongyloides ratti, Nippostrongylus braziliensis, Trichinella spiralis, T. colubriformis* and *T. muris* in mice, rats and guinea pigs (see Fig. 9). Inspite of the availability of several interesting data, the precise mechanisms of this expulsion is not very clear and has been extensively reviewed by Befus & Binenstock [9, 10]; Larsh & Race [64]; Miller [82]; Mitchell [84]; Ogilvie & Love [89]; Soulsby [123,124]; Wakelin [137, 138, 139, 141]. Density-dependent effects delay spontaneous cure as do pregnancy, lactation and concurrent infections (Fig. 10). Recently, host genes have been identified that influence the capacity to develop and express immune responses against parasites and these clearly influence host specificity (see Chapter 15). Obviously, immunological findings (see Tables 2 and 3; Behnke [11]; Leid [65]) have revolutionised the entire concept of parasite-host specificity and much remains to be discovered.

Ecologic Specificity

A parasite and its host represent a biologic system whose maintenance depends on ecological factors. It is not easy to generalise the causes of ecological specificity since multiple factors are involved in the process. These factors not only vary in extra-host situations but also at inter- as well as intra-host levels. For example, the swim bladder fluke *Isoparochis hypselobagri* normally occurs in fish but it has been reported from the bile duct of pig and humans. *Artyfechinostomum sufrartyfex* though parasitises intestine

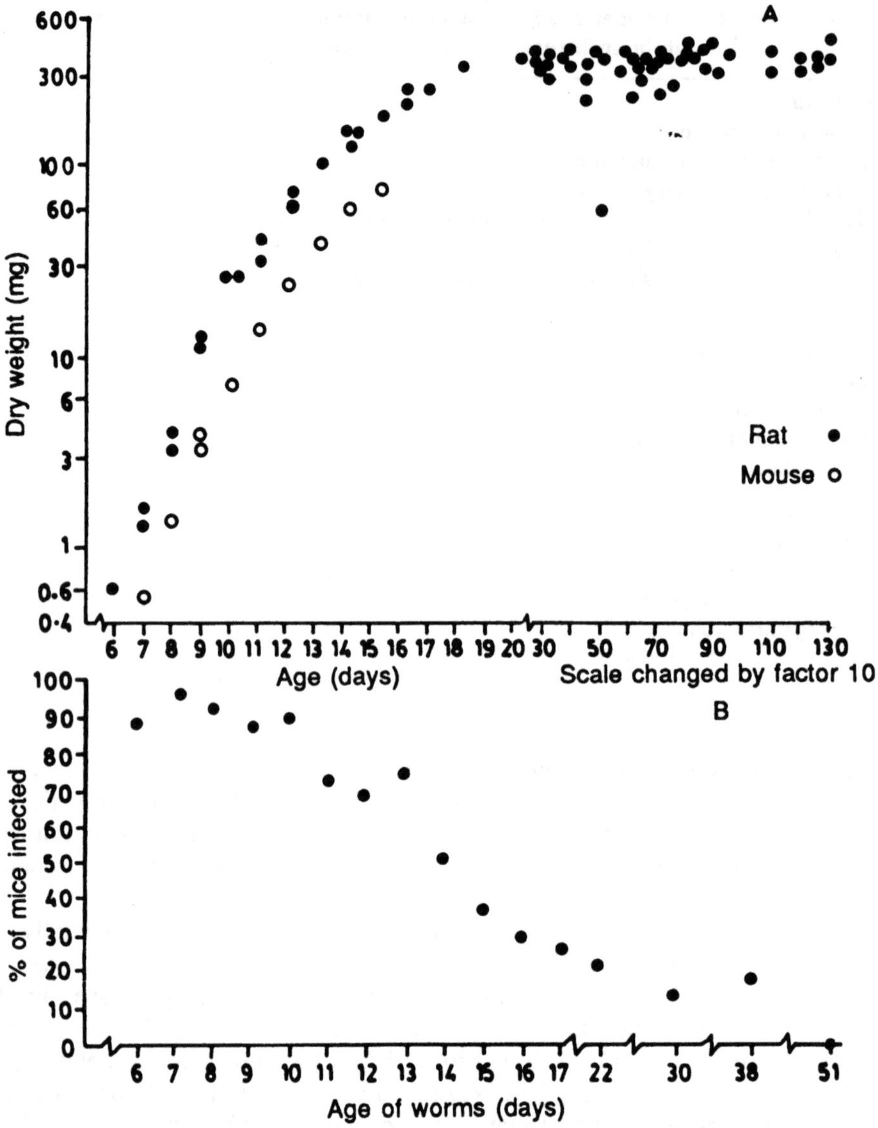

Fig. 8. (A)—A comparison of the growth rate of *Hymenolepis diminuta* in rats and mice. Destrobilated worms in mice are not included; (B)—A survivor-ship plot of *H. diminuta* in single worm infections in male mice (Both after Hopkins et al, 1972).

of humans and pig and less commonly of dog and cat also occurs in honey-badger, rat, spiny-tailed lizard and palm civet (see Appendix; Chapter 3).

Contrary to this, the vast majority of pinworms occur in a single species, genus of higher texa if those are monophylectic i.e., narrow host range. Obviously, this type of specificity has a strong intra- or extra-host component [3]. Adamson [3] argued that closely related hosts differ fundamentally so far ecology of the gut is concerned and, therefore, sympatric hosts frequently harbour the same parasite even if they are distantly related. On the other hand, closely related hosts may harbour distinct parasite populations even if they (parasite populations) occur in sympatry. Inglis [56] thus commented, "when the host population

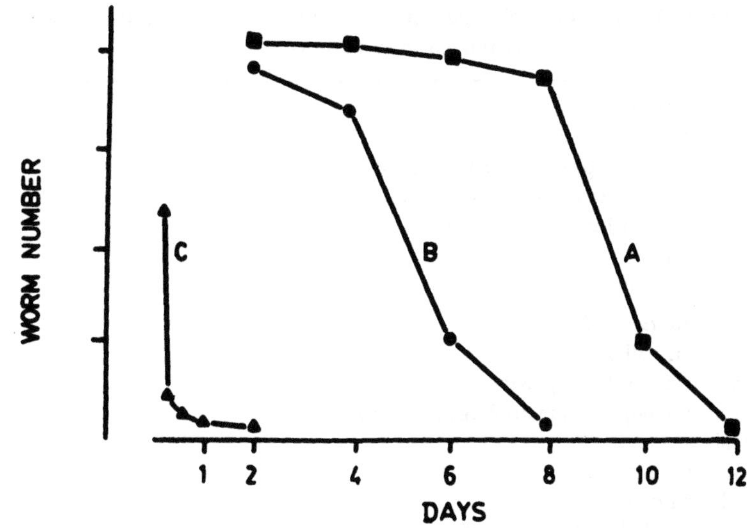

Fig. 9. Time course of infections with *Trichinella spiralis* in NIH mice. A—Primary infection, showing spontaneous cure; B—Challenge infection given 21 days after priamry and C—Challenge infection given 16 days after primary, showing rapid expulsion (After Wakelin, 1987).

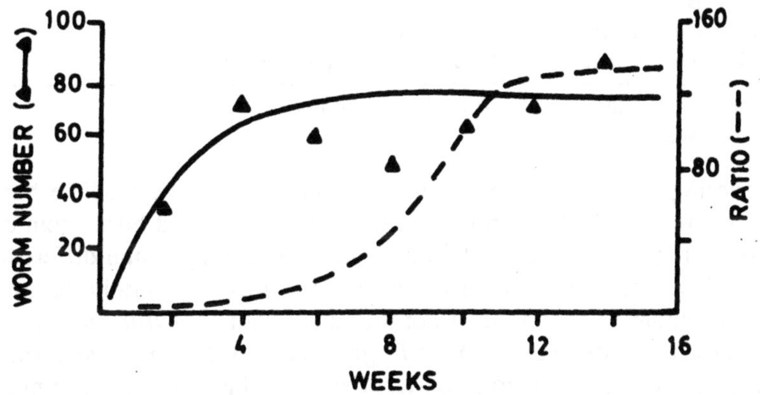

Fig. 10. Mean worm recoveries and changes in variance-to-mean ratio of the number of worms/mouse in male MF_1 mice given 52 larval *Nematospiroides dubius* every 2 weeks. (After Wakelin, 1987; Redrawn from Keymer, 1985).

re-unite the parasite species cross infest, compete and stabilize as distinct species if there is a niche available". Host specificity is also explicable on the basis of intra-host competition due to multiple invasions. Smyth [116] discussed the changing concept in micro- and macroecology with particular reference to hydatid disease. The intimate relationships of *E. granulosus* or *E. multilocularis* within the microenvironment of their definitive and intermediate hosts, essentially represents the microecological features of the life cycle (Fig. 11a, b). In Russia, the maturation of *E. granulosus* in dogs varies between 39–54 and 35–49 days during summer and winter months, respectively [143]. Indeed, another startling finding in India was that hydatid cysts from the liver and lungs of the same host (buffalo) become adult in dogs with different prepatent periods (W.A. Nizami, unpublished data), a situation has been explained as due to variation in "nutritional reserves" in protoscoleces or "inbuilt" (?) differences in the organ or host (see [116]).

Table 3. Potential changes in intestinal structure and function induced by parasitic infections

Physiological changes
Increased secretion of mucus
Increased vascular and epithelial permeability
Increased net fluid secretion across mucosa
Fluid accumulation in lumen
Increased motility
Decreased transit time
Cellular and structural changes
Altered epithelial cell kinetics
Villous atrophy, crypt hyperplasia
Infiltration by eosinophils, macrophages, neutrophils
Mastocystosis
Increased goblet cells
Increased plasma cells
Increased intra-epithelial lymphocytes
Membrane changes in epithelial cells
Dedifferentiation
Mediator Changes
Increased levels of Ig in lumen
Increased levels of myeloperoxidase and phospholipase
Increased levels of histamine, 5-HT, leukotrienes, prostaglandins

Note: Ig, immunoglobulin; 5-HT, 5-hydroxytryptamine
Adapted from Wakelin [140] .

Macroecology, covering both ecology of potential definitive and intermediate hosts, is a vast area beyond the scope of this review. Isolates from the same host or different hosts from different areas are known to exhibit morphological, biochemical, nutritional and developmental differences [79, 80]. It is now a well established fact that *E. granulosus* has many(?) strains even within the same host and same country and possibly the situation is similar, if not the same, with *E. multilocuris* [119]. The reason(s) for the appearance of new range of strains of both *E. granulosus* and *E. multilocularis* is (are) not clear (see Fig. 12) but it is presumed to be due to cross-fertilization, which induces recombination of genome resulting in intraspecific variation.

Intraspecific Host Specificity

It is well recognised that parasites resort to formation of new strain(s) by several methods like self fertilization, polyembryony, asexual reproduction (Fig. 13), clone formation and isolation.

Intraspecific differences in host specificity of parasites presumably results from interactions of genetical, immunological, behavioural and ecological characters of host-parasite system(s) which have been discussed earlier by Inglis [56] and Nelson [87] among others. Strain differences in helminths have been extensively reviewed by Bryant & Flockhart [22]; Thompson & Lymbery [132]; Smyth & McManus [119]; (see also Hurd [54]; Rollinson & Anderson [107]; Lie et al [68]; Richards & Shade [100]; Minchella [83]; McLennan & Brooks [78]; Chapters 5 and 15).

Human Factor

Under natural eco-systems, the specificity of any parasite-host system is favourably maintained. Human

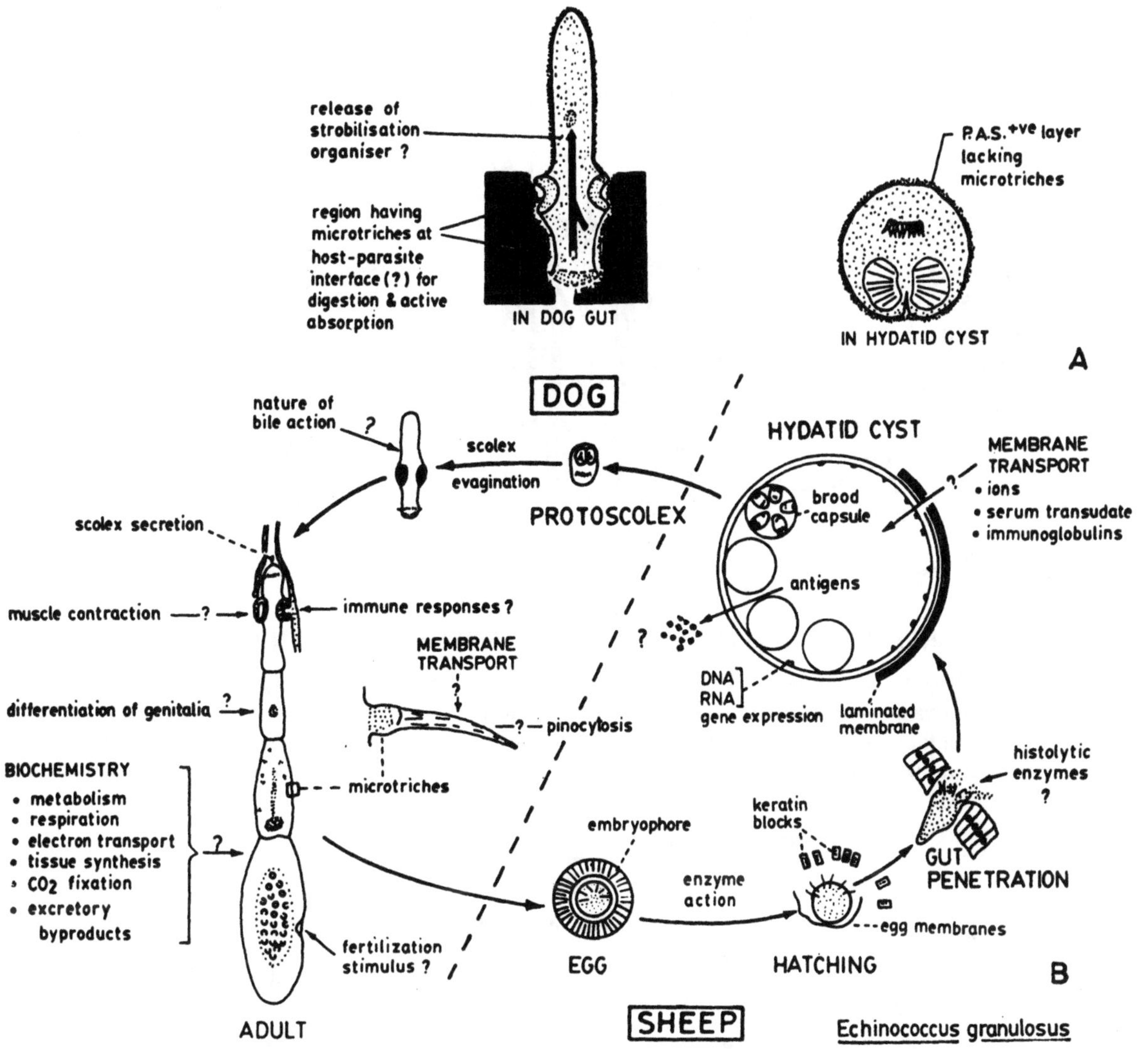

Fig.11. (A)—Comparison of the protoscolex of *Echinococcus granulosus* in a hydatid cyst and, after evagination, in the dog gut. In the latter position, the exposed microtriches make contact with the protein surface and the cestode cells are stimulated to differentiate in a strobilar direction. It is speculated that an "organiser" is involved, the release of which may be initiated via a neurosecretion (Adapted from Smyth et al, 1967) (From Physiology of Cestodes. Courtesy: Longman Group, U.K.); (B)—microecological aspects of the life cycle of *E. granulosus*. The question marks indicate areas of particular importance where further research is needed (Adapted from Smyth, 1987) (Reprinted by permission of Kluwer Academic Publishers).

intervention is one of the causes which alters the natural compatibility between parasite and host; the end result being the break down of specificity. The introduction of *F. hepatica* to Australian continent has been discussed earlier. An excellent account is given by Baugh [8] on the occurrence of human urinary schistosomiasis in India. During World War II visits by soldiers, traders and pilgrims returning from the African continent and Arab countries, in fact, introduced the disease. The fluke *S. haematobium* consequently became adapted to a new local snail, *Ferrissia tenuis* in Maharashtra (see Fig. 1; Chapter 3).

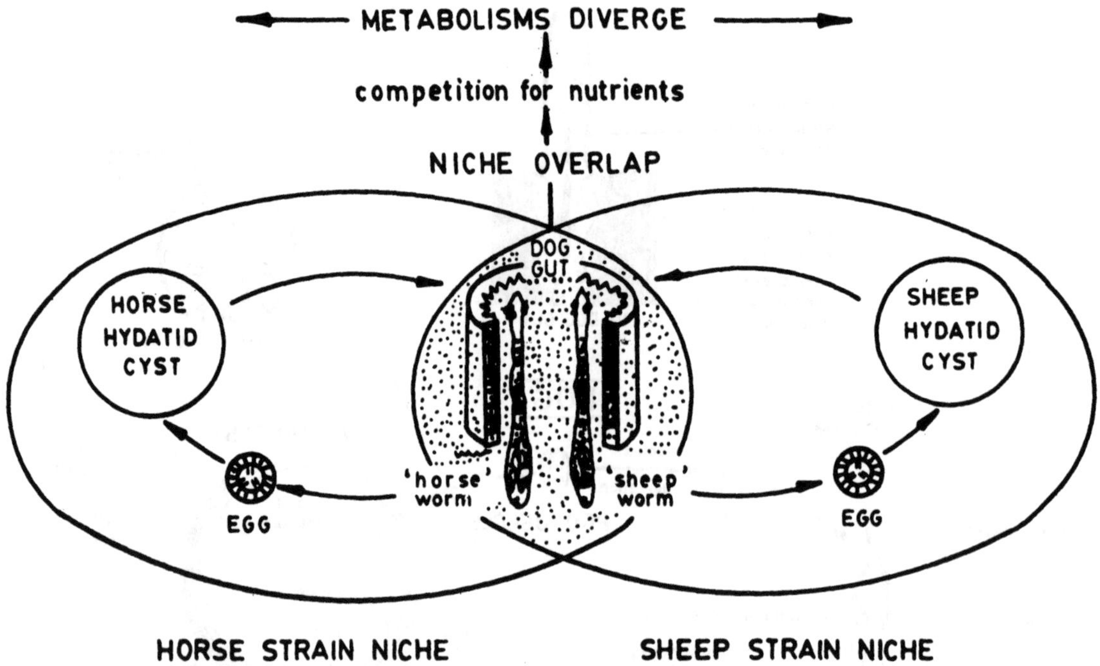

Fig. 12. Gause's Hypothesis applied to the sheep and horse "strains" of *Echinococcus granulosus*. The niches of these two strains overlap at one point in their life cycles in that they both inhabit the dog gut. It has been shown that these two strains have evolved different metabolic pathways, so that competition for the same nutrients is likely to be reduced (After Smyth, 1987). (Reprinted by permission of Kluwer Academic Publishers).

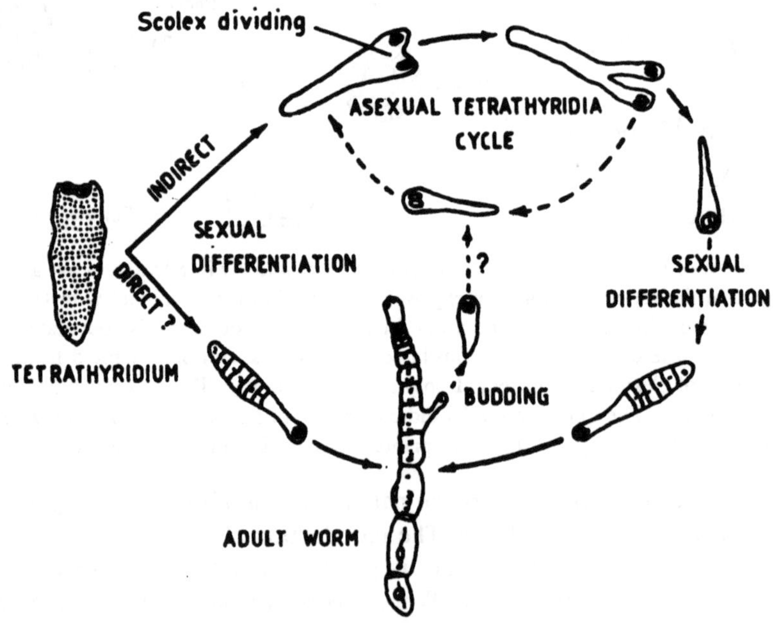

Fig. 13. *Mesocestoides corti*: simplified diagram of the sexual and asexual processes occurring in the intestine of the definitive host (cat/dog) (After Smyth & McManus, 1989).

COMPARATIVE HOST SPECIFICITY

Trematodes

Adult trematodes inhabit different internal organs of their hosts but host specificity is not well marked in all families of digenetic trematodes though it exists to a considerable extent. Interestingly, different members of the same family sometime can exhibit spectacular differences in the degree of specificity [74]. Thus, schistosomes are an exceptional group of trematodes which demonstrate certain unique features of specificity. For example, the host range of human schistosome, *S. mansoni* is restricted to humans, monkeys, rat and mouse while *S. japonicum* is a successful parasite of humans, dog, cat, rat, mouse, cattle, buffalo, pig, horse, sheep, goat, and other carnivores. Similarly, *S. haematobium* develops well in monkeys, baboon, albino mice, hamster but grows poorly in cat, albino rat, cotton rat, guinea pig and goat and is refractory in rabbit and dog.

The malady "nasal schistosomiasis" is a disease entity which occurs exclusively in Indian subcontinent and is due to *S. nasale* which affects cattle and buffalo and rarely sheep and goat. The disease is invariably accompanied by cauliflower-like growths in the nasal septa of cattle only (see Chapter 3). Identical conditions have been reproduced both in cattle and buffaloes by cross transmission experiments thereby eliminating (?) the possibility of strain difference of the parasite [61]. The blood fluke can develop and reach in nasal veins of experimental kids, but not in lambs, mice, guinea pig and rabbit [109]. *S. incognitum* naturally occurs in pig, dog and sometimes in rodents in India, Thailand and Indonesia but experimentally this blood fluke can be grown in large numbers in monkeys, rabbit, white rat, guinea pig, squirrel, cat and very young kid but only a few worms develop in cattle, sheep and adult goat [4, 115]. *S. spindale* is found mainly in ungulates (sometime in rodents) whereas *Orientobilharzia dattai* exclusively occurs in ruminants. Albino mice appear to be a slightly more suitable experimental hosts of *S. indicum* than rabbit or guinea pig but white rat is not very susceptible host and females do not attain maturity both somatically and sexually. In contrast, the females of *Bilharziella polonica* and *S. douthitti* reach sexual maturity even in the absence of males. Surprisingly, unmated males and females of *S. japonicum* remain immature indicating that the species of *Schistosoma* and *Bilharziella* are the least evolved genera of the family Schistosomatidae. Some of the special features in *Schistosoma* spp. are undoubtedly genetically controlled which have been elegantly reviewed in Chapter 15.

In tropical Africa, *Paramphistomum microbothrium* occurs commonly in sheep, cattle and goat but experimental studies revealed that cattle is a preferred host [53]. Similarly, each species of schistosome reported from Indian subcontinent is known to have a host preference even among ruminants (see Chapter 3; Table 6).

Detailed experimental studies with *S. indicum* in India have demonstrated that though miracidial infection of *Indoplanorbis exustus, Gyrulus convexiusculus, L. luteola and L. acuminata* took place in different snails but actual development of the parasite with subsequent production of cercariae occurred in some specimens of *I. exustus* only. The miracidium of *O. dattai*, however, is attracted to *L. luteola* and not to *I. exustus* [128]. Likewise, the miracidium of *O. felineus* is attracted to the snail *Bithynia leachi* but not to the closely related *B. tentaculata* which occurs in the same locality. Exceptionally, however, the carcariae of *S. indicum* and *S. spindale* which are enzootic in India often utilize the same species of snail, *I. exustus* as their intermediate host.

It is now evident that host specificity among digenetic trematodes is often greater at their larval forms as compared to adult stage. Thus two species of fluke can live together as adults in the intestines of avian or mammalian hosts but require different species of snails to complete their life cycles.

Cestodes

The range of host specificity in cestodes varies considerably. The high host specificity and distribution of particular group of cestodes in particular host groups show that the hosts and the cestodes evolved in parallel lines. Cameron [25] believed that "parasitism" in vertebrates began in Palaeozoic period; transition to land animals started during Mesozoic era. The birds first became parasitised during that time due to sudden population explosion and their tapeworms became group specific. The majority of advanced type of cyclophyllidean tapeworms became parasites of birds and exhibit considerably high host specificity.

Among cestodes, the host specificity is more dependent on phylogenetic relationhips than that on ecological segregation of their hosts. The species of genera *Dipylidium, Taenia* and *Echinococcus*, for example, are found in the carnivores while *Moniezia, Avitellina* and *Stilesia* are restricted to ruminants. In many instances, host specificity of adult cestode is dependent on food habits of the host rather than its macro- or microenvironments. Thus, tapeworms of the genus *Hymenolepis* are widespread among insectivorous birds and mammals but at species level the host range becomes very narrow. Similarly, Anoplocephalan tapeworms adopt insects or mites as intermediate hosts and have a wider host range which, however, is restricted to generic level. Considerable specificity is seen in adult *T. saginata* and *T. solium* and in either case, man is the only known definitive host. Apart from man, *T. solium* can be developed experimentally in lar gibbon but *T. saginata* cannot be developed in chimpanzee, baboon, Java and rhesus monkeys [23, 62]. A spectacular specificity is, however, demonstrated by *T. solium* which cannot develop in rhesus monkeys (with or without immunosuppressants) though their larval forms "react" and attain maturity in the same host as in man. The adult of this cestode though develop in hamsters but remain somatically and sexually immature [110].

Larval tapeworms frequently tolerate a much wider host range. The metacestodes of *Taenia* and *Echinococcus* have many kinds of hosts for successful completion of life cycles (see Chapter 3; Tables 10 and 12). The factors determining host specificity in coracidia in copepods are not clear. The nature of digestive juices, structure of alimentary canal, nature of haemocoele, including the reactions of the phagocytic or encapsulating cells are believed to play definite role (Smyth & McManus [119]; see also Phares [94]).

Acanthocephala

The adult acanthocephalans occur in the intestine of animals and wild anseriforms and show high degree of host specificity. Specificity in these worms is directly related to the type of their intermediate host(s). Interestingly, these worms do not have free-living stages and cystacanths like spirurids, may re-encyst in the vertebrate hosts, other than their normal definitive hosts following their ingestion. Indeed, adequate basic information on the morphologic and physiologic specificity of this group is lacking.

Nematodes

The parasitic nematodes display a startling variety of life histories. Though majority of them invade and parasitise any tissue of their hosts, a large number of them remain in the gut. The host specificity among zoo-parasitic namatodes is perplexing because of multitude of associations between this diverse group of parasites and their hosts from entirely free living stages to obligatory parasites (Figs. 14 and 15).

Inglis [57] stated that though there is little host specificity (exists only at level of classification) in nematodes, every group is host restricted or exhibits specificity inspite of their wide host range. Undoubtedly, the parasitism began a number of times during evolution of zoo-parasitic nematodes and caused not only the emergence of certain distinct and unrelated groups but also diversified and/or narrowed their host range.

In subclass Enoplida, the species of genera *Capillaria, Trichuris* and *Trichinella* form an old group which is believed to have been derived from stylet-bearing dorylaimid ancestors ([45]; see discussion [2]). The genus *Capillaria* is widely distributed in vertebrates ranging from mammals to fishes, whereas

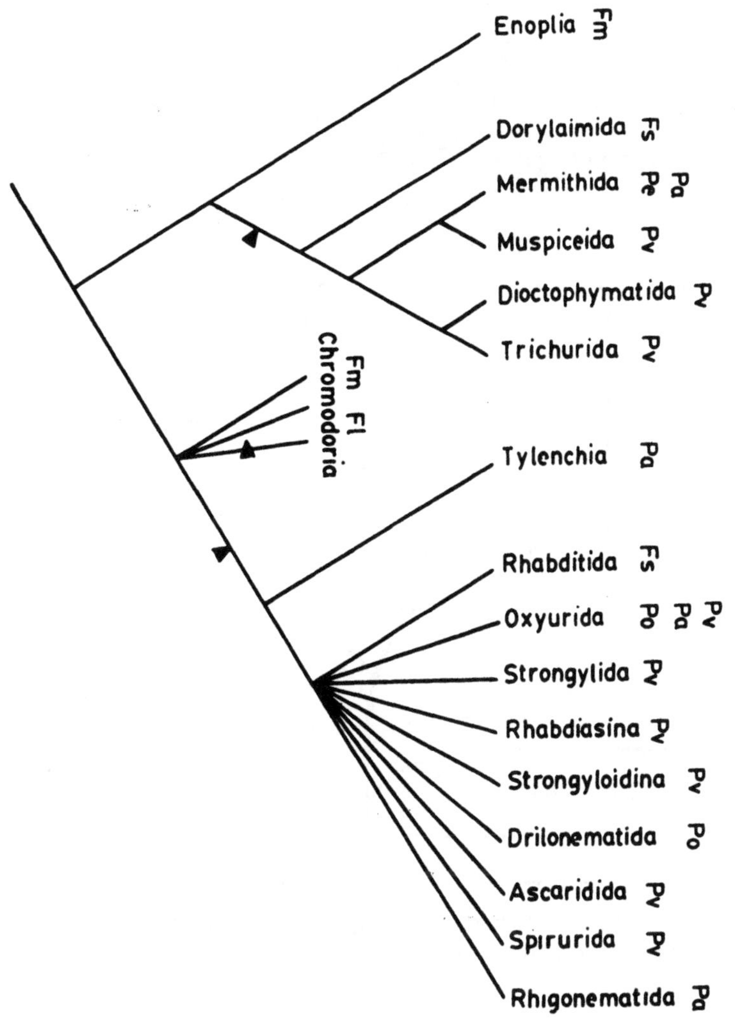

Fig. 14. Cladogram indicating presumed relationships among major groups of Nematoda. Letters above terminal Texa indicate habitat: Fl = free-living; Fm = free-living marine; Fs = free-living soil and freshwater; Pa = parasitic in arthropods; Po = parasitic in oligochaetes; Pe = parasitic in echinoderms; Pv = parasitic in vertebrates (After Adamson, 1989).

Trichuris is restricted to mammals including marsupials, artiodactyles, carnivores, rodents, insectivores and primates. The genus *Trichinella* normally occurs in mammals but experimentally develops in chicken, golden hamster and only in young Chinese hamster [103].

It is strongly felt that the present day Ascarids are of marine origin and Anisakidae is their ancestor which parasitises sea mammals, birds and elasmobranchs and, therefore, advantageously requires an intermediate host to complete the life cycle. The members of the Ascaridae are widely distributed in mammals, birds, reptiles and amphibians (see [127]). Though some members of this group have a direct life cycle, a few still demonstrate their phylogenetic reminiscence by the abnormal migration in their hosts which is nothing but a "recapture" (see [56]). Thus, while *Ascaridia galli* of birds has almost lost this instinct, *T. canis* of carnivores has modified this to an *in utero* infection of the young ones. *T. vitulorum*

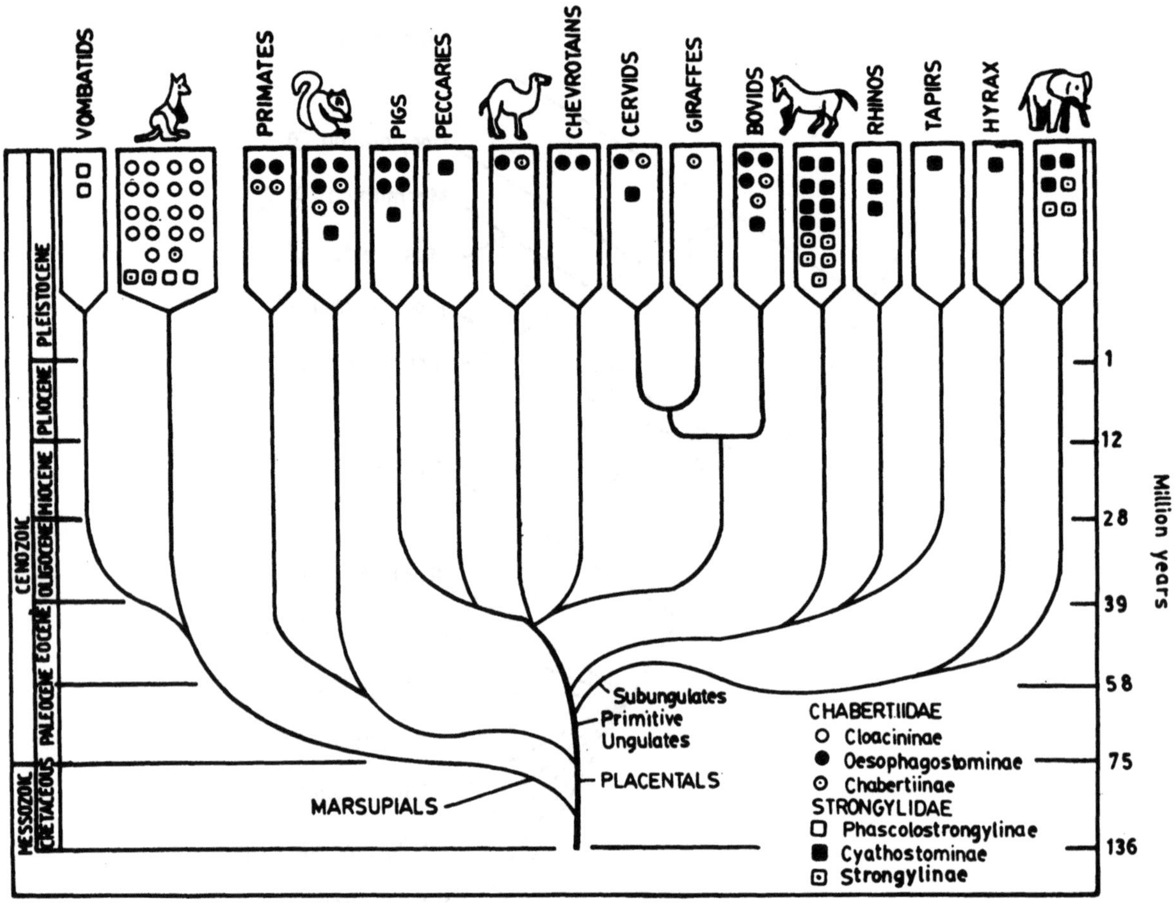

Fig. 15. Dendrogram of evolution of mammals showing number of genera of each strongyloid family and subfamily in each host group (After Lichtenfels, 1979).

of bovines further advanced it to follow the transmammary route to become adult in the new borns. *Parascaris equorum*, a member of this group is very host specific developing only in equines. *Ascaris suum*, however, grows in both sheep and goat and *A. lumbricoides* in sheep, buffalo and Malayan giant squirrel beside their normal hosts, pig and humans. *T. vitulorum* reaches adult stage in goats in addition to its known hosts—cattle and buffalo (see also reviews by Otto [90]; Adamson [2] and Chapter 3).

The Oxyurida has a distinct history of evolution and is the only group which has radiated both in vertebrate and invertebrate hosts (see review [3]), while the genus *Oxyuris* quite atypical morphologically has adapted life to perissodactyles and paenungulates. The atypical oxyurids do not occur in birds but these hosts sometime become parasitised by the members of the genus *Subulura* which has, beside an arthropod host, unrelated definitive hosts like primates and monkeys.

Cameron [25] argued that there is no evidence suggesting that the genus *Strongyloides* is related to the bursate nematodes, the largest group of zoo-parasitic nematodes which essentially has similar free-living stages. This genus, like oxyuridans, is widely distributed in mammals and birds in addition to reptiles and amphibians.

Recently, Lichtenfels [67] reviewed classification and phylogeny of strongylida 1982 through 1985 (see Figs. 15, 16 and 17). It has been suggested that the members having Y-shaped ovejectors and dorsal rays with six branches are parasitic in perissodactyles while those possessing J-shaped objectors are found in artiodactyles, marsupials, rodents and primates (see Lichtenfels [66]; Beveridge [12]; Durette-Desset [40]; Gibbons & Khalil [46]).

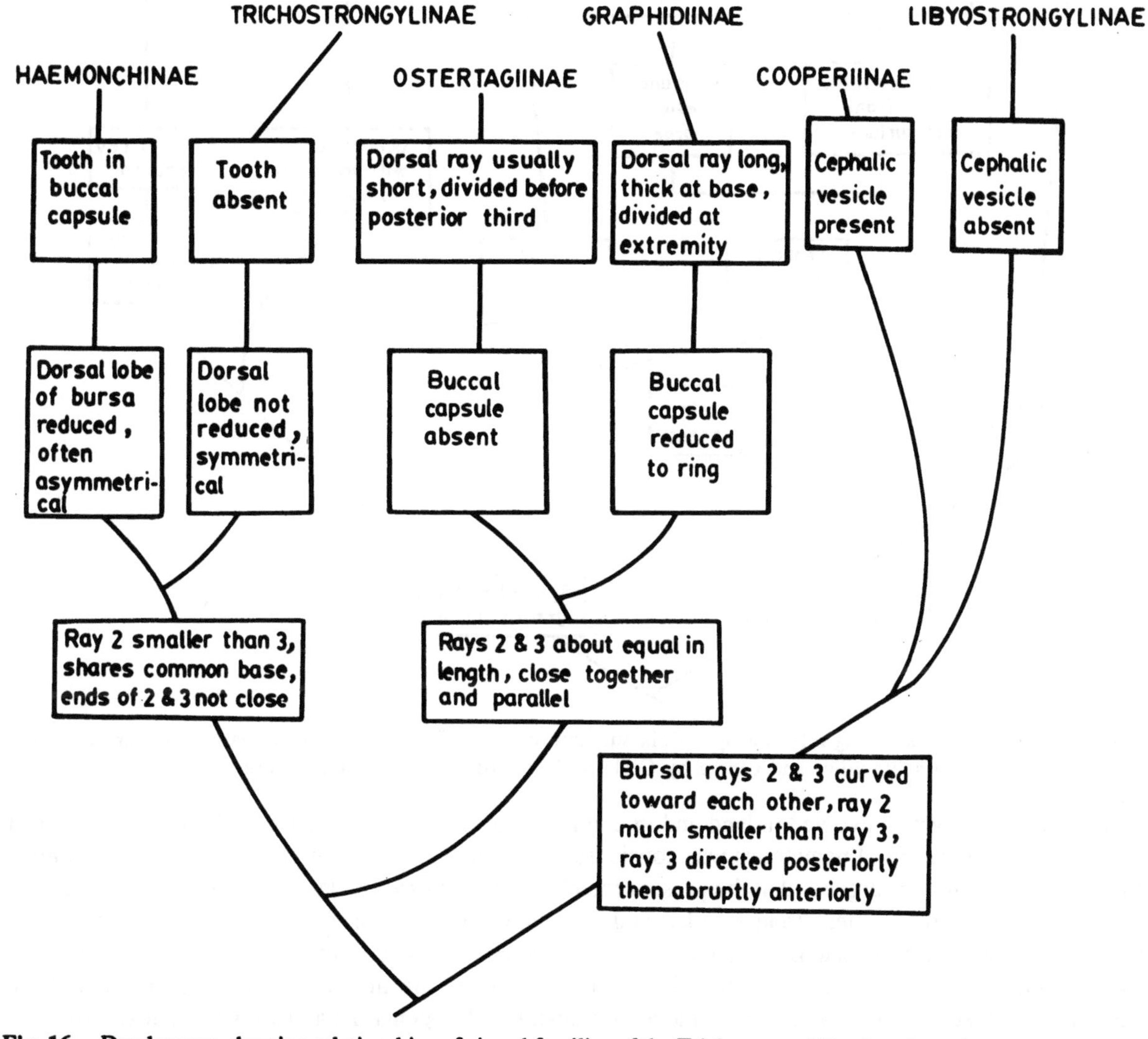

Fig. 16. Dendrogram showing relationships of six subfamilies of the Trichostrongylidae based on characters described in a key to subfamilies by Durette-Desset (1983) (original figure).

The members of the Trichostrongylidae, usually small nematodes with rudimentary buccal capsule, parasitise all classes of vertebrates beside amphibians (see review [41]). *T. axei* is the only species which develops experimentally in rabbit, hamster and guinea pig.

The Metastrongylidae is another group of small nematodes with a rudimentary buccal capsule and poorly-developed bursa and is confined to mammals. Anderson [5] believed that heteroxenity and

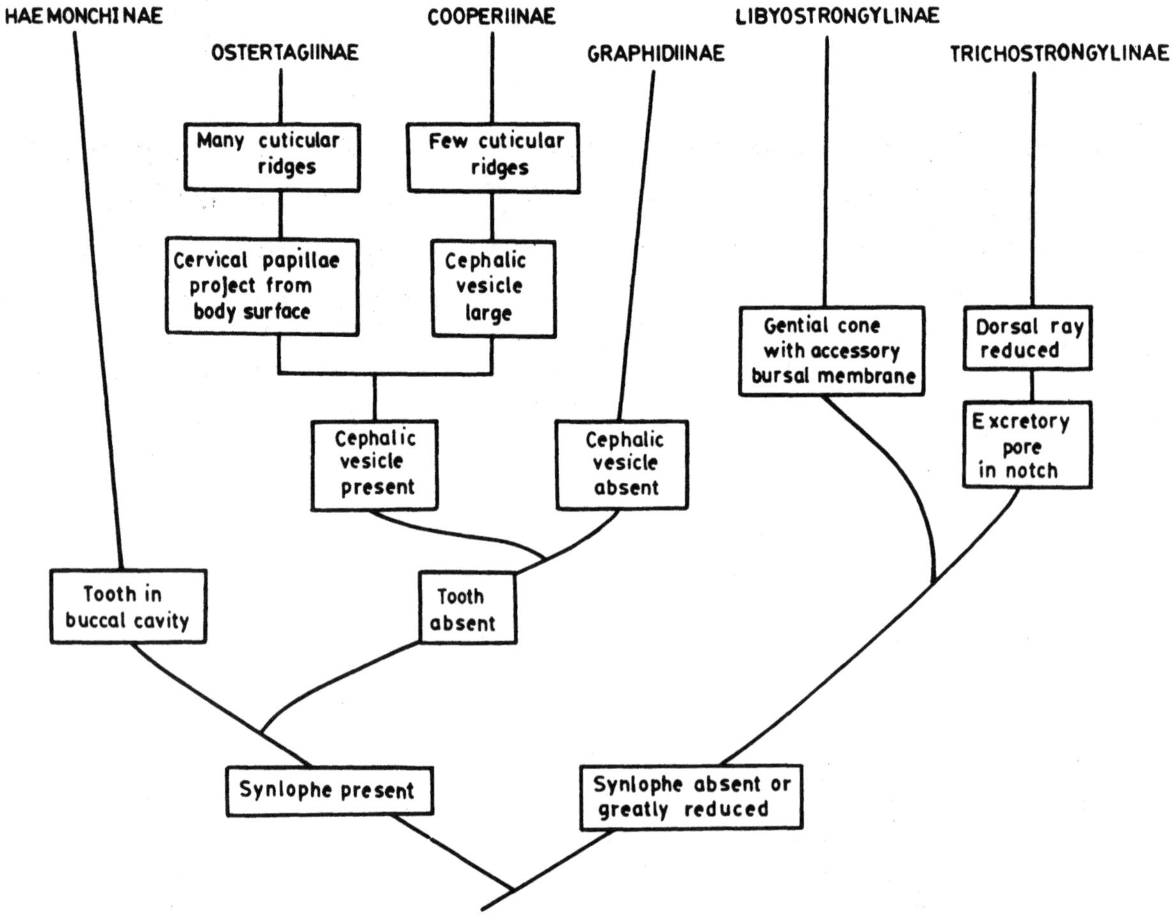

Fig. 17. Dendrogram showing relationships of six subfamilies of the Trichostrongylidae based on characters used in a key to subfamilies and genera by Gibbons & Khalil (1982) (original figure).

paratenesis in this group evolved in land and in carnivores before the appearance of the Pinnipedia. Host specificity in this group of nematodes was considered to be developed to a high degree in the final host. Pryadko [98], however, argued that the present members of Protostrogylidae, parasitic in ruminants and lagomorphs are more ancient than reptiles and mammals, whereas original Metastrongyloidea were parasitic in amphibia of the late Devonian era. Thus, Dougherty [37] had rightly remarked, "the strongylina by no means preserved strict host-specificity, but enough of the ancestral pattern appears to have survived to make possible certain reasonable conclusions of a general nature on the pattern of their evolution".

The spirurids commonly found in the stomach, are not very host specific and are parasitic in mammals, birds, reptiles and amphibians. Some authors believe that this group has evolved from Ascaridida (see [27]). Maggenti [72] suggested that spirurids were originally parasites of pre-historic dinosaurs which later became extinct and their worms subsequently adapted to ecologically analogous mammals, birds, and fishes. As a result of extinction of insectivorous dinosaurs and disappearance of free-living and monoxenous ancestors of spirurids, this group finally became a "taxonomic island".

The filarids are essentially somatic parasites with a secluded life style both in the intermediate as well as wide spectrum of definitive hosts like mammals, birds (see Chapter 3; Table 13), reptiles and amphibians. Contrary to spirurids, this group often exhibits high host specificity. Thus, *Wuchereria bancrofti* has been experimentally developed in Mongolian gerbil and Taiwan monkey and, *Brugia malayi* and *B. pahangi* in the jird; however, the jird could not be infected with *W. bancrofti, Loa loa loa, Onchocerca volvulus* and *Mansonella ozzardi* (see [130]).

CONCLUSIONS

Parasites constitute good model systems for studying general evolutionary principles as they have evolved along with their hosts. Direct evolutionary correlation has been observed in various groups of helminths and their hosts. Host-parasite relationship is usually a stable and balanced state on which selection continues to act. Coevolutionary patterns in parasite-host associations indicate that both classes of organisms show a phylogenetic retention of several traits that are sufficient to follow reconstruction of their historical relationships. Examples of coevolutionary association have been observed in different groups of parasitic helminths. A coherent explanation of the integrated data from ecology, systematics and developmental biology to study coevolutionary process has long been the goal of parasitologists.

Parasites have evolved from their free-living ancestors. The digenetic trematodes possibly were parasites of molluscs, where they became sexually mature. It has been postulated that the digeneans (and oncophoreans (monogeneans and cestodes) had separate origins from free-living rhabdocoele-like ancestors. However, more information is needed, particularly, about cestodes from marine habitats before a complete pattern of cestode ontogeny becomes evident. In making phylogenetic inferences from stages in the life cycle of the main groups of parasitic flatworms, the digeneans, monogeneans and cestodes, it will be necessary to attempt to settle the major difference of opinion about their life cycles. The acanthocephala probably originated as parasites of freshwater fish, with a single invertebrate host—an arthroped. Parasitic nematodes are descendants of several independent parasitic events that have given rise to lineages differing greatly in number of species.

Our knowledge of parasitism is limited and there is dearth of basic information on host-parasite relationship, particularly, about the process which regulate/influence infection, migration and development of infective stages. Undoubtedly, there is enough gap in our understanding of host-parasite interactions at molecular level. Though some information is now available on hormones and growth factors, much yet remains to be known on the mode of action of ecdysteroids reported in helminths and their intermediate hosts. Finally, understanding of the nature of parasitism will depend on characterization of the genetic dependence of parasites on their hosts.

Host specificity is one of the fundamental characteristics of parasitism. Phylogenetic specificity is dependent on ecological or physiological factors or a combination of both. Intraspecific differences of helminths in host specificity are receiving renewed impetus. Among trematodes, host specificity is greater at their larval forms as compared to adult stage. Unlike trematodes, the range of host specificity in adult cestodes varies considerably, it being very strict in some cases and unexpectedly low in other instances. However, larval tapeworms frequently tolerate a much wider host range. In acanthocephalans, the specificity is directly related to the type of their intermediate host(s). The host specificity among zoo-parasitic nematodes is perplexing because of multitude of associations between this diverse group of parasites and their hosts from entirely free-living stages to obligatory parasites.

The genesis of host specificity of helminths is now subjected to the discovery of newer facts from the upcoming fields of molecular biology and genetic engineering.

ACKNOWLEDGEMENTS

The first author is grateful to several authors and to the following editors and publishers for granting permission to use material: Academic Press Inc. (London) Ltd.; Cambridge University Press, U.K.; S. Karger AG, Basel; Journal of Helminthological Society, Washington; American Society of Parasitologists and Australian Academy of Science in addition to those acknowledged in the legends. Special thanks to Professor Emeritus J.D. Smyth for the permission to use several important figures which have considerably improved this Chapter. Thanks are also due to Professor S.P. Ahuja, PAU, for the suggestions in preparation of Figs. 2 and 11; Dr. S.S. Saini, Assistant Professor, College of Veterinary Science, PAU and Dr. S.K. Dey, Assistant Professor, College of Agriculture, PAU, for critically going through the final manuscript.

REFERENCES

1. Adamson, A. Phylogenetic analysis of the higher classification of the nematoda. Can. J. Zool. **65**: 243–255 (1987).
2. Adamson, M.L. Modes of transmission and evolution of life histories in zooparasitic nematodes Can. J. Zool. **64**:1375–1384 (1986).
3. Adamson, M.L. Evolutionary biology of the Oxyurida (Nematode): biofacies of a haplodiploid taxon. Adv. Parasitol. **28**:175–228 (1989).
4. Agrawal, M.C. and Shah, H.L. A review on *Schistosoma incognitum* Chandler, 1926. Helminthol. Abstr. **58**: 239–251 (1989).
5. Anderson, R.C. Host-parasite relations and evolution of the Metastrongyloidea (Nematoda). Mem. Mus. Natl. Hist. Nat. Ser. A. Zool. **123**: 129–132 (1982).
6. Anderson, R.C. The origins of zooparasitic nematodes. Can. J. Zool. **62**: 317–328 (1982).
7. Anderson, R.C. Nematode transmission patterns. J. Parasitol. **74**: 30–45 (1988).
8. Baugh, S.C. A century of schistosomiasis in India: human and animal. Riv. Iber. Parasitol. **38**: 435–472 (1978).
9. Befus, A.D. and Bienenstock, J. Factors involved in symbiosis and resistance at mucosa parasite interface. Prog. Allergy **31**: 76–177 (1982).
10. Befus, A.D., Lee, T. Ernst, P. Egwang, T., McElory, P., Gauldie, J. and Bienenstock, J. Unique characteristics of local responses in host resistance to mucosal parasitic infections. Vet. Parasitol. **20**: 175–194 (1986).
11. Behnke, J.M. Evasion of immunity by nematode parasites causing chronic infections. Adv. Parasitol. **26**: 1–71 (1989).
12. Beveridge, I. Evolution of strongyloid nematodes of Australian marsupials. Mem. Mus. Natl. Hist. Nat. Ser A. Zool. **123**: 87–91 (1982).
13. Beveridge, I. Coevolutionary relationships of the helminth parasites of Australian marsupials. In: *Coevolution and systematics* (A.R. Stone, and D.L. Hawksworth eds), Clarendon Press: Oxford, U.K. (1986). South Australian Dep. Agric., Adelaide, South Australia.
14. Bianco, A.E. and Maizels, R.M. Parasite development and adaptive specialization. *Symposia of the British Society for Parasitology*, Vol. 26: Parasitology **99**: S 113– S 123 (1989).
15. Brooks, D.R. Testing the context and extent of host-parasite coevolution. Syst. Zool. **28**: 200–307 (1979).
16. Brooks, D.R. Analysis of host-parasite co-evolution. Int. J. Parasitol. **17**: 291–297 (1987).
17. Brooks, D.R. macroevolutionary comparisons of host and parasite phylogenies. Ann. Rev. Ecol. Syst. **19**: 235–259 (1988).
18. Brooks, D.R. The phylogeny of the cercomeria (Platyhelminthes: Rhabdocoela) and general evolutionary principles. J. Parasitol. **75**: 606–616 (1989).
19. Brooks D.R.: *Parasites as model systems in comparative evolutionary biology*. Sminthsonian Institution University Press: Spring (1993).
20. Brooks, D.R. and O'Grady, R.T. Crocodilians and their helminth parasites: Microevolutionary considerations. In: *Symposium on the biology of the crocodilia* (V.A. Lance, organizer), Ann. Meets. Am. Soc. Zool., Dec. 27–30, 1987, New Orleans, LA. Am. Zool. **29**: 873–883 (1989).

21. Brooks, D.R., O'Grady, R.T. and Glen, D.R. The phylogeny of the cercomeria Brooks, 1982 (Platyhelminths) Proc. Helm. Soc. Wash. **52**: 1–20 (1985).
22. Bryant, C. and Flockhart, H.A. Biochemical strain variation in parasitic helminths. Adv. Parasitol. **25**: 275–319 (1986).
23. Cadigan, F.C., Jack, S.S., Prayoth, T. and Verachart, C. The lar gibbon as definitive intermediate host of *Taenia solium*. J. Parasitol. **53**: 844: (1967).
24. Cameron, T.W.M. The species of *Enterobius* Leach in primates. J. Helminthol. **7**: 161 192 (1929).
25. Cameron, T.W.M. Host specificity and the evolution of helminth parasites. Adv. Parasitol. **2**: 1–34 (1964).
26. Capron, A. and Dessaint, J.P. Molecular basis of host-parasite relationship: towards the definition of protective antigens. Imunol. Rev. No. 112, 27–48 (1989).
27. Chabaud, A.G. Keys to the genera of the superfamilies Cosmocercoidea, Seuratoidea, Heterokoidea and Subuluroidea. In: *CIH Keys to the nematode parasites of vertebrates* No. 6. Commonwealth Agricultural Bureaux: Buckinghamshire, England (1978).
28. Chowdhury, N.: Effects of fat soluble vitamins (Vitamin A, D_3 and E) on axenically *in vitro* growth of *Hymenolepis microstoma*. Z. Parasitenkunde. **56**: 29–38 (1979).
29. Chowdhury, N., Korenaga, M., Mimori, T. and Tada, I. *In vitro* culture of *Strongyloides ratti* larvae. Jpn. J. Parasitol. **34**: 229–235 (1985).
30. Chowdhury, N. and Rycke, P.H. de. Structure, formation and functions of calcareous corpuscles in *Hymenolepis microstoma*. Z. Parasitenkunde. **53**: 159–169 (1977).
31. Chowdhury, N. and Rycke, P.H. de. Effects of pancreatin on axenically *in vitro* growth of *Hymenolepis microstoma*. Jpn. J. Parasitol. **28**: 261–269 (1979).
32. Chowdhury, N., Rycke, P.H. de and Van Grembergen, G. Effects of ox bile on the growth and organogenesis of *Hymenolepis microstoma in vitro*. Z. Parasitenkunde. **70**: 73–80 (1984).
33. Chowdhury, N. and Singh, A.I. Role of calcareous corpuscles in the embryonic organization in *Raillietina* spp. Z. Parasitenkunde. **56**: 309–312 (1978).
34. Chowdhury, N. and Singh, R. Distribution of zinc in parasitic helminths. J. Helminthol. **63**: 149–152 (1989).
35. Clegg, J.A. and Smyth, J.D. Growth, development and culture methods: parasitic platyhelminths. In: *Chemical zoology*. Vol. II. (M. Florkin and B.T. Scheer eds), Academic Press: London (1968).
36. Combes, D.E., Fournier, A., Jourdane, J., Mone, H., Morand, S. and Theorn, A. Schistosomatidae (Trematoda): some views on their origin and evolution. Rev. Iber. Parasitol. **51**: in press (1992).
37. Daugherty, E.C. Evolution of zooparasitic groups in the phylum nematoda, with special reference to host distribution. J. Parasitol. **37**: 353–378 (1951).
38. Davis, G.M. Snail hosts of asian *Schistosoma* infecting man: evolution and coevolution. Malacol. Rev. (Suppl.) **2**: 195–238 (1980).
39. Dick, T.A., Daugherty, D.A. and Wassom, D.L. *Trichinella spiralis* infections of inbred mice: genetics of the host response following infection with different *Trichinella* isolates. J. Parasitol. **74**: 665–669 (1988).
40. Durette-Desset, M.C. Keys to genera of superfamily Trichostrongyloidea. In: *CIH Keys to the nematode parasites of vertebrates*. No. 10. (R.C. Anderson, A.G. Chabaud and S. Willmott eds), Commonwealth Agricultural Bureau: Buckinghamshire, England (1983).
41. Durette-Desset, M.C. Trichstrongyloid nematodes and their vertebrate hosts: reconstruction of the phylogeny of a parasitic group. Adv. Parasitol. **24**: 239–306 (1985).
42. Ehlers, U. Phylogenetic relationships within the platyhelminths. In: *The origins and relationships of lower invertebrates*. (S. Conway Morris, J.D. George, R. Gibson and H.R. Platt eds), Oxford University Press: Oxford (1985).
43. Flisser, A., Espinoza, B., Tovar, A., Plancarte, A., and Correa, D. Host-parasite relationship in cysticercosis: immunologic study in different compartments of the host. Vet. Parasitol. **20**: 95–102 (1986).
44. Freeman, R.S. Ontogeny of cestodes and its bearing on their phylogeny and systamatics. Adv. Parasitol. **11**: 481–557 (1973).
45. Fülleborn, F. On the larval migration of some parasitic nematodes in the body of the host and its significance. J. Helminthol. **7**: 15–26 (1929).

46. Gibbons, L.M. and Khalil. L.F. A key for the identification of genera of the nematode family Trichostrongylidae Leiper, 1912. J. Helminthol. **56**: 185–233 (1982).
47. Gibbs, H.C. Hypobiosis in parasitic nematodes—an update. Adv. Parasitol. **25**: 129–174 (1986).
48. Glen, D.R. and Brooks, D.R. Parasitological evidence pertaining to the phylogeny of the hominoid primates. Biol. J. Linn. Soc. **27**: 331–354 (1986).
49. Harvell, C.D. The evolution of iducible defence. *Symposia of the British Society for Parasitology*, Vol. **27**: Parasitology **100**: S53–S61 (1990).
50. Haslewood, G.A.D. *Bile salts*, Methuen: London (1967).
51. Hopkins, C.A. and Andreassen, J. Inhibition of growth of a tapeworm *Hymenolepsis diminuta* in its normal host (rat). Int. J. Parasitol. **21**: 47–55 (1991).
52. Hopkins, C.A., Subramanian, G. and Stallard, H. The development of *Hymenolepis diminuta* in primary and secondary infections in mice. Parasitology **64**: 401–412 (1972).
53. Horak, I.G. Paramphistomiasis of domestic ruminants. Adv. Parasitol. **9**: 33–72 (1971).
54. Hurd, H. Physiological and behavioural interactions between parasites and invertebrate hosts. Adv. Parasitol. **29**: 271–318 (1990).
55. Hyde, J.E. *Molecular parasitology*. Open University Press: Milton Keynes (1990).
56. Inglis, W.G. Patterns of evolution in parasitic nematodes. In: *Evolution of parasites* (A.E.R. Taylor ed), Blackwell Scientific Publications: Oxford (1965).
57. Inglis, W.G. Speciation in parasitic nematodes. Adv. Parasitol. **9**: 185–223 (1971).
58. Jackson, H.C. The role of blood in helminth nutrition. Helminthol. Abstr. **56**: 427–434 (1987).
59. Keymer, A.E. and Read, A.F. The evolutionary biology of parasitism. Parasitol. Today **6**: 2–3 (1990).
60. Klein, J. *Immunology*: *The science of self-non-self discrimination*. Willey Interscience: New York (1982).
61. Koshy, T.J., Achuthan, H.N. amd Alwar, V.S. Cross transmissibility of *Schistosoma nasale* (Rao, 1933) infection between cattle and buffaloes. Indian vet. J. **52**: 216–218 (1975).
62. Kumar, V., Geerts, S. and Mortelmans, J. Failure of 74 days old *Cysticercus bovis* to develop in anthropoid apes, monkeys and hamsters. Ann. Soc. belge Med. trop. **57**: 181–184 (1977).
63. Lapage, G. *Veterinary Parasitology*. Oliver & Boyd: Edinburg (1956).
64. Larsh, J.E. and Race, G.J. Allergic inflamation is a hypothesis for the expulsion of worms from tissues: a review. Exp. Parasitol. **37**: 251–266 (1975).
65. Leid, R.W. Parasites and complement. Adv. Parasitol. **27**: 131–168 (1988).
66. Lichtenfels, J.R. A conventional approach to classification of the strongyloidea, nematode parasites of mammals. Am. Zool. **19**: 1185–1194 (1979).
67. Lichtenfels, J.R. Phylogenetic inference from adult morphology in the nematoda; with emphasis on the bursate nematodes, the strongylida; advancements (1982–1985) and recommendations for further work. Int. J. Parasitol. **17**: 269–279 (1987).
68. Lie, K.J., Jeog, K.H. and Heyneman, D. Acquired resistance to *Schistosoma mansoni* in *Biomphalaria glabrata*. Int. J. Parasitol. **13**: 301–304 (1983).
69. Llewellyn, J. The evolution of parasitic platyhelminths, In: *Evolution of parasites* (A.E.R. Taylor ed), Blackwell Scientific Publications: Oxford (1965).
70. Llewellyn, J. Phylogenetic inference from platyhelminth life-cycle stages. Int. J. Parasitol. **17**: 281–289 (1987).
71. Mackiewicz, J.S. Caryophyllidea (Cestoidea): evolution and classification. Adv. Parasitol. **19**: 139–206 (1981).
72. Maggenti, A. *General nematology*. Springer Verlag: Heidelberg (1981).
73. Malviya, H.C. The susceptibility of mammals to *Fasciolopsis buski*. J. Helminthol. **59**: 19–22 (1985).
74. Manter, H.W. Host specificity and other host relationships among the digenetic trematodes of marine fishes. In: *First symposium on host specificity among parasites of vertebrates*. Institute of Zoology, University of Neuchatel (1957).
75. May, R.M. and Anderson, R.M. Parasite-host coevolution. *Symposia of the British Society for Parasitology*, Vol. **27**: Parasitology **100**: S89–S101 (1990).
76. McInnis, A.J. A general theory of parasitism. Proc. Third Int. Congr. Parasitol. **3**: 1511–1512 (1974).
77. McLaren, D.J. Disguise as a strategem of parasitic organisms. Parasitology **88**: 597–611 (1984).
78. McLennan, D.A. and Brooks, D.R. Parasites and sexual selection: a macroevolutionary perspective. Quart. Rev. Biol. **66**: 255–286 (1991).

79. McManus, D.P. and Symth, J.D. Difference in the chemical composition and carbohydrate metabolism of *Echinococcus granulosus* (horse and sheep strains) and *E. multilocularis*. Parasitology **77**: 103–109 (1978).
80. McManus, D.P. and Smyth, J.D. Changing concepts in epidemiology and speciation. Parasitology **84**: 351–366 (1986).
81. Mercer, J.G. Developmental hormones in parasitic helminths. Parasitol. Today **1**: 96–100 (1985).
82. Miller, H.R.P. The protective muscosal response against gastrointestinal nematodes in ruminants and laboratory animals. Vet. Immunol. Immunopath. **6**: 167–259 (1984).
83. Minchella, D.J. Host life-history variation in response to parasitism. Parasitology **90**: 205–216 (1985).
84. Mitchell, G.F. Response to infection with metazoan and protozoan parasites in Mice. Adv. Immunol. **28**: 451–511 (1979).
85. Mitchison, N.A. The evolution of acquired immunity to parasites. Parastitology **100**: S27-S34 (1990).
86. Mitter, C. and Brooks, D.R. Phylogenetic aspects of coevolution. In: *Coevolution* (D.J. Futuyma and N. Slatkin eds), Sunderland, M.A., Sinauer Associates (1983).
87. Nelson, G.S. The epidemiological significance of intra-specific variations in helminths of medical importance; with particular reference to *Trichinella* and *Schistosoma*. In: *H.D. Srivastava commemoration volume* (K.S. Singh and B.K. Tandon eds), Indian Veterinary Research Institute, Izatnagar (1970).
88. Nicholas, W.L. The biology of Acanthocephala. Adv. Parasitol. **2**: 205–237 (1967).
89. Ogilvie, B.M. and Love, R.J. Cooperation between antibodies and cells in immunity to a nematode parasite. Transplant. Rev. **19**: 147–168 (1974).
90. Otto, G.F. Development of parasitic stages of nematodes. In: *Biology of Parasites* (E.J.L. Soulsby ed), Academic Press: New York (1966).
91. Parkhouse, R.M.E. Parasite evasion of the immune response. *Symposium of the British Society for Parasitology*, Vol. **21**: Parasitology **88**: 571–682 (1984).
92. Pearson, J.C. A phylogeny of life-cycle patterns of the Digenea. Adv. Parasitol. **10**: 153–189 (1972).
93. Petronijevic, T. and Rogers, W.P. Gene activity and the development of early parasitic stages of nematodes. Int. J. Parasitol. **13**: 197–199 (1993).
94. Phares, C.K. Plerocercoid growth factor: a homologue of human growth hormone. Parasitol. Today **3**: 346–349 (1987).
95. Price, P.W. General concepts on evolutionary biology of parasites. Evolution **31**: 405–420 (1977).
96. Price, P.W. *The evolutionary biology of parasites*. Princeton University Press: Princeton, New Jersey (1980).
97. Price, P.W. Evolution in parasite communities. Int. J. Parasitol. **17**: 209–213 (1987).
98. Pryadko, E.I. (The genesis of the prostrongylid fauna.) In: *Gel 'minty sel' skokhozyaistvennykh i okhotnich' epromyslivykh zhivotnykh*. (M.D. Sonin ed), "Nauka": Moscow, USSR (1984).
99. Rhode, K. Phylogenetic origin of the trematodes. Parasitol. Schriftenr. **21**: 17–27 (1971).
100. Richards, C.S. and Shade, P.C. The genetic variation of compitability in *Biomphalaria glabrata* and *Schistosoma mansoni*. J. Parasitol. **73**: 1146–1151 (1987).
101. Rickard, M.D. Larval taeniid cestodes—models for research on host-parasite interactions. In: *Parasite lives* (M. Cremin, C. Dobson and D.E. Moorhouse eds), University of Queensland Press: St. Lucia, Australia (1986).
102. Rickard, M.D. and Williums, J.F. Hydatidosis/cysticercosis, immune mechanisms and immunization against infection. Adv. Parasitol. **21**: 229–296 (1982).
103. Ritterson, A.L. Age dependency of tissue resistance to *Trichinella spiralis* in Chinese hamsters. J. Parasitol. **53**: 652–653 (1967).
104. Rogers, W.P. *The Nature of Parasitism. The relationship of some metazoan parasites to their hosts*. Academic Press: New York (1962).
105. Rogers, W.P. The interaction of juvenile hormone and carbon dioxide in the hatching of nematode eggs: its significance in the physiology of infective stages. Comp. Biochem. Physiol. **64A**: 77–80 (1979).
106. Rogers W.P. and Sommerville, R.I. The infectious process, and its relation to the development of early parasitic stages in nematodes. Adv. Parasitol. **6**: 327–372 (1968).
107. Rollinson, D. and Anderson, M.R. (eds). *Ecology and genetics of host-parasite interactions*. Academic Press: London (1985).

108. Roza, L. (Host-parasite relationships: coevolution or sequential evolution?). Parasitol. Hungarica **22**: 29–33 (1989).
109. Sahay, M.N. and Sahai, B.N. Studies on the susceptibility of the laboratory animals, kids, and lambs to experimental infestion with *Schistosoma nasale* Rao, 1933. J. Parasitol. **64**: 1135–1136 (1978).
110. Saleque, A., Chowdhury, N., Iyer, P.K.R. and Baruah, G. Induced *Taenia solium* cysticercosis in rhesus monkeys (*Macaca mulatta*): a clinico-pathological study. Ann. Trop. Med. Parasitol. **82**: 103–105 (1988).
111. Schwabe, C.W. and Kilejian, A. Chemical aspects of ecology of platyhelminths. In: *Chemical zoology*. Vol. II (M. Florkin and B.T. Scheer eds), Academic Press: London (1968).
112. Schwabe, C.W., Luttermoser, G.W., Koussa, M. and Ali, S.R. Serial passage of fertile hydatid cysts of *Echinococcus granulosus* in the absence of the definitive host. J. Parasitol. **50**: 260: (1964).
113. Simpson, A.J.G. (ed) Parasites and molecular biology: application of new techniques. *Symposia of the British Society for Parasitology*, Vol. **23**: Parasitology: **92**: S1: (1986).
114. Singh, A.I. and Chowdhury, N. Shape, size and pattern of distribution of calcareous corpuscles in some avian and mammalian cestodes. Proc. 8th. Wld Vet. Conf., Sydney, Australia (1977). MVSc. Thesis, Assam Agric. Univ., Guwahati, 1976.
115. Sinha, P.K. and Srivastava, H.D. Studies on *Schistosoma incognitum* Chandler, 1926 on the host specificity of the blood fluke. Indian vet. J. **42**: 335–341 (1965).
116. Smyth, J.D. Changing concepts in microecology, macroecology and epidemiology of hydatid disease. In: *Helminth Zoonoses* (S. Geerts, V. Kumar and J. Brandt eds), Martinus Nijhoff Publishers: Dordrecht (1987).
117. Smyth, J.D. and Halton, D.W. *The physiology of trematodes*. 2nd edn. Cambridge University Press: Cambridge (1983).
118. Smyth, J.D. and Haslewood, G.A.D. The biochemistry of bile as a factor in determining host specificity in intestinal parasites, with special reference to *Echinococcus granulosus*. Ann. N.Y. Acad. Sci. **113**: 234–260 (1963).
119. Smyth, J.D. and Mcmanus, D.P. *The physiology and biochemistry of cestodes*. Cambridge University Press: Cambridge (1989).
120. Smyth, J.D., Miller, H.J. and Hawkins, A.B. Further analysis of the factors controlling strobilization, differentiation and maturation of *Echinococcus granulosus in vitro*. Exp. Parasitol. **21**: 31–41 (1967).
121. Soulsby, E.J.L. *Helminths, arthropods and protozoa of domesticated animals*. 6th edn. Bailliere Tindall: London (1968).
122. Soulsby, E.J.L. The control of parasites: the role of the host. Proc. Helm. Soc. Wash. **44**: 28–43 (1977).
123. Soulsby, E.J.L. The immune system and helminth infection in domestic species. Adv. Vet. Sci. Comp. Med. **23**: 71–102 (1979).
124. Soulsby, E.J.L. (ed). Immune response in parasitic infections: *immunology, immunopathology and immunoprophylaxis*. Vol. I. Trematodes and Cestodes. C.R.C. Press Inc.: Boca Raton, Florida (1987).
125. Spasski, A.A. Anoplocephalate tapeworms of domestic and wild animals. In: *Essentials of cestodology*. (K.I. Skrjabin ed), Vol. I. (Transl. A. Birron and Z.S. Cole. Israel Program for Scientific Translation, 1961.) (1951).
126. Sprent, J.F.A. Parasitism, immunity and evolution. In: *The Evolution of living organisms* (G.W. Leeper ed), Melbourne University Press; Melbourne (1962).
127. Sprent, J.F.A. Host parasite relationships of ascaroid nematodes and their vertebrate hosts in time and space. Mem. Mus. Natl. Hist. Nat. Ser A. Zool. **123**: 255–263 (1982).
128. Srivastava, H.D. and Dutt, S.C. Studies on *Schistosoma indicum*. Indian Council of Agricultural Research Series No. 34. I.C.A.R.: New Delhi (1962).
129. Stunkard, H.W. Studies on north American blood flukes. Bull. Am. Mus. Natl. Hist. **48**: 165–221 (1923).
130. Suswillo, R.R., Nelson, G.S., Muller, R., McGreevy, P.B., Duke, B.O.L. and Denham, D.A. Attempt to infect jirds (*Meriones unguiculatus*) with *Wuchereria bancrofti, Onchocerca volvulus, Loa loa loa* and *Mansonella ozzardi*, J. Helminthol. **51**: 132–133 (1977).
131. Terry, R.J. Parasites and molecular biology. Experientia **47**: 115–118 (1991).

132. Thompson, R.C.A. and Lymbery, A.J. The nature, extent and significance of variation within the genus *Echinococcus*. Adv. Parasitol. **27**: 209–258 (1988).
133. Ulmer, M.J. Site finding behaviour in helminths in intermediate and definitive hosts. In: *Ecology and Physiology of parasites* (A.M. Falis ed), Adam Hilger, Ltd.: London (1971).
134. Vickerman, K. The impact of future research on our understanding of parasitism. Int. J. Parasitol. **17**: 731–735 (1987).
135. Von Brand, T. *Biochemistry of parasites*. 2nd edn. Academic Press Inc.: London (1973).
136. Von Brand, T. *Biochemistry and physiology of endoparasites*. Elsevier/North Holland Biomedical Press:Amsterdam (1979).
137. Wakelin, D.Immunity to intestinal parasites. Nature (Lond.) **273**: 617–620 (1978).
138. Wakelin, D. Immunity to helminths and prospects for control. *In: Critical reviews in tropical medicine.* Vol. 2 (R.K. Chandra ed), Plenum Press: New York (1984).
139. Wakelin, D. *Immunity to parasites. How animals control parasitic infections.* Edward Arnold: London (1984).
140. Wakelin, D. Immunity to helminths. EOS Riv. Immunol. Immunofarmacol. 6: 35–46 (1986).
141. Wakelin, D. The role of the immune response in helminth population regulation. Int. J. Parasitol. **17**: 549–557 (1987).
142. Warren, K.S. The evolution of parasitology. In: *Parasitic infections* (J.H. Leech, M.A. Sande and K.H. Root eds), Churchill Livingstone: New York (1988).
143. Zhuravets, S.K. Periods of maturation of *Echinococcus granulosus* in dogs. Byull. vses. Inst. gel'minth. K.I. Sktyabina. **32**: 28–32 (1982).
144. Zuk, M. Reproductive strategies and disease susceptibility: an evolutionary view point. Parasitol. Today **6**: 231–233 (1990).

2. Epidemiology of Some Helminth Infections of Domesticated Animals in the Tropics with Emphasis on Fasciolosis and Parasitic Gastroenteritis

S.N. Chiejina
Department of Veterinary Parasitology and Entomology
Faculty of Veterinary Medicine
University of Nigeria, Nsukka, Nigeria

INTRODUCTION

Helminthosis remains one of the world's most prevalent and economically important parasitoses of man and his domesticated animals. This is particularly the case in the developing countries, much of which lies within the tropical regions of the world and where, for most part, systems of livestock production, environmental and socio-economic conditions are highly conducive for the development, maintenance and transmission of infection.

The word 'tropics' may be defined simply as the region between the Tropics of Cancer and Capricorn. It embraces several territories and countries in the continent of Africa, including its off-shore Islands, Central and South America , the Carribean, the Indian sub-continent, western and south-eastern Asia, northern Australia and the Oceanic Islands. These areas cover a variety of topographies, sub-climates and vegetation complexes, montane forests, humid equatorial and tropical rain-forests. Associated with the varieties of subclimates and vegetation complexes are variations in short-term weather systems between and, sometimes, within the different territories that make up this vast geographical entity. Payne [217] therefore considered it 'meaningless' to talk about a 'typical tropical climate'.

Nevertheless, the definition proposed by Payne, [216], which is based on the climatic-ecological concept of Blumenstock [36] is useful and will be adopted in this work. According to Payne [216], the tropics comprises all those areas north and south of the equator, within the bounds of the 18°C isotherm for the coolest month of the year. The tropics, therefore, extend northwards to a maximum of 30° latitude and southwards to a maximum of 38° latitude. Also in this area the diurnal temperature range is greater than the annual temperature range. Day length is virtually constant throughout the region, the variability being least, about 30 minutes, at the equator and greatest, approximately 60+ minutes, at the Tropics [217].

Parasitologically, climate may be defined as the synthesis of the day-to-day values of the meteorological elements affecting a locality, expressed as seasonal or annual values, while weather describes the day-to-day meteorological conditions which constitute the climate [280]. Thus, climate influences the general epidemiological pattern, while weather determines the timing and magnitude of specific events that make up that pattern.

From the standpoint of epidemiology of helminthosis, the tropical environment may further be subdivided into two broad climatological-ecological zones [252] namely, that in which environmental conditions are permanently or continuously favourable for development and/or transmission of infection and that in which environmental conditions are intermittently favourable. The former situation is characteristic of the wet, subhumid and humid (rainforest) zones while the latter occurs in some highland regions and in the arid, semi-arid and drier parts of the subhumid tropics. The unique climatic conditions found at very high altitudes (montane climate) in some parts of the tropics [217] are also intermittently favourable for parasite development and transmission but will not be considered here since they are more 'temperate' than 'tropical' and overall support relatively small populations of domesticated animals.

This chapter presents and analyses available information on the epidemiology of two major helminth infections of livestock, other than poultry, in the tropics namely, fasciolosis and parasitic gastroenteritis (PGE). The choice of these two helminthoses has been influenced partly by the author's assessment of their overall importance in the livestock industries and economy of the region and partly by constraints of space. Consequently, an exhaustive review has not been attempted and several other important helminths and their associated diseases have been left out. Nevertheless, it is hoped that these two examples and the information provided will serve to illustrate and highlight the uniqueness of the tropical climate, livestock production systems and husbandry practices and how these and other epidemiological determinants influence the transmission dynamics and incidence of helminthoses in general.

The standardized nomenclature of animal parasitic diseases (SNOAPAD) proposed by Kassai et al [150] has been adopted in this work.

FASCIOLOSIS

Introduction

Fasciolosis has been rightly described by Fabiyi [97] as the most important single helminth infection of cattle. This is not only because of its high prevalence rates but also on account of its enormous production losses in this animal which, in some countries such as Burma [124] and Nigeria [98], run into millions of US dollars annually. The disease in some areas is no less important in buffaloes and small ruminants (sheep and goats), acute disease being commoner in small than in large ruminants [39, 123, 124, 135, 247, 262].

Estimates of its prevalence rate in tropical livestock are largely conjectural but available figures suggest that infection rates between 30% and 90% are common in the highly endemic areas of East, West and Central Africa [20, 81,161, 173, 198]. The situation appears to be much the same in tropical Asia, particularly in some parts of Bangladesh [224], Burma [124], India [255], Indonesia [93, 261] and Thailand [265]. However, although a good deal is known about the prevalence and potential economic importance of the infection, few detailed studies appear to have been done on its epidemiology and control in the major climatic-ecological zones of the tropics, particularly, in host species other than cattle and sheep. Therefore, the information provided applies mainly to these two species and to some extent also to goats and buffaloes.

Aetiology, distribution and method of transmission

Fasciolosis, also referred to as liver fluke disease and liver rot, is caused by several species of fluke belonging to the genera *Fasciola* Linnaeus, 1758 and *Fascioloides* Ward, 1917. Only the former occurs naturally in the tropics while the latter is found predominantly in north America [40, 262]. The two known species of *Fasciola* which are parasitic in domesticated animals and, sometimes, man in the tropics are *F. gigantica* and *F. hepatica*. Their major domestic animal hosts are sheep, goat, cattle and buffalo. The horse, donkey, mule and camel are less susceptible but epidemiologically important while the domestic pig is naturally resistant to infection, especially with *F. hepatica* [39, 262].

Three other species of *Fasciola* namely, *F. jacksoni*, *F. nyanzae* and *F. tragelaphi* have been described from the Indian elephant, hippopotamus and the Situatunga antelope, respectively [40, 262].

F. hepatica is restricted in its dstribution to the more temperate high altitude (Highland) regions [88, 90, 142, 164, 255, 287]. *F. gigantica*, on the other hand, is a more widely distributed enzootic particularly in Africa, Indian subcontinent, parts of China, South East Asia, Phillipines and Hawaii [40, 164, 193, 194, 254, 255, 262, 287]. However, in some parts of Africa and the Indian subcontinent overlapping infections with both species may occur [39, 86, 128, 154].

The life-cycle of *Fasciola* species involves (1) the egg (2) a free-swimming larva, the miracidium (3) three other larval stages namely, sporocyst, redia and cercaria, which develop in the snail intermediate host and (4) the encysted cercaria or metacercaria, which is the infective stage and is present on herbage or vegetation. Metacercariae which encyst on the surface of water may trap air bubbles and float for long distances [39]. Infection is acquired through the consumption of herbage and or water containing metacercariae.

Factors influencing the life-cycle and epidemiology

Biology of Snail Vectors

Successful maintenance and transmission of fasciolosis depend to a large extent on the occurrence of the snails belonging to the family Lymnaeidae, which act as the intermediate hosts of the trematode. Information on the several species of lymnaeids which can fulfil this role, their geographical distribution and susceptibility to various species of *Fasciola* were discussed by Kendall [155] and Boray [38–40]. The snail vector of *F. gigantica* in tropical Africa is *Lymnaea natalensis*. The equivalent species in the Indian subcontinent are *L. auricularia* s.s, *L. rufescens* and *L. acuminata* and in Malaysia, *L. rubiginosa* [262, 265, 287, 305]. The primary snail vector of *F. hepatica* in these areas is *L. truncatula*, although *L. diaphena*, *L. columella*, *L. viridis* group [39] and *L. auricularia* s.s [255] may also be involved in the life-cycle in specific situations (see also Chapter 3).

The snail intermediate hosts of *F. gigantica* are truly aquatic and require large, permanent, still or slowly-flowing, clear bodies of water with abundant vegetation but little decaying organic matter. Therefore, amphibious conditions in water courses or other aquatic habitats which disappear during dry spells are unsuitable. Thus, under natural conditions, the ideal habitats are found in borders of permanent lakes, streams, irrigation channels, flood-prone areas and low-lying marshes [287]. By contrast, the snail intermediate hosts of *F. hepatica* are amphibious. Typical habitats include badly-drained land and shallow marshy areas, temporary springs and watercourses, drainage ditches, areas of seepage of springs and broken drains. According to Soulsby [262], even hoof prints of animals in clay soils as well as muddy areas in gateways and around drinking troughs may provide temporary habitats for this species of *Fasciola*.

In general, the formation, size and suitability of habitats are influenced by several biological, chemical and physical factors in the environment [209] which include (1) topography, soil type and geological formation of the area. These influence water drainage and retention (2) pH and salinity of the aquatic habitat (3) availability of food for the snails and (4) presence of man-made drainage and water management schemes. For example, *L. tomentosa*, the major snail vector of *F. hepatica* in Australia [39], thrives at a pH range of 6.0 to 7.0 and salinity range of 25 to 160 ppm [39, 62]. A drastic reduction in snail population occurs with increasing salinity following evaporation [39].

Detailed observations on the biology of the snail vectors of *F. hepatica* in Australia [37, 39] and Britain [208, 209, 210] have shown that in temperate climates maintenance of wet conditions with adequate balance between rainfall and evapotranspiration during dry periods, and plentiful food supply enhance the reproductive capacity, growth and survival of snails. The better the state of nutrition of snails, the larger is their size and that of the larval fluke burden which they can support, large snails harbouring almost ten times the number of developmental stages found in smaller, malnourished ones [153]. Hence the size of snails, and not the number, is the important factor in their vectorial capacity [156].

The ability of *L. truncatula* and other amphibious snails to survive in a state of dormancy during the winter (hibernation) and prolonged drought (aestivation) is another important factor in their biology and efficiency in transmission of infection [39, 262]. In temperate climates, hibernating and aestivating snails may survive in frozen or dried mud, as the case may be, for several months. On the return of favourable environmental conditions such snails quickly repopulate their habitat due to their high biotic potential. This is also accompanied by increased production and release of cercariae as a result of enhanced development of the larval stages which had accumulated in the snails, thereby creating large herbage populations of metacercariae [39, 262]. Equally important are short spells of drought followed by wet conditions and intermittent, sudden marked fluctuations in climatic factors like temperature and moisture in the habitat of dormant snails. These influences can lead to massive emissions of cercariae and heavy fluke burdens in animals [39].

The extent and significance of aestivation in tropical species of lymnaeids are less well documented. Boray [39] has suggested that all lymnaeids may aestivate for a considerable time during dry periods, in which case the phenomenon may also be of epidemiological importance in the tropics. On the other hand, Soulsby [262] stated that the snail vectors of *F. gigantica* can survive an amphibious existence but can aestivate for only very short periods.

Dispersal of snails by active and passive migration is an important aspect of their population biology and distribution within and between habitats [39, 304]. Lowered oxygen tension is apparently one of the factors which can initiate active migration of some species of amphibious lymnaeids out of stagnant pools [174] while their speed of migration depends on water temperature and stream gradient of the habitat. Studies on the Australian lymnaeid, *L. tomentosa*, showed that at temperatures of 4.5 to 6.2°C, it migrated up to 150 m in approximately 12 weeks. At higher temperatures, it travelled up to 150 m in 8 weeks [39]. Snails can also migrate passively by attaching their feet to the water surface and drifting with the water current for long periods. Flooding of habitats following heavy tropical rainfall also encourages passive migration and dispersal of snails. In large-scale flooding, snails can be carried to new habitats hundreds of miles away from their origin [39]. This aids re-colonisation of habitats and the establishment of new habitats and foci of infection, an important consideration in the planning of molluscicide applications in flood-prone areas.

There is good evidence that several biological agents, including some species of algae, bacteria, flies, fungi, nematodes and plants can adversely affect the reproduction and/or growth of snails and attempts have been made to apply this concept in the biocontrol of snail vectors of trematodes [39]. However, although some of these agents such as the larvae and adults of the sciomyzid flies, *Sepedon macroplus, Dichaetophora hendeli* and *D. biroi* are reported to be 'efficient killers' of snails [39] and, theoretically, could influence the population dynamics of the vector; there is little evidence that they are of practical value under field conditions. Indeed, the presence of sciomyzid larvae in snail habitats is probably coincidental, their numbers being determined by the availability of food, which includes snails. Similarly, birds which may consume substantial numbers of snails in certain habitats, are not considered to be epidemiologically important [39].

Biology of Fasciola

There are at least four crucial stages in the preparasitic development of *Fasciola*, each of which is directly influenced by climatic factors and must be successfully completed before transmission of infection can take place. These are: (1) development and hatching of eggs (2) infection of snail intermediate host by the free-swimming miracidium (3) parthenogenetic development of larval stages in the snail and (4) cercarial emission from the snail and formation of metacercariae on herbage.

The bionomics of *F. hepatica* eggs and miracidia have been studied by several workers [214, 241] and available data were summarised by Boray [39], Ollerenshaw & Smith [211] and Soulsby [262]. Relevant data on *F. gigantica* was provided by Dinnik & Dinnik [88, 90] from studies carried out in the Kenya Highlands, at an altitude of 6,800 ft. In general, fluke eggs are protected from desiccation and other adverse climatic factors by faeces. However, no significant development occurs as long as they are in contact with visible faecal material. Therefore, break up of host faeces and liberation of the eggs are essential for development and hatching to occur, provided also that a film of moisture is maintained over the surface of the freed eggs [209, 211].

Eggs of *F. hepatica* do not develop at temperatures below 10°C and hatching of fully-developed eggs takes up to 90–100 days at temperatures between 1°C and 11°C [39, 241]. There is an increased rate of development at temperatures ranging from 10 to 26°C, development time decreasing from 60 days at 12°C to 12 days at 26°C. Development is again inhibited at temperatures above 37°C. Thus, the eggs of this trematode are unable to develop and hatch during drought, ideal conditions being provided under cool temperate climates. During summer, the development and hatching times may also increase in stagnant pools with plenty of organic matter [39].

Development and hatching of *F. gigantica* eggs occur at higher temperatures. Development time is approximately 17 days at a constant temperature of 26°C, which is longer than for *F. hepatica* eggs at this temperature [39, 241]. Development and hatching of *F. gigantica* eggs must take place entirely under water. These processes are, therefore, largely controlled by the water temperature and are highly susceptible to desiccation and also to low temperatures. Development times from egg to miracidium during the warm and coldest seasons in the Kenya Highlands were 52 to 70 days and 90 days, respectively and hatching of eggs did not take place when the lowest minimum and highest maximum temperatures were 5.5°C and 19.5°C, respectively [88]. However, under optimum conditions, such as exist in permanent habitats in more typical (Lowland) tropical areas where temperature is not a limiting factor, development and hatching of eggs take place faster and all the year round [287].

Infection of snails also must take place in an aquatic environment, soon after the emergence of the miracidium and under warm conditions. It has been suggested that the initial attachment and penetration of the epithelial cells of *L. truncatula* is performed by the miracidium while the final entry of the larva into the snail is accomplished by the young sporocyst [80].

Parthenogenetic development of the subsequent larval stages is closely linked with the biology of the snail vector. The latter provides an ideal environment for rapid multiplication of the parthenitae through polyembryony. The reproductive potential of digenetic trematodes is so enormous that a single miracidium is theoretically capable of giving rise to millions of cercariae [262]. Boray [40] has estimated that approximately 4 to 5 generations of cercariae can develop in a year in tropical climates and about 4 in tropical highlands and subtropical climates. This contrasts with about 1 to 2 generations in temperate climates [262].

Accumulation of larval stages in aestivating and hibernating snails also provides an effective means of parasite survival during adverse climatic conditions and a guarantee of successful transmission when environmental conditions improve, particularly, in the case of *F. hepatica* in cool temperate and subtropical climates.

Dinnik & Dinnik [90] have estimated that development of the larval stages of *F. gigantica* in *L. natalensis* under field conditions in Kenya would be completed in 69 to 197 days, depending on the seasonal temperature. In aquaria maintained at a constant and more suitable temperature of 26°C, it took a minimum of 33 days. Since the development from egg to miracidium under similar conditions took 17 days [88], the whole process from egg to cercaria at this temperature was approximately 50 days.

Cercarial output of snails increases with rising temperature [155] but this may be partly offset by a corresponding increase in mortality of snails and by reduced infectivity of metacercariae [39]. Marked fluctuations in climatic factors in the snail habitat also induce spontaneous emission of cercariae, especially by aestivating snails [39]. The resultant sudden increase in herbage metacercarial infestation may in some situations give rise to major disease outbreaks. Cercariae encyst within a few hours of their leaving the intermediate host and the process of encystment requires abundant vegetation and warm, wet conditions. Cercariae of *F. gigantica* , in particular, encyst on fully submerged vegetation or herbage but occasionally encystment may occur on the water surface. In the latter situation some of the metacercariae may sink to the bottom of the body of water, becoming available when stirred up by the host as it wades into the water to drink [262].

The longevity of metacercariae under field conditions is limited as they are susceptible to desiccation and direct sunlight [41, 62, 213]. The longevity of *F. gigantica* metacercariae on grass blades is about 3 to 6 months [287] and some can survive on stored fodder such as rice straw for up to 4 months [262]. However, this appears also to depend on relative humidity (R.H) and the moisture content of the stored product. This suggests that in some situations stored fodder may constitute a potential source of *F. gigantica* infection.

In contrast, metacercariae of *F. hepatica* survive at relatively low temperatures [41], which is consistent with the greater adaptability of this trematode species to temperate climates. In laboratory studies by Boray [39] 100% of metacercariae survived and were infective after exposure to a temperature of 12 to 14°C for 6 months. Survival time, which was dependent on R.H was progressively shorter at lower or much higher temperatures. For example, a high proportion of metacercariae were viable and infective for at least 36 days at 25°C but only for 14 days at 30°C. Thus, in cool temperate and subtropical climates, the longevity of *F. hepatica* metacercariae on herbage is much greater in winter than in summer [208].

Metacercariae accidentally ingested by snails are passed out in faeces, unaffected by the latter's digestive processes and are fully infective for the definitive host [39, 305]. Although this is one possible method of re-cycling and passive dispersal of metacercariae within and between habitats, its epidemiological significance remains to be determined. A commoner and possibly more important method of passive dispersal of metacercariae is when cercariae which encyst on water surface are transported for long periods and distances by water current.

Livestock Husbandry Practices, Systems of Production and Patterns of Infection

The method of flock and herd management in any locality has a major influence on whether or not the host comes into contact with infection, if this is present, as well as on the frequency and duration of that contact. This has been little studied with regard to fasciolosis under tropical systems of livestock management and production and, therefore, only general statements on its epidemiological significance based on the limited available information are possible.

In the tropics, areas of high endemicity of fasciolosis correspond to the major livestock raising areas such as the dry savanna and Sahel zones of Africa. During the wet season there are abundant grazing and alternative sources of drinking water. This reduces the need for animals to graze near, and to drink from, particular permanent water holes. Moreover, snail habitats and pastures are constantly flooded and, therefore, snails and free-living larval stages of *Fasciola* are regularly flushed out and disseminated over a large area. In some cases animals are deliberately pastured away from wet marshy areas and permanent pools [287]. Even when snail habitats are localised, as in the case of *F. hepatica* endemic areas, animals would normally prefer to graze or browse on dry pasture and thereby aviod 'flukey' parts of the pasture [39], as long as this choice of grazing or browse is available to them. Consequently, only light infections are likely to be acquired during the wet season in most cases. In contrast, when the rains cease and the dry season sets in, there is a reduction of the available grazing areas and sources of drinking water for stock. This increases the need for animals to graze near, and drink from permanent dams, water holes and other snail habitats which are usually the only sources of grazing and drinking water at that time of the year. The attendant over-crowding and increased contamination of these areas with fluke eggs soon give rise to heavy infections in snails, a significant increase in herbage metacercarial density and disease outbreaks during the dry season and sometimes extending to the start of the succeeding rainy season, depending on the locality [20, 247, 287]. In those parts of West Africa, in which the dry season lasts from the end of October to March or April, *L. natalensis* populations start to increase around October, reaching a peak during November/December. This peak is maintained throughout the dry season [287]. A marked increase in herbage metacercarial density occurs by January and peak infestations during February/March. Consequently, outbreaks of acute fasciolosis are generally observed from late February and subacute and chronic disease a little later in the season and during the early part of the succeeding rainy season [171, 173, 247, 287]. A broadly similar pattern of infection has been described in Malawi by Mzembe and Choudhry [193, 194].

Nomadic herds and flocks which are constantly on the move usually run relatively little risk of heavy infection except when they accidentally graze in or drink from heavily contaminated areas at some point in their migratory circuit. However, widespread heavy infection is one of the consequences of disruption of normal migratory patterns of nomadic pastoralists by prolonged drought and other natural disasters which result in migration to unusually distant and unfamiliar territories, prolonged stay and over-crowding at few available water holes. These, coupled with malnutrition, climatic stress, concurrent infections and general lowering of body resistance were identified as being responsible for the unusually high incidence of helminthoses during the 1973 to 1974 drought in northern Nigeria [149, 245]. Transhumant flocks and herds also acquire significant infection from permanent snail habitats during the dry season but acquisition of infection ceases when they leave the permanent water holes at the end of the season [287].

The risk of fasciolosis is greater in intensively managed sedentary, than in migratory, herds and flocks. However, this risk may be considerably reduced where a high standard of grazing management and other prophylactic measures are routinely applied. Such managerial practices as avoidance of permanent snail habitats, provision of alternative watering places during the dry season, use of correct stocking rate, regular and strategic anthelmintic treatment of stock and, to some extent, various methods of snail control have a significant influence on the life-cycle of *Fasciola* and on the epidemiology of the disease [39, 40, 72]. In some situations mixed grazing of sheep and cattle may also influence the incidence of fasciolosis, especially in the former more susceptible host, as the less susceptible host would ingest and destroy numerous metacercariae which would otherwise become available to sheep [39].

The increasing use of flood plains, irrigated land, man-made lakes and dams for livestock rearing has major implications for the spread and incidence of fasciolosis and other snail-transmitted trematodiases [170, 247], particularly, since such places may be used for the various Government—sponsored schemes for the resettlement of nomadic pastoralists. In the permanently favourable environment existing in these areas infection would be present throughout the year and the optimum number of *F. gigantica* cercariae per annum [40] would be expected to be produced.

Relationship Between Fasciola and Snails

The susceptibility of lymnaeids to *Fasciola* and the adaptability of the latter to its intermediate host are important considerations in the epidemiology of fasciolosis. Of the two species of *Fasciola* parasitic in tropical livestock, *F. gigantica* has been shown in field observations and laboratory experiments to be capable of attaining full development in a variety of snails, including *L. truncatula*, the most important intermediate host of *F. hepatica* in the tropics. Some of the relevant information was summarised by Boray [39]. Therefore, the possibility of transmission of *F. gigantica* by alternative snail vectors should always be considered in those tropical areas in which such snails occur contemporaneously with the more natural vector.

Larval stages of *Fasciola* are true parasites of, and may cause severe pathology in, their snail host, producing acute, subacute and chronic disease states which, respectively, correspond to early infection, developmental and accumulation phases in the development of the parthenitae [39]. Each of these phases can result in snail mortality which, if significant, can influence the availability of infected snails and hence the incidence and intensity of fasciolosis.

It has been shown that a number of biological agents in the snail's habitat have a modulating influence on the balance between *Fasciola* and its snail vectors, with possible epidemiological implications. For example, the larvae of trematodes belonging to the family Echinostomatidae Poche, 1926, which also parasitise lymnaeids, feed on sporocysts and rediae of *Fasciola* and other trematodes, thereby inhibiting the latter's larval development and diminishing its pathogenicity for snails [39]. Consequently, they are considered as possible biological control agents for *Fasciola* species. Similarly, the worm, *Chaetogaster limnaei* (Oligochaeta), which feeds on miracidia is potentially capable of influencing the intensity of fluke infection in snails and, therefore, in definitive hosts. However, neither the occurrence nor the epidemiological significance of these biological agents has been adequately evaluated under tropical conditions.

Host-parasite Relationships

Boray [39] has classified some common domestic and synanthropic animal hosts of *Fasciola* spp. into 3 broad groups on the basis of their susceptibility and response to infection. This simple classification is useful in the consideration of the role of different host-parasite systems in the epidemiology of fasciolosis. The first group, the early resistance group is naturally resistant to the parasite, as their tissues are unsuitable for the establishment and survival of the parasite which, as a result, is quickly eliminated with minimal or no pathogenic effects on the host. Animals in this group include the domestic pig, dog and cat. The infection in these animals is, therefore, self-limiting.

The second group, the delayed resistance group, is moderately susceptible to infection which, however, is controlled in the early stages by 'mechanical resistance' and in the later stage of the infection by delayed host fibrous tissue reaction. The profound alteration in liver architecture associated with this reaction ultimately leads to the elimination of any parasites reaching the bile ducts, by imposing a 'mechanical barrier' to their feeding. The host, nevertheless, may show severe liver pathology resulting in high morbidity and mortality, particularly in young and malnourished animals. Therefore, the parasite readily completes its life-cycle and can be successfully transmitted to other susceptible hosts. The infection in this group is also self-limiting. Members of the group include cattle, buffalo, horse, donkey, mule, ass and man.

The third group, the low resistance group, is highly susceptible and shows severe early (acute) and delayed (chronic) tissue reactions but with insufficient hepatic and biliary fibrosis and cellular response to interfere with the development and survival of the parasite. Usually all ages of animal are susceptible and light infections may survive as long as the hosts. Both the acute and chronic stages of the disease are often accompanied by marked clinical symptoms and significant mortality. Some members of this group are sheep, goat, rabbit, rat and mice.

Boray [39] regarded the relationship between *Fasciola* spp. and most of their domestic animal hosts as basically 'very unbalanced' and considered it 'futile' attempting to single out the most successful host-parasite system. Moreover, some overlapping between groups certainly exists. Nevertheless, the above three broad categories give an indication of the potential roles of each group of hosts in the life-cycle, maintenance and transmission of infection. Thus, while most of the hosts in group one are largely irrelevant in the transmission of infection, those in groups two and three are crucial to the successful life-cycle and transmission of the parasite. The high prevalence rate [81, 124, 161, 198, 255, 265] and the pathogenic effects [32, 136, 138, 190, 242, 253] in such hosts as sheep, goat, cattle and buffalo account for the world-wide economic importance of the parasite. Other hosts in these two groups such as the rabbit, horse, donkey and camel may, in specific situations, act as important reservoir hosts. The role of the rabbit as a reservoir host with very long-lived infection is well documented [39].

The egg production capacity and life-span of *Fasciola* in the definitive host are a reflection of the susceptibility and resistance status of the host. Flukes normally live for about 9 months in sheep before they die or are expelled from the host but longevities of 8 to 11 years have been reported [39, 262]. The mechanism responsible for this prolonged survival is unknown. None of the various 'avoidance mechanisms' which could possibly enable the parasite to survive in its host such as acquisition of host antigens, turnover of parasite glycocalyx to prevent antibody attachment and/or cell adherence and production by the parasite of a substance or substances which can prevent host effector cells from functioning, appear to be implicated [140].

Boray [39] estimated the daily faecal egg output in light subclinical *F. hepatica* infections in sheep to be 0.5 million. Animals with moderate infections shed 2.5 to 3.0 million eggs daily each in their faeces. Thus, the level of pasture contamination is considerable, even in subclinical infections. This high fecundity of the parasite, the high reproductive capacity of the snail vector and polyembryony of larval flukes in snails are responsible for the rapid recolonisation of habitats and contribute to the successful maintenance and spread of infections in endemic areas. Furthermore, in those situations in which highly susceptible hosts are exposed to repeated waves of infection, lack of synchrony in attainment of maturity of flukes derived from the succession of infective stages and the prolonged parasite survival in the host would tend to prolong very considerably the duration of patency and increase the intensity of pasture contamination. The same lack of synchrony in the maturation of flukes also occurs in single infections [262].

Egg production by *Fasciola* spp. in cattle is less persistent than in sheep and goats due to limited patent period of infections (7–9 weeks in moderate infections and 13–15 weeks in heavy infections), lower fecundity and more rapid elimination of flukes [39, 82]. Under Australian conditions, the daily faecal egg production of cattle chronically infected with *F. hepatica* and having a faecal egg count of only 5 eggs per gram (EPG) was approximately 20 times less than that of similarly infected sheep [39]. Nevertheless, in cattle, buffalo and other large animal hosts such low EPG is epidemiologically important and can contribute significantly to pasture contamination during the patent period, in view of the relatively large volume of faeces produced by these hosts.

These relationships between *Fasciola* and its definitive host are strongly influenced by several factors, the most important of which are the resistance status, age and nutritional plane of the host [39, 140, 245] and concurrent infections [228, 245]. In rats and mice the sex and genotype of the host also appear to be important [39, 140].

Laboratory animals, notably rats and mice are able to control primary infections with *F. hepatica* and subsequently develop strong resistance to reinfection (reviewed by Boray [39] and Hughes [140]). This resistance can be transferred to recipient naive animals by immune serum or cells [17, 71, 159, 160]. In rats the resistance wanes with time and after about 6 months the animals are again susceptible to infection. Sensitised rats can kill orally administered metacercariae, newly excysted juvenile flukes and adults which have been transplanted into the recipient's peritoneal cavity. The killing occurs at the gut wall within 48 hours of challenge [132, 133] and there is evidence that antigens common to the three stages of fluke are involved in the protective immunity.

Although the evidence for acquired resistance to infection in small laboratory animals is incontrovertible, this is far from the case in domesticated livestock. So far, it has not been possible to establish a role for immunologically mediated resistance in the control of *Fasciola* infections, particularly in sheep and goats where attempts to induce immunity through the administration of irradiated metacercariae, *Fasciola* antigens and other procedures have been unsuccessful [140]. The current situation was summarised by Hughes [140] who stated that 'no worker or group of workers have to date been able to unequivocally prevent reinfection in sheep by immunological means'.

On the other hand, it is true that cell-mediated immune response, assessed by means of *in vitro* lymphocyte proliferation test and interleukin (IL_2) production, has been demonstrated in bovine fasciolosis [207] and encouraging results have been reported by several workers who have attempted to immunise this host using irradiated metacercariae [18, 33, 195, 308], transfer of antiserum or lymphoid cells [71, 79], there is also insufficient evidence to show that resistance to infection, which undoubtedly occurs in both experimental and field infections in this host, is immunologically mediated. On the contrary, it is generally believed that the resistance is associated with the marked alterations in liver and bile duct architecture resulting from damage of the liver by migrating flukes and of the bile ducts by adult flukes [39, 140]. However, Dargie et al [79] have shown that in monozygotic twin calves resistance to re-infection can develop in the absence of liver damage. Nevertheless, one must conclude from available evidence that immunologically mediated (acquired) resistance is unlikely to play an important role in the epidemiology of naturally occurring bovine fasciolosis but that non-immunological resistance may have a limited role in this and other naturally relatively resistant hosts, through its influence on parasite fecundity, growth and survival.

In cattle, heavy infections and clinical disease occur most commonly in calves and resistance to challenge infection, which usually develops following either single primary or continuous low level infections [39] is stronger in adult than in young animals. This contrasts with the situation in sheep and goats, adults of which are fully susceptible to infection and readily develop clinical disease [199, 247].

Nutritional plane of the host has broadly the same effect on body resistance as age in most susceptible hosts. Experiments in which groups of lambs fed different rations were infected with a standard dose of metacercariae showed that those on a low plane of nutrition died much sooner after heavy infection and developed more profound anaemia than those on a high plane of nutrition [39]. This was attributed to increased 'state of resistance' of the well-nourished animals. Schillhorn van Veen [245], working in northern Nigeria, reached a similar conclusion based on field observations on nomadic and semi-nomadic cattle during severe drought. His observations highlighted the interactive influences of malnutrition, long distance trekking by migrant herds and concomitant parasitic and non-parasitic infections on the incidence of fasciolosis.

Little information is available on the role of concurrent infections of the epidemiology of fasciolosis in the tropics. The best known examples of naturally occurring associations between *Fasciola* and other pathogens in livestock are the interaction between *Fasciola* and *Clostridium oedematiens novyi* [262], in the causation of necrotic hepatitis (Black disease), that between *F. hepatica* and chronic *Salmonella dublin* infections in cattle and the *F. hepatica-Oestertagia ostertagi* complex in some temperate climates [228]. In these associations, fasciolosis usually enhanced the pathogenecity and/or the persistence of the other pathogens in the host, except that in fasciolosis-ostertagiosis complex there appears to be minimal interaction between the two parasites. It is noteworthy that experimentally concurrent infections of cattle with *Fasciola* and *Schistosoma bovis* appear to elicit heterologous resistance. Yagi et al [306] have shown that mature *F. gigantica* infections in zebu calves can induce significant resistance to challenge infection with *S. bovis* cercariae and vice versa. Some non-immunologically mediated resistance to *Fasciola* has also been observed in sheep concomitantly infected with *Cysticercus tenuicollis* [45, 85].

However, the latter observation was not confirmed by Hughes and his co-workers [141]. In view of the well known immunomodulatory properties of helminth and protozoan parasites and the fact that multiple parasitic infections are the more natural forms of parasitism in tropical livestock [52, 245], it would be surprising if concurrent infections did not have some modulatory influence on the course of fasciolosis in some natural host-parasite systems.

PARASITIC GASTROENTERITIS

Introduction

Domesticated ruminants namely, cattle, sheep, goats and, in some areas, buffaloes and camels are the dominant group of livestock in the tropics, generating substantial revenue and providing employment, animal protein for human consumption, fuel, manure and, in some places, alternative means of traction and transportation. Their health status and productivity, and hence their ability to fulfil this multipurpose function, are generally suboptimal largely due to disease. Gastrointestinal nematodes have been identified as one of the causes of the production losses, which arise primarily through mortality, severe weight loss, poor meat, milk and wool production, carcase and offal condemnation and impaired reproductive performance [3, 4, 46, 52, 70, 95, 121, 124, 186, 244, 291].

Nematode parasites have two features of epidemiological importance which are also found in some cestodes but are lacking in trematodes of domestic animals. No multiplication of their preparasitic stages occurs and their infective stage, whether acquired directly by the host or indirectly via intermediate hosts, always gives rise to a single adult male or female parasite in the definitive host. Therefore, the number of infective stages present in the host environment at any given period is related to the number of warm eggs passed by the host and largely determines the number of parasites potentially capable of establishing in a susceptible host population. However, as will be discussed later, this, in turn, is influenced to varying degrees by a number of environmental as well as host- and parasite-dependent factors.

Aetiology and Methods of Transmission

PGE is a gastroenteropathy caused by mixed infections with several nematodes belonging to three main families namely, Ancylostomatidae Loos, 1905, Trichonematidae Witenberg, 1925 and Trichostrongylidae Leiper, 1912, with the occasional involvement of members of Ascarididae Baird, 1953, Strongyloididae Chitwood and McIntosh, 1934 and Trichuridae Railliet, 1915 (Table 1). In its typical form, it is a disease of young animals and may be acquired through a variety of ways [167, 189, 262, 287, 298], depending on the species of nematode (Table 1). *Toxocara vitulorum* is unique in being parasitic exclusively in young animals [162, 262], particularly, young buffalo calves [124] and patent infections can only be acquired through prenatal and lactogenic routes.

Factors Influencing Epidemiology

Bionomics of Free-living Stages

Pasture, which in its broadest meaning is the complex of herbage, faeces, soil and mat (the layer of organic matter on the surface of the soil) [74] provides the micro-environment in which free-living or preparastic development and survival must take place, worm eggs and larvae being distributed in an unpredictable manner between these sites [74, 280]. The number of infective stages which eventually become available to the host depends on the independent and interactive influences of several factors in the macro- and micro-environment. These include temperature, moisture, humidity, sunlight and oxygen

Table 1. Nematodes which cause parasitic gastroenteritis in domesticated ruminants and camels in the tropics, their main hosts and methods of transmission

	Family	Species	Main host	Infective stage	Method/route of infection
1.	Ascarididae	*Toxocara vitulorum*	Cattle and buffalo	Egg containing L_2;L_3 in maternal tissues	(a) Ingestion of infective stage, followed by somatic migration by L_2 in neonates, juveniles and adults. (b) Lactogenic transmission from population of L_3 in tissues. Larvae develop into adult worms in small intestine in about 4 weeks. (c) Prenatal transmission. L_3 from maternal tissues reach amniotic fluid via placenta. Foetus infected orally.
2.	Ancylostomatidae (Hookworms)	*Bunostomum phlebotomum*	Cattle	Third stage larvae	Percutaneous and oral routes
		B. trigonocephalum	Sheep and goat	"	"
		Gaigeria pachyscelis	Sheep and goat	"	Percutaneous route only
		Agriostomum vryburgi	Cattle	"	Percutaneous and oral routes
3.	Strongyloididae	*Strongyloides papillosus*	Cattle, sheep and goat	Third stage larvae	Lactogenic, oral, percutaneous and prenatal routes
4.	Trichonematidae	*Chabertia ovina*	Sheep, goat and cattle	Third stage larvae	Oral route
		**Oesophagostomum asperum (O. indicum)*	Goat and sheep	"	"
		O. columbianum	Sheep, goat and camel	"	"
		**O. multifoliatum*	Sheep and goat	"	"
		O. radiatum	Cattle and buffalo	"	"
		O. venulosum	Sheep, goat and camel	"	"
5.	*Trichostongylidae*	*Trichostrongylus axei*	Sheep, goat and cattle	Third stage larvae	Oral route
		T. colubriformis	Sheep, goat, cattle and camel	"	"
		**T. longispicularis*	Sheep and cattle	"	"
		**T. probolurus*	Sheep, goat and camel	"	"
		Camelostrongylus mentulatus	Camel, sheep and goat	Third stage larvae	Oral route

(*Contd.*)

Family	Species	Main host	Infective stage	Method/route of infection
	Cooperia curticei	Sheep and goat	"	"
	C. pectinata	Cattle	"	"
	C. punctata	Cattle	Third stage larvae	Oral route
	**C. spatulata*	Cattle and sheep	"	"
	**Cooperiodes kenyensis*	Cattle, sheep and goat	"	"
	Haemonchus contortus	Sheep, goat and cattle	"	"
	H. longistipes	Camel and dromedary	"	"
	H. placei	Cattle	"	"
	**Impalaia nudicollis*	Camel, sheep and goat	"	"
	Mecistocirrus digitatus	Buffalo, cattle sheep and goat	"	"
	**Marshallagia marshalli*	Sheep and goat	"	"
	**Nematodirus spathiger*	Sheep, goat and dromedary	"	"
	Paracooperia nodulosa	Buffalo	"	"
6. Trichuridae	**Trichuris cameli*	Camel	Egg containing first stage larvae	Oral route
	**T. discolor*	Cattle, buffalo, sheep and goat	"	"
	T. globulosa	Camel, sheep goat and cattle	"	"
	T. ovis	Goat, sheep and cattle	"	"

Table compiled from data from the following sources: Mosgovoi & Shakhmatova [189]; Warren [298]; Schillhorn van Veen et al [249]; Anon [14]; Soulsby [262] and Troncy [289] (see also Chapter 3).

*These species are rare, and/or relatively non pathogenic.

supply [10, 11, 83, 235, 259, 295, 296], structure of the soil, growth and composition of pasture herbage [63, 111, 157], as well as size and consistency of faeces [57, 58, 237, 238, 259, 307]. Of these, temperature and moisture exert the dominant influence and are the most conveniently measured parameters. Consequently, considerable amount of data is currently available from both controlled laboratory experiments [9, 11, 31, 74, 236, 259, 295, 296] and observations on small artificial experimental grass plots [83, 111, 114–116, 166] on the response of eggs and larvae to these two meteorological factors. These data have been reviewed by Kates [151, 152] Wallace [294], Crofton [74], and Levine [165]. However, their relevance and applicability to field conditions have been criticised on the grounds that the experiments from which they were obtained were invariably carried out under static, artificial conditions that bear only superficial resemblance to the highly complex, dynamic micro-environment in which the free-living stages are found [180]. Furthermore, very few of such studies were carried out in the tropics or under conditions simulating those in the tropics.

Partly in order to circumvent the drawback associated with laboratory-based experiments and partly as a result of a recognition of the importance of parasite-pasture-grazing animal interaction in the epidemiology of helminthoses, several investigators have carried out their experiments and field observations under more natural conditions in which pastures or paddocks were either naturally or artificially contaminated with worm eggs and, in some cases, subsequently grazed at short intervals by a succession of tracer animals. Emphasis in such studies, many of which have been reported from tropical and subtropical climates [8, 22, 48, 56–58, 87, 89, 104, 105, 144, 145, 200, 204–206, 220, 307], were first, on the trends in the populations of L_3 in pasture and, secondly, on the overall pattern of infection and desease incidence in animals. The parasitological data were then related to the prevailing weather conditions. Environmental conditions were considered suitable for free-living development and transmission of infection if a large population of L_3 was available in pasture and/or if significant infection developed in tracer animals exposed to the herbage infestation. Thus, no attempt was made to assess the precise effect(s) of specific macro- or micro-climatic factors on either the individual free-living stages or on the overall trend of parasite population in pasture and in animals. Where such observations were carried out over several consecutive seasons or years it was usually possible to obtain information on seasonal and annual variations in infection patterns in the locality. An important assumption in this appoach, the validity of which has been questioned [180], is that there is a regular relationship between macro- and micro-climate which is sufficiently close to permit the use of measurements in the former to interprete, and even predict parasite population patterns in pasture and animals [23, 211, 212, 280, 281].

Despite the obvious limitations of the laboratory and field studies it is generally accepted that many of them [9, 10, 74, 89, 114, 166, 237, 238, 259, 295, 307] have yielded very valuable information that has led to important conclusions which have influenced current thinking and knowledge of climate-free- living stage relationship and its relevance to parasite epidemiology. Some of these information and conclusions may by summarised as follows:

1. Free-living stages of nematode parasites of grazing animals differ in their responses to environmental conditions, notably temperature and moisture. These differences exist within and between nematode species. In general, L_3 and embryonated eggs are the least susceptible to adverse environmental conditions, followed by unembryonated eggs, L_1 and L_2 in that order.

2. High moisture levels, particularly surface soil moisture, which is determined by the amount and distribution of precipitation, rate of evapotranspiration from the soil as well as by vegetation and soil type [262] are a general requirement for larval development and migration [83, 307]. Hookworm larvae generally require moist, slightly sandy soil and temperature range of 23°C to 30°C for optimum development and activity [262, 287]. Transcutaneous infection with these larvae occurs most readily in a muddy environment, as they are apparently incapable of penetrating the skin in a 'very wet' environment [287] and are not usually found in large numbers on herbage. On the other hand, excessive moisture or precipitation

associated with sustained torrential rainfall characteristic of the wet season in the tropics adversely affects development of eggs and pasture larval densities through rapid disintegration of faeces and washing away of eggs and larvae by rain and flood water [44, 105, 144, 259, 307] and also possibly through reduced oxygen tension in water-logged faeces [256, 259]. By contrast, larval development and survival, particularly in faeces of cattle and other large ruminants, may be considerably enhanced during prolonged dry weather [22, 237, 238, 260, 307]. The smaller the faecal mass, the faster are its rate of moisture loss (Fig. 1) and the rate of mortality of free-living stages, hence there is hardly any larval development in faecal pellets of small ruminants, in unshaded pasture, during the hot dry season [58, 89].

3. Warm conditions and a continuous film of moisture are required for successful translation of L_3 onto herbage [258, 262]. However, L_3 are capable of limited lateral and vertical migration from soil and faeces, even under optimum conditions. There appears to be a log decrease in the number of L_3 in pasture as the distance from faecal pat increases. In some situations, larval dispersal in pasture may be aided by various mechanical agents such as farm machinery, implements and foot-wear of farm personnel. Biotic factors such as some species of fungi which have been shown to transport the larvae of *Cooperia punctata* and *Trichostrongylus colubriformis* [34] and psychodid flies, which can mechanically transport L_3 of *Oesophagostomum* and *Ostertagia* spp. [148, 283] may also be involved in larval dispersal in pasture in some situations.

4. Once on herbage the longevity of L_3 is extremely limited under hot dry tropical conditions [58, 89]. In the Trichuridae and Ascarididae, which do not have free-living infective larvae, the infective egg

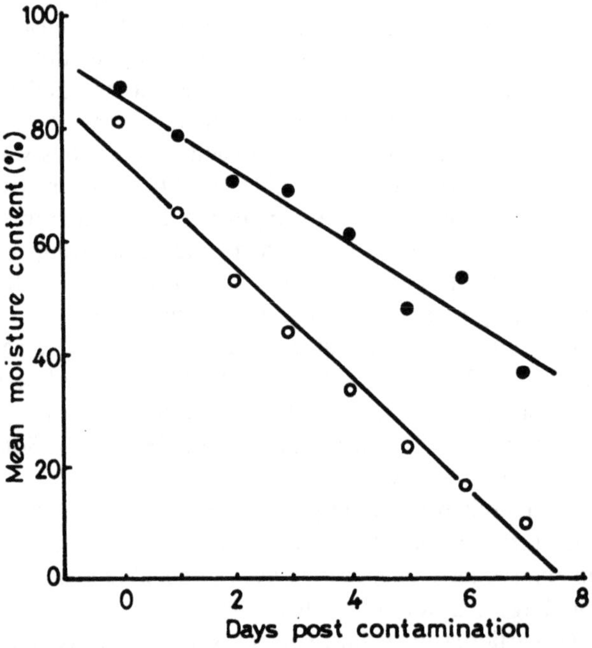

Fig. 1. Mean moisture content of 1.0 kg (●) and 200 g (○) bovine faecal pats deposited in unshaded pasture at Nsukka during the 1984/85 dry season (November to March). The regressions of moisture content on day post contamination for the two sizes of pats during the first week following deposition are : $Y = 86.06 - 6.55X$: $r = -0.9724$, $P < 0.001$ and $Y = 75.35 - 9.86X$: $r = -0.9911$, $P < 0.001$, respectively. Reproduced with permission from Chiejina and Fakae [57].

can survive in a warm humid environment for several years [287] but their longevity is also severely curtailed during the hot dry season and by direct sunlight. Surviving eggs in soil have been shown to be a source of infection for animals in subsequent years.

Seasonal Patterns of Infection

The close relationship between climate and the incidence of helminthoses in grazing animals is aptly summed up by Gordon [118] in his much-quoted statement that 'worms work by the weather' [280]. A practical application of this concept is the increasing use of standard meteorological data to predict the pattern of availability of L_3 in pasture and the occurrence and incidence of disease outbreaks [23, 211, 212, 215, 280, 281].

In general, two distinct patterns of pasture infestation may be encountered in the tropics, depending largely on the distribution of rainfall. The first is seen in the dry tropics and in those parts of the subhumid zone which have clearly defined favourable and unfavourable seasons for preparasitic development. In both climatic zones the latter season is sufficiently hot, dry and prolonged to cause complete cessation of free-living development and survival. Consequently, except in irrigated or other permanently wet pastures, transmission of PGE in such areas is restricted to the wet season [49, 163, 185, 200] and the only means of carry-over of infection from one rainy season to another is through animals harbouring adult worms and/or arrested (hypobiotic) larvae [59, 130, 163, 185, 200, 201, 264].

However, in the less dry parts of the subhumid and wet tropics faecal reservoirs of L_3 are also a very important additional source of the early rains herbage infestation in dry season-contaminated cattle pastures but not in pastures contaminated by sheep and goats [57, 58, 99]. Such larvae develop and accumulate inside faecal pats throughout the dry season but do not appear on herbage until adequate moisture conditions return at the start of the succeeding rainy season, when a spontaneous and synchronous translation of the entire faecal larval population takes place, usually within 24 to 48 hours of the first substantial rainfall of the season [57, 104]. In spite of the high mortality of eggs and larvae in faecal pats soon after deposition (Fig. 2), large numbers of L_3 invariably survive inside the dry faecal masses to ensure massive early rains herbage infestation [44, 57, 104]. A similar phenomenon has been described in temperate and sub-tropical climates where it is of considerable epidemiological importance [8, 22, 55, 92, 237–239, 307].

With the return of the rains and the re-establishment of favourable environmental conditions, there is a continuous cycle of infection between the host and pasture for as long as these conditions last. Consequently, grazing animals and pasture herbage are constantly infected during this period. However, pasture larval densities and worm populations in animals fluctuate considerably throughout the season [56,105, 144], in response to variations in the size of contamination, grazing intensity and frequency, short term fluctuations in weather conditions and host responses to infection. In the derived savanna zone of Nigeria, for example, each episode of artificial or natural contamination during the 7-month rainy season, produces a rapid build-up of herbage infestation to a peak. This is followed by an equally rapid disappearance of infestation (Figs. 3 and 4) and frequent contaminations are required to maintain a consistently high level of herbage infestation throughout the season. The relatively short (6–8 weeks) survival of L_3 on herbage during the wet season is the net result of several contributory factors. The most important is the constant heavy rainfall which causes accelerated disintegration of faecal pats and disappearance of free-living stages from pasture. In some situations, rapid break up and burying of faeces in the soil by dung beetles, earthworms and termites [44, 48, 101, 126] also contribute to the fast depletion of pasture larval population. Predacious fungi, which play a similar role in temperate climates [131] have not been identified in faeces of tropical livestock.

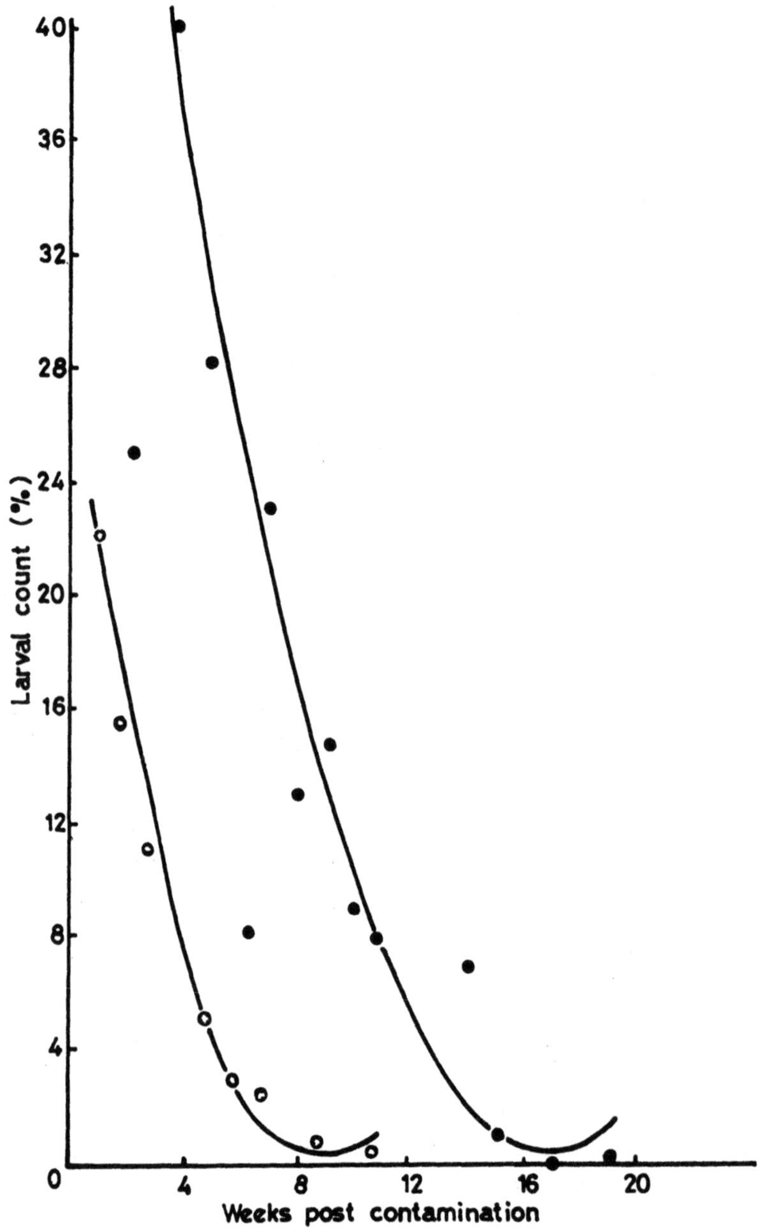

Fig. 2. The relationship between faecal larval count and time post contamination in two grass paddocks, P2(●) and P4(○), contaminated in December and February (early and late dry season, respectively). The faecal larval counts were expressed as a percentage of the total initial population of strongyle eggs in each paddock. The curves were fitted using the following quadratic equations: $Y = 62.07 - 7.30X + 0.22X^2$ ($F = 85.67$, $P < 0.001$) for December contamination and $Y = 27.74 - 6.02X + 0.33X^2$ ($F = 256.02$, $P < 0.001$) for February contamination. Reproduced with permission from Chiejina and Fakae [57].

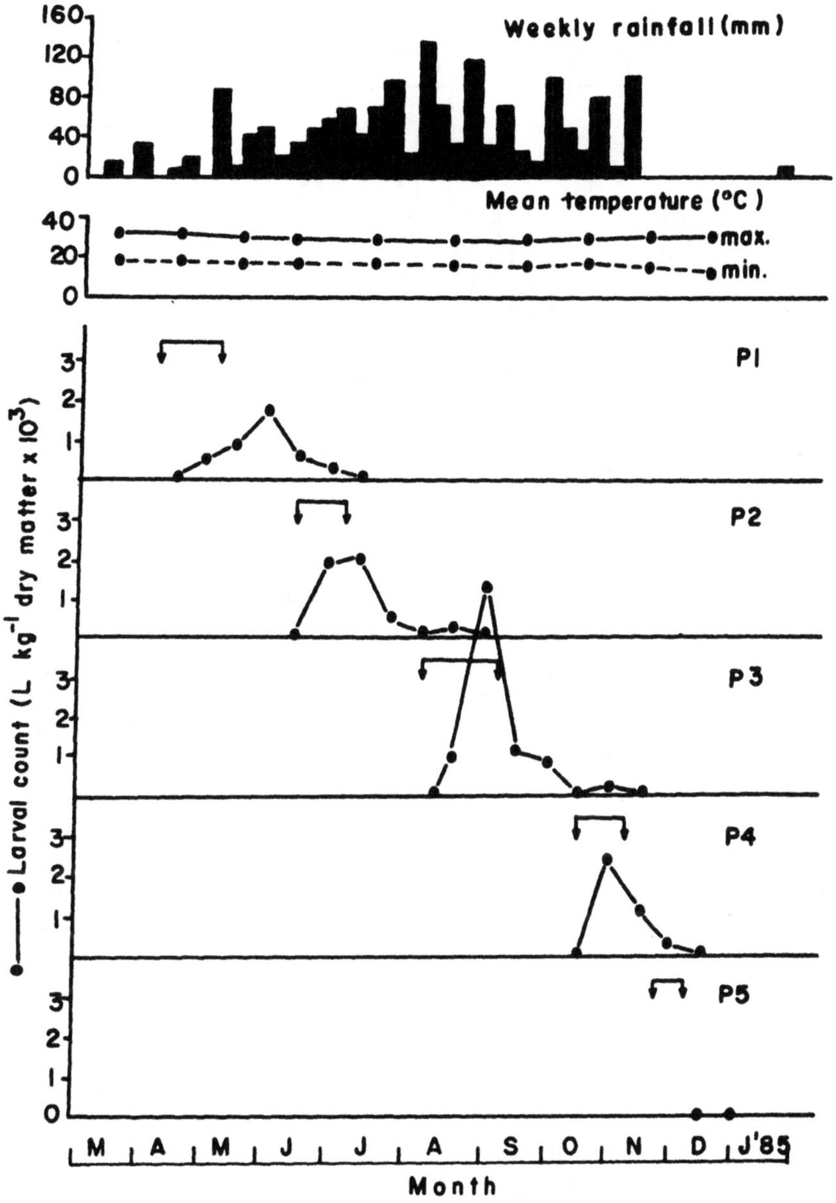

Fig. 3. The herbage trichonstrongylid larval populations in five experimental paddocks, P1 to P5, sequentially contaminated with faeces of infected cattle at Nsukka, in the Nigerian derived savanna, between March and November 1984 and the weekly rainfall and mean monthly temperatures for Nsukka during March 1984 to January 1985. ↓‾‾↓ indicates period of contamination. Each episode of contamination was followed, after a lag period of 1 to 3 weeks, by rapid rise and decline in herbage infestation. The longevity of L_3 on herbage was shortest following the contamination carried out at the peak of the rains in July/August. Reproduced with permission from Fakae and Chiejina [105].

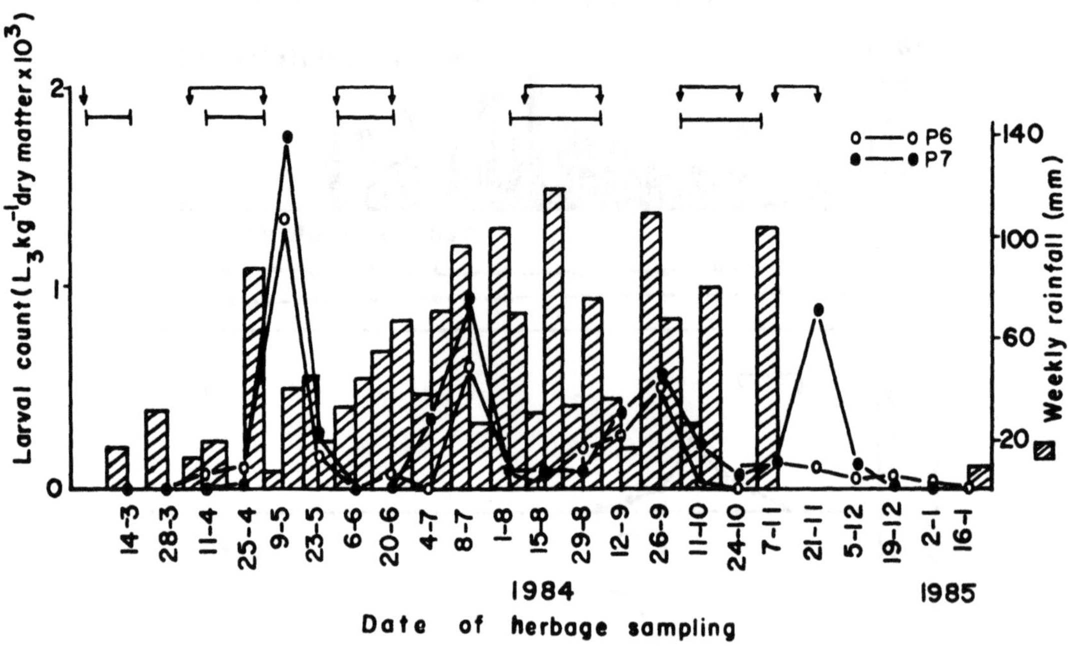

Fig. 4. Weekly rainfall at Nsukka during March 1984 to January 1985 and the trichostrongylid larval population in two paddocks, P6 and P7. The former was artificially contaminated at monthly intervals with faeces of infected cattle. The latter was grazed by the same infected cattle at approximately 4–6 weekly intervals. Both patterns of contamination produced 3 to 4 coincident and relatively short-lived waves of herbage infestation during the 7-month rainy season. ↓‾‾↓ indicates the periods of contamination of P6. ⊢⊣ indicates the periods of contamination of P7. ↑ indicates time of turn-out of grazing animals. Reproduced with permission from Fakae and Chiejina [105].

As a result of this pattern of free-living development and survival, pastures which are repeatedly contaminated during the wet season show a multi-peak larval population curve (Fig. 4) which is the summation of the descrete waves of infestation resulting from the individual contaminations. The number of such waves and of their associated peaks are in turn determined by the length of the favourable season, frequency and duration of contamination and the generation times [74] of the nematodes [105]. Thus, the number of larval peaks is a good indication of the number of generations of the nematodes which can develop in pasture and in susceptible animals during the favourable season. It has been estimated that a minimum of about 3 to 4 generations of *Haemonchus* and *Trichostrongylus* spp. can develop in small ruminants during the 6–7 month rainy season in the Nigerian subhumid zone [95, 105] while approximately 3 generations of *Haemonchus* and *Trichostrongylus* spp. can develop in goats in the humid zone of Malaysia [144]. These contrast with a maximum of two generations of ovine trichostrongylids under temperate conditions of North-east England [42].

Although many species of nematodes may be represented in each of the generations, particular species predominate in particular generations, and there is a definite order of succession of dominant species [74, 95, 96].

The second major infection pattern is non-discontinuous and occurs under the continuously favourable climatic conditions in the humid tropics. In this zone there is no clear-cut seasonality in the pattern of larval availability in pasture [127, 144, 204] and several larval peaks and generations of parasites develop in pasture and animals, respectively throughout the year, actual pasture larval densities, parasite burdens and disease incidence being determined more by livestock husbandry practices and agricultural systems, host age, genotype, nutrition and immune responses than by meteorological factors.

Acquisition of infective eggs of *T. vitulorum* may also follow a seasonal pattern in the dry tropics while in the permanently warm and humid zone infection may be constantly present both in the environment and in calves in endemic areas. However, the pattern of lactogenic and congenital infections is necessarily determined primarily by the dynamics of populations of dormant L_3 in maternal tissues, the reproductive state of the dam and methods of calf husbandry [234, 298]. Somatic larvae can probably survive in maternal tissues for several months or years [262].

Livestock Production Systems and Husbandry Practices

Just as meteorological and other environmental factors largely determine the size of the overall population of infective stages of nematode parasites available in the environment at any given time, livestock husbandry systems and managerial practices have a major influence on the transmissibility of that infection to a susceptible host population. The methods of management of domesticated ruminants in the tropics are varied [51, 75, 106, 112, 146, 196, 216, 223, 225, 267, 271, 288] and are inextricably linked with systems of agricultural production as well as with local traditions and beliefs [196, 216]. Table 2 summarises some of the more commonly practised traditional methods of management of small ruminants in the tropics. In addition to these methods of livestock management, modern intensive methods are also practised in large Government and some privately-owned commercial and Institutional farms. However, animals in the latter establishments account for a relatively small proportion of the small ruminant population in most tropical countries.

Table 2. Some common traditional methods of management of small ruminants in the tropics

Type of management	Method of management and its features
1. Permanent confinement with or without restricted grazing:	Flocks are small and sedentary. Involves either tethering of dams in paddocks, fallow land or near hedgerows during the day and housing at night, or total confinement indoors in sheds throughout the year. While indoors, animals are fed on tree foliage, cut fodder and browse, harvest litter and sometimes left-over kitchen wastes and crop residues.
2. Semi-permanent confinement with some free grazing:	Flocks are small and sedentary. May have access to one or more fenced paddocks of varying sizes during the day. Animals are housed in simple sheds or huts at night. Some fodder, browse and left-over kitchen wastes are usually available when housed. Confinement may be seasonal, related to cropping/planting season.
3. Free-range system with no confinement:	Small to medium-sized sedentary, village flocks, roam by day but usually return to their owner's premises at night for shelter. Involves extensive and unrestricted grazing and access to bushes, fallow land, roadside verges, hedgerows, garbage dumps and occasionally crop residues and other agricultural by-products.
4. Nomadism and transhumance:	Flocks are large and are grazed extensively. Grazing areas constantly change as the pastoralists and their animals migrate in search of fresh pasture and water. Transhumant flocks are sedentary during period of abundant food supply but also migrate like true nomads, though to less distant locations, during periods of food and water shortage, returning to their village base when food supply situation improves. Many nomadic and transhumant flocks are managed alongside similar herds of cattle and some camels, horses and donkeys.

It has been estimated that approximately half of all the large ruminants in the tropics are raised in non-monsoonal Asia where they are used by many subsistence farmers for work purposes [216]. Nomadism and transhumance are also important methods of cattle husbandry and land utilisation in the drier parts of East, West and Central Africa as well as in Indian peninsula and Southern Arabia, while commercial ranching for beef production is common in Central and South Africa, the Carribean, East Africa, Australia, some Pacific Islands and the Phillipines. Most ranches are located in savanna highland and plateau areas but some are in the humid zones. As in nomadic herding, ranching involves the use of relatively large grazing areas which are used rather more intensively. Ranchers, unlike nomadic herdsmen, practise some orderly grazing and supplementary feeding and have breeding and selection programmes.

Clearly, opportunities for acquisition of worm infections and the risk of disease outbreaks will vary according to the type of husbandry and livestock production system in the particular area. In some traditional systems, which utilise extensive grazing (Table 2) faecal contaminations, and hence infective stages, are thinly spread over a large territory. Therefore, heavy infections rarely occur, especially where animals are able to select their feed of browse. The same is true in village herds and flocks that are necessarily permanently housed and fed on cut fodder and foliage which have little chance of being contaminated with infective stages. In such village production systems, infections, when present, are usually chronic and subclinical in nature. On the other hand, housed animals maintained entirely on zero-grazing or other cut-and-carry feeding regimens [232] sometimes acquire significant worm infections if the fodder is contaminated with infective stages and litter is allowed to accumulate for long periods in warm, humid sheds [2, 102, 112]. In a study of traditionally managed small ruminants confined and zero-grazed in sheds throughout the cropping (rainy) season but allowed free roam during the harvest (dry) season. Fakae [102] observed escalating worm burdens and faecal egg counts during the former and low prevalence and intensity of infections during the latter season (Fig.5).

Similarly, nomadic, and to a lesser extent transhumant, herds and flocks usually harbour low level of infections, but if such animals camp in one locality for an extended period they are liable to create significant infection, particularly in their gathering pens, paddocks and the more frequently grazed areas near water holes, streams and other water courses, particularly during drought. Confinement of large numbers of young animals in unhygienic and wet environments also predisposes them to heavy infections particularly with skin-penetrating nematodes like *Bunostomum* and *Strongyloides* spp. [143, 229, 287]. Moreover, long distance trekking and forced migration of nomadic and transhumant herds and flocks to unfamiliar territories, apart from exposing them to new species and strains of parasites and to inadequate feed supply, frequently lead to loss of vitality and body resistance. This has been linked with high incidence of parasitism in general and PGE in particular in such animals [245]. In contrast, losses due to PGE are relatively common in intensively managed herds and flocks since high standards of grazing management and effective deworming programmes which are essential in such situations are frequently lacking [52, 95, 147, 248, 264].

In areas with marked seasonality in the availability of infective stages in pasture, the period of year in which animals are born and their subsequent management may also influence the epidemiology of PGE in such animals during their first year of life [50, 181]. Young animals are unlikely to acquire significant infection with trichostrongylid nematodes from pasture for as long as they rely on their dam's milk for nourishment. As they get older and more independent and begin to graze or browse, their worm burden increases in proportion to the level of pasture infestation. As a result, calves and lambs which are weaned during the dry season acquire lighter infections than those weaned during the rainy season when pasture infestation is generally high [50, 279]. However, since tropical breeds of ruminants are non-seasonal breeders and few traditional livestock owners have organised breeding programmes, the

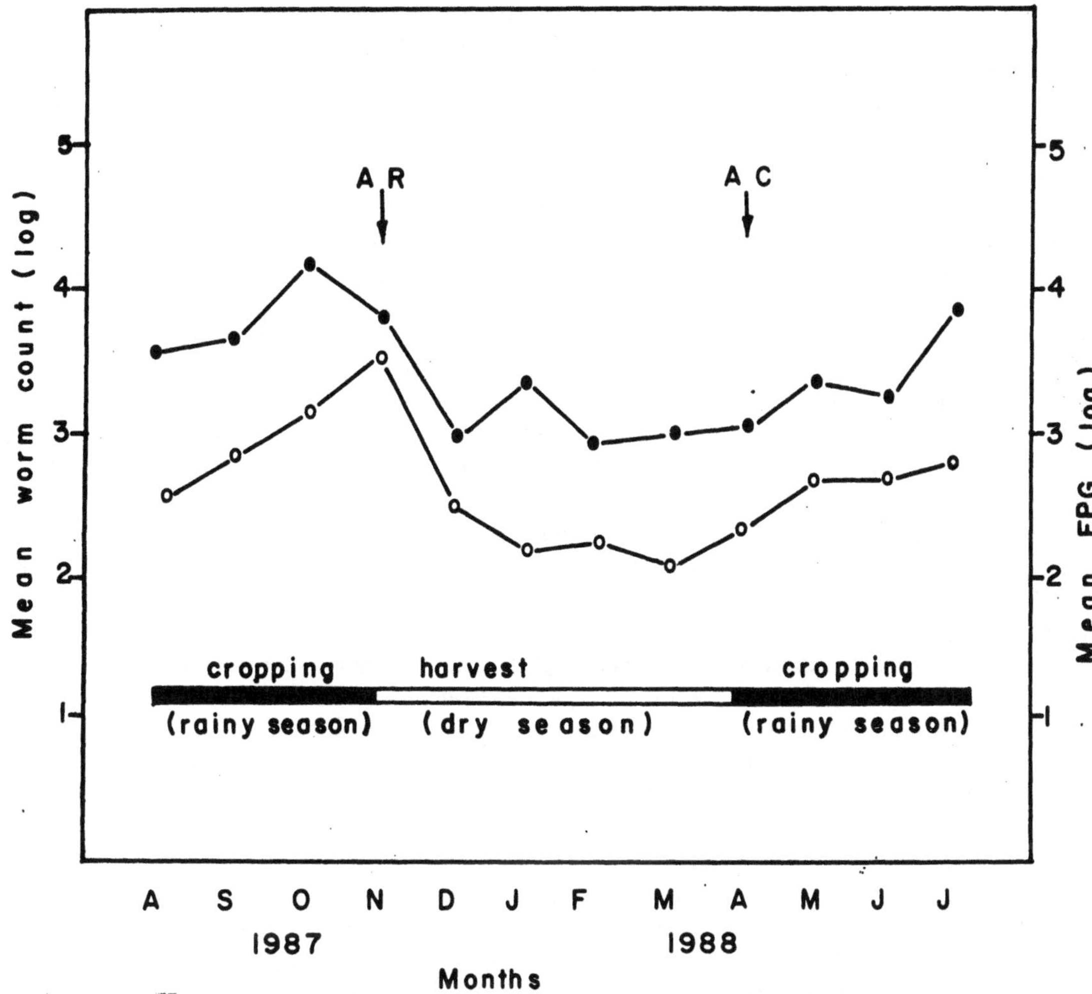

Fig. 5. The mean strongyle burden (—O—O—) and the mean monthly faecal worm egg count (—●—●—) of traditionally managed West African Dwarf sheep and goats during the harvest (dry) and cropping (rainy) seasons in the Nigerian derived savanna. The animals were confined during the latter, but were on free-roam during the former, seasons. AR indicates when they were released and AC when they were confined. Reproduced with permission from Fakae [102].

epidemiological importance of the timing of parturition and weaning is often difficult to determine, except that in the case of *T. vitulorum* parturition and lactation have a major influence on the occurrence and intensity of infections in neonates and juveniles.

Nutrition

Host nutrition is known to influence the outcome of nematode infections in man and animals [67, 76, 110, 230, 272, 284]. Faults in food intake and quality, particularly of protein, minerals and vitamins, lead to the lowering of host resistance to infections which in turn results in increased establishment, survival and pathogenicity of the parasite and hence in increased morbidity and mortality of the host [1, 24, 113, 119, 245, 272]. Gastrointestinal nematodes for example, exert these effects through a variety of pathogenic mechanisms which include (1) anorexia and post-absorptive disturbances in protein metabolism [21, 69, 129, 226, 262, 272]. (2) some, usually localised, impairment of digestive and absorptive functions [108, 191, 231, 257, 263, 273, 276], (3) endocrinological disorders [129, 275, 282] (4) disorders of gastrointestinal smooth muscle reactivity and gut motility [122]. Many of these effects have been observed in subclinically infected lambs [68, 69, 274] and, paradoxically, also in adult resistant cattle [21, 226].

The crude protein content of most native tropical grasses is adequate for only moderate levels of animal production for a few months of the year when the grasses are young [232]. Even when young grasses are plentiful, they consist of very few legume forages and more of tall, stemmy and lignified grasses of rapidly diminishing quality [2] and require protein and other dietary supplementations to maintain statisfactory health and production [202]. Consequently, there is usually severe seasonal food shortage leading to wide-spread malnutrition and sometimes heavy parasitism, particularly in the semi-arid and dry savanna zones [49, 245]. The situation is worsened by indiscriminate bush and grass burning, incessant drought and desert encroachment, which have become a major problem in some parts of the dry tropics. Although transhumant and other traditional systems of livestock rearing which rely heavily on unpredictable sources of natural grazing or on cut-and-carry fodder and browse are mostly at risk, animals raised intensively in large modern farms are also sometimes badly affected since very few of the latter establishments pay adequate attention to fodder conservation and dry season supplementary feeding. Moreover, in the humid tropics, silage making and other forms of fodder preservation are problematical because of inclement weather [202].

The full extent of malnutrition and its impact on gastrointestinal nematode infection in tropical livestock are impossible to assess accurately. However, it is reasonable to conclude, based on the widespread occurrence of infections and undernutrition in animals and the clearly demonstrated adverse production effects of even low-grade infections in other parts of the world [21, 68, 69, 226, 274] that host nutrition also enhances host susceptibility as well as pathogenicity, incidence and spread of infection in the tropics.

Host Age, Acquired Resistance and Genotype

Age also influences the susceptibility of the host to helminth parasites and the pathogenicity of the latter. This influence has an immunological basis. For example, neonates are generally incapable of responding immunologically to nematode parasites. In sheep, the ability to respond to *T. colubriformis* and *H. contortus* infections is not fully developed until at least 3 and 6 months of age, respectively [29, 53, 169]. Hence, under field conditions disease outbreaks are more likely to occur in young than in adult ruminants. The latter usually harbour chronic, low-level infections and, therefore, act as a constant source of infection for more susceptible animals. In ovine trichurosis, there is a strong age resistance to infection from 8 months of age [262]. Age resistance is most effective in *T. vitulorum* infections in which expulsion of adult worms commences as early as 38 days after birth [262] and is completed at 3 to 6 months of age. Consequently, adult worms are rare in older cattle and buffaloes [162, 262].

Traditionally managed flocks/herds contain a disproportionate number of old, largely female, stock which are in constant contact with their young from birth till the next pregnancy or sometimes parturition, a situation which favours maintenance and transmission of infection. This could offset any benefits likely to accrue from dilution of herbage infestation levels by the large volume of faeces passed by the adult resistant stock and from the ingestion, by the adults, of large numbers of infective stages which would otherwise become available to their offspring. Moreover, the popular view that clinical PGE is restricted to young animals requires some revision since there are several reports of major disease outbreaks in adult small and large ruminants [6, 96, 139, 245, 246, 251, 299]. Such outbreaks usually occurred either under nutritional stress and intercurrent infection [245] or as a result of poorly developed and waning immunity. In the tropics, animals protected from infection through zero-grazing, non-availability of infective stages during the dry season, intensive prophylactic dosing or other managerial interventions in early life are also likely to remain susceptible to infection, as adults, but more so in their second year of life.

In well established host-parasite relatioships, parasite populations in the host are regulated primarily by mechanisms and processes associated with either natural or acquired resistance, the latter being under genetic control [293]. It is as a result of natural resistance that some nematode parasites which are well adapted to one host species may be unable to establish and reproduce successfully in another host. This is the basis for using mixed and alternate grazing of different species of animals for the control of animal helminthoses [19]. Acquired resistance, on the other hand, occurs more widely and helps to control parasite populations in field infections. Of its many manifestations [54, 176], resistance to the establishment of new infections, inhibition of ovulation and worm rejection are of epidemiological significance.

The best-known immunologically-mediated worm rejection mechanism in field infections is the self-cure response [6, 130, 163, 268–270, 290], which periodically results in spontaneous exponential rejection of worm infections and is sometimes also accompanied by protection from further challenge infection [120]. However, self-cure is not always immunologically-mediated, as it has been observed in non-immune *Haemonchus*-infected sheep grazing worm-free lush pasture. The stimulus responsible for inducing the latter type of response in sheep is believed to be either an anthelmintic or an allergic substance present in freshly growing grass or as yet undetermined physiological alterations in the abomasum [5]. Michel [175, 177] has also described a non-immunologically-mediated mechanism for regulating burdens of *Ostertagia* spp. in calves in which worm populations are maintained by a balance between loss of existing worms and the acquisition of new infection. However, this mechanism may not apply to all host-parasite systems [53, 54].

It has been suggested that the characteristic rise in faecal strongyle egg output of ewes beginning in late pregnancy and reaching a peak in early lactation, known as periparturient rise [43, 65, 73, 188, 277, 290] is due to a temporary relaxation or suppression of host immunity [262], which allows maturation of hypobiotic larvae, increase in fecundity of female worms and enhanced establishment of new infections. The most important strongyle nematodes of sheep contributing to the rise in the tropics are *Haemonchus* and *Trichostrongylus* spp. [290] while *Ostertagia*, *Trichostrongylus* and occasionally *Haemonchus* spp., are mostly involved in temperate climates [43, 65]. The net result is a substantial increase in host worm burden and pasture contamination with worm eggs. The rise, which also occurs, but to a lesser extent, in cattle [181], goats [203] and possibly other ruminants, is of considerable epidemiological importance in sheep in temperate climates where it is sometimes the sole source of the first major wave of worm infections in lambs during the summer [42, 134]. Its epidemiological significance in the tropics is less well known.

Extensive literature is currently available on the genetic basis of resistance to parasitic, especially gastrointestinal nematode infections of laboratory animals [25–28, 78, 292]. Early work on the heritability of resistance to trichostrongylid infections in sheep [301, 302] and in zebu cattle [240] had suggested that individual or strain, rather than breed, differences in susceptibility existed in the former host. However, subsequent studies, reviewed by Dargie [78], confirmed the occurrence of differences in the resistance of tropical [125, 221, 222] and non-tropical [158, 250, 285] breeds of small ruminants, particularly to *H. contortus*. This subject was reviewed by Piper and Barger [219] with respect to parasites of sheep and by Frisch [109] with respect to parasites of cattle in the tropics. In general, indigenous tropical breeds are more resistant than their exotic counterparts, while crosses between the two are intermediate in their response, particularly with regard to establishment of worms in the host, faecal worm egg output and pathogenicity of infections, all of which are of epidemiological importance (see Chapter 15). The recognition of the potential value of this hitherto unexploited genetic characteristic in livestock selection and improvement schemes and in parasite control has prompted the search for suitable markers such as haemoglobin polymorphisms, lymphocyte antigen types [7, 84, 91, 233, 266, 303] and, more recently, DNA markers [Windon, personal communication] for predicting resistance to nematode parasites.

Some workers have suggested that resistance of indigenous breeds of small ruminants to haemonchosis is the result of natural genetic selection [221] and that this is dependent on the degree of selection pressure in the host environment to develop resistance to this nematode. In the case of goats, which are browsers in their natural environment, this pressure would be either absent or minimal since they would, as a result of their browsing habit, avoid significant infection normally present in succulent grasses available to natural grazers like sheep [52, 146]. This may have resulted in the failure of goats to evolve effective resistance mechanisms against their common gastrointestinal nematodes [262] and hence in their apparently high susceptibility to PGE under intensive grazing systems [52].

Inhibition of Development (Hypobiosis)

Many nematode parasites of domestic animals are capable of temporarily interrupting their parasitic development at a precise larval stage characteristic of the particular nematode. This temporary cessation of development, also known as inhibition of development or hypobiosis is important in many gastrointestinal nematodes of domesticated ruminants in which the fourth larval stage (L_4) or sometimes the L_3 is mostly affected [179, 243]. The factors which are known to induce or influence larval hypobiosis include (1) age of host, acquired resistance, pregnancy, lactation and the presence of adult worms, as for example in *O. ostertagi* infections in calves [178, 184], (2) exposure of L_3 to darkness or diminishing photoperiod, as in *H. contortus* infections in sheep [66, 201], (3) the strain of parasite, as in *O. ostertagi* [107] and *H. contortus* [47], (4) exposure of L_3 to adverse environmental conditions such as very low temperatures and hot dry conditions, as in *O. ostertagi* and *Cooperia oncophora* infections in calves [16, 182, 183]. Of these, exposure to adverse environmental conditions is the most important stimulus for the induction of hypobiosis in gastrointestinal nematodes of ruminants.

This is consistent with the observation that the phenomenon occurs most commonly in cold temperate [12, 35, 64, 66, 94, 182, 183, 227, 297] and in hot dry climates [15, 59, 130, 200, 201, 246]. In both climatic zones arrested development is highly seasonal in its incidence and sets in just before the commencement, and terminates at the end, of the unfavourable season, which corresponds to the autumn/winter and the dry season in temperate and tropical climates, respectively. In the dry savanna zone of Nigeria, for example, the highest and lowest incidences are observed during the dry and wet seasons, respectively (Fig. 6). Consequently, hypobiosis is considered by many [64, 278, 297] to be an adaptation by the affected parasites which enables them to survive in their hosts at a developmental stage which is unaffected by the host immunological responses and at a time when environmental conditions are unfavourable for free-living development and survival. Thus, it serves to synchronise the development of the parasite with events in the host and the environment [262]. Reports from Kenya [6, 15], India [127], Malaysia [145] and Nigeria [59, 103] indicate that only limited hypobiosis or none at all occurs in those climatic zones where environmental conditions are permanently suitable for free-living development and survival. However, Capitini et al [47] believed that although hypobiosis in *H. contortus* in North America is also a survival strategy, it is genetically-determined, requiring no environmental stimulus to induce.

In the dry tropics hypobiosis is the most important means of survival and carry-over of infection in the host, with species such as *Haemonchus* and *Cooperia*, from the end of one rainy season to the beginning of another [130, 200, 201]. Its onset also serves to limit further increase in the number of adult worms [243], which are the more pathogenic stages, at a time when pasture larval density is still high. However, synchronous development of large numbers of hypobiotic larvae can sometimes occur well before the

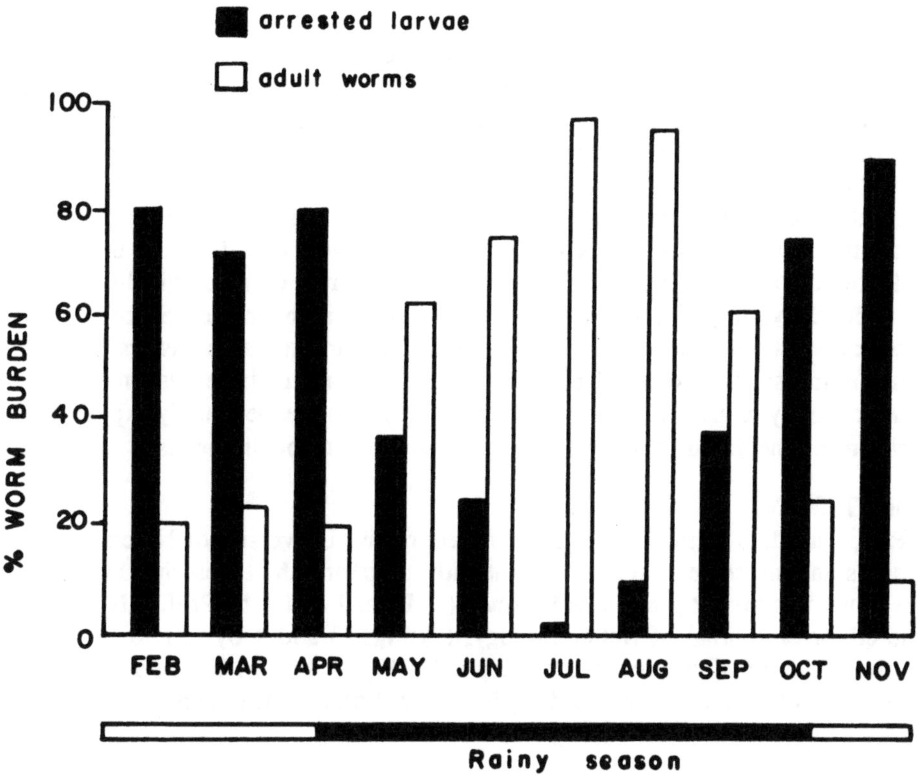

Fig. 6. Inhibition of development of naturally acquired *Haemonchus contortus* in goats of northern Nigeria. Inhibition commenced during the late rainy season and reached a peak during the early dry season. The lowest percentage inhibition was observed at the peak of the rains, from June of August. Reproduced with permission from Chiejina et al [59].

end of the dry season, thereby resulting in a sudden increase in the numbers of pathogenic stages which may lead to dry season outbreaks of PGE [100, 246].

Concurrent Infections

The capacity of an animal to control a given parasitic infection may be markedly compromised by concomitant infection with another related or unrelated parasite or pathogen [27, 197, 218, 289]. This phenomenon has been extensively studied in laboratory animal models utilising protozoal and helminth parasites, some of which are powerful suppressors of heterologous and, sometimes also, homologous immunity [30, 60, 61, 117, 172, 187, 192, 218, 289, 300]. The phenomenon is subject to host strain-dependent variability which is, in turn, genetically-determined (see Chapter 15).

These laboratory findings have been confirmed by field observations on concurrent *Trypanosoma congolense-H. contortus* infections in indigenous and exotic breeds of goats in East Africa [125]. This and other species of trypanosomes have also been shown to suppress host immune response to a number of important bacterial vaccines used in domesticated animals, including livestock, in the tropics [13, 137, 168]. Although the extent of immunodepression in concurrent infections in tropical livestock is unknown, it is likely to be of epidemiological importance in trypanosome-endemic areas where concurrent trypanosome- helminth infections are common [52] and are frequently associated with nutritional and climatic stresses, which are known to influence host resistance to infections [1, 77, 119, 245, 286].

CONCLUSION

A great deal remains to be learnt about the epidemiology of helminthoses of tropical livestock. This is particularly true of infections in buffaloes, camels, goats and other less popular but regionally important farm animals. Available information shows that the epidemiology of fasciolosis, parasitic gastroenteritis and, no doubt, other important helminthoses of animals differs in many important respects from that in temperate and other climates. Thus, while the considerable amount of epidemiological data available from temperate climates, in particular, are useful for comparative studies, they are not always directly applicable to the tropical environment. Therefore, there is an urgent need for more original research data derived from laboratory and field studies carried out in the different climatological-ecological zones and localities and under the variety of livestock production systems in the tropics. Since the vast majority of tropical livestock are owned and managed by nomadic and semi-nomadic pastoralists and subsistence farmers, greater emphasis should be placed on studies relevant to these systems of production and management. Also worthy of greater attention is the possible role of the potentially exploitable phenomenon of genetic resistance to nematode parasites among tropical breeds of ruminants.

ACKNOWLEDGEMENT

I thank Dr. R.N. Anyadike of the Department of Geography, University of Nigeria, Nsukka, for very useful discussions and my secretarial staff for patiently preparing the typescript. Permission to use the figures in this work was kindly granted as follows: (1) Figs. 1 and 2 by Prof. S.N. Chiejina, Dr. B.B. Fakae and the editor of Journal of Helminthology, (2) Figs. 3 and 4 by Dr. B.B. Fakae, Prof. S.N. Chiejina and the editor of Veterinary Parasitology, (3) Fig. 5 by Dr. B.B. Fakae and the editor of Research in Veterinary Communications and (4) Fig. 6 by Prof. S.N. Chiejina, Dr. B.B. Fakae, Dr. B.O. Eze and the editor of Veterinary Parasitology.

REFERENCES

1. Abbot, E.M., Parkins, J.J. and Holmes, P.H. Influence of dietary protein on the pathophysiology of haemonchosis in lambs given continuous infections. Res. Vet. Sci. **45**: 41–49 (1988).
2. Ademosun, A.A. Appropriate management systems for the West African Dwarf goat in the humid tropics. In: *Goat Production in the Humid Tropics*. Proceedings of a Workshop, University of Ife, Nigeria, pp. 21–28 (1987).
3. Akerejola, O.O., Schillhorn van Veen, T.W. and Njoku, C.O. Ovine and caprine diseases in Nigeria: a review of economic losses. Bull. Anim. Hlth. Prod. Afr. **27**: 65–70 (1979).
*4. Allonby, E.W. Annual report of the sheep and goat development project, Kenya (1975).
5. Allonby, E.W. and Urquhart, G.M. Self-cure of *Haemonchus contortus* infections under field conditions. Parasitology **66**: 43–53 (1973).
6. Allonby, E.W. and Urquhart, G.M. The epidemiology and pathogenic significance of haemonchosis in a Merino flock in East Africa. Vet. Parasitol. **1**: 129–143 (1975).
7. Altaif, K.I. and Dargie, J.D. Genetic resistance to helminths. The influence of breed and haemoglobin type on the response of sheep to re-infection with *Haemonchus contortus*. Parasitology **77**: 177–187 (1978).
8. Altaif, K.I. and Yakoob, A.Y. Development and survival of *Haemonchus contortus* larvae on pasture in Iraq. Trop. Anim. Hlth. Prod. **19**: 88–92 (1987).
9. Andersen, F.L. and Levine, N.D. Effect of desiccation on survival of the free-living stages of *Trichostrongylus colubriformis*. J. Parasitol. **54**: 117–128 (1968).
10. Andersen, F.L., Levine, N.D. and Boatman, P.A. Survival of third stage *Trichostrongylus colubriformis* larvae on pasture. J. Parasitol. **56**: 209–232 (1970).
11. Andersen, F.L., Wang, G.T. and Levine, N.D. Effect of temperature on survival of the free-living stages of *Trichostrongylus colubriformis*. J. Parasitol. **52**: 713–721 (1966).

12. Andersen, N., Armour, J., Jennings, F.W., Ritchie, J.D. and Urquhart, G.M. Inhibited development of *Ostertagia ostertagi*. Vet. Rec. **77**: 146–147 (1965).
13. Anene, B.M., Chukwu, C.C. and Anika, S.M. Immunosuppression of humoral immune response in canine trypanosomiasis. Microbios Letters. **40**: 37–46 (1989).
14. Anon, *Handbook on Animal Diseases in the Tropics* (A. Robertson ed). 3rd edn., pp. 212–259. British Veterinary Association (1976).
15. Armour, J. The epidemiology of helminth diseases in farm animals. Vet. Parasitol. **6**: 7–46 (1980).
16. Armour, J. and Bruce, R.G. Inhibited development in *Ostertagia ostertagi* infections—a diapause phenomenon in a nematode. Parasitology **69**: 161–174 (1974).
17. Armour, J. and Dargie, J.D. Immunity to *Fasciola hepatica* in the rat: successful transfer of immunity by lymphoid cells and serum. Exp. Parasitol. **35**: 381–388 (1974).
*18. Armour, J., Dargie, J.D., Doyle, J.J., Murray, M., Robinson, P. and Rushton, B. Immunisation against fascioliasis. Proc. Third Int. Congr. Parasitol., Munich, Vol. 1. p. 494. Facta Publications (1974).
19. Arundel, J.H. and Hamilton, D. The effect of mixed grazing of sheep and cattle on worm burdens in lambs. Aust. Vet. J. **51**: 436–439 (1975).
20. Babalola, D.A. and Schillhorn van Veen, T.W. Incidence of fascioliasis in cattle slaughtered in Bauchi (Nigeria). Trop. Anim. Hlth. Prod. **8**: 243–247 (1976).
21. Barger, I.A. and Southcott, W.H. Trichostrongylosis and wool growth. 3. The wool growth response of resistant grazing sheep to larval challenge. Aust. J. Exp. Agric. Anim. Husb. **15**: 167–172 (1975).
22. Barger, I.A., Lewis, R.J. and Brown, G.F. Survival of infective larvae of nematode parasites of cattle during drought. Vet. Parasitol. **14**: 143–152 (1984).
23. Barnes, E.H., Dobson, R.J., Donald, A.R. and Waller, P.J. Predicting populations of *Trichostrongylus colubriformis* infective larvae on pasture from meteorological data. Int. J. Parasitol. **18**: 767–774 (1988).
24. Bawden, R.J. Relationships between *Oesophagostomum columbianum* infections and the nutritional status of sheep. III. Serum and tissue protein changes. Aust. J. Agric. Res. **20**: 965–970 (1964).
25. Bell, R.G. Genetic analysis of expulsion of adult *Trichinella spiralis* in NFS, C3H/He and BIO. BR mice. Exp. Parasitol. **66**: 57–65 (1988).
26. Bell, R.G. and Liu, W.M. *Trichinella spiralis*: quantitative relationships between intestinal worm burden, worm rejection and the measurement of intestinal immunity in inbred mice. Exp. Parasitol. **66**: 44–45 (1988).
27. Bell, R.G., Adams, L.S. and Ogden, R.W. *Trypanosoma musculi* with *Trichinella spiralis* or *Heligmosomoides polygyrus*. Concomitant infections in the mouse. Exp. Parasitol. **58**: 8–18 (1984).
28. Bell, R.G., Adams, L.S. and Ogden, R.W. *Trypanosoma musculi* and *Trichinella spiralis*: Concomitant infections and selection for resistance genotypes in mice. Exp. Parasitol. **58**: 19–26 (1984).
29. Benitez-Usher, C., Armour J., Duncan, J.L., Urquhart, G.M. and Gettinby, G.A. A study of some factors influencing the immunisation of sheep against *Haemonchus contortus*—attenuated larvae. Vet. Parasitol. **3**: 327–342 (1977).
30. Behnke, J.M. Hannah, J. and Pritchard, D.I. *Nematospiroides dubius* in the mouse: evidence that adult worms depress the expression of homologous immunity. Parasite Immunol. **5**: 397–408 (1983).
31. Berberian, J.F. and Mizelle, J.D. Developmental studies on *Haemonchus contortus* Rudolphi (1803). Am. Midl. Natrl. **57**: 421–439 (1957).
32. Berry, C.I. and Dargie, J.D. Pathophysiology of ovine fascioliasis: the influence of dietary protein and iron on the erythrokinetics of sheep experimentally infected with *Fasciola hepatica*. Vet. Parasitol **4**: 327–339 (1978).
33. Bitakaramire, P.K. Preliminary studies on the immunisation of cattle against fascioliasis using gamma-irradiated metacercariae of *Fasciola gigantica*. In: *Isotopes and Radiation in Parasitology*, Vol. 3, p. 23. IAEA, Vienna (1973).
34. Bizzell, W.E. and Ciordia, H. Dissemination of infective larvae of trichostrongylid parasites of ruminants from faeces to pasture by the fungus *Pilobolus* spp. J. Parasitol. **51**: 184 (1965).
35. Blitz, N.M. and Gibbs, H.C. Studies on the arrested development of *Haemonchus contortus* in sheep. I. The induction of arrested development. Int. J. Parasitol. **2**: 5–12 (1972).

36. Blumenstock, D.K. Distribution and characteristics of tropical climates. Proc. Ninth Pacific Sci. Congr., Bangkok **20**: 3–24 (1958).
37. Boray, J.C. The ecology of *Fasciola hepatica* with particular reference to its intermediate host in Australia. Proc. Seventeenth Wld. Vet. Congr., Hannover, Vol. 1. pp. 709–715 (1963).
38. Boray, J.C. Studies on the relative susceptibility of some lymnaeids to infection with *Fasciola hepatica* and *F. gigantica* and on adaptation of *Fasciola* spp. Ann. Trop. Med. Parasitol. **60**: 114–124 (1966).
39. Boray, J.C. Experimental fascioliasis in Australia. Adv. Parasitol. **7**: 95–210 (1969).
40. Boray, J.C. Fascioliasis. In: *Handbook Series on Zoonoses. Section C: Parasitic Zoonoses.* Vol. III (G.V. Hillyer and C.E. Hopla eds), pp 71–88. CRC Press Inc.: Florida, USA (1982).
41. Boray, J.C. and Enigk, K. Laboratory studies on the survival and infectivity of *Fasciola hepatica* and *F. gigantica* metacercariae. Z.Tropenmed. Parasitol. **15**: 324–333 (1964).
42. Boag, B. and Thomas, R.J. Epidemiological studies on gastrointestinal nematode parasites of sheep. Infection patterns on clean and autumn-contaminated pasture. Res. Vet. Sci. **12**: 132–139 (1971).
43. Brunsdon, R.V. and Vlassoff, A. The post-parturient rise: a comparison of the pattern and relative generic composition of strongyle egg output from lactating and non-lactating ewes. N.Z. Vet. J. **19**: 19–25 (1971).
44. Bryan, R.P. and Kerr, J.D. Factors affecting the survival and migration of the free-living stages of gastrointestinal nematode parasites of cattle in Central Queensland. Vet. Parasitol. **30**: 315–326 (1989).
45. Campbell, N.J., Kelly, J.D., Townsend, R.B. and Dineen, J.K. The stimulation of resistance in sheep to *Fasciola hepatica* by infection with *Cysticercus tenuicollis*. Int. J. Parasitol. **7**: 347–351 (1967).
46. Carmichael, I.H. Helminthiasis in domestic and wild ruminants in Botswana—Preliminary investigations. Trop. Anim. Hlth. Prod. **4**: 175–181 (1972).
47. Capitini, L.A., McClure, K.E. and Herd, R.P. Effect of environmental stimuli on pre-infective stages of *Haemonchus contortus* in the northern United States for the induction of hypobiosis. Vet. Parasitol. **35**: 281–293 (1990).
48. Catto, J.B. Longevity of infective larvae of gastrointestinal nematodes of cattle in the Pantanal, Brazil. Pesquisa Agropecuaria, Brasiliera, **22**: 847–854 (1987).
49. Charles, T.P. Seasonal prevalence of gastrointestinal nematodes of goats in Pernambuco State, Brazil. Vet. Parasitol **30**: 335–343 (1989).
50. Chiejina, S.N. Bovine parasitic gastroenteris in Nsukka. Worm egg output and herbage infestation following dry season calving and the performance of calves during their first grazing season, Trop. Vet. 2: 128–132 (1984).
51. Chiejina, S.N. The epizootiology and control of parasitic gastroenteritis of domesticated ruminants in Nigeria. Helminthol. Abstr. **55**: 413–429 (1986).
52. Chiejina, S.N. Some parasitic diseases of intensively managed West African Dwarf sheep and goats in Nsukka, eastern Nigeria. British Vet. J. **143**: 264–272 (1987).
53. Chiejina, S.N. and Sewell, M.M.H. Experimental infections with *Trichostrongylus colubriformis* (Giles, 1892) Loos, 1905 in lambs: worm burden, growth rate and host resistance resulting from prolonged escalating infections. Parasitology **69**: 301–314 (1974).
54. Chiejina, S.N. and Sewell, M.M.H. Worm burdens, acquired resistance and liveweight gains in lambs during prolonged daily infections with *Trichostrongylus colubriformis* (Giles, 1892) Loos 1905. Parasitology **69**: 315–327 (1974).
55. Chiejina, S.N. and Clegg, F.G. Some observations on the epidemiology of ostertagiasis in calves in Britain: an analysis of laboratory diagnostic and field data for 1974 to 1978. British Vet. J. **134**: 541–550 (1978).
56. Chiejina, S.N. and Emehelu, C.O. Seasonal changes in pasture populations of infective larvae of gastrointestinal nematodes of cattle in eastern Nigeria. Res. Vet. Sci. 37: 144–147 (1984).
57. Chiejina, S.N. and Fakae, B.B. The ecology of infective larvae of bovine gastrointestinal tichostrongylids in dry season contaminated pastures in the Nigerian derived savanna. J. Helminthol. **63**: 127–129 (1989).
58. Chiejina, S.N., Fakae, B.B. and Eze, P.I. Development and survival of free-living stages of gastrointestinal nematodes of sheep and goats on pasture in the Nigerian derived savanna. Vet. Res. Com. **13**: 103–112 (1989).
59. Chiejina, S.N., Fakae, B.B. and Eze, B.O. Arrested development of gastrointestinal trichostrongylids in goats in Nigeria. Vet. Parasitol. **28**: 103–113 (1988).

60. Chowaniec, W., Wescott, R.B. and Congdon, L.L. Interaction of *Nematospiroides dubius* and influenza virus in mice. Exp. Parasitol. **32**: 33–44 (1972).
61. Clayton, C.E., Sacks, D.L., Ogilvie, B.M. and Askonas, B.A. Membrane fractions of trypanosomes mimic the immunosupressive and mitogenic effects of living parasites on the host. Parasite Immunol. **1**: 241–249 (1979).
*62. Clunies Ross, I. and Mckay, A.C. Bull. Coun. Scient. Ind. Res. Melb., No. 43, 66 pp. (1929).
63. Collis-George, N. The physical environment of soil animals. Ecology **40**: 550–557 (1959).
64. Connan, R.M. The seasonal incidence of inhibition of development in *Haemonchus contortus*. Res. Vet. Sci. **12**: 272–274 (1971).
65. Connan, R.M. Ovine parasitic gastro-enteritis. The spring rise in faecal egg count. In: *Helminth Diseases of Cattle, Sheep and Horses in Europe*. Proceedings of a workshop, University of Glasgow Veterinary School, 1973, pp. 36–41 and 52–53 (1975).
66. Connan, R.M., Inhibited development in *Haemonchus contortus*. Parasitology **71**: 239–246 (1975).
67. Cook J.C. The clinical significance of gastrointestinal helminths—a review. Trans. R. Soc. Trop. Med. Hyg. **80**: 675–685 (1986).
68. Coop, R.L. and Angus, K.W. The effect of continuous doses of *Trichostrongylus colubriformis* larvae on the intestinal mucosae of sheep and on liver vitamin A concentration. Parasitology **70**: 1–9 (1975).
69. Coop, R.L. and Angus, K.W. How helminths affect sheep. In: *Practice*. Suppl. Vet. Rec. **3**: 4–11 (1981).
70. Copeman, D.B. and Hutchinson, G.W. The economic significance of bovine gastrointestinal nematode parasitism in North Queensland. (W.A. Geering, R.J. Roe and L.Chapman eds), pp. 383–389. Aust. Govt. Publ. Ser. (1979).
71. Corba, J., Armour, J., Roberts, R.J. and Urquhart, G.M. Transfer of immunity to *Fasciola hepatica* infection by lymphoid cells. Res. Vet. Sci. **12**: 292–295 (1971).
72. Coyle, T.J. Experiments in the diagnosis and control of fascioliasis in Ugandan cattle. Bull. epiz. Dis. Afr. **6**: 255–272 (1958).
73. Crofton, H.D. Nematode parasite populations in sheep on lowland farms. I. Worm egg counts in ewes. Parasitology **44**: 465–477 (1954).
74. Crofton, H.D. *Nematode parasite populations in sheep and on pasture*. Tech. Commun. Commonw. Bur. Helminthol. No. 35. 104 pp. St. Albans: England (1963).
*75. Croix, St. F.W. *The Fulani of northern Nigeria*. Government Printers: Lagos, Nigeria (1944).
76. Crompton, D.W.D. Nutritional aspects of infection. Trans. R. Soc. Trop. Med. Hyg. **80**: 697–705 (1986).
77. Cummins, A.G., Duncombe, V.M., Bolin, T.D., Davies, A.E. and Yong, J. The response of the small intestine of the protein-deficient rat to infection with *Nippostrongylus brasiliensis*. Int. J. Parasitol. **17**: 1445–1450 (1987).
78. Dargie, J.D. The influence of genetic factors on the resistance of ruminants to gastrointestinal nematode and trypanosome infections. In: *Animal Models in Parasitology* (D.G. Owen ed), pp. 17–51. McMillan Press: London (1982).
79. Dargie, J.D., Armour, J. and Urquhart, G.M. Studies on immunity to *Fasciola hepatica*. Parasitology **67**: 25–32 (1973).
80. Dawes, B. A study of the miracidium of *Fasciola hepatica* and an account of the mode of penetration of the sporocyst into *Lymnaea truncatula*. In: *Libro Hmenaje Dr. Eduardo Caballero of Caballero*. pp. 95–111. Escuela Nacional de Ciencias Biologicas: Mexico (1960).
81. Diaw, D.T., Seye, M. and Sarr, Y. Epidemiology of trematodiases in livestock in the Kolda region, Casamance (Senegal). Rev. Elev. Med. Pays Vet. Trop. **41**: 257–264 (1988).
82. Dickson, K.E. The relative suitability of sheep and cattle as hosts for the liver fluke, *Fasciola hepatica* L. J. Helminthol. **38**: 203–212 (1964).
83. Dinaburg, A.G. Development and survival under outdoor conditions of eggs and larvae of the common ruminant stomach worm *Haemonchus contortus*. J. Agric Res. **69**: 421–433 (1944).
84. Dineen, J.K. and Windon, R.G. The effect of sire selection on the response of lambs to vaccination with irradiated *Trichostrongylus colubriformis* larvae. Int. J. Parasitol. **10**: 189–196 (1980).
85. Dineen, J.K., Kelly, J.D. and Campbell, N.J. Further observations on the nature and characteristics of cross

protection against *Fasciola hepatica* produced in sheep by prior infection with *Cysticercus tenuicollis*. Int. J. Parasitol. **8**: 173–176 (1978).

86. Dinnik, J.A. and Dinnik, N.N. Report of the East African Veterinary Research Organisation, 1956–1957. p. 50 (1950).
87. Dinnik, J.A. and Dinnik, N.N. Observations on the development of *Haemonchus contortus* larvae under field conditions in the Kenya Highlands. Bull. epiz. Dis. Afr. **6**: 11–21 (1958).
88. Dinnik, J.A. and Dinnik, N.N. Effect of the seasonal variations of temperature on the development of *Fasciola gigantica* eggs in the Kenya Highlands. Bull. epiz. Dis. Afr. **7**: 357–369 (1959).
89. Dinnik, J.A. and Dinnik, N.N. Observations on the longevity of *Haemonchus contortus* larvae under field conditions in the Kenya Highlands. Bull. epiz. Dis. Afr. **9**: 193–208 (1961).
90. Dinnik, J.A. and Dinnik, N.N. Effect of seasonal variation of temperature on the development of *Fasciola gigantica* in the snail host in the Kenya Highlands. Bull. epiz. Dis. Afr. **11**: 197–207 (1963).
91. Douch, P.G.C. and Outteridge, P.M. The relationship between ovine lymphocyte antigens and parasitological and production parameters in Romney sheep. Int. J. Parasitol. **19**: 35–41 (1989).
92. Durie, D.H.: Parasitic gastroenteritis of cattle: the distribution and survival of infective strongyle larvae on pasture. Aust. J. Agric. Res. **12**: 1200–1211 (1961).
93. Edney, J.M. and Muchlis, A. Fascioliasis in Indonesian livestock. Comm.Vet. Bogor. **6**: 49–62 (1962).
94 Eysker, M. Inhibition of the development of *Trichostrongylus* spp. as third stage larvae in sheep. Vet. Parasitol. **4**: 29–33 (1978).
95. Eysker, M. and Ogunsusi, R.A. Observations on epidemiological and clinical aspects of gastrointestinal helminthiasis of sheep in northern Nigeria during the rainy season. Res. Vet. Sci. **28**: 58–62 (1980).
96. Fabiyi, J.P. Seasonal fluctuations of nematode infestations in goats in the savanna belt of Nigeria. Bull. epiz. Dis. Afr. **21**: 277–286 (1973).
97. Fabiyi, J.P. Production losses and control of helminths in ruminants of tropical regions. Int. J. Parasitol. **17**: 435–442 (1986).
98. Fabiyi, J.P. and Adeleye, G.A. Bovine fascioliasis on the Jos Plateau, northern Nigeria, with particular reference to economic importance. Bull. epiz. Dis. Afr. **30**: 41–43 (1982).
99. Fabiyi, J.P., Copeman, D.B. and Hutchinson, G.W. Abundance and survival of infective larvae of the cattle nematodes *Cooperia punctata, Haemonchus placei* and *Oesophagostomum radiatum* from faecal pats in a wet tropical climate. Aust. Vet. J. **65**: 229–231 (1988).
100. Fabiyi , J.P., Oluyede, D.A. and Negedu, J.O. Late dry season outbreak of clinical haemonchosis and cooperiasis in cattle of northern Nigeria. Vet. Rec. **105**: 399–400 (1979).
101. Fakae, B.B. Aspects of the epidemiology of bovine parasitic gastroenteritis in the derived savanna area of eastern Nigeria. MSc. Dissertation, University of Nigeria, Nsukka, Nigeria. xi+131 pp. (1986).
102. Fakae, B.B. The epidemiology of helminthosis in small ruminants under the traditional husbandry system in eastern Nigeria. Vet. Res. Comm. **14**: 381–391 (1990).
103. Fakae, B.B. Seasonal changes and Hypobiosis in *Haemonchus contortus* infection in the West African Dwarf sheep and goats in the Nigerian derived savanna. Vet. Parasitol. **36**: 123–130 (1990).
104. Fakae, B.B. and Chiejina, S.N. Relative contributions of late dry season and early rains pasture contaminations with trichostrongyle eggs to the wet season herbage infestation in eastern Nigeria. Vet. Parasitol. **28**: 115–123 (1988).
105. Fakae, B.B. and Chiejina, S.N. Further studies on the development and availability of infective larvae of bovine gastrointestinal trichostrogylids on pasture in eastern Nigeria. Vet. Parasitol. **28**: 143–152 (1988).
106. Francis, P.A. Livestock and farming systems in south-eastern Nigeria. In: *Goat Production in the Humid Tropics*. Proceedings of a workshop, University of Ife, Nigeria, pp. 159–169 (1987).
107. Frank, G.R., Herd, R.P., Marbury, K.S., Williams, J.C. and Willis, E.R. Additional investigations on hypobiosis of *Ostertagia ostertagi* transferred between northern and southern USA. Int. J. Parasitol. **18**: 171–177 (1988).
108. Franklin, M.C., Gordon, H. McL. and MacGregor, C.H. A study of nutritional and biochemical effects in sheep of infestation with *Trichostrongylus colubriformis*. J. Coun. Sci. Ind. Res. **19**: 46–60 (1946).
109. Frisch, J.E. Factors affecting resistance to ecto— and endo-parasites of cattle in tropical areas and the implications for selection. In: *Genetic and Nutritional Influences on the Immunology. and Pathogenesis of Parasitic Infections in Ruminants*. IAEA, Vienna (1981).

110. Frye, W.W. Nutrition and intestinal parasitism. Ann. N.Y. Acad. Sci. **63**: 175–185 (1955).
111. Furman, D.P. Effects of environment upon the free-living stages of *Ostertagia circumcincta* (Stadelmann) Trichostrongylidae. II. Field experiments. Am. J. Vet. Res. **5**: 147–153 (1944).
112. Gatenby, R.M. and Handayani, Sri. W. Would Java goat and sheep houses be useful in Nigeria? In: *Goat Production in the Humid Tropics*. Proceedings of a workshop, University of Ife, Nigeria, pp. 170–173 (1987).
113. Gibson, T.E. Studies on *Trichostrongylus axei*. IV Factors in the causation of pathogenic effects by *T. axei*. J. Comp. Path. **81**: 493–498 (1955).
114. Gibson, T.E. and Everett, G. The ecology of free-living stages of *Trichostrongylus colubriformis*. Parasitology **57**: 533–547 (1967).
115. Gibson, T.E. and Everett, G. The ecology of free-living stages of *Ostertagia circumcincta*. Parasitology **64**: 451–460 (1970).
116. Goldberg, A. Develoment, migration and survival on pasture of gastrointestinal nematode parasites of cattle: summer contamination. Proc. Helm. Soc. Wash. **37**: 166–169 (1970).
117. Goodwin, L.G. Green, D.G., Guy, M.W. and Voller, A.: Immunosuppression during trypanosomiasis. British J. Exp. Path. **53**: 40–43 (1972).
118. Gordon, H. McL.: The epidemiology of parasitic diseases with special reference to studies with nematode parasites of sheep. Aust. Vet. J. **24**: 17–44 (1948).
119. Gordon, H. McL. Studies on resistance to *Trichostrongylus colubriformis*. Influence of a quantitative reduction in the ration. Aust. Vet. J. **40**: 55–61 (1964).
120. Gordon, H. McL. Self-cure reaction. In: *The Reaction of the Host to Parasitism* (E.J.L. Soulsby ed), pp. 174–190. Elwert: Marburg (1968).
*121. Graber, M.: Helminthes and Helminthiasis faisant obstacle a al'ameleration de la production ovine en republique de Tchad. Report Lab. de Farcha. L.E.M.V.T. Fort Lamy, Chad (1965).
122. Gregory, P.C., Wenham, G., Poppi, D.,Coop, R.L., CaeRae, J.C. and Miller, S.J. The influence of chronic subclinical infection of *Trichostrongylus colubriformis* on gastrointestinal motility and digesta flow in sheep. Parasitology **91**: 381–396 (1986).
123. Gretillat, S. Note preliminaire sur l'epedemiologie de la distomatose bovine en Senegal. Rev. Elev. Med. Pays Vet. Trop. **14**: 283–291 (1961).
124. Griffiths, R.B. Report to the Government of Burma on parasitic diseases of livestock. Report No. 684. 127 pp. FAO, Rome (1957).
125. Griffin, L., Allonby, E.W. and Preston, J.M. The interaction of *Trypanosoma congolense* and *Haemonchus contortus* in two breeds of goat. I. Parasitology. J. Comp. Path. **91**: 85–95 (1981).
126. Gronvold, J.: Field experiments on the ability of earthworms (Lumbricidae) to reduce the transmission of infective larvae of *Cooperia oncophora* (Trichostrongylidae) from cow pats to grass. J. Parasitol. **73**: 1133–1137 (1987).
127. Gupta, R.P., Yadav, C.L. and Chaudhri, S.S. Epidemiology of gastrointestinal nematodes of sheep and goats in Haryana, India. Vet. Parasitol. **24**: 117–127 (1987).
*128. Guralp, N., Ozcan, C. and Simms, B.T.: *Fasciola gigantica* and fascioliasis in Turkey. Am. J. Vet. Res. **25**: 196–210 (1964).
129. Hammerberg, B. Pathophysiology of nematodiasis in cattle. In: *The Veterinary Clinics of North America. Food Animal Practice. Parasites: Epidemiology and Control* (H.C. Gibbs, R.P. Herd and K.D. Murrell eds), Vol. 2, pp. 225–234. W.B. Saunders: Philadelphia, London & Toronto (1986).
130. Hart, J.A. Observations on the dry season strongyle infestations of zebu cattle in northern Nigeria. British Vet. J. **120**: 87–95 (1964).
131. Hashmi, H.A. and Connan, R.M. Biological control of ruminant trichostrongylids by *Arthrobotrys oligospora*, a predacious fungus. Parasitol. Today **5**: 28–30 (1989).
132. Hayes, T.J. and Mitrovic, M. The early expression of protective immunity to *Fasciola hepatica* in rats. J. Parasitol. **63**: 584–587 (1977).
133. Hayes, T.J. Further evidence for the early expression of immunity to *Fasciola hepatica* in rates. *J. Parasitol.* **64**: 374–376 (1978).
134. Heath, G.B.S. and Michel, J.F, A contribution to the epidemiology of parasitic gastroenteritis in lambs. Vet. Rec. **85**: 305–308 (1969).

135. Henderson, W.W. Diseases of sheep and goats: Liver fluke (fascioliasis). Annual Report, Veterinary Department of Nigeria, 1935. p. 15 (1936).
136. Holmes, P.H., Dargie, J.D., MacLean, J.M. and Mulligan, W. The anaemia of fascioliasis: Studies with ^{51}Cr-labelled red cells. J. Comp. Path. **78**: 415–420 (1968).
137. Holmes, P.H., Mammo, E., Thomson, A., Knight, P.A., Lucken, R., Murray, P.K., Murray, M., Jennings, F.W. and Urquhart, G.M. Immunosuppression in bovine trypanosomiasis. Vet. Rec. **95**: 86–87 (1974).
138. Hope Cawdery, M.J. The effects of fascioliasis on ewe fertility. British Vet. J. **132**: 568–575 (1976).
139. Hotson, I.K. Ostertagiosis in cattle. Aust. Vet. J. **43**: 383–387 (1967).
140. Hughes, D.L. *Fasciola* and *Fascioloides*. In: *Immune Responses in Parasitic Infections: Immunology, Immunopathology and Immunoprophylaxis*. Vol. II, Trematodes and Cestodes (E.J.L. Soulsby ed), pp. 91–114. CRC Press: Florida, USA (1987).
141. Hughes, D.L., Harness, E. and Doy, T.G. Failure to demonstrate resistance in goats, sheep and cattle to *Fasciola hepatica* after infection with *Cysticercus tenuicollis*. Res. Vet. Sci. **25**: 356–359 (1978).
142. Hunter, A.G. and Heath, P.J. Ovine internal parasitism in the Yemen Arab Republic. Trop. Anim. Hlth. Prod. **16**: 95–106 (1984).
143. Ikeme, M.M. *Strongyloides papillosus* and *Neoascaris vitulorum*: naturally acquired mixed infestations of calves in the Plateau area of northern Nigeria and the treatment given. Bull. epiz. Dis. Afr. **18**: 339–345 (1971).
144. Ikeme, M.M., Fatimah, I. and Lee, C.C. Seasonal incidence of infective nematode larvae on pasture and nematode worm populations of goats in Selangor. Kaj. Vet. **18**: 69–76 (1986).
145. Ikeme, M.M., Iskander, F. and Chong, L.C. Seasonal changes in the prevalence of *Haemonchus* and *Trichostrongylus* hypobiotic larvae in tracer goats in Malaysia. Trop. Anim. Hlth. Prod. **19**: 184–190 (1987).
146. International Livestock Centre for Africa. *Small Ruminant Production in the Humid Tropics*. Systems study No. 3. ILCA, Addis Ababa: Ethiopia (1979).
147. Irfan, M. Trials on the efficacy of thiabendazole against *Haemonchus contortus* in West African Dwarf sheep and goats. Bull. epiz. Dis. Afr. **15**: 369–371 (1967).
148. Jacobs, D.E., Tod, M.E., Dunn, A.M. and Walker, J. Farm-to-farm transmission of porcine oesophagostomiasis. Vet. Rec. **82**: 57 (1968).
149. Jiya, M. Report on the effect of drought conditions on livestock production in North-western State. Nig. J. Anim. Prod. **1**: 17–23 (1974).
150. Kassai, T., Cordero Del Campillo, M., Euzeby, J., Gaafar, S., Hiepe, Th. and Himonas, C.A. Standardized nomenclature of animal parasitic diseases (SNOAPAD) . Vet. Parasitol. **29**: 299–326 (1988).
151. Kates, K.C. Survival on pasture of free-living stages of some common gastrointestinal nematodes of sheep. Proc. Helm. Soc. Wash. **17**: 39–58 (1950).
152. Kates, K.C. Ecological aspects of helminth transmission in domesticated animals. Am. Zoologist **5**: 95–130 (1965).
153. Kendall, S.B. Nutritional factors affecting the rate of development of *Fasciola hepatica* in *Lymnaea truncatula*. J. Helminthol. **53**: 179–190 (1949).
154. Kendall, S.B. Fascioliasis in Pakistan Ann. Trop. Med. Parasitol. **48**: 307–313 (1954).
155. Kendall, S.B. Relationships between the species of *Fasciola* and their molluscan hosts. Adv. Parasitol. **3**: 59–58 (1965).
156. Kendall, S.B. and Ollerenshaw, C.B. The effect of nutrition on the growth of *Fasciola hepatica* in its snail host. Proc. Nutr. Soc. **22**: 41–46 (1963).
157. Knapp, S.F. Relationship of different species of forage to the survival and infectivity of *Haemonchus contortus* in lambs. J. Parasitol **50**: 144–148 (1964).
158. Knight, R.A., Vegirs, H.H. and Glimp, H.A. Effects of breed and date of birth of lambs on gastrointestinal nematode infections Am. J. Vet. Res. **34**: 323–327 (1973).
159. Lang, B.Z. Host-parasite relationships of *Fasciola hepatica* in the white mouse. II. Studies on acquired immunity. J. Parasitol. **53**: 21–30 (1967).
160. Lang, B.Z., Larsh, J.E. Jr., Weatherby, N.F. and Goulson, T.H. Demonstration of immunity to *Fasciola hepatica* in recipient mice given peritoneal exudate cells. J. Parasitol. **53**: 208–209 (1967).
161. Le Roux, P. Report to the Government of Rhodesia and Nyasaland on the control of parasitic diseases of livestock. Report No. 696. pp. 47, FAO: Rome (1957).

162. Lee, R.P. Bovine ascariasis. In: *Symposium on Helminthiasis in Domestic Animals*, Nairobi, Kenya. Publication No. 549, pp. 131–145. Science Council of Africa South of the Sahara (1959).
163. Lee, R.P., Armour, J. and Ross, J.G. The seasonal variations of strongyle infestations in Nigerian zebu cattle. British Vet. J. **116**: 34–46 (1960).
164. Lemma, B., Gabre-Ab, F. and Tedla, S. Studies on fascioliasis in four selected sites in Ethiopia. Vet. Parasitol. **18**: 29–37 (1985).
165. Levine, N.D. Weather, climate and the bionomics of ruminant nematode larvae. Adv. Vet. Sci. **8**: 215-261 (1963).
166. Levine, N.D. and Anderson, F.L.: Development and survival of *Trichostrongylus colubriformis* on pasture. J. Parasitol. **59**: 147–165 (1973).
167. Lyons, E.T., Drudge, J.H. and Tolliver, S.C. *Strongyloides* larvae in the milk of sheep and cattle. Mod. Vet. Pract. **51**: 65–68 (1970).
168. Mackenzie, P.K.I., Boyt, W.P., Emslie, V.W., Lander, K.P. and Swanepoel, R. Immunosuppression in ovine trypanosomiasis. Vet. Rec. **97**: 452–453 (1975).
169. Manton, V.J.A., Peacock, R., Poynter, D., Silverman, P.H. and Terry, R.J. The influence of age on naturally acquired resistance to *Haemonchus contortus*. Res. Vet. Sci. **3**: 308–313 (1962).
170. McCanley, E.H., Majid, A.A., Tayeb, A. and Bushara, H.O. Clinical diagnosis of schistosomiasis in Sudanese cattle. Trop. Anim. Hlth. Prod. **15**: 129–136 (1983).
171. McCullough, F.S. *Lymnaea natalensis* and Fascioliasis in Ghana. Ann. Trop. Med. Parasitol. **58**: 320–326 (1965).
172. McElroy, P.J. and Befus, A.D. *Nippostrongylus brasiliensis*—induced depression of heterologous immune responses: Effects of the histamine receptor antagonist, cimetidine. J. Parasitol. **73**:1158–1163 (1987).
173. Megard, J.P. *La fasciolose en Afrique noire*. Merck, Sharp and Dohme an service de l'Elevage, Paris (1976).
*174. Mehl, S. *Arb. bayer, Landesanst. BflBau*, No. 10. 177 pp. (1932).
175. Michel, J.F. The phenomena of host resistance and the course of *Ostertagia ostertagi* in calves. Parasitology **53**: 63–84 (1963).
176. Michel, J.F. Immunity to helminths associated with tissues. In: *Immunity to Parasites. Symposium of the British Society for Parasitology* (A.E.R. Taylor ed), pp. 67–89. Blackwell Scientific Publications: London (1968).
177. Michel, J.F. The regulation of populations of *Ostertagia ostertagi*. Parasitology **61**: 435–447 (1970).
178. Michel, J.F. Adult worms as a factor in the inhibition of development of *Ostertagia ostertagi* in the host. Int. J. Parasitol. **1**: 31–36 (1971).
179. Michel, J.F. Arrested development of nematodes and some related phenomena. Adv. Parasitol. **12**: 279–366 (1974).
180. Michel, J.F. The epidemiology and control of some nematode infections of grazing animals. Adv. Parasitol. **14**: 355–397 (1976).
181. Michel, J.F., Lancaster, M.B. and Hong, C. The epidemiology of gastro-intestinal nematode infection in the single-suckled calf. Vet. Rec. **91**: 301–306 (1972).
182. Michel, J.F. Lancaster, M.B. and Hong, C. Studies on arrested development of *Ostertagia ostertagi* and *Cooperia oncophora*. J. Comp. Path. **84**: 539–554 (1974).
183. Michel, J.F. Lancaster, B.M. and Hong, C. Arrested development of *Ostertagia ostertagi* and *Cooperia oncophora*. Effect of temperature on the free-living third stage. J. Comp. Path. **85**: 133–138 (1975).
184. Michel, J.F., Lancaster, M.B. and Hong, C. The effect of age, acquired resistance, pregnancy and lactation on some reactions of cattle to infection with *Ostertagia ostertagi*. Parasitology **79**: 157–168 (1978).
185. Misra, S.C. and Ruprah, N.S. Incidence of helminths in goats at Hisar. J. Res. Punjab Agric. Univ. **5**: 279–286 (1968).
186. Mitchell, J.R. The epidemiology and economic importance of *Oesophagostomum radiatum* in Swaziland. Bull. epiz. Dis. Afr. **22**: 95–98 (1974).
187. Molinari, J.A., Cypes, R.H. and Ebersole, J.L. Effect of *Trichinella spiralis* infection on the cell-mediated immune response to BCG. Int. Archs Allergy Appl. Immunol. **47**: 483–487 (1974).
188. Morgan, D.O. and Sloan, J.E.N. Researches on helminths in hill sheep with reference to seasonal variations in worm egg output. Scottish Agric. **27**: 28–35 (1947).

189. Mozgovoi, A.A. and Shakhmatova, V.I. On the study of the life cycle of *Neoascaris vitulorum* (Ascaridata: Anisakidae), nematode of ruminants. Trud. Gel'mint. Lab. Akad Nauk. USSR. **17**: 95–103. (1969).
190. Murray, M. Fascioliasis: Pathology. In: *Helminth Diseases of Cattle, Sheep and Horses in Europe* (G.M. Urquhart and J. Armour eds), pp. 81–144. MacLehouse: Glasgow, Scotland (1973).
191. Murray, M., Jennings, F.W. and Armour, J. Bovine ostertagiasis: structure, function and mode of differentiation of the bovine gastric mucosa and kinetics of the worm loss. Res. vet. Sci. **11**: 417–427 (1970).
192. Murray, P.K., Jennings, F.W., Murray, M. and Urquhart, G.M. Immunosuppression in trypanosomiasis. In: *Parasitic Zoonoses, Clinical and Experimental Studies*. Proc. Sixth Int. Conf. Wld Assoc. Adv. Vet. Parasitol. Vienna (E.J.L. Soulsby ed), pp. 133–150 (1973).
193. Mzembe, S.A.T. and Chaudhry, M.A. The epidemiology of fascioliasis in Malawi: I. The epidemiology in the intermediate host. Trop. Anim. Hlth. Prod. **11**: 246–250 (1979).
194. Mzembe, S.A.T. and Chaudhry, M.A. The epidemiology of fascioliasis in Malawi. Part II. Epidemiology in the definitive host. Trop. Anim. Hlth. Prod. **13**: 27–33 (1980).
195. Nansen, P. Resistance in cattle to *Fasciola hepatica* induced by X-ray attenuated larvae: results of a controlled field trial. Res. Vet. Sci. **19**: 278–283 (1975).
196. Ndamukong, K.J.N., Sewell, M.M.H. and Asanji, M.F. Management and productivity of small ruminants in the North-west province of Cameroon. Trop. Anim. Hlth. Prod. **21**: 109–119 (1989).
197. Nichol, C.P. and Sewell, M.M.H. Immunosuppression by larval cestodes and *Babesia microti*. Ann. Trop. Med. Parasitol. **78**: 228–233 (1984).
198. Ogambo-Ongoma, A.H. Fascioliasis survey in Uganda. Bull. epiz. Dis. Afr. **20**: 35–41 (1972).
199. Ogunrinade, A.F. Infectivity and pathogenicity of *Fasciola gigantica* in West African Dwarf sheep and goats. Trop. Anim. Hlth. Prod. **16**: 161–166 (1984).
200. Ogunsusi, R.A. Pasture infectivity with trichostrongylid larvae in the northern Guinea savanna of Nigeria. Res. Vet. Sci. **26**: 320–323 (1979).
201. Ogunsusi, R.A. and Eysker, M. Inhibited development of trichostrongylids of sheep in northern Nigeria. Res. Vet. Sci. **26**: 108–110 (1979).
202. Okeke, G.C. and Oji, U.I. The nutritive value of grass ensiled with cassava peel and poultry excreta. In: *Goat production in the Humid Tropics*. Proceedings of a workshop, University of Ife, Nigeria, pp. 101–106 (1987).
203. Okon, E.D. Effect of parturition on faecal strongyle egg output in Nigerian goats. Bull. epiz. Dis. Afr. **28**: 155–158 (1980).
204. Okon, E.D. and Akinpelu, A.I. Development and survival of nematode larvae on pasture in Calabar, Nigeria. Trop. Anim. Hlth. Prod. **14**: 23–25 (1982).
205. Okon., E.D. and Enyenihi, U.K. Infectivity of *Haemonchus contortus* and *Trichostrongylus colubriformis* larvae on pasture at Ibadan. Bull. Off. Int. epiz. Paris. **83**: 1139–1144 (1975).
206. Okon, E.D. and Enyenihi, U.K. Devlopment and survival of *Haemonchus contortus* larvae on pasture in Ibadan. Trop. Anim. Hlth. Prod. **9**: 7–10 (1977).
207. Oldham, G. and Williams, L. Cell-mediated immunity to liver fluke antigens during experimental *Fasciola hepatica* infection in cattle. Parasite Immunol. **7**: 503–516 (1985).
208. Ollerenshaw, C.B. The ecology of the liver fluke (*Fasciola hepatica*). Vet. Rec. **71**: 957–963 (1959).
209. Ollerenshaw, C.B. Forecasting Liver fluke disease. In: *The Effects of Meteorological Factors upon Parasites. Symposium of the British Society for Parasitology* (A.E.R. Taylor and R. Muller eds), pp. 33–52. Blackwell Scientific Publications: Oxford, England (1974).
210. Ollerenshaw, C.B. and Rowlands, W.T. A method of forecasting the incidence of fascioliasis in Anglesey. Vet. Rec. **71**: 591–598 (1959).
211. Ollerenshaw, C.B. and Smith, L.P. Meteorological factors and forecasts of helminth disease. Adv. Parasitol. **7**: 283–323 (1969).
212. Ollerenshaw, C.B., Graham, E.G. and Smith, L.P. Forecasting the incidence of parasitic gastroenteritis in lambs in England and Wales. Vet. Rec. **103**: 461–465 (1978).
213. Olsen, O.W. Longevity of metacercariae of *Fasciola hepatica* on pastures in the upper coastal region of Texas and its relationship to liver fluke control. J. Parasitol. **31**: 36–42 (1947).
214. Pantelouris, E.M. *The Common Liver Fluke*. Pergamon Press: Oxford, England (1965).

215. Paton, G., Thomas, R.J. and Waller, P.J. A prediction model for parasitic gastroenteritis in lambs. Int. J. Parasitol. **14**: 439–445 (1984).
216. Payne, W.J.A. *Cattle production in the tropics*, Vol. 1. Longman Group Ltd.: London (1970).
217. Payne, W.J.A. Systems of beef production in developing countries. In: *Beef Cattle Production in Developing Countries* (A.J. Smith ed), pp. 118–131. Lewis Reprints Ltd.: Tonbridge, Scotland (1976).
218. Phillips, R.S., Selby, G.R. and Wakelin, D. The effect of *Plasmodium berghei* and *Trypanosoma brucei* on the immune expulsion of the nematode, *Trichuris muris* from mice. Int. J. Parasitol. **4**: 409–415 (1974).
219. Piper, L.R. and Barger, I.A. Resistance to gastrointestinal strongyles: Feasibility of a breeding programme. In: *Proceedings of Third World Congress on Sheep and Beef Cattle Breeding, Paris*. Vol. 1, pp. 593–611 (1988).
220. Prats,V.M.V. and Fuentos, R.N. A study of the infective larvae of gastrointestinal nematodes of sheep in a humid sub-tropical climate. Tec. Pec. Mex. **25**: 25–31 (1987).
221. Preston, J.M. and Allonby, E.W. The influence of breed on the susceptibility of sheep and goats to a single experimental infection with *Haemonchus contortus*. Vet. Rec. **103**: 509–512 (1978).
222. Preston, J.M. and Allonby, E.W. The influence of breed on the susceptibility of sheep to *Haemonchus contortus* infection. Res. Vet. Sci. **26**: 134–139 (1979).
223. Pullan, N.B. Productivity of White Fulani cattle on the Jos Plateau, Nigeria. III. Disease and management factors. Trop. Anim. Hlth. Prod. **12**: 77–84 (1980).
224. Rahman, M.A., Ali, A.K. and Rahman, A. Incidence of disease of cattle in Mymensingh. Bangladesh Vet. J. **6**: 25–30 (1972).
225. Rains, A.B. Institute for Agricultural Research, Samaru, Zaria, Nigeria, Miscellaneous Paper No. 1, pp. 24–26 (1963).
226. Randall R.W. and Gibbs, H.C. Effects of clinical and subclinical gastrointestinal helminthiasis on digestion and energy metabolism in calves. Am. J. Vet. Res. **42**: 1730–1734 (1981).
227. Reid, J.F.S. and Armour, J. Seasonal fluctuations in inhibited development of gastro-intestinal nematodes of sheep. Res. Vet. Sci. **13**: 225–229 (1972).
228. Reid, J.F.S., Armour, J., Jennings, F.W., Kirkpatrick, K.S. and Urquhart, G.M. The fascioliasis-ostertagiasis complex in young cattle. A guide to diagnosis and therapy. Vet. Rec. **80**: 371–374 (1967).
229. Reinecke, R.K. and Thomas, R.J. Epizootiology of nematode parasites with special reference to animal husbandry methods. In: *Symposium on Helminthiasis in Domestic Animals*. IACED Publication No. 49, pp. 194–215. Nairobi: Kenya (1959).
230. Reveron, A.E. and Topps, J.H. Nutrition and gastrointestinal parasitism in ruminants. Outlook Agric. **6**: 131–136.
231. Reveron, A.E., Topps, J.H. and Gelman, A.L. Mineral metabolism and skeletal development of lambs affected by *Trichostrongylus colubriformis*. Res. Vet. Sci. **14**: 310–319 (1974).
232. Reynolds, L. and Adediran, S.O. The effects of browse supplementation on the productivity of West African Dwarf sheep over two reproductive cycles, *In: Goat production in the Humid Tropics*. Proceedings of a workshop, University of Ife, Nigeria, pp. 83–91 (1987).
233. Riffkin, G.G. and Dobson, C. Predicting resistance of sheep to *Haemonchus contortus* infection. Vet. Parasitol. **5**: 365–378 (1979).
234. Roberts, J.A. *Toxocara vitulorum*: treatment based on the duration of the infectivity of buffalo cows (*Bubalus bubalis*) for their calves. J. Vet. Pharm. Therap. **12**: 5–13 (1989).
235. Rogers, W.P. The effects of environmental conditions on the accessibility of third stage trichostrongyle larvae to grazing animals. Parasitology **32**: 208–225 (1940).
236. Rogers, W.P. and Sommerville, R.I. The infective stage of nematode parasites and its significance in parasitism. Adv. Parasitol. **7**: 109–177 (1963).
237. Rose, J.H. Some observations on the free-living stages of *Ostertagia ostertagi*, a stomach worm of cattle. Parasitology **51**: 295–307 (1961).
238. Rose, J.H. Further observations on the free-living stages of *Ostertagia ostertagi* in cattle. J. Comp. Path. Ther. **72**: 11–18 (1962).
239. Rose, J.H. Parasitic gastroenteritis in cattle. Factors influencing the time of increase in the worm population of pastures. Res. Vet. Sci. **11**: 199–208 (1970).

240. Ross, J.G., Armour, J. and Lee, R.P. Further observations on helminthiasis in Nigerian zebu cattle: natural and acquired resistance. British Vet. J. **117**: 485–490 (1961).
241. Rowcliffe, S.A. and Ollerenshaw, C.B. Observations on the bionomics of the egg of *Fasciola hepatica*. Ann. Trop. Med. Parasitol. **54**: 172–181 (1960).
242. Rushton, B. and Murray, M. Hepatic pathology of a primary experimental infection of *Fasciola hepatica* in sheep. J. Comp. Path. **87**: 459–470 (1977).
243. Schad, G.A. The role of arrested development in the regulation of nematode populations. In: *Regulation of Parasite Populations. Proceedings of Symposium sponsored by American Microscopical Society and American Society for Parasitology, New Orleans* (G.W. Esch ed), pp. 111–167. Academic Press: London (1977).
244. Schillhorn van Veen, T.W. Small ruminant health problems in northern Nigeria with emphasis on the helminthiasis. Nig. Vet. J. **2**: 26–31 (1973).
245. Schillhorn van Veen, T.W. Drought, malnutrition and parasitism. Nig. J. Anim. Prod. **1**: 231–236 (1974).
246. Schillhorn van Veen, T.W. Haemonchosis in sheep during the dry season in the Nigerian savanna. Vet. Rec. **102**: 364–365 (1978).
247. Schillhorn van Veen, T.W. Ovine fascioliasis (*F. gigantica*) on the Ahmadu Bello University Farm. Trop. Anim. Hlth. Prod. **11**: 151–156 (1979).
248. Schillhorn van Veen, T.W. and Brinckman, W.L. An evaluation of anthelmintic drenching schemes for lambs on wet season grazing. Samaru Agric. News **17**: 70–74 (1975).
249. Schillhorn van Veen, T.W., Shonekan, R.A.O. and Fabiyi, J.P. Host-parasite check-list of helminth parasites of domestic animals in northern Nigeria. Bull. epiz. Dis. Afr. **23**: 269–288 (1975).
250. Scrivner, L.H.: Breed resistance to ostertagiasis in sheep. J. Am. Vet. Med. Assoc. **144**: 883–887 (1964).
251. Selman, I.E., Reid, J.F.S., Armour, J. and Jennings, F.W. Type II ostertagiasis in adult cattle. Vet. Rec. **99**: 141–143 (1976).
252. Sewell, M.M.H. The role of management in the control of helminth diseases. In: *Beef Cattle Production in Developing countries* (A.J. Smith ed), pp. 138–149. Lewis Reprints Ltd.: Tonbridge, Scotland (1976).
253. Sewell, M.M.H., Hammond, J.A. and Dinning, D.C. Studies on the aetiology of anaemia in chronic fascioliasis in sheep. British Vet. J. **124**: 160–170 (1968).
254. Sharma, A.K. and Lal, S.S. A note on the trematode infection in buffalo (*Bos bubalis*) in Meerut. India. Vet. Res. J. **6**: 93–95 (1983).
255. Sharma, R.L., Dhar, D.N. and Raina, O.K. Studies on the prevalence and laboratory transmission of fascioliasis in animals in the Kashmir Valley. British Vet. J. **145**: 57–61 (1989).
256. Shorb, D.A. Survival on grass plots of eggs and larvae of the stomach worm *Haemonchus contortus*. J. Agric. Res. **68**: 317–324 (1944).
257. Shumard, R.F., Bolin, D.W. and Eveleth, D.F. Physiological and nutritional changes in lambs infected with the nematodes *Haemonchus contortus, Trichostrongylus colubriformis* and *Nematodirus spathiger*. Am. J. Vet. Res. **18**: 330–337 (1957).
258. Silangwa, S.M. and Todd, A.C. Vertical migration of trichostongylid larvae on grasses. J. Parasitol. **50**: 278–285 (1964).
259. Silverman, P.H. and Campbell, J.A. Studies on parasitic worms of sheep in Scotland. I. Embryonic and larval development of *Haemonchus contortus* at constant temperatures. Parasitology **49**: 23–37 (1959).
260. Smeal, M.G., Robinson, G.G. and Fraser, G.C. Seasonal availability of nematode larvae on pasture grazed by cattle in New South Wales. Aust. Vet. J. **56**: 74–79 (1980).
261. Soesetya, R.H.B. The prevalence of *Fasciola gigantica* infection in cattle in East Java, Indonesia. Malayasian Vet. J. **6**: 5–8: (1975).
262. Soulsby, E.J.L. *Helminths, Arthropods and Protozoa of Domesticated Animals*. 7th edn. Bailliere Tindall: London (1982).
263. Spedding, C.W.R. Effect of a subclinical worm burden on the digestive efficiency of sheep. J. Comp. Path. **64**: 5–14 (1954).
264. Sprent, J.F.A. Some observations on the bionomics of *Bunostomum phlebotomum*, a hookworm of cattle. Parasitology **37**: 202–210 (1946).
265. Srikitjakarn, L. Control of helminthoses of calves and fascioliasis of adult buffaloes in north-

eastern Thailand. Inaugural dissertation. Fachbereich Veterinarimedizin der Freien Universitat Berlin. 70 pp. (1986).

266. Steer, M.J., Tierney, T.J., Baldock, F.C., Brown, S.C., Nicholas, F.W. and Rudder, T.H. Class I antigens of the bovine major histocompatibility system are weakly associated with variation in faecal worm egg counts in naturally infected cattle. Anim. Genet. **19**: 115–122 (1988).
267. Stenning, D.J. *Savanna nomads*. Oxford University Press: Oxford, England (1959).
268. Stewart, D.F. Studies on resistance of sheep to infestation with *Haemonchus contortus*, and *Trichostrongylus* spp. and on the immunological reactions of sheep exposed to infestation. V. The nature of the 'self-cure' phenomenon. Aust. J. Agric. Res. **4**: 100–117 (1955).
269. Stoll, N.R. The occurrence of self-cure and protection in typical nematode parasitism. J. Parasitol. **15**: 147–148 (1928).
270. Stoll, N.R. Studies with the strongyloid nematode, *Haemonchus contortus*. I. Acquired resistance of hosts under natural reinfection conditions out-of-doors. Am. J. Hyg. **10**: 384–418 (1939).
271. Sumberg, J.E. and Mack, S.D. Village production of West African Dwarf goats and sheep in Nigeria. Trop. Anim. Hlth. Prod. **17**: 135–140 (1985).
272. Sykes, A.R. The effect of subclinical parasitism in sheep. Vet. Rec. **102**: 32–34 (1978).
273. Sykes, A.R., Coop, R.L. and Angus, K.W. Experimental production of osteoporosis in growing lambs by continuous dosing with *Trichostrongylus colubriformis* larvae. J. Comp. Path. **85**: 549–559 (1975).
274. Sykes, A.R., Coop, R.L. and Angus, K.W. The influence of chronic *Ostertagia circumcincta* infection on the skeleton of growing sheep. J. Comp. Path. **87**: 521–529 (1977).
275. Symons, L.E.A. and Hennesy, D.R. Cholycystokinin and anorexia in sheep infected by the intestinal nematode *Trichostrongylus colubriformis*. Int. J. Parasitol. **11**: 55–58 (1981).
276. Symons, L.E.A. and Steel, J.W. Pathogenesis of the loss of production in gastrointestinal parasitism. In: *The Epidemiology and Control of Gastrointestinal Parasites of Sheep in Australia* (A.D. Donald, W.H. Southcott and J.K. Dineen eds), pp. 9–22. Commonwealth Scientific and Industrial Research Organisation, Australia (1978).
277. Taylor, E.L. Seasonal fluctuation in the number of eggs of trichostrongylid worms in the faeces of ewes. J. Parasitol. **21**: 175–179 (1935).
278. Taylor, E.L. and Michel, J.F. The parasitological and pathological significance of arrested development in nematodes. J. Helminthol. **27**: 199–205 (1953).
279. Thomas, R.J. The survival of pasture infestation with nematode parasites of sheep in a summer rainfall area. J.S. Afr. Vet. Med. Assoc. **38**: 181–183 (1967).
280. Thomas, R.J. The role of climate in the epidemiology of nematode parasitism in ruminants. In: *The Effects of Meteorological Factors upon Parasites. Symposium of the British Society for Parasitology* (A.E.R. Taylor and R. Muller eds), pp. 13–32. Blackwell Scientific Publications: Oxford, England (1974).
281. Thomas, R.J. and Starr, J.R. Forecasting the peak of gastrointestinal nematode infection in lambs. Vet. Rec. **103**: 465–469 (1978).
282. Tichen, D.A. Hormonal and Physiological changes in helminth infestation. In: *Biology and Control of Endoparasites* (L.E.A. Symons, A.D. Donald and J.K. Dineen eds), pp. 257–275. Academic Press: New York (1982).
283. Tod, M.E., Jacobs, D.E. and Dunn, A.M. Mechanisms for the dispersal of parasitic nematode larvae. 1. Psychodid flies as transport hosts. J. Helminthol. **45**: 133–137 (1971).
284. Todd, A.C. Parasitism related to host nutrition. Congreso Mundial de Medicina Veterinaria Zootecnia (XIX), Mexico City. Vol. 1, pp. 74–77 (1971).
285. Todd, K.S. Maisfield, M.E. and Levine, N.D. *Haemonchus contortus* infections in Targhee and Targhee—Barbados Blackbelly cross lambs. Am. J. Vet. Res. **39**: 865–866 (1978).
286. Tromp, S.W. Biometeorology. The Impact of the Weather and Climate on Humans and their Environment (Animals and Plants). In: *Heyden International Topics in Science* (L.C. Thomas ed), Heyden & Son Ltd.: Philadelphia, USA (1980).
287. Troncy, P.M. Helminths of livestock and poultry in tropical Africa. In: *Manual of Ttropical Veterinary Parasitology*. English Ed. pp. 3–175. CAB International: Wallingford, Oxon, England (1989).

288. Upton, M. Goat Production in the humid tropics. Actual and potential contribution to agricultural development. In: *Goat Production in the Humid Tropics*. Proceedings of a workshop, University of Ife, Nigeria, pp. 11–20 (1987).

289. Urquhart, G.M., Murray, M. Murray P.K., Jennings, F.W. and Bate, E. Immunosuppression in *Trypanosoma brucei* infections in rats and mice. Trans. R. Soc. Trop. Med. Hyg. **4**: 528–535 (1973).

290. Van Geldorp, P.J.A. and Schillhorn van Veen, T.W. Periparturient rise in faecal helminth egg counts of Udah sheep in the Zaria area of Nigeria. Vet. Parasitol. **1**: 265–269 (1976).

291. Vassillades, G. The helminthic infestations among cattle in Senegal river basin. Bull. epiz. Dis. Afr. **22**: 69–74 (1974).

292. Wakelin, D. Genetic control of immunity to Parasites. Infection with *Trichinella spiralis* in inbred and congenic mice showing rapid and slow responses to infection. Parasite Immunol. **2**: 85–98 (1980).

293. Wakelin, D. *Immunity to Parasites. How Animals Control Parasite Infections*. Edward Arnold: London (1988).

294. Wallace, H.R. The bionomics of the free-living stages of zoo-parasitic and phyto-parasitic nematodes—a critical survey. Helminthol. Abstr. **30**: 1–22 (1961).

295. Waller, P.J. and Donald, A.D. The response to desiccation of eggs of *Trichostrongylus colubriformis* and *Haemonchus contortus*. Parasitology **61**: 195–204 (1970).

296. Waller, P.J. and Donald, A.D. Effects of changes in temperature and saturation deficit on the survival of eggs of *Trichostrongylus colubriformis* (Nematoda: Trichostrongylidae). Int. J. Parasitol. **2**: 439–447 (1972).

297. Waller, P.J. and Thomas, R.J. Field studies on inhibition of *Haemonchus contortus* in sheep. Parasitology **71**: 285–291 (1975).

298. Warren, E.G. Observations on the migration and development of *Toxocara vitulorum* in natural and experimental hosts. Int. J. Parasitol. **1**: 85–90 (1971).

299. Wedderburn, J.F. Ostertagiasis in adult cattle: a clinical report of an outbreak in the field. N.Z. Vet. J. **18**: 168–170 (1970).

300. Wedderburn, N. Immunodepression produced by malarial infection in mice. In: *Parasites in the Immunised Host: Mechanisms of survival* (R. Porter and J. Knight eds), pp. 123–135. Associated Scientific Publishers: New York (1974).

301. Whitlock, J.H. A study of the inheritance of resistance to trichostrongylidosis in sheep. Cornell Vet. **45**: 422–439 (1955).

302. Whitlock, J.H. and Masden, H. The inheritance of resistance to trichostrongylidosis in sheep. II. Observations on the genetic mechanism in trichostrongylidosis. Cornell Vet. **48**: 134–135 (1958).

303. Windon, R.G., Dineen, J.K. and Kelly, J.D. The segregation of lambs into 'responders' and 'non-responders': response to vaccination with irradiated *Trichostrongylus colubriformis* larvae before weaning. Int. J. Parasitol. **10**: 65–73 (1980).

304. Woolhouse, M.E.J. Passive dispersal of *Biomphalaria globosus*. Ann. Trop. Med. Parasitol. **82**: 315–317 (1988).

305. Yadav, S.C. and Gupta, S.C. On the viability of *Fasciola gigantica* metacercariae ingested by *Lymnaea auricularia*. J. Helminthol. **62**: 303–304 (1988).

306. Yagi, A.I., Younis, S.A., Haroun, E.M., Gameel, A.A., Bushara, H.O. and Taylor, M.G. Studies on heterologous resistance between *Schistosoma bovis* and *Fasciola gigantica*. J. Helminthol. **60**: 55–59 (1986).

307. Young, R.R. and Anderson, N. The ecology of the free-living stages of *Ostertagia ostertagi* in a winter rainfall region. Aust. J. Agric. Res. **32**: 371–388 (1981).

308. Younis, S.A., Yagi, A.I., Haroun, E.M., Gameel, A.A. and Taylor, M.G. Immunisation of Zebu calves against *Fasciola gigantica* using irradiated metacercariae. J. Helminthol. **60**: 123–124 (1986).

*Cited by other workers.

3. Helminths of Domesticated Animals in Indian Subcontinent

N. Chowdhury
Department of Parasitology, College of Veterinary Science
Punjab Agricultural University, Ludhiana, India

INTRODUCTION

India has the largest livestock population (0.692 billion; subcontinent: 1.227 billions; see [18]) in the world which contributes nearly 7% towards its national income. While cattle and buffaloes alone generate 54% of energy for agricultural operations, they provide 32% of all the energy required for rural economy in India [172]. Unfortunately, the productivity of domesticated animals is appallingly poor. The reasons for this are many as well as varied, such as a large animal population with rapidly diminishing grazing areas and consequent over-stocking, poor nutritional standard and traditional husbandry practices compounded by clinical or subclinical often with multiple worm infections. It is estimated that more than 300 species of helminths parasitise livestock in India [185] and new species are being frequently discovered and added to this already swollen list.

Even though some knowledge pertaining to helminths was known in ancient India [75], the credit for laying the scientific foundation of helminthology in general and that of Veterinary and Medical Helminthology in particular, goes to a few enthusiastic officers of the British Veterinary and Medical Services who worked in India during the second half of the 19th century. Thus, the first worthwhile report of helminths from British India was that of Cobbold's who in 1869 gave information on helminth parasites of horses, elephants and dolphins. In 1910, immediately before the first world war, Gaiger published his first check-list of helminths of domesticated animals of India which he revised in 1915. Likewise, Sewell [176] published a monograph, which is now considered a classic on Indian cercariae. In mid-thirties, another monograph was published by Bhalerao [26] based on updated informations on helminths of Indian livestock which still remains indispensable. Indeed, the compilations of helminths of Indian subcontinent in four volumes by Southwell [191] and Baylis [25] also remain a landmark. In undivided India, Srivastava [194] first surveyed the helminth parasites of domestic animals in N.E.F. and Sind provinces. At least two subsequent surveys by Moghe [98] and later by Thapar [201] during the post-independent period have provided invaluable data on helminths of livestock in India. Deo [51], Bhatia & Chauhan [30] and recently, Parija [117] have compiled excellent informations on different aspects of helminths of both domesticated animals and man of the subcontinent of India. Within India a few very informative bibliographies and check-lists are now available [10, 14, 17, 143, 185]. Anantaraman & Anantaraman [13] have reviewed the work on acanthocephala described from this country. Similarly, Seneviratna [174] and Rahman & Mandal [138] have compiled important informations on helminths of Sri Lanka and Bangladesh, respectively.

The topics discussed in the text are restricted to prevalence, geographical distribution, available information on pathology and epidemiology of selected helminths of the subcontinent. However, relevant reviews concerning diverse aspects have been cited in the text, tables or elsewhere (see Table 3; Appendix; Parija [117]; Chapters 12, 13 and 14). Because of space limitation, scientific names of wild animals could not be included and may be seen in Prater [134] and *Encyclopaedia Britanica.* In this updated account, only those older references are cited which are still considered to be relevant and hence useful.

DISEASES

Fascioliasis

This is one of the most serious and widely prevalent of all helminthic diseases of domestic animals in Indian subcontinent. Unquestionably, the combined losses from mortality, wasting and debility and reduction in the production of meat, milk and wool due to this disease may reach a staggering figure. To cite a few examples; the total annual loss from condemnation of liver alone in Bangladesh was estimated to be Taka 0.588 million [35]. Likewise, combined losses due to rejection of both sheep and goat livers in local slaughter houses in Bareilly, India was calculated to be Rs. 4.178 million [22].

Parasite

Varma [201] considered the common liverfluke of Indian ruminants, hitherto recognised as *Fasciola gigantica* to be a new species, *Fasciola indica*. The validity of *F. indica* was, however, not accepted; hence became a synonym of *F. gigantica* [85]. Coincidentally, certain morphological variations in *F. gigantica* were later reported from Egypt (see [37]); Chapter 5).

The susceptibility of *F. gigantica* was studied in a number of laboratory animals e.g. monkey, rabbit, guinea pig, mouse, hamster and rat. The fluke became sexually mature in 117 days in monkeys and in 143 days in rabbits (Chandra, personal communication).

Transmission

Thapar & Tandon [202] while working on the life cycle of *F. gigantica* in India, incriminated the snail *Lymnaea acuminata* (=*L. auricularia rufescens*) to be its principal intermediate host. Soon, it was realised that the same species of snail also plays this role in the plains of Pakistan [85]. Recently, yet another species of snail, *L. andersoniana* has been discovered to act as an intermediate host of *F. gigantica* in Sikkim [82].

In the hill-tracts of Pakistan, the snail *L. truncatula* is reported to act as intermediate host of *F. hepatica* [85]. On the other hand, in the higher altitudes of Kashmir both the flukes coexist, but *L. auricularia rufescens* is conspicuous by its absence there. Instead, a different variety of the snail i.e. *L. auricularia sensu stricto* is widely prevalent. Possibly this species might be serving as the intermediate host for both the flukes. This contention needs further investigation. Not much is known about the intermediate host(s) of *F. gigantica* or *F. hepatica* from elsewhere in the subcontinent (see below).

Prevalence

The results of various surveys carried out (Table 1) in different regions of the subcontinent (Fig. 1) have established the predominance of *F. gigantica* which is widely distributed.

On the highlands (above 1,200 metre) of Baluchistan and places with similar altitudes in Pakistan, *F. gigantica* has been replaced by *F. hepatica*. In border areas, intermingling/overlapping of both parasites and intermediate snail hosts is a common occurrence [85]. Somewhat similar situation also exists in Kashmir. Indeed, the true picture of all the desert and semi-desert areas of the subcontinent is not very clear though malady was recorded twice in 1949 and 1953 in Chittoorgarh district, Rajasthan, India.

Presence of *F. hepatica* from the plains of India has been reported from cattle in Orissa [201] and

Table 1. Prevalence (%) of fascioliasis in Indian subcontinent

Countries/States	Sheep		Goat		Cattle		Buffalo	
PAKISTAN								
Peshawar	—	—	—	—	—	40.0-80.0*	—	(?)
INDIA								
Assam	—	—	5.0	(123)	17.0	(53)	70.0	(7)
Bihar	—	—	—		60.0	(550)	40.0	(550)
Orissa	—	—	89.5	(39)	91.5	(200)	95.0	(300)
Andhra Pradesh	0.15	(4,407)	—	—	0.02	(4,217)	—	—
Tamil Nadu	4.42	(?)	8.0	(?)	—	—	—	—
Karnataka	—	—	—	—	20.69	(87)	66.7	(56)
Maharashtra	—	—	—	—	3.76	(1,225)	7.85	(2,451)
Madhya Pradesh	25.3	(20)	54.7	(150)	18.5	(66)	65.4	(70)
Gujarat	—	—	—	—	12.32	(113)	18.40	(87)
Uttar Pradesh	81.44	(97)	54.30	(89)			39.0	(605)
Haryana	—	—	—	—	1.27	—	39.61*	(1,286)
Himachal Pradesh	8.12–65.80	(1,793)	—	—	—	—	—	—
Jammu & Kashmir								
Gulmarg/Srinagar	61.8	(55)	16.4	(61)	100.0	(5)	—	—
Sikkim	13.7	(22)	30.0	(10)	25.0	(8)	—	—
BANGLADESH	27.0	(178)	18.0	(245)	59.3	(526)	78.7	(315)

Figures in parentheses indicate number of necropsy examination.
*Mixed/unspecified figures of host/parasite.

once in a horse in Maharashtra. Two isolated cases of human involvement, one with *F. hepatica* and the other with *F. gigantica* have also been recorded [41].

The existence of both *F. gigantica* (=*F. indica*) and *F. hepatica* was known in Sri Lanka in the South [174]. In the hill-tracts of Chittagong and Sylhet in Bangladesh, only *F. gigantica* was reported. However, in Burma both *F. gigantica* and *F. hepatica* were known to coexist.

Outbreaks of fascioliasis have been reported both from the plains and hill-tracts of the subcontinent. At higher altitudes, acute outbreaks are usually known to occur in winter months, between January to February. In the plains, on the other hand, outbreaks frequently occur during October to March with higher incidence between March to May. However, sporadic outbreaks of fascioliasis occur almost throughout the year [96, 153].

Animals with subclinical infections were shown to play a definite role in acute outbreaks. Imported sheep were more susceptible with very high mortality than local breeds. Susceptibility and mortality rates were more in calves than in adult cattle. Surprisingly, in two outbreaks, one in goats in Punjab and another in buffaloes in Haryana, morbidity was 100 percent [70, 73]. The mortality due to fascioliasis was high in cattle and buffaloes and was low in sheep and goats in coastal areas. However, in the plains of northern India such species difference is not evident [32, 48].

Pathology

Most of the observations on the pathology of fascioliasis in the subcontinent are largely confined to *F. gigantica*. Even, here, detailed and experimental studies on pathogenesis in different domestic animals are lacking. Likewise, little is known about *F. hepatica* from this region (see [183]).

Sheep and Goat
Several excellent reports are available describing clinical picture of acute ovine and caprine fascioliasis [82, 106, 108]. An isolated case of clinical fascioliasis in naturally-infected Black Bengal goats was reported from Bangladesh [108].

In a critical study on acute outbreaks in sheep in Sikkim, in addition to various lesions in the affected liver the lungs revealed foci of haemorrhages, alveolar oedema, and occasionally sub-acute bronchitis and cloudy swelling in myocardium.

In natural fascioliasis in sheep, the main histopathological changes observed were hypertrophy and hyperplasia of bile duct epithelium with periductal fibrosis, thickening of hepatic arterioles, egg granuloma and well formed lympho-folicular hyperplastic aggregates in the portal areas. In a rare experimental study, venous thrombosis, cholengio-hepatitis, extensive parenchymal necrosis were also recorded [186]. In naturally infected goats, acute hepatitis with microabscesses, infiltration of eosinophils and some plasma cells in the liver parenchyma were also seen. The blood serum globulin and bilurubin were increased significantly [108].

Cattle and Buffalo
In a more recent study in naturally infected buffaloes with *F. gigantica*, a decrease in total serum protein and albumin and albumin-globulin ratio and a significant increase in α- and β-globulins was observed. Increased activity of several serum enzymes was also reported [87, 88].

In cattle and buffaloes, the important lesions observed in natural cases were, marked perilobular fibrosis in the affected liver with varying degrees of degenerative and inflammatory changes. In a comparative study of affected livers of cattle and buffalo, a greater degree of eosinophilic infiltration was observed in the former than in the latter [161]. The affected livers in buffaloes were found to contain significantly low fat and high protein contents; PAS-positive substances were extensively depleted.

Wild Animals
A few reports of natural fascioliasis in wild animals also exist [34, 147]. While studying the liver lesions in spotted deer and black buck, periductular and perivascular fibrosis were observed to be more pronounced in the spotted deer than in the black buck. Extensive calcification of common bile duct and densely stained eosinophilic club-shaped radiating structures on the external surface of the egg was encountered only in deer, were other rare findings of the authors [147].

In a rare case of fascioliasis of an Indian rhino calf, the important changes described were acute perihepatitis, multiple adhesions and focal necrosis in the hepatic parenchyma, periarteritis and mesarteritis. The epithelial cells in the bile duct were anaplastic in nature with a tendency to form bile duct-like adenomatous structures. There was total or partial disappearance of the hepatic cord cells and atrophy of the hepatic cells more marked in the involved Glisson's capsul [34].

Epidemiology
In Indian subcontinent, the agro-climatic conditions differ considerably. Consequently, fascioliasis is also likely to vary from region to region and even within a region. However, detailed epidemiological studies are few. In the last three decades agro-climatic conditions of many States of India have drastically changed mainly due to the construction of dams, canals and improvised irrigation facilities for intensive cultivation. One of the aftermaths of these changes, is that snail-borne diseases like fascioliasis, amphistomiasis and schistosomiasis are on the increase.

In the plains, buffalo appeared to be the principal host among domestic animals. However, in Kashmir, the prevalence was found to be the highest in cattle, moderate in sheep and the lowest in goats. These data obviously reflected that the grazing habits of animals in the region can directly be related to the prevalence of the fluke. The infection rate in such areas is reported to increase from autumn (42%) and reach its peak in winter (54%) and then decline in spring (48%) with lowest infection being in summer (33%) [115].

Interestingly, in Tarai region of Uttar Pradesh, the prevalence of fascioliasis in buffaloes occurred in three peaks-September/October, January and June. In this area, the highest incidence of fascioliasis in buffaloes was reported during the rainy season and in sheep and goat in winter [32].

In many areas in the plains and sometimes in hill tracts for example, in swamps of Sikkim with stagnant or shallow water that gets accumulated in the unlined channels which get overgrown with aquatic weeds, are ideal conditions for the growth, persistence and propagation of *L. auricularia*. In such a situation, outbreaks of fascioliasis could occur under zero grazing system [73]. Lack of fodder in certain months of the year and feeding of animals on aquatic vegetations could also lead to infection of stall-fed animals.

Kendall [85] outlined the bionomics and some ecological requirements of *L. auricularia rufescens* in Pakistan and the then East Pakistan (now Bangladesh), and *L. trancatula*, the amphibious snail which occurred above 1,500 metre and only in Pakistan. Subsequently, *L. auricularia rufescens* with special emphasis on its ecology and distribution was studied by others in the Punjab region of Pakistan. The geographical distribution, environment, habits of *L. auricularia rufescens* has been reported from Orissa [125]. Similar investigations from other endemic areas of Indian region are desired.

Apart from domestic ruminants, high incidence (10–60%) of *F. gigantica* has been reported from black buck and several species of wild deers in Orissa and Uttar Pradesh [69, 149]. Evidently, *F. gigantica* appeared to be a well established parasite of both deer and black buck in the subcontinent and these might be playing a significant role as reservoirs of this disease. The fluke is also known to occur in rhinoceros in Assam. In a recent survey in Sikkim, 2.1% of 225 yaks were found positive for *F. gigantica* [139]. The relationship of fascioliasis between domestic animals and its wild counterparts particularly where sancturies and forests are in proximity to inhabited localities is an important area of epidemiological study.

Paramphistomiasis

Immature amphistomiasis is a major disease of sheep, goat, cattle and buffalo in Indian subcontinent. It ranks next to fascioliasis. The acute form of the disease is due to immature stages in the duodenal and pyloric wall of abomasum and is characterised by dark foetid diarrhoea associated with gastroenteritis, rapid emaciation, general weakness and high morbidity. Though economic losses due to it must be enormous, no accurate figures are available. Various aspects of the disease has been reviewed by Alwar [8] , Horak [76], Dutt [57] and Gupta [71] and hence will not be discussed here. However, important informations have been summarised in Table 2.

Parasitic Gastroenteritis

Parasitic gastroenteritis (PGE) is world-wide but is most prevalent in hot and humid climate of tropics. Considering the high mortality, morbidity and serious but unaccounted, economic losses resulting from reduction and/or spoilage of wool, meat and milk, this condition can justifiably be ranked as a major helminthic disease of sheep, goat, cattle and buffalo in the subcontinent.

Parasites

SHEEP AND GOAT

Helminths of these animals, associated with PGE in the subcontinent, have been reported by several

Table 2. Some common paramphistomes of Indian subcontinent

Parasite	Definitive host	Intermediate host	Geographical distribution
**Paramphistomum cervi*[1]	Sheep, goat, cattle, buffalo, camel	*Indoplanorbis exustus*	Pakistan, India, Sri Lanka, Burma
**Gigantocotyle explanatum*[2]	Buffalo, cattle, sheep, goat, sambar, dog	*Gyraulus convexiusculus*	Pakistan, India, Sri Lanka, Nepal, Burma
**Cotylophoron cotylophorum*	Cattle, buffalo, sheep, goat, spotted deer, black buck	*I. exustus*	Pakistan, India, Bangladesh, Nepal
C. indicum	Sheep, goat, cattle,	*I. exustus*	Pakistan, India
Palamphistomum dutti	Sheep, goat, buffalo	*G. convexiusculus*	India
P. lobatum	Sheep, goat, buffalo	*G. convexiusculus*	India
Orthrocoelium scolicoelium	Cattle, buffalo, sheep, goat	*Bithynia pulchella*	India
O. streptocoelium	Cattle, buffalo, sheep, goat	*Bulimus tentaculata*	India, Sri Lanka, Burma
Srivastavaia indica?	Sheep, goat	*I. exustus*	India
Olveria indica	Cattle, buffalo, sheep, goat	*G. convexiusculus*	India
**Gastrothylax crumenifer*	Cattle, buffalo, sambar, spotted deer, black buck	*G. convexiusculus*	Pakistan, India, Sri Lanka, Bangladesh, Nepal, Burma
**Fischoederius elongatus*	Buffalo, cattle, sheep, goat, sambar, barking deer	*Lymnaea luteola*	India, Sri Lanka, Nepal
Gastrodiscus secundus	Horse, mule, ass, elephant	*I. exustus*	India
Gastrodiscoides hominis	Man, pig, monkey	*Helicorbis coenosus*	India
Pseudodiscus collinsi	Horse, mule, elephant, cattle	*I. exustus*	India, Burma

1. Occurs in rumen and reticulum; sometime in bile duct, gall bladder and liver.
2. Found in bile duct, liver, gall bladder; sometime in duodenum, rumen or pancreatic duct. All others occur in rumen/reticulum or both.

*Associated with outbreaks in sheep, goat, cattle, buffalo.

Adapted partly from Dutt [57]; Dutt & Srivastava [59]; Peter [128]; Peter & Srivastava [129]; Tripathi & Srivastava [206].

workers [30, 31, 97, 173, 174, 201]. The main species involved were: *Haemonchus contortus, Ostertagia circumcincta, O. mentulata, Mecistocirrus digitatus, Trichostrongylus axei, T. colubriformis, Bunostomum trigonocephalum, Gaigeria pachyscelis, Strongyloides papillosus, Oesophagostomum columbianum, O. asperum, Nematodirus filicollis, N. spathiger, Cooperia punctata, Chabertia ovina, Skrjabinema ovis, Trichuris globulosa, T. ovis, Moniezia expansa, Stilesia* spp., *Cysticercus tenuicollis*, immature amphistomes and *Schistosoma indicum*.

CATTLE AND BUFFALO

The helminths responsible for PGE in these animals have been reported by many workers in different regions of the subcontinent [30, 99, 138, 168, 179, 201].

Frequently encountered helminths were: *M. digitatus, H. contortus, H. placei, O. ostertagi, Skrjabinagia boevi, Paracooperia nodulosa, C. oncophora, C. punctata, C. pectinata, C. laterouniformis, T. colubriformis T. longispicularis, B. phlebotomum, G. pachyscelis, Agryostomum vryburgi, O. columbianum, O. radiatum, S. papillosus, Toxocara vitulorum, C. ovina, Capillaria bilobata, T. globulosa, T. discolor* in addition to *Moniezia, Stilesia* and *Avitellina* spp., immature amphistomes, *S. indicum* and *S. spindale*.

In Indian subcontinent, PGE is a disease complex due to multiple helminthic infections as haemonchosis, ostertagiosis, trichostrongylosis, mecistocirrosis, ascariasis, strongyloidiasis, bunostomiasis, oesophagostomiasis, paracooperiasis, cooperiasis, cestodiasis, immature amphistomiasis, hepato-intestinal schistosomiasis. Some of these cause serious economic losses both from morbidity and mortality and have been discussed by Bhatia & Chauhan [30] at length, hence will not be repeated here (see Appendix; Table 3).

Table 3. Helminths incriminated in some of the outbreaks in Indian subcontinent

Host	Parasite*	Locality
Mixed infection		
Sheep	*Haemonchus contortus, Oesophagostomum columbianum* or *Trichuris ovis*	Rajasthan
Sheep	*H. contortus, Bunostomum trigonocephalum, Nematodirus filicollis, Trichostrongylus colubriformis, T. axei, Ostertagia* sp., *Chabertia ovina, Ogmocotyle indica*	Sikkim
Sheep	Similar to above [8]	
Goat	*H. contortus, O. columbianum, T. ovis, T. globulosa, T. colubriformis*	Orissa
Duck	*Amphemerus anatis, Amoebotaenia sphenoides, Ascaridia galli, Polymorphus boschadis, Filicollis anatis*	Kashmir
Duck	*A. galli, Hymenolepis collaris, Fimbriaria fasciolaris, Psilochasmus oxyurus*	Tamil Nadu
Monospecific infection		
Goat	*Cysticercus tenuicollis*	Punjab
Sheep	*Coenurus* sp.	Uttar Pradesh
Buffalo	*C. macracantha* (= *Multiceps macracantha*) [2]	Punjab
Cattle	*Mecistocirrus digitatus*	Uttar Pradesh
Sheep	*H. contortus*	Rajasthan
Goat	*Dictyocaulus filaria*	Delhi
Cattle	*D. viviparus*	Madhya Pradesh
Pig	*Metastrongylus salmi* [3]	West Bengal
Pig	*M. apri* [16]	Tamil Nadu
Sheep	*Protostrogylus rufescens*	Uttar Pradesh
Buffalo	*Stephanofilaria zaheeri*	Assam
Cow Calves	*Trichuris globulosa*	Bangladesh
Cow calves	*T. discolor*?	Uttar Pradesh
Elephant	*S. srivastavai*?	Assam
Poultry	*Hymonolopis carioca*	Bihar
Poultry	*Dyspharynx nasuta*	Bihar
Duck	*H. megalops*	Bihar

*In order of preference in mixed infections; figures in parenthesis indicate number of outbreaks/ attacks.

Haemonchosis

Haemonchus contortus is the most pathogenic species of this genus and it is widespread in the entire subcontinent. The different aspects of the disease have been reviewed by Sood [190] and Bhatia & Chauhan [30]. Additional information has been provided in Tables 3 and 4 and appropriate sections of this part.

Ostertagiosis

Due to lack of adequate surveys in different regions of the subcontinent, limited information is available on the incidence, geographical distribution and pathogenecity of different species involved. *Ostertagia mentulata* is the commonest species of this genus known to parasitise 50% of sheep and goats in Lahore, Pakistan. *O. ostertagi* was recorded in hill goats in Indo-Pakistan subcontinent while *O. circumcincta* was encountered sporadically in sheep and goats in India. *O. ostertagi* is mostly seen in cattle than in buffaloes in India whereas *Skrjabingia boevi* is more common in buffaloes than in cattle and was found in nearly 50% of buffaloes examined in Uttar Pradesh [111].

Trichostrongylosis

Trichostrongylus axei is found in the hill-tracts of Pakistan. *T. probolurus* is reported to be very common in sheep and goats in the plains of West Punjab. *T. vitrinus* and *T. longispicularis* are restricted to Peshawar region. The latter three species are not known to occur elsewhere in the subcontinent.

T. colubriformis is widely distributed in the Indo-Pakistan subcontinent. Infection in sheep and goats with *T. colubriformis* occurs in certain areas of Uttar Pradesh and Punjab, India, and is reported to go as high as 90–100%.

Mecistocirrosis

In an old survey in Pakistan, 20% cattle and 11% buffaloes were found to harbour *Mecistocirrus digitatus*.

In India, it was recorded in an outbreak in 1931 in cattle in Uttar Pradesh (Table 3). Subsequently, it has been reported in buffaloes and goats in Assam, Orissa, Madhya Pradesh, Gujarat, Maharashtra, Karnataka and Kerala. In Sikkim 45% of 20 yaks examined were found positive for this species.

M. digitatus is widespread in Sri Lanka and is also known to occur in Bangladesh.

M. digitatus sometimes observed to be more pathogenic than *H. contortus*. The mucosa of the affected abomasum becomes enormously swollen with eosinophilic oedmatous fluid. Calves usually show signs of diarrhoea, anaemia, bottle jaw and dehydration. The associated changes observed were areas of degeneration, haemorrhage, necrosis, desquamation or formation of gastric ulcers and atrophy of parietal or oxyntic cells [61, 66]. Liver and heart of such affected animals also showed necrosis and cloudy swelling.

Strongyloidiasis

Patnaik & Pande [124] considered *Strongyloides papillosus* second to *Toxocara vitulorum* in frequency and pathogenecity in young calves in India under 5 weeks of age. In Rajasthan, its incidence in lambs was as high as 100% [121] whereas in farm-bred goats in Kerala it was 73% [171] and was reported to be common in sheep in Tarai region and Gujarat. Recently, this species was found to be associated with mortality in buffalo calves in Punjab. In Uttar Pradesh 57% of buffalo calves upto one year old were found to harbour this infection and was more common in 3 to 6 months old calves [44].

In Sri Lanka, *S. papillosus* is widespread in cattle, buffaloes, sheep and goats with heavy infections predominantly recorded from 10 days to 8 months old buffalo calves [179]. It has been recorded in cattle in Bangladesh.

Bunostomiasis

Information on bunostomiasis in domestic animals is scanty in the subcontinent. *Bunostomum phlebotomum* was studied by Patnaik and Pande [124]. In Rajasthan, both *B. trigonocephalum* and *Gaigeria pachyscelis* were encountered in 100% of 442 adult farm-bred and exotic sheep whereas in the plains of Aligarh and Mhow 50–58% of sheep harboured *B. trigonocephalum* and prevalence was only

11% in the humid areas of Kerala [121, 171]. In Sikkim, 50% of 20 viscera of yaks were found positive for *B. phlebotomum*.

Oesophagostomiasis

Oesophagostomum columbianum is widely distributed in the plains of the subcontinent and often reported to occur as mixed infections with *H. contortus*, *T. colubriformis*, *T. ovis* and *T. globulosa* (Table 3). In Kerala, India, 57% of indigenous farm-goats harboured this infection. Chauhan [43] found that the prevalence of *O. radiatum* infection was as high as 71% in buffaloes in Uttar Pradesh.

O. columbianum is reported to be widespread with heavy infection in goats in dry zone areas of Sri Lanka. In Bangladesh, 46.6% of 13,595 goats examined were found infected with *Oesophagostomum* spp. [135].

Fatal peritonitis and intussusception in sheep and goats due to *O. columbianum* have occasionally been reported from Indian region [109, 121]. A rare case of acute colitis in a calf due to *O. radiatum* has been recorded [103]. In addition to hyperplastic changes and cellular proliferation, histologically three types of nodules have been described in sheep and goats. The changes in nodules appear to differ in different hosts. The formation and the number of nodules is believed to be dependent on the number of times host got exposed to the infection and usually associated with necrosis, caseation and calcification.

Paracooperiasis

Alwar [9] considered *Paracooperia nodulosa* highly pathogenic to buffalo calves in India and ranked it next to *T. vitulorum*. Though *P. nodulosa* was first recorded in buffalo calves in Bareilly, India and later on in Pakistan and other states of India (see review [101]). In Uttar Pradesh, *P. nodulosa* was found in buffalo calves from 5 weeks to $2^1/_2$ years of age. The occurrence of this species in cattle, sheep and goats in areas where buffaloes are found infected needs investigations. This infection is very common in Tamil Nadu where in a livestock farm, 8 of 22 buffalo calves once reported to have died due to paracooperiasis. The reported prevalence of *P. matoffi* in certain areas also needs further investigations.

Chauhan and Pande [46] described lesions of paracooperiasis in detail.

Cooperiasis

Few reports are available on the incidence, geographical distribution and pathogenesis of *Cooperia* spp. *C. punctata* and *C. pectinata* were found in cattle in Tamil Nadu, India, which also occurred frequently in hill cattle and sheep at Mukteswar and Bareilly, India. *C. pectinata* was recorded in 14% of 100 sheep examined in Madhya Pradesh whereas *C. laterouniformis* was encountered in 4.3% buffalo calves at Mathura [124].

C. Punctata and *C. pectinata* are suspected to cause death in cattle at Mukteswar. *C. punctata* is believed to cause nodular lesions with centrally necrosed area. Detailed informations on *C. laterouniformis* and *C. pectinata* are lacking.

Mortality and Morbidity

Sheep and Goat

Minnett [95] while investigating the mortality in sheep and goats, observed 76% deaths due to haemonchosis and 67% due to immature amphistomiasis in sheep in Hissar and Monghyr sheep breeding farms. High mortality was also attributed in Bhamburda (Poona) farm to haemonchosis, oesophagostomiasis and amphistomiasis and in Hosur farm (Tamil Nadu), to haemonchosis and mecistocirrosis.

Atleast three outbreaks that swept through Central Rajasthan in early fifties were probably the worst outbreaks ever documented from the subcontinent in which thousands of sheep and goats perished (see Table 3). The 'culprits' responsible were *H. contortus* and *O. columbianum*. In Central Sheep and Wool Research Institute, Rajasthan where 6,000 sheep of different genetic groups were being maintained, a total of 336 animals died during the period, 1971–1977 despite the fact that these were range-managed with 6–8 h grazing and regularly dehelminthiased [92].

In October 1988, in Mattewara sheep farm, Punjab, an outbreak of PGE occurred which accounted for 40% deaths of 250 imported Pol dorset sheep mainly due to haemonchosis; the morbidity rate in the same outbreak was even higher (90%) and the loss incurred was estimated to be about Rs. 1 lakh (Gupta, personal communication). In Gujarat 15% of total deaths in farm-bred sheep were believed to be due to PGE with highest death rate in the age group of 1–3 years.

In 2 goat farms in Sri Lanka, 28% of 2,530 kids died before reaching the age of 6 months mainly due to PGE. The main helminths responsible were: *Haemonchus* spp., *Trichostrongylus* spp., *Oesophagostomum* spp. and tapeworms [146].

Among the goats belonging to the goat farm of Bangladesh Agricultural University, Mymensingh, and its neighbouring villages, 25% of 214 kids and 43% of 115 adults of BAU farm, and 30% of 227 kids and 5% of 487 adults of these villages died due to PGE [178].

CATTLE AND BUFFALO

Dhanda and Khera [52], on the basis of a country-wide survey of Govt. farms, reported that the gastrointestinal disorders in calves were about 45%. In Orissa, 33% of calf mortality was attributed to digestive disorders alongwith ascariasis, haemonchosis and amphistomiasis. Published reports indicated that in Gujarat, 24% deaths in cattle and 29% that in buffaloes were due to parasitic diarrhoea/dysentery for which mecistocirrosis was held responsible [145]. Gastroenteritis resulting from combined parasitic and bacterial infections was responsible for about 11% deaths in cattle in Karnataka. Chronic parasitic hepatitis accounted for another 21% and 66% deaths among local and cross-bred cattle and buffaloes, respectively in the same region. Of 2,835 calves both cow and buffalo, in Tamil Nadu, atleast 27% died due to PGE. The specific parasitic diseases identified were: haemonchosis, bunostomiasis, mecistocirrosis, ascariasis, trichuriasis, monieziasis and immature amphistomiasis [49].

Sub-clinical cases of PGE in dairy herds were reported from Karnataka. The infection rate was 65% in cattle and 70% in buffaloes. These infections were common in those animals that were over 8 years old [56].

Epidemiology

Comprehensive data on epidemiology of PGE, that cover the entire subcontinent, are at best and at worst lacking. It is, however, well recognised that seasons, rainfall and atmospheric temperature play definite role in gastrointestinal parasitism in the subcontinent which vary according to different agro-climatic conditions, different types of grazing, rearing, housing and other prevailing husbandry practices (see [90, 91]; Chapter 2).

SEASON AND RAINFALL

Sheep and goats often pick up infections while grazing on contaminated pastures. The mid-eastern semi-arid Rajasthan experiences an average rainfall of 37-40 cm/year in 30 to 45 rainy days. Most of the precipitation occurs in July-September months which are fairly warmer months. Gastrointestinal parasitism in Rambouillet sheep in the region flared up in clinical form during post-monsoon period, while it remained subclinical in winter, spring and summer months. Tripathi [209], on the basis of larval counts and worm-egg output, observed that in the plains of Uttar Pradesh during post-monsoon season, prevalence of infective larvae of gastrointestinal nematodes in the grazing plots was in order of *Haemonchus* > *Strongyloides* > *Trichostrongylus* > *Oesophagostomum* spp. and in winter it was in order of *Strongyloides* > *Trichostrongylus* > *Haemonchus* > *Oesphagostomum* spp. In Gujarat, *M. digitatus* infection was fairly high throughout the year, while *Trichostrongylus, Ostertagia* and *Cooperia* spp. remained at low level.

In the semi-arid Rajasthan, *Haemonchus* infection, that had been reported frequently in post-monsoon period, sometimes extend to early winter depending upon rainfall during the period. Further, hot summer months have adverse effect on the development of larvae on pasture and onset of rains triggered the development process of infection rapidly [121]. The prevalence of *Haemonchus* infection in other areas of India was observed in monsoon and post-monsoon months which may reach as high as 100% in sheep and goats ([207]; see review [190]). The prevalence of *G. pachyscelis* in goats in Uttar Pradesh was, however, at its maximum during post-monsoon months and winter, moderate in monsoon and minimum in the pre-monsoon months [21]. In Uttar Pradesh, the prevalence of *O. radiatum* was maximum during summer months. The occurrence of adults of *P. nodulosa* in the lumen was 57% in winter and 15% in summer months whereas nodules in buffalo calves were found 100% in summer, 85% in rainy season and 60% in winter season ([43]; see [30]).

Humidity and Temperature

Little information is available on the effect of environmental conditions on the distribution of nematode larvae on the pasture. Survival of larvae often has been correlated with climatological data. Both atmospheric temperature and moisture were found essential for the development of eggs and larvae of gastrointestinal nematodes in the plains of northern India, but dryness caused early deaths of eggs on one hand and permitted larvae to migrate faster in sandy soil and live longer in loamy soil, on the other hand (see [207, 208]).

Tripathi [209] further observed that larval presence in the grazing plot was directly proportional to the prevailing humidity and temperature in the ambience. However, changes in relative humidity markedly effect the upward migration of *H. contortus* larvae than changes in temperature. Likewise, moderate temperature (20–29°C) and moderate relative humidity (52–80%) led to increased larval population [97]. Temperature is believed to influence survival of larvae of *Haemonchus* on herbage which led to higher level of infection in autumn and summer (see review [190]).

In the plains of Uttar Pradesh, optimum temperature for the development of *Oesophagostomum* larvae was 40°C, during which the third-stage larvae were produced in maximum numbers in 82 hours. A high humidity was required for normal development of eggs of both *O. venulosum* and *O. columbianum* whereas a high humidity and comparatively low (non-freezing) temperature was found best for survival of infective larvae (see [91]).

Altitude

Topography appears to influence considerably the geographical distribution of gastrointestinal nematodes. In Kashmir valley, *T. axei* and *T. colubriformis* were the two predominant species in sheep which is in sharp contrast to the plains of India where *Haemonchus, Oesophagostomum* and *Bunostomum* infections were predominant ([121, 201]). Both *N. spathiger* and *N. filicollis*, though not known to occur in plains, were common in sheep in Jammu and Kashmir, and Sikkim (Fig. 1).

Susceptibility Age

Age appeared to be an important factor in the epidemiology of gastrointestinal nematodiasis. In Gujarat, young sheep were observed relatively more susceptible to PGE; the acquired resistance that developed in rams was not absolute and could easily be broken down due to malnutrition. In Tarai region of Uttar Pradesh and Rajasthan, both being climatically contrasting regions, kids and lambs were found to suffer equally heavily from strongyloidiasis with 100% infection [121]. Incidence of bursate nematodes was maximum in 6 to 9 months old calves, slightly less in 3 to 6 months old and the least in 9 to 12 months old calves [44]. Young buffalo calves were also found to be more susceptible to *Trichostrongylus* infection than older animals (see [30]) .

BREED

One of the most important aspects of epidemiology of PGE is that imported sheep were usually found to be highly susceptible to *H. contortus* which sometimes occurred as outbreaks. Thus, Patnaik et al [121] observed that while lambs were not susceptible, adult Rambouillet sheep were most vulnerable to *Haemonchus* infection in comparison to cross-bred and local breeds. Rambouillet sheep were also highly susceptible to *O. columbianum* with higher incidence than in local breeds and the infection persisted in sub-clinical form throughout the year [109, 121].

POST PARTURIANT FAECAL OUTPUT AND HYPOBIOSIS

Isolated reports on spontaneous elimination of worms, post-parturient rise in faecal output and hypobiotic behaviour of nematode larvae in certain geographical areas are available. In Maharashtra, adults of *M. digitatus* were encountered from middle of June to first week of November reaching peak infection between mid-July to mid-September. Spontaneous elimination of worms took place from November onward and none were found from December to January. Rise in strongyle egg counts occurred in March and April coinciding with lambing season in Kashmir valley. Lambs showed rise in faecal egg output from May onwards reaching the peak in September. The exact cause of post parturiant rise needs further investigations if this is due to resumption of development of arrested worms or due to an increase in number of newly acquired worms by the host subsequent to loss of resistance. Similarly, hypobiotic behaviour of *H. contortus* larvae was observed in Tamil Nadu which led to sudden increase in egg and worm counts in March inspite of the fact that larval burden on pasture and e.p.g. in February was minimum [165]. It is believed that the larvae developed from the eggs of previously inhibited adults which gave rise to another peak e.p.g. and worm counts in June-July.

HUSBANDRY PRACTICES

Incidence of PGE, particularly in sheep, was very high in Gujarat during October. The main reason for this might be overstocking of pastures which facilitated the infection of animals during rainy season. In Bangladesh, mortality due to PGE in goats under farm conditions was higher than in villages [136]. The major factors claimed for this mortality were concentration of animals in farm houses, grazing in a small pasture and minimal individual care.

WILD ANIMALS

A large number of wild/zoo animals (Table 4) are known to harbour several important nematodes [43, 67, 149] that are associated with PGE in domestic animals. But no adequate data are available to evaluate the potential of these nematodes in causing clinical diseases in wild life and in assessing their role in epidemiology of PGE in domestic ruminants. There is, thus, a pressing need for extensive surveys in different localities and in different seasons of the year to understand the implications of these helminthiases in the wild life.

Table 4. Prevalence of some gastrointestinal nematodes of domestic ruminants in wild/zoo animals in India (cf. text)

Parasite	Host
Skrjabinema ovis	Black buck
Strongyloides papillosus	Black buck, wild deer?
Oesophagostomum radiatum	Deer?
O. asperum	Nilgai
O. venulosum	Nilgai
Bunostomum trigonocephalum	Black buck, deer?
B. phlebotomum	Yak, Indian hare
Gaigeria pachyscelis	Black buck, nilgai
Trichostrongylus axei	Black buck
T. colubriformis	Black buck, Indian hare
Haemonchus contortus	Black buck, spotted deer, swamp deer, Nilgiri tahr
H. bubalis	Black buck
H. placei	Yak
Paracooperia nodulosa	Black buck
Cooperia laterouniformis	Black buck
Mecistocirrus digitatus	Yak
Trichuris ovis	Black buck, deer?
T. globulosa	Yak, nilgai
T. discolor	Nilgai

Ascariasis

Calf-ascariasis due to *Toxocara vitulorum* is widespread in the subcontinent as elsewhere in Asia and Africa. In the entire subcontinent 'parasitic white scour' is the chief helminthic condition in neonatal calves and considered to be a major limiting factor in calf rearing in organised farms, particularly in dairy farms, because it leads to high mortality, morbidity and loss of production. Besides, it comes as an impediment in effective implementation of breeding programmes. Despite the realisation of its economic implications in different regions [30, 178], there is hardly any factual information on annual losses incurred due to this disease at national level.

Parasite

The biology and zoonotic potential of the parasite has recently been discussed [210]. *T. vitulorum* is more lethal to buffalo calves than cow calves. Obviously, detailed studies are necessary to determine if this is due to strain variation of the parasite (see Chapter 5). *T. vitulorum* has also been recorded in goats in Kerala, India and Sri Lanka [141, 178], and Rajamohan et al [141] successfully transmitted it to experimental goats. In Bangladesh, ova of *Ascaris lumbricoides* has been recovered from 3 goats and 27 sheep. Significantly, in Kashmir valley sheep/man cycle of *A. lumbricoides* is reported to exist [53]. There is, therefore, an urgent need to ascertain the species/strain(s) of *Ascaris* recorded in sheep, goat and man in different geographical regions in view of the epidemiology of the disease and zoonotic potential of the parasite.

Prevalence

Ascariasis in calves is very common in the entire subcontinent. Its prevalence, however, is far more common in buffalo calves than in cow calves but differ in different regions due to animal husbandry practices followed and few other undetermined local factor(s).

In a recent survey in Tirupati, India, its prevalence was shown to be 5% in cow and 18% in buffalo calves. At Hissar (Haryana), 60% of 320 cases of cow and buffalo calves brought to clinic had ascariasis. On the other hand, two field surveys revealed 52% of 1,151 and 27.5% of 316 buffalo calves had ascariasis. In an extensive survey involving four livestock farms in Uttar Pradesh, its incidence was shown to be as high as 46% in buffalo and 54% in cow calves that were below 1 year [44]. The only survey conducted in Rajasthan revealed that 14% of 409 pure exotic calves had ascariasis.

In Sri Lanka, ascariasis had been reported to be widespread in buffalo calves in southern, north central and central provinces [174]. During a survey in government owned farms 63 of 66 buffalo and 14 of 66 cow calves, respectively were found positive for *T. vitulorum* infection [178].

In Bangladesh, almost all buffalo calves were reported to be infected [94].

Pathology

In Uttar Pradesh, India as many as 400 adult *T. vitulorum* were once recovered from a buffalo calf less than 3 months old [197]. Such heavily infected calves usually die after a bout or colic or with variable combination of symptoms. Besides, in 3 other calves the worms had migrated and localised in the liver after perforating the intestinal wall. In natural infection in buffalo calves haemoconcentration, toxaemia, pyrexia, lymphocytosis, increased blood histamine and activity of transaminases has been observed. Excellent accounts of pathomorphological aspects of ascariasis in buffalo calves are now available ([192]; see [30]).

Epidemiology

The only study on seasonal incidence of ascariasis in calves is that of Chauhan et al [44] according to whom it was the highest (55% and 37%) and the lowest (7% and 4%) in cow and buffalo calves in summer and rainy seasons, respectively.

Milk samples from buffaloes have been shown often to contain third-stage larvae of *T. vitulorum* [23] suggesting transcolostral route of infection. This has also been demonstrated experimentally [94].

In Tamil Nadu, the mortality in calves due to ascariasis alone was about 4% whereas in Uttar Pradesh it was about 11%. It was even higher in the Tarai region of Uttar Pradesh which has humid climate. In Orissa, India and Bangladesh, the calf mortality has been reported to be as high as 80%, the tropical monsoon climate prevailing in the latter area. Further studies are necessary to determine the local factor(s), if any, influencing such a high toll of calves in these regions.

In Sikkim, more than 10% of 225 yaks were found to harbour *T. vitulorum* but its relation to other related domestic stock is obscure.

Schistosomiasis

Schistosomiasis in domestic animals and in man is by and large confined to tropical countries of the world. Ironically, no other tropical country in the world has so many species of schistosomes parasitising domestic animals as in Indian subcontinent (see further).

For a long time, this disease complex had taken a heavy toll of animals because of indifferent attitute towards it. Schistosomiasis is now well recognised as the fifth major helminthiasis of domestic animals in the subcontinent in terms of both morbidity and mortality in cattle and buffaloes. Recently, in three excellent reviews, different aspects of the disease complex have been discussed by Kumar and Burbure [89], Agrawal & Shah [4] and Shah & Agrawal [177] and hence, will not be discussed here. Additional and pertinent information, however, has been furnished in Table 6. Looking to the peculiar geographical distribution and epidemiological complexity of some aspects of nasal schistosomiasis, it was thought necessary to discuss here*.

---

*See also "Nasal Schistosomiasis: A review" by Agrawal, M.C. and Alwar, V.S. Helminthol. Abstr. **61**: 373–384 (1992).

Nasal Schistosomiasis

This malady has long been recognised as one of the most debilitating diseases of cattle in the subcontinent and colloquially known at different regions as 'Busa Rogan', 'Sun Suna' or 'Pinas' meaning either snoring or rotten nose. The name of Jay Singh Raj is associated with this disease who first documented it in 1910 from Tamil Nadu, but the casual parasite, however, remained a mystery until 1931. Earlier, the condition was erroneously thought to be of actinomycotic origin, rhinosporidiosis or due to some unknown protozoan parasite.

Prevalence

Data on the prevalence of nasal schistosomiasis in different domestic animals and its geographical distribution is provided in Table 5 (see Fig. 1). The disease, though widespread in the plains of Indo-Pakistan subcontinent, has not been reported from Himachal Pradesh, Punjab, Rajasthan, major parts of Western Uttar Pradesh and Delhi.

Table 5. Prevalence (%) of nasal schistosomiasis in Indian subcontinent

Countries/States	Cattle		Buffalo		Sheep	Goat
PAKISTAN	(?)	(85)	48.9	(?)		
INDIA						
Assam	6.0–15.0*	(68)	2.0–40.0*	(65)		
West Bengal	100.0	(500)*	100.0	(500)*		
Bihar	47.5	(601)	—	(103)		(59)
Orissa			50.0	(50)		
Andhra Pradesh			88.8	(36)		
Tamil Nadu	15.5–53.0	(?)	24.0–60.0	(?)	1.8 (87)	41.4 (88)
Karnataka	24.7	(286)	51.2	(41)		
Kerala			59.5	(227)		
Madhya Pradesh	1.0	(357)	0.9	(107)	— (79)	— (274)
Gujarat	13.3	(15)			4.0 (50)	
Uttar Pradesh	—	(125)	13.45	(?)		
SRI LANKA	12.6	(1,393)				
BANGLADESH	12.0	(224)	—	(5)	— (20)	3.0(118)

Figures in parentheses indicate number of animals examined.
*Mixed data; — indicate surveys in which negative cases or inconclusive results were recorded.

Pathology

Recently, nasal schistosomiasis has been studied in naturally infected cattle in Sri Lanka [65, 214]. The various lesions encountered have been classified into 6 specific types; the type 5 is believed to be the cause of symptoms of snoring. Interestingly, epithelioid cell granulomas containing live and dead eggs within the severly inflammed mucosa were reported for the first time. Beside typical ulcers, globule leucocytes and goblet cell hyperplasia were also detected in the areas where eggs penetrated the mucosa.

Epidemiology

Several important aspects of epidemiology of nasal schistosomiasis still remain unclear. Nothing substantial is known regarding the factor(s) which limit the spread of infection beyond the subcontinent and also in some areas within it.

Table 6. Schistosomes of animals in Indian subcontinent

Geographical distribution	Definitive host	Location	Intermediate host	Pathology
Schistosoma nasale				
Pakistan, India, Sri Lanka, Bangladesh, Burma	Cattle, buffalo, sheep, goat, rarely horse	Veins of nasal mucosa, sometime elsewhere	*Indoplanorbis exustus*	endarteritis, pseudoactinobody around egg, rhinitis, granulomatous growth, hypergammaglobulinaemia
S. spindale				
Pakistan, India, Sri Lanka, Bangladesh	Buffalo, cattle, sheep, goat, horse, donkey, rodent, rarely dog	Portal, mesenteric, pancreatic, gastric, splenic, renal veins or elsewhere	*I. exusus*	verminous thrombophlebitis, hepato-intestinal pseudotubercle, peri-portal cirrhosis, colitis, patchy enteritis
S. indicum				
Pakistan, India, Bangladesh	Sheep, goat, cattle, buffalo, camel, horse, donkey, elephant	Portal, mesenteric, hepatic veins, pulmonary artery or elsewhere	*I. exustus*	chronic venous congestion, 'Hoeppli body' formation, nodular portal cirrhosis, hypergammaglo-bulinaema
S. incognitum				
India	Pig, dog	Portal, hepatic, mesenteric veins or elsewhere	*Lymnaea luteola*	intrahepatic thrombophlebitis, angeiomatoid formation, bilharzial pseudotubercle, persistent dysentery
Orientobilharzia turkestanicum				
Pakistan, India	Cattle, sheep, goat	Mesenteric, portal veins	*Lymnaea* spp.	not reported
O. dattai				
India	Buffalo, cattle, sheep, goat	Portal, hepatic, mesenteric, gastric, pancreatic veins or elsewhere	*L. luteola*	bilharzial pseudotubercle, haemorrhage
O. bomfordi				
India	Cattle	Portal, mesenteric veins	?	not known
O. harinasuta				
India	Buffalo	?	?	not known
Bivitellobilharzia nairi				
India, Burma	Elephant	Portal, mesenteric veins	?	biliary pseudotubercle ? anorexia, constipation, diarrhoea

Principal host preceded in definitive host list.
Rabbit, guinea pig, mice and albino rat are susceptible laboratory models.
Adapted mainly from Agrawal & Shah [4]; Ahluwalia [5]; Kumar & Burbure [89]; Shah & Agrawal [177]; Srivastava [195].

No satisfactory explanation so far is given on the occurrence of *Schistosoma nasale* only in buffaloes and not in cattle in areas far remote from those where nasal schistosomiasis is endemic in cattle [99]. In a survey in Pakistan nearly half of the buffaloes had nasal schistosomiasis but none of the cattle in the same survey was found infected [1].

The nasal granuloma is reported to be higher in cattle than in buffaloes; surprisingly, though incidence of the disease is more common in buffaloes than in cattle. Infected cattle in certain areas were observed to excrete two or three times more eggs in nasal mucus than buffaloes [154] whereas in other regions eggs were seen in a fair number of buffaloes but only in stray cases in cattle. The nasal discharge was profuse in some buffaloes, but snoring was not pronounced nor there was any evidence of marked enlargement of nasal septa. Like buffaloes, a large number of cattle in endemic areas were reported to remain asymptomatic [58].

Cross transmission between cattle and buffaloes revealed no difference in the morphology of eggs, larval stages or adults, but accompanying pathology, however, differs in either case (see review [89]).

It is believed that endemically the buffalo is a "latent reservoir" [11] or this host is reported to have an "innate resistance" [89]. However, research is needed to establish if the host-parasite behaviour is due to an extrinsic factor(s), partial immunity from an earlier exposure to more than one species, "complete compatibility" with host or there is strain difference on lines known to exist in some other schistosomes (see review [37]).

It is generally believed that the disease is more prevalent in older animals than to young stock but sufficient data are needed from different geographical regions to substantiate the age susceptibility of the disease.

In some areas in Karnataka, the incidence of nasal schistosomiasis in cattle and buffaloes was reported to be high from October to December and at other places from June to July (102). Both *Lymnaea luteola* and *Indoplanorbis exustus* were thought to act as intermediate hosts of *S. nasale*, but subsequent studies showed that *I. exustus* was the only natural and experimental host of *S. nasale*, the other species of snails were refractory to this infection [58].

Parasitic Bronchitis

Parasitic bronchitis has long been recognised as a major helminthic infection in both temperate and tropical countries of the world and has been reviewed by Poynter & Selway [130] and Poynter [131]. In Indian subcontinent too, it is one of the important helminth infections particularly in sheep and cattle and is associated with considerable morbidity and mortality in young and growing flocks (see Table 3).

Prevalence

Verminous pneumonia has been reported from different regions of the subcontinent [99, 133, 174, 180, 189]. The geographical distribution of different species of lungworms and their prevalence in domestic animals and wild counterparts have been presented in Table 7 (see Fig. 1).

In North-West Frontier and Sind provinces of Pakistan, the prevalence of *Dictyocaulus filaria* and *Protostrongulus rufescens* was to the extent of 16% and 24% in sheep and 14% and 17% in goats, respectively [194].

According to an old estimate, at least 77–92% sheep in Himalayan region were found infected with lungworms [189]. In Kashmir valley, 29% sheep had *Varestrongylus* sp., 21% *D. filaria* and 7% had both. Contrary to this, in Garhwal Hills of Uttar Pradesh 31% sheep and 19% goats were infected only with *D. filaria*. Similarly, in Nepal maximum prevalence of *D. filaria* was 45% in sheep and 24% in goats [200].

Bovine dictyocauliasis was recorded in 17% of 3,000 lungs examined in a survey in Madhya Pradesh

Table 7. Some lungworms of domestic animals in Indian subcontinent

Parasite	Definitive host	Geographical distribution
Dictyocaulus filaria	Sheep, goat, buffalo	Pakistan, India, Nepal, Burma
D. viviparus	Cattle, Buffalo?	India, Sri Lanka, Bangladesh, Burma
D. arnfieldi	Donkey, horse	India
Varestrongylus pneumonicus	Sheep, goat	Pakistan, India, Nepal
**V. capricola*? (= *V. pneumonicus*)	Siberian wild goat	Pakistan
Metastrongylus apri	Pig, sheep, goat	India, Sri Lanka
M. salmi	Pig	India
Protostrongylus rufescens	Sheep, goat, Buffalo?	Pakistan, India
**P. indicus*?	Siberian wild goat	Pakistan
**P. ovis*?	Sheep	Pakistan
Muellerius spp.	Sheep, goat, spotted deer	India

?*Species inquirenda*/occurrence doubtful.
*Conf. Boev [36]; Sarwar [167].

and is reported to be common in dry zone of Sri Lanka [174]. In Burma, the prevalence of lungworm is usually light in buffaloes but sometime heavy in cattle.

Other species of lungworms have not been studied to that extent in the subcontinent. Of the 28 lungs of donkeys examined at Mathura, Uttar Pradesh only 2 had *D. arnfieldi* and in contrast, 13 ponies were free from this infection [81]. In Kashmir valley, 1.6% of 248 and in Sikkim, 5% of 300 sheep harboured *Muellerius* sp. Significantly, the prevalence of this species was as high as 31% in 197 deers (?) examined in three national parks in India [144].

Pathology

Considerable data are available from histopathological studies in different domestic animals including sheep [24], goat [24], cattle [133], buffalo [80], donkey [81], and pig [33]. In sheep and goats mixed infection with *D. filaria*, *P. rufescens* and *V. pneumonicus* is very common. However, detailed experimental studies on the pathogenicity of this disease are rare (see [28]).

Epidemiology

Sharma et al [180] briefly described the epidemiology of lungworm infections in sheep and goats in India. No evidence of an age resistance in sheep with *D. filaria* was observed in part of Punjab, Pakistan, while in India aged sheep were found more resistant than young ones to this infection. Goats were reported to be more susceptible to *D. filaria* than sheep. The incidence of *D. filaria* is reported to be more common in female than in male goats in the age group of 3–5 years. Further study is necessary to substantiate these informations.

D. filaria, *V. pneumonicus* and *Muellerius* spp. occur in hill sheep and their wild counterparts but no cross transmission attempts between them, have been made in order to assess their respective role in dissemination of these infections.

Both in Madhya Pradesh and Karnataka, deaths due to parasitic bronchitis had been recorded in old cattle (see Table 3).

Macrochlamys (= *Euastenia*) *cassida* acts as intermediate host of *V. pneumonicus* in India [27]. The earthworm, *Megascolex mauritii* was incriminated as an intermediate host for *Metastrongylus apri* in

Tamil Nadu [199] whereas, other species of earthworms viz. *Perionyx excavatus* and *Dichogaster bolaui, Drawida pellucida* var *pallida, lampito mauritii* had been shown experimentally to play the role for *M. salmi* in West Bengal and Kerala, respectively [33, 203]. Nothing is known, however, about intermediate host(s) of *P. rufescens* and *Muellerius* spp. in any part of the subcontinent.

Interestingly, the lungworms hitherto believed to be confined to certain cold hill-tracts of northern India, have now successfully adapted themselves to warmer southern plains. This fact calls for detailed investigations on their bionomics for effective implementation of control measures those might be devised against them.

Echinococcosis/hydatidosis

Hydatidosis is a serious zoonotic disease of global dimensions and is caused by the members of the genus *Echinococcus*. What is alarming is the fact that it is spreading to new foci. Various aspects of the disease have been reviewed by WHO/FAO Expert Group on Zoonoses [15], Rickard & Williams [158] and Thompson [204].

Hydatidosis/Echinococcosis has great importance in agriculture-based countries with a large rural population. In India, and for this matter, the entire subcontinent, ideal conditions exist for establishment, propagation and dissemination of hydatidosis both in man and livestock. Yet, some standard text books do not mention India as one of the countries where hydatidosis is widely prevalent and is endemic [118].

The exact status of hydatidosis in the subcontinent, however, is far from clear except periodic surveys on prevalence in animals and occasional clinical reports both in man and animals. Its economic impact, which must be of considerable magnitude, has not been assessed at national level. This is despite the fact that according to a recent estimate, the condemnation of buffalo livers at Aligarh, Uttar Pradesh alone accounted for a loss of Rs. 131,400 per annum [78].

Parasite

Consensus of parasitological opinion accepts four species of *Echinococcus* as taxonomically valid [187]. Of these, only two are of public health importance viz. *E. granulosus* and *E. multilocularis*. The latter species is distributed in the countries of northern hemisphere from Canada and Alaska to Japan, Siberia, northern Europe and U.S.S.R. (see review [156]) and it is not likely to be encountered in India. However, two reports [6, 86] do exist that claim its presence in northern India although it needs verification. In the meantime, only *E. granulosus* has relevance in India.

Cameron [38] was first to point out the existence of a sub-species within *E. granulosus* in Canada. Later works have provided substantial evidence for the existencee of 3 other sub-species of *E. granulosus*. Apart from all other factors, all the four sub-species differed from each other on the basis of their host assemblage.

It is now well established that several (and perhaps many) strains of *E. granulosus* (and *E. multilocularis*) exist but all of which may not be infective to man ([93]; see also [188]). Thus, Gill & Rao [68] and Pandey [114] showed, on the basis of their studies on morphology and developmental stages of *E. granulosus* of buffalo-dog and goat-dog origins, that did not fall into "classical" description of this tapeworm. They, therefore, concluded that it might be a "mutant strain" developed in the process of evolution (see Chapter 5). Similarly, morphological variations had also been observed in adult worms by feeding protoscolices from cysts taken from the liver and lungs of buffaloes [77]. In Sri Lanka, hydatid cysts in animals usually has a large number of scolices in brood capsules which are believed to be the characteristic of a different ecological strain ([166]; see review [205]).

No transmission experiments have so far been conducted to determine the host-specificity of buffalo, cattle, sheep, goat, pig and camel strains in Indian subcontinent. Nothing is known about the *in vitro* growth and physiological characteristics of many of these strains.

Prevalence

Hydatidosis

Hydatidosis has been recorded in different domestic animals in most regions of the subcontinent so far surveyed (Table 8, Fig. 1).

Table 8. Prevalence (%) of hydatidosis in food animals in Indian subcontinent

Locality	Sheep		Goat		Cattle		Buffalo		Pig	
PAKISTAN	4.6	(?)	2.1	(?)	27.0	(62)	19.0	(9,000)	—	—
INDIA										
N.E. Region	—	—	7.8	(203)	9.7	(227)	10.6	(57)	2.8	(106)
West Bengal	—	—	—	—	17.82	(2,048)	24.08	(2,686)	—	—
Bihar	17.72	(79)	10.74	(773)	19.70	(462)	14.42	(312)	—	—
Tamil Nadu	0.89	(2,869)	—	—	41.0	(163)	—	—	—	—
Karnataka	21.75	(1,177)	16.96	(731)	60.94	(192)	—	—	7.7	(65)
Kerala	2.55	(705)	1.05	(1,045)	35.47	(761)	32.96	(91)	—	—
Rajasthan	2.27	220)	5.83	(360)	—	—	—	—	—	—
Uttar Pradesh	6.0	(109)	2.0	(1,208)	—	—	36.0	(1,556)	1.0	(559)
Haryana	4.66	(386)	3.03	(362)	—	—	11.03	(928)	2.27	(357)
Punjab	5.1	(328)	2.8	(315)	7.5	(79)	8.9	(156)	—	—
Sikkim	50.0	(44)	33.0	(12)	25.0	(20)	—	—	—	—
SRI LANKA	2.4*	(?)	—	—	0.5	(?)	—	—	—	—
BANGLADESH	7.3	(55)	2.6	(115)	53.6	(155)	48.6	(35)	—	—

Figures in parentheses indicate number of animals examined.
*Mixed data.

SHEEP AND GOAT

Prevalence of hydatidosis in the subcontinent varies according to surveys carried out in different regions. Sheep is regarded as the best disseminator of hydatid infection. Even though sheep-rearing is not a major occupation in the subcontinent, sheep population is still considerable i.e. 80,331 million in India alone (see [18]). In hill-tracts of northern India, in particular, sheep-rearing is widely practised but data on the prevalence of hydatidosis in them is lacking.

Rao [151] recorded hydatid infection in sheep to be 65 in lungs, 35 in liver and 4% in spleen in Madras city whereas in Kurnool, the incidence in lungs was 66 in sheep and 80% in goats and in liver 19 in sheep and 10% in goats [157]. Significantly, human hydatidosis in Tamil Nadu and Andhra Pradesh is endemic.

Data on the viability of fertile cysts in the subcontinent are scarce. In a survey in Punjab, 96 and 84% sheep and goat cysts, respectively were found fertile, whereas in Rajasthan ovine cysts, were sterile but goat cysts were 100% fertile ([50]; see [78]). In Sri Lanka, the fertility of goat cysts was high whereas in Bangladesh it was low in comparison to sheep cysts.

Clinical hydatidosis in live sheep and goats is seldom documented (see review [172]). However, Varma & Malviya [213] recently recorded a rare case in a Jamunapari goat that died from functional atrophy of both lungs and heart due to mechanical pressure by a huge secondary cyst in the pleural cavity.

CATTLE AND BUFFALOES

Cattle cysts, like in buffaloes, are observed to be more or less uniformly distributed in liver and lungs. However, in a survey in Karnataka, 86% cysts were found in the lungs of cattle. Cattle cysts are often believed to be sterile but in Punjab 39% of lung and 42% liver cysts were found fertile [50]. In Sri Lanka, cattle cysts as a rule were fertile as they contained numerous scolices in their brood capsules. A rare case of hepatic hydatidosis of prenatal origin was reported in a 19 day-old suckling calf from Uttar Pradesh (Gaur, personal communication).

Hydatid cysts in buffaloes have been described in spleen, kidney, heart, omentum, brain in addition to liver and lungs with the later organs being the more common site. Gill & Rao [69] observed 90% fertility of buffalo cysts in Bareilly whereas in Trichur and Ludhiana, the fertility was 73% and 40%, respectively [2, 50].

Multilocular hydatid cyst is rare in goat, less common in buffalo but frequently found in cattle. Calcified and suppurative cysts, which are occasionally found in sheep and goats have been reported in buffaloes. Unlike other hosts, secondary hydatid cysts is very common in buffaloes and frequently found encapsulated in lungs, diaphragm, pleural cavity, subcutaneous tissue, abdominal cavity and omentum. Singh et al [184] have documented an exceptionally high infection in a buffalo in which the liver had 105 cysts on the dorsal and 89 cysts on the ventral surface. Two of the primary cysts had 61 secondary cysts internally and 36 externally. Surprisingly, in India prevalence of hydatidosis had similar sex difference in buffaloes as humans, it is being higher in females than in males. Like cattle, sheep and goats, the prevalence of hydatidosis is reported to be higher in older buffaloes. The parasite, however, appears to have well adapted in buffaloes, a long-lived intermediate host. The fertility of cysts in buffaloes increased dramatically with increasing cyst size though viability of protoscolices is not consistent [78]. It must, however, be noted that the primary cyst of buffalo sometimes reaches 90–100% fertility, it has exceptional ability to produce secondary cysts which are consistently 100% fertile and in general 81% viable, have greater infectivity and establishment rate. The parasite is observed to have similar *in vivo* requirements to that of human material!

YAK

In Sikkim, 80% of yaks were found infected, the incidence of hydatid cysts was 35% in the liver and 5% in lungs [139]. The cysts were unilocular. Neither fertility rate nor clinical manifestations in this species has been studied so far.

EQUINES

Hydatidisis in equines in the subcontinent have rarely been documented. During the course of a post-mortem examination of a 7 year old horse, the author observed massive involvement of both liver and lungs. At least 20–30 cysts of different sizes (4–8 cm) were recovered which had replaced 30–40% of parenchyma of both organs. Majority of the cysts were unilocular, but their fertility was not ascertained.

SWINE

Limited surveys have been carried out for hydatidosis in pigs in the subcontinent. The highest (17%) prevalence had been recorded in Madhya Pradesh. The cysts were unilocular and sterile. However, 29% cysts from pigs in Uttar Pradesh and 76% from Bihar were fertile.

CAMEL

An overall prevalence of 71% hydatidosis have been reported from Indian camels. Cysts were found in 85% lungs, 75% liver, 5% spleen and 5% rumen. The fertility rate varied from 15 to 100% in different geographical areas [7, 78].

Echinococcosis

Prevalence of echinococcosis in different geographical regions of the subcontinent of India has been reviewed by Sharma & Chitkara [181]. More recent informations have been provided in Table 9. Unfortunately, up-to-date data on echinococcosis are not available in the Indian region of the subcontinent.

Table 9. Prevalence of *Echinococcus granulosus* infection in dogs in Indian subcontinent

Locality	Year	Number of dogs examined	Source	Percentage
PAKISTAN				
Karachi	1965	200	City	7.0
Faisalabad	1991	100	City	13.0–18.0
INDIA				
Amritsar	1963	287	City	10.8
		173	Village	3.46
Madras	1967	49	City	4.0
Kurnool	1968	150	Town	33.3
Mhow/Jabalpur	1969	74	Town/City	10.81
Bareilly	1969	51	City	5.8
Patna	1971	20	City	10.0
Calcutta	1973	219	City	0.46
Bangalore	1974	10	City	10.0
Pantnagar	1980	53	Town	1.86
Srinagar ?	1988	400	city	11.0–17.0
SRI LANKA				
Colombo	1960	3	City	+
BANGLADESH				
Dacca	1977	267	City	61.0*
Mymensingh	1980	125	City	62.52*

+ Two of the dogs examined harboured several hundred worms
*cf. text

In a very recent survey in Pakistan, 19 out of 100 jackals examined were found to harbour *E. granulosus* infection [79]. In Aligarh, Uttar Pradesh, it has been estimated that there are more than 40 street dogs/ per sq km; one of the dogs known to scavange from buffalo abattoir when autopsised, was found to harbour a patent *E. granulosus* infection [78]. Natural infections of *E. granulosus* in Indian wolf, jackal and fox had been recorded in Orissa, India and Sri Lanka ([54, 149, 150]; see [147]).

In Bangladesh, echinococcosis in dogs is very common; the highest (61–75%) incidence was recorded in dogs near the slaughter house and the lowest (6–35%) in those caught away from it. The incidence in dogs of urban area was 51.2% but comparable data for those of rural areas are not available.

Pathology

Schwabe [172] has recently reviewed the clinical aspect of hydatidosis in cattle and buffaloes in India. Pathoanatomy of hydatidosis has been studied in different domestic animals including sheep, goat, cattle, buffalo, camel, pig and yak. The main lesions observed were similar in all animals (see [126, 139]). Significantly, in the subcontinent mild to acute catarrhal enteritis with or without severe anaemia has been documented in experimental dogs and naturally infected wolf [150, 151]. However, in the literature echinococcosis in dogs is reported to be non-pathogenic.

Epidemiology

Hydatidosis in the subcontinent is far more common in domestic animals than is realised. Several factors including biological, cultural, educational, socio-economical, agricultural, environmental and faunatic contribute to the transmission of infection between man and domestic animals and has been reviewed by a number of workers ([157,159, 204]; see also [54, 78, 79]).

Seasonal fluctuations of echinococcosis in dogs has been determined in Bangladesh; the lowest (47%) prevalence was between January to March and the highest (78%) between October to December (Islam, personal communication). Comparable data, however, lacking for other regions of the subcontinent.

Pulmonary hydatidosis in both man and animals is common [157, 159]; it is, however, surmised that majority of human infections occurred through consumption of infected vegetables and by drinking contaminated water rather than man to dog contact [215].

In the subcontinent, animal hydatidosis in different regions is believed to be biocoenotically related. Consequently, hydatidosis is more common in buffaloes in northern alluvial plains, in zebu cattle in eastern region, in goats in Tamil Nadu, Andhra Pradesh and Madhya Pradesh and in wild asses and camels in Gujarat, Rajasthan and *run of Kutch* bordering Pakistan (see Fig.1). Apart from dogs, black ants (?) have been incriminated in semi-arid regions of the subcontinent of India to disseminate *E. granulosus* eggs on shrubs [60]. High prevalence of human/animal hydatidosis in Tamil Nadu, Andhra Pradesh and Karnataka where existence of a dog-black ant-goat/man cycle is a possibility. In Kashmir, Himachal Pradesh, hilly regions of Uttar Pradesh, dog-sheep/man cycle probably predominates. In Sri Lanka, primarily a Jackal-Sambhur, deer, wild buffalo, wild boar cycle is reported to exist ([54]; see also [79]).

A wide range of both captive and wild animals (Table 10) are known to act as intermediate hosts in the subcontinent. Two cysts from a great Indian squirrel once caught from Malabar Hills, were found fertile, so also from a giant flying squirrel and a spotted deer in Nandan Kanan Zoo in Orissa. Undoubtedly, there is fairly high prevalence of human hydatidosis in the subcontinent [159] for the propagation of which, not only a domestic cycle consisting of sheep, goat, buffalo, camel might be responsible, but also a sylviatic cycle involving some of the wild carnivores and ruminants. The latter aspects need investigation.

Cysticercosis

Cysticercosis is not only of great public health significance but its impact on agricultural economy in tropical countries is enormous. In most developing countries, cysticercosis often occurs in hyperendemic form (see review [16]) and the subcontinent of India is no exception. However, accurate basic data on the prevalence of bovine and porcine cysticercosis is not available. Likewise, the economic losses due to ovine as well as caprine cysticercosis, if any, have yet to be assessed in the entire subcontinent.

An older estimate in undivided Punjab indicated that among the cattle slaughtered, 5.5% to 6% had *Cysticercus bovis*. More recent information on the prevalence of *Taenia solium* and *T. saginata* cysticercosis has been furnished in Table 11 (see Fig. 1). The metacestodes and their adults (*Taenia* spp.) have been reported from a wide variety of mammalian hosts in the subcontinent (see Table 12).

Pathology

Neurocysticercosis in man is one of the major diseases in the subcontinent of India. The condition is frequently associated with violent manifestations. Various aspects of this disease had been reviewed by Flisser et al [62, 63, 64] and Pawlowski & Schultz [127] (see also [117, 118]).

The exact cause(s) of cerebral symptoms in man is obscure. Apart from critical anatomical location of cysts, clinical symptoms are believed to be associated with acuteness or chronic state. Degeneration,

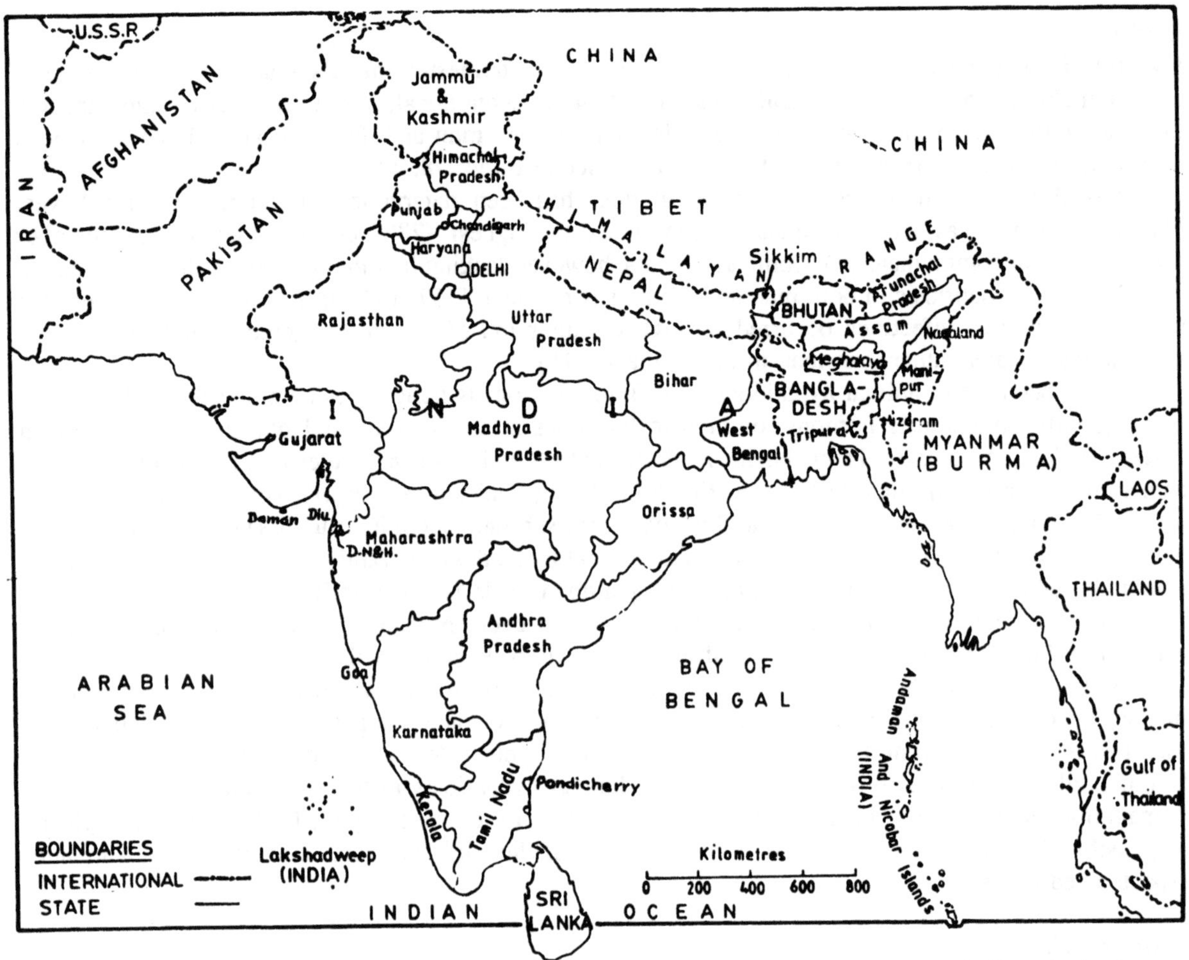

Based upon Survey of India map with the permission of the Surveyor General of India. © Government of India copyright 1991.
The boundary of Meghalaya shown on this map is as interpreted from the North-Eastern Areas (Reorganisation) Act, 1971, but has yet to be verified.

The administrative headquarters of Chandigarh Haryana and Punjab are at Chandigarh.

The territorial waters of India extend into the sea to a distance of twelve nautical miles measured from the appropriate base-line.

Fig. 1 Map of Indian subcontinent

absorption, cicatrisation and calcification of cysts are also reported. Recently, "glycocalyx", a major antigenic component of surface membrane of the cyst wall also felt to be responsible for clinical manifestations [42].

Experimental infections in monkeys, however, showed that high dose of eggs (12,000 eggs/animal) induced cerebral syndromes even before metacestode had matured and caused death in monkeys [164] (see Table 12). Delayed hypersensitivity reaction manifested by granuloma formation in several internal organs in monkeys has been demonstrated experimentally [163].

Table 10. Prevalence of hydatidosis in captive and wild animals in Indian subcontinent (cf. text)

Host	Organs affected	Geographical distribution
Rabbit	Central nervous system	India
Spotted deer	Lung	India
Four-horned antelope	Lung	India
Ceylon sambhur	Lung	Sri Lanka
Blue bull	Different sites?	India
Wild buffalo	?	Sri Lanka
Yak	Liver, lung	India
Giraffe	Lung	India
Wild ass	?	India
Indian pangolion	Lung	India
Common giant flying squirrel	Lung	India
Great Indian squirrel	Lung	India
Indian elephant	Different sites?	India
Wild boar	Peritoneal cavity	Pakistan, India, Sri Lanka
Lion*	Liver	India
Monkey	Liver, lung, spleen	India, Sri Lanka
Nilgiri langur	Lung, spleen	India

* ? Infection suspected from feeding entrails of infected dogs.

Table 11. Prevalence (%) of *Taenia solium* and *T. saginata* cysticercosis in food animals in India

Locality	T. solium		T. saginata					
	Pig		Goat		Cattle		Buffalo	
Army Command	1.40	(14,027)	—	—	1.40	(1,212)	—	—
N.E. Region	20.8	(106)	—	—	5.3	(227)	5.3	(57)
Assam	9.23	(65)	2.87	(209)	7.62	(118)	—	—
Mizoram	7.14	(42)	—	(36)	3.79	(79)	—	—
Nagaland	4.68	(64)	4.61	(24)	4.87	(82)	—	—
Meghalaya	19.03	(289)	—	—	—	—	—	—
West Bengal	—	—	—	—	3.43	(3,640)	2.33	(4,075)
Orissa	—	—	—	—	0.35	(556)	—	—
Tamil Nadu	7.61	(65)	—	—	3.57	(448)	0.75	(267)
Uttar Pradesh	8.0	(200)	—	—	—	—	—	—
Haryana	0.09	(2,163)	—	—	—	—	—	—

Figures in parentheses indicate number of animals examined.

Epidemiology

In India, three cases of generalised cysticercosis, one in a cross-bred boar in West Bengal and two in indigenous pigs in Haryana have been documented. Neurocysticercosis to the extent of 1.4% and 3.1% in stray dogs was observed in Tamil Nadu and Karnataka, respectively. In Uttar Pradesh, none of the 758 animals including sheep, goats, cattle, buffaloes was found positive for *C. cellulosae* [119]. The reported development of *C. cellulosae* in a lamb but not in kids [212] needs confirmation; so also the occurrence of this metacestode in rodents [140] and *C. bovis* in goats [160].

In spite of sporadic reports of human taeniasis/cysticercosis from different regions of the subcontinent,

Table 12. Metacestodes of Indian subcontinent

Geographical distribution	Hosts*	Location	Pathology
Cysticercus cellulosae (Taenia solium) India, Sri Lanka	Man, hamster (DH); domestic pig, wild boar, dog, man, monkey, sheep, goat, rodent (IH)	Voluntary and involuntary muscles, internal organs, fascia, omental fat, mesentery, brain, spinal cord	Neuronal degeneration, cerebral necrosis, monoplegia/diplegia; lymphoproliferative changes in internal organs; nodulations in skeletal muscles; pulmonary hyperplasia
C. bovis (T. saginata) Pakistan?, India, Sri Lanka, Bangladesh	Man (DH); cattle, buffalo, goat (IH)	Voluntary and involuntary muscles, liver	Arteriosclerosis, granulomatous reaction in heart; atrophy, caseation, calcification of muscles
C. tenuicollis (T. hydatigena) Pakistan, India, Sri Lanka, Burma	Dog, jackal (DH); sheep, goat, cattle, buffalo, camel, pig, rabbit, deers, Beisa orynx, gibbon (IH)	Mesentery, liver, abdominal cavity, cervix, omental fat, rarely in lungs, rectal wall, brain	Erythrocytopaenia, hypohaemoglobinaemia; granuloma in Liver, degenerative changes in hepatic cells; pneumonia
C. ovis (T. ovis) Pakistan, India	Dog (DH); sheep, goat (IH)	Skeletal muscles, lungs, stomach wall, kidney, urinary bladder, tunica vaginalis	Atrophy of seminiferous tubules, neoplastic transformation?
C. pisiformis (T. pisiformis) Pakistan, India Bangladesh	Dog, domestic cat, jackal, wild cat, tiger (DH); rabbit (IH)	Liver, mesentery, abdominal cavity, intestine	Not reported
C. fasciolaris (T. taeniaeformis) Pakistan, India, Sri Lanka, Bangladesh, Burma	Cat, tiger, clouded leopard (DH); rat, mice, golden hamster, bandicoot, palm squirrel, gerbal (IH)	Liver, urinary bladder, mesentery, body cavities	Necrosis, cloudy swelling/atrophy of hepatic cells, depletion of glycogen and increase in carbohydrate complexes in hepatic tissue
Coenurus cerebralis (Multiceps multiceps) Pakistan, India, Sri Lanka, Bangladesh, Burma	Dog, jackal (DH); sheep, goat, cattle, buffalo, camel, horse, pig, rabbit (IH)	Brain, heart, placenta, lymph node, diaphragm, thigh muscles, rarely in other organs	Convulsion, severe meningitis, granuloma in brain; fatty changes in liver; coagulative necrosis in kidneys; myocardial haemorrhage
C. gaigeri (M. gaigeri) Pakistan, India, Sri Lanka, Bangladesh	Dog (DH); goat, sheep, pig (IH)	Intramuscular connective tissue, peritoneum, pubic symphysis, other internal organs, rarely in nervous tissues	Toxaemia, glomerular atrophy, necrosis; calcification in brain
C. serialis (M. serialis) Pakistan, India, Sri Lanka	Dog (DH); goat, pig, rabbit, hare (IH)	Subcutaneous tissue, heart	Not reported

*DH: Definitive hosts; IH: Intermediate hosts; adult tapeworms are cited in parentheses. Adapted partly from Chakraborty et al [40]; Gupta & Chowdhury [72]; Parija [117]; Rao & Anantaraman [152]; Saleque et al [164]; Seneviratna & Kulasiri [175].

there are a few reports on the incidence in animals; for example, incidence of *Taenia saginata*/*T. solium*, is documented for Kashmir and Sikkim but there is no comparable information available on bovine and porcine infections due to their larval stages. In an extensive survey in Calcutta, 51% of cattle and 50% of buffalo cysticerci were found to be viable which undoubtedly pose a potential source of infection for perpetuation of taeniasis in man (see also review in [117]).

In Tarai region of Uttar Pradesh, the prevalence of *C. tenuicollis* in sheep and goats was observed to be the highest in rainy season and the lowest in summer. Aged animals harboured more metacestodes than young animals. The reported higher incidence in goats as compared to sheep in Uttar Pradesh obviously either due to their different grazing habits in the region or varying susceptibility of the two hosts. This, however, needs verification.

The seasonal fluctuation of 'gid' in goats with higher prevalence in rainy and post-monsoon seasons, similar to *C. tenuicollis*, has been observed in Assam. However, no data on the prevalence or age susceptibility of coenuriasis in sheep and goats are available. Availability of moisture probably is a pre-requisite for eggs of these taeniid cestodes since cysticerciasis/coenuriasis sometime assumes epidemic form in high rainfall regions of the subcontinent such as eastern India, Bangladesh and Uttar Pradesh (Table 3).

Filariasis

Filarioids, widespread in tropical countries, also occur in Indian subcontinent. Despite the fact that these nematodes are associated with negligible mortality, they yet cause significant, but hitherto unestimated economic losses by way of reducing working capacity of draft animals and by damaging the hides. Some of these nematodes are suspected to be of public health importance an aspect probably least studied. This group comprises of 98 species under 37 genera of 4 families of Filarioidea parasitising Indian animals and birds including their wild counterparts. The important filarial parasites of the subcontinent and their vectors are listed in Tables 13 and 14. Some of the diseases due to them have been reviewed by Bhatia & Chauhan ([30]; see also review in [117]).

Dirofilariasis

Prevalence

Cunningham as early as 1873 recorded the incidence of *Dirofilaria immitis* in India. Yet, the data on its prevalence are meagre. In Calcutta, 13% of 1,245 dogs examined during 1962–1972 were found positive for this worm [166]. The disease is reported to be common in dogs in foot-hills of Himalayas. In Orissa, an endemic area of filariasis, 57% dogs and 50% cats harboured this infection [122].

D. repens had been recorded in dogs in Bihar and Maharashtra. Again, in Orissa 14% dogs and 50% cats were found infected, whereas in Colombo, Sri Lanka over 50% dogs harboured this infection [175].

Pathology

A few studies have been made on the clinicopathological aspect of canine dirofilariasis [39, 148, 166]. The important changes in dogs observed were, cardiac dilatation, myocardial degeneration, congestive spleenomegaly, pneumonia and enlargement of liver. In a rare case in an Indian fox, the associated lesions observed were more severe and included endarteritis, multiple pulmonary infarcts and embolic granular nephritis [148].

Epidemiology

Little information is available on the epidemiology of the disease. In Tarai region of Uttar Pradesh, dirofilariasis is common in aged Tibetan mastif and Alsatian dogs hinting the presence of both age and breed susceptibility to dirofilariasis. No cross immunity, however, existed since dogs harboured multiple infections of *D. immitis, D. repens* and *D. reconditum.*

Parafilariasis

Gulati was the first to report "Filaria haemorrhagica" in working bullocks in Lahore, undivided Punjab in 1934 due to *Parafilaria,* as slightly raised nodules which periodically bursted and bled. He attributed this condition to *Parafilaria multipapillosa.* Subsequently, it was shown that two species of *Parafilaria, P. bovicola* and *P. sahaii* infect cattle and buffaloes [162, 196]. However, Patnaik and Pande [123] maintained that *P. sahaii* was a synonym of *P. bovicola.* The reported occurrence of *P. bovicola* with a larger microfilaria than its known range in buffaloes needs investigation.

Prevalence

Data on the prevalence and geographical distribution of parafilariasis in the subcontinent are scanty. In Orissa, 3% of 81 buffaloes were found infected with *P. bovicola* [122] whereas in Maharashtra, 5% of 160 cattle carried microfilariae in the blood [182]. Haemmorhagic dermatitis due to *P. bovicola* and *P. sahaii* locally known as "Seraun Phutana" is fairly high in Jammu and Kashmir and in certain pockets of Bihar and Uttar Pradesh (see [162]).

Pathology

The nodular lesions in parafilariasis are mainly due to accumulation of oedematous fluid in the interstitial areolar connective tissue and panniculus muscularis. In mules, *P. multipapillosa* infection did not cause nodule formation; instead, there were subcutaneous granulomatous foci, thickened arteries, perivascular eosinophilic infiltration and induration of local arterioles.

Epidemiology

Parafilariasis is prevalent both at higher altitudes and in plains of the subcontinent. However, the condition is seasonal and prevalent during summer and rainy months and usually not encountered after the onset of winter. The disease may persist as latent infection as filarids are known to be long lived. Mostly adult cattle were observed to be affected by the disease indicating the existence of age susceptibility in parafilariasis.

Elaeophoriasis

Varma in 1953 first reported occurrence of *Elaeophora poeli* from Indian buffaloes in Assam though the specimens examined by him were collected by Buckley in 1934. Almost two decades later, a survey revealed 19% of 256 aortae of buffaloes in Uttar Pradesh had this filarid worm [132].

Pathology

Prasad & Bhalla [132] ascribed pathology of aortitis due to *E. poeli* and categorised the gross changes into three types: (1) The "nodulations" on the intima frequently found associated with aneurysm, (2) occasional fibro-hyalinisation and (3) partial to complete elastolysis in the media, perivascular fibrosis and hyperplasia of adventitial vasa vasorai.

Epidemiology

No data on the epidemiology of elaeophoriasis is available. *E. poeli* probably has much wider host range and geographical distribution than is believed. Sustained search and differentiation of it's lesions from those of aortic onchocerciasis, a more prevalent infection in Indian domestic animals are likely to yield valuable information.

Setariasis

Though von Linstow described *Setaria digitata* from Zebu (*Bos indicus*) in Sri Lanka, it was Place who in 1911 first observed the high prevalence of *Filaria papillosa* (= *S. digitata*?) in Burma (now Myanmar) and suspected it to be the possible cause of "Kumree" in horses.

Parasite

Enormous confusiom exists on the taxonomic position of each species (Table 13) of *Setaria* and *Artionema* prevalent in the subcontinent. The validity of the species has been a subject of controversy for many years (see review [19]).

Table 13. Some filarial parasites of Indian subcontinent

Parasite	Location	Host	Geographical distribution
Dirofilaria immitis	Heart, pleural cavity	Dog, domestic cat, Indian fox, jackal, tiger	India, Sri Lanka, Bangladesh, Nepal, Burma
D. repens	Subcutaneous tissues	Dog, domestic cat, golden cat, Indian fox, man	India, Sri Lanka
Brugia ceylonensis	Lymphatics	Dog	Sri Lanka
B. buckleyi	Heart, blood vessels	Hare	Sri Lanka
B. malayi	Lymphatics	Cat, dog, leaf monkey, other domestic animals?	India, Sri Lanka
Wuchereria bancrofti	Lymphatics	Man	India, Sri Lanka
Parafilaria bovicola	Skin, subcutanecus tissues, eye, or other sites	Cattle buffalo?	Pakistan, Indian
P. sahaii	Skin	Buffalo, cattle?	India
P. multipapillosa	Skin, subcutaneous tissue	Horse, donkey, mule, cattle?	*Pakistan?, India
Indofilaria pattabirmani	Subcutaneous tissue	Elephant	India
I. elephantis?	Unknown	Elephant	India
Bhalfilaria badamii,	Heart	Domestic fowl	India
Gallifilaria mhowensis	Heart	Domestic fowl	India
Cardiofilaria nilesi	Heart	Domestic fowl	Sri Lanka
Elaeophora poeli	Aorta	Buffalo	India
Dipetalonema evansi	Spermatic and pulmonary arteries, cerebrospinal fluid	Camel, horse	*Pakistan?, India
D. reconditum	Vascular system, lungs, liver, kidney, spleen, brain	Dog, antilope	Pakistan, India
D. dracunculoides	Peritoneal cavity	Dog	India
D. grassii	Subcutaneous tissue	Dog	India
Dipetalonema sp.?	Vascular system	Cattle, buffalo	India
Tetrapetalonema sp.?	Stomach	Bear cat	India
Setaria equina	Peritoneal cavity, thoracic cavity, circulatory system, liver, testes, eye, spinal column	Horse, mule, cattle	Pakistan, India, Sri Lanka, Bangladesh?, Burma

(Contd.)

Parasite	Location	Host	Geographical distribution
S. labiato-papillosa	Peritoneal cavity, pleural sac, subcutaneous fasciae, pericardium, eye, subdural space	Buffalo, cattle, sheep, goat, sambar, nilgai	India, Burma
S. digitata	Peritoneal cavity, kidney, urinary bladder, eye	Cattle, buffalo, horse, sheep, goat	Pakistan, India, Sri Lanka, Bangladesh?, Burma
S. cervi (=S. axis?)	Peritoneal cavity, subserosa, liver, scrotum, eye, cranial cavity, spinal column	Buffalo, cattle, goat, sheep, horse, antilope, ceylonese spotted deer	India, Sri Lanka
S. marshalli	Peritoneal cavity	Cattle	Sri Lanka, Burma
S. m. pandei?	Peritoneal cavity	Buffalo	India
S. bernardi	Peritoneal cavity	Pig	Burma
Artionema indica?	Peritoneal cavity	Sheep	India
A.harturchi	Peritoneal cavity	Black buck	India
Papillosetaria veversi	Peritonenl cavity	Black buck	India
Stephanofilaria assamensis	Skin	Cattle, buffalo, goat, elephant, rhinoceros	India, Bangladesh
S. jaheeri	Ear	Buffalo, cattle?, nilgai	India, Bangladesh
S. andamani?	Skin	Buffalo	India
S. srivastavai?	Skin	Domestic elephant	India
Onchocerca gibsoni	Subcutaneous tissue, muscle	Cattle, buffalo, camel	India, Bangladesh
O. reticulata	Tendon (foot), connective tissue	Horse	India
O. gutturosa	Ligamentum nuchae	Buffalo,cattle	India, Bangladesh
O. armillata	Aortic intima	Cattle, buffalo, sheep, goat, donkey, pony, four-horned antilope	India, Bangladesh
O. cervicalis	Ligamentum nuchae	Horse	India
O. indica?	Subcutaneous tissue	Cattle, Buffalo	India, Sri Lanka
O. fasciata	Subcutaneous tissue	Camel	*Pakistan?, India
O. linealis?	Under pectoral muscles	Cattle	India,*Sri Lanka?
O. pandei?	Ligamentum nuchae	Buffalo	India
O. sweetae?	Subcutaneous tissue	Buffalo	India
O. cebei?	Skin	Buffalo	India
Onchocerca sp.?	Subcutaneous tissue	Buffalo, cattle	Burma
O. volvulus	Subcutaneous nodule	Man	India

? Species inquirenda/occurrence doubtful.
* Undivided India.

Prevalence

In a survey on animal filariasis in west Godavari district of Andhra Pradesh, the only filarial worm encountered was *S. digitata*. Its prevalence was shown to be 4.6% of 5,571 cattle and 5.4% of 13,244 buffaloes examined [170]. In Bihar, *S. digitata* was by far the most prevalent species in cattle and *S. cervi*, in buffaloes.

Microfilariasis due to *S. cervi* in very common in Tarai region of Uttar Pradesh where about one fourth of the buffalo population was suspected to suffer from this disease. Significantly, the incidence was determined to be 23% more in lactating buffaloes than non-lactating ones and was the least in heifers.

The occurrence of *S. labiato-papillosa* has been reported from several states of India. In Maharashtra, its prevalence was as high as 80% [182]. In Orissa, 82.3% buffaloes, 19.3% local cattle, 12.5% exotic cattle, 4.2% goats and 2.3% sheep were found infected [122]. In Tamil Nadu, this species was reported to be less common in cattle than in buffaloes.

In Sri Lanka, *S. marshalli* was least common among the species of *Setaria* which, however, were encountered in 75% of cattle slaughtered at Kandy [174].

The data on the prevalence of other species is lacking.

Pathology

Though cattle and buffaloes are natural hosts of *S. digitata*, animals with high density of microfilariae in blood manifest pyrexia, agalactia, diarrhoea, oedema of legs and udder and occasionally accompanied by mortality in untreated cases [170]. Sheep, goats and horses being abnormal hosts, the common symptoms encountered in them were weakness, lameness, incoordination or complete paralysis of hind legs. The disease 'Lumber paralysis' has been studied in different regions of the subcontinent and earlier work on the subject has been systematically reviewed amongst others by Mohiyuddeen [100] and Sarwar [169].

Table 14. Some vectors of filarial parasites in Indian subcontinent

Parasite	Intermediate host
Dirofilaria immitis	Mosquitoes, lice
D. repens	Mosquitoes
Wuchereria bancrofti	Mosquitoes
Brugia malayi	Mosquitoes
B. ceylonensis	Mosquitoes
Parafilaria bovicola	*Musca vitripennis*
Indofilaria pattabiramani	*Tabanus* sp.?, Mosquitoes ?
Cardiofilaria nilesi	Mosquitoes
Dipetalonema reconditum	Mosquitoes
Setaria equina	Masquitoes
S. labiato-papillosa	Mosquitoes
S. digitata	Mosquitoes
S. cervi	Mosquitoes
Stephanofilaria assamensis	M. conducens
S. zaheeri	M. planiceps, M. automnalis

S. digitata is reported to cause fatal peritonitis in goats, whereas nodular lesions on the serosa of buffaloes is caused by both *S. cervi* and *S. labiato-papillosa*. Interestingly, eczematous dermatitis or weeping wounds over the body in buffaloes caused by microfilariae of *S. cervi* has been reported only from Uttar Pradesh (see [30]).

Epidemiology

Ansari [20] observed the highest prevalence of *S. cervi* in buffaloes in the rainy season with maximum density of microfilariae in the blood during August and September and the lowest in winter months. Lumber paralysis in epidemic form, however, occurred between September to December.

In Orissa, the Beetal goats and their graded progeny of all age groups and in Uttar Pradesh, all pure and cross-bred sheep and goats were affected by setariosis *digitata* indicating thereby breed and age susceptibility to it.

Since microfilariae of *S. digitata* had been recovered from milk of affected cattle and buffaloes [170], the contention of Anantaraman [12] that oral transmission of filarial infections in man is not a farfetched idea but a possibility. Therefore, intensive epidemiological study of *S. digitata* in endemic areas is necessary.

Stephanofilariasis

Since Agrawal & Shah [3] have reviewed stephanofilarial dermatitis in India, the topic will not be discussed here in detail. The dermatitis due to *Stephanofilaria assamensis* is of particular interest to eastern India and Bangladesh. It is estimated that loss of millions of *Takas* in foreign currency incurs in Bangladesh from exporting lower quality hides and skins alone where prevalence in cattle in some areas is reported to be as high as 91.5%. Further studies in the region have revealed that the occurrence of disease was significantly higher in certain breeds, aged males, working bullocks beside milch cattle in institutional dairy farms. The lesions were noticed in at least 13 different sites of the body [107].

Onchocerciasis

Lingard in 1905 first recorded *Onchocerca armillata* in equines and cattle in India.

Prevalence

From the available data, aortic onchocerciasis appeared to have a limited geographical distribution in the plains of the subcontinent. In Orissa, 24.5% local and 12.5% exotic cattle, 35.3% buffaloes and 1.4% goats harboured this infection [122]. Its prevalence has been recorded in Maharashtra and Uttar Pradesh (see [30]). In Bangladesh, 54% aortae of slaughtered cattle were found infected with *O. armillata* [137].

In a survey in Uttar Pradesh, 66 to 68% of ligamentum nuchae of buffaloes and cattle, respectively, were found to harbour *O. gutterosa* and in Maharashtra the infection was 83%.The only record of occurrence of *O. cervicalis* in the ligamentum nuchae of Indian horses is that of Srivastava [193].

Shastry [182] observed that 40% of cattle and 66% of buffaloes in Maharashtra had subcutaneous and intradermal nodules due to *O. gibsoni* and *O. cebei*, respectively. Likewise, the prevalence of both *O. gibsoni* and *O. gutterosa* was 10% of 400 superficial skin nodules examined in Bangladesh.

O. volvulus once recovered from a subcutaneous nodule on the forehead of a woman in Assam, India.

Pathology

Epileptiform fits, periodic opthalmia and blindness in cattle are clinical manifestations recorded in aortic onchocerciasis [120].

The aortic lesions due to *O. armillata* have been studied by several workers in different domestic animals including sheep [29], goats [48, 83], cattle [104, 120], buffaloes [198], equines [113] and pigs [142].

Pande & Ahluwalia [110] and Chauhan & Pande [45] are probably the only workers who described the lesion in the ligamentum nuchae of cattle and buffaloes due to *O. gutterosa* and subcutaneous nodules caused by *O. gibsoni* in India.

Epidemiology

Buffalo is believed to be a reservoir host of *O. armillata* [45]. The incidence of aortic onchocersiasis in adult cattle is reported to be as high as 98% to 99% in certain areas whereas 3 to 6 month-old calves in the same area were found free of infection [104, 120]. Evidently, older cattle also serve as reservoirs of this infection. However, onchocerciasis is a rare condition in goats and recovery of adult nematodes from this host has not been reported so far. The vector of this filarid worm still remains unidentified, hence, hampering the experimental studies.

High microfilarial count in the peripheral blood was observed only in Red Sindhi and Haryana bulls which suffered from ocular onchocerciasis [120] suggesting that some breeds of cattle were more susceptible to *O. armillata* than others. Further study, however, is necessary to substantiate this claim.

Concurrent infections of *O. armillata, S. digitata, S. libiato-papillosa* and *Stephanofilaria* sp. in cattle and buffaloes, and *O. armillata* and *Spirocerca lupi* in goats had been reported [118, 122]. It would be of interest to obtain immunological data of these infections in more natural hosts like cattle and buffaloes than in goats which might be unnatural hosts. Epidemiological data on other species of *Onchocerca* are either fragmentary or lacking particularly with regard to those associated with cutaneous nodules. The reported occurrence of *O. armillata* (?) in atheromatous lesions of pigs in India also needs further investigations. The zoonotic potentials of filarial infections of animal origin in man as pointed out by Dissanaike ([55]; see also review [105]) needs renewed emphasis.

Thelaziasis

Thelaziasis has been recently recognised as an important helminthic infection responsible for considerable economic losses from eye infection of zebu cattle and buffaloes in India [112]. Though its prevalence is very high in certain areas, the exact magnitude of the disease produced by it is yet to be ascertained.

Prevalence

Thelazia rhodesii is the only species reported from the subcontinent. Pande et al [112] in a study of 1,112 exenerated eye-balls of cattle and buffaloes found that beside *T. rhodesii* and *T. skrjabini, T. gulosa* was also associated with serious eye lesions in these animals in Uttar Pradesh. This infection has also been reported from Andhra Pradesh and Madhya Pradesh. *T. bubalis*, originally recovered from the eye of a 6 month-old buffalo calf in Tamil Nadu, has been synonymised with *T. gulosa. T. rhodesii* was once recorded in a nilgai [43].

Pathology

The pathogenesis of the disease is not well studied. Conjunctivitis, abscess in the eye lids, corneal opacity, keratitis are common lesions, in addition to complications from secondary bacterial infections ([112]; see [30]). The worms are believed to secrete a toxin which needs experimental confirmation.

Epidemiology

The highest incidence of thelaziasis was found in winter (59.4%) followed by rainy (44.2%) and summer (41.3%) seasons; *T. skrjabini* is the commonest species encountered in Uttar Pradesh [47]. Transmission of the infection is imperfectly known but *Musca domestica* and *M. viscina* are believed to act as intermediate host(s) of *T. gulosa* [74] and an unknown species of *Thelazia* [155].

CONCLUSION

From the foregoing review it is evident that enormous confusion exists on the occurrence, geographical distribution and taxonomic position of a number of helminths reported from Indian subcontinent; some are difficult to resolve. For example, among the three species of *Opisthorchis* known to parasitise domestic and wild carnivores which one of these was exactly reported from man by McConnell in 1876 is difficult to ascertain in retrospect. Considerable controversy also prevails on the taxonomic validity of these three species. Similarly, among several species of *Paragonimus, P. compactus* was first documented in India in a mongoose by Cobbold in 1859. Subsequently, *P. kelliotti, P. edwardsi* and *P. westermanii* were also reported, the last two species were encountered in wild felidae from Himalayan regions that connect Assam geographically to endemic South-East Asia (Fig. 1). Except for *P. westermanii* in India, *P. compactus* and *P. westermanii* in Sri Lanka, presently the other species have become obscure (misidentified!?). *Nanophyetus salmincola*, the cause of 'salmon poisoning' complex elsewhere, was recorded only once in four pure and cross-bred dogs in Uttar Pradesh and this species has not been encountered for the second time from any part of the subcontinent. The life cycles of few important flukes viz. *P. westermanii, Dicrocoelium dendriticum, Eurytrema pancreaticum* and *Ogmocotyle indica* are still unknown; the snail host of the first species once reported to exist in the gangiatic plain!

In the subcontinent, few species of *Diphyllobothrium* recorded at different times from dog, cat, wild felidae and cattle; their taxonomic position, however, remained uncertain though most of these are believed to be *Spirometra* spp. The occurrence of *Raillietina celebensis* in *Mus decumanus, Rattus rattus* and *Bandicota bengalensis* in India and, *R. madagascariensis* in *R. rattus* in Sri Lanka are known. Human infection with these cestodes has never been documented though these parasites have been recorded in man elsewhere.

Among nematodes, life histories of several species have been partially studied; little is known about the biology of the important lungworm, *Protostrongylus rufescens* which is widely prevalent in the plains of Pakistan and northern India. The subcontinent has a number of animal filarioids, the vector(s) of only few species are known.

During the last several years, the list of amphistomes and filarioids in the subcontinent has considerably swollen. To resolve the basic problem of taxonomic confusion, in addition to transmission experiments and studies on host-parasite relationships, ultrastructural investigations and karyotype analyses are helpful. Application of rcent molecular and biochemical techniques including studies of DNA sequence and isoenzyme patterns would remove many such anomalies. Though considerable data on the prevalence of fascioliasis, amphistomiasis, nasal schistosomiasis and ascariasis have been accumulated, there is dearth of information on epidemiology of several important helminths. Within the subcontinent existence of "strains" of helminths is known. Detailed experimental studies are, therefore, necessary to ascertain their effect(s) on different species/breeds of animals from different geographical areas. In India, few systematic studies have been undertaken on zoonotic potentials of helminths of livestock. For the past several years, this country is under constant threat of deforestation, and under the disturbed eco-system, some helminths of wild life might be emerging as zoonoses. There is, therefore, a need to keep vigil on human habitations in proximities of forests and wild sanctuaries.

The subcontinent has dubious distinction of possessing an extremely rich helminth fauna and consequently, is a fertile ground for research on both fundamental and applied helminthology. Indeed, comparative studies on different groups of helminths in areas like evolution, host-specificity, ultrastructure, physiology, biochemistry, genetics, pathology and immunology can provide much needed information.

ACKNOWLEDGEMENTS

I am grateful to Professor S.P. Ahuja, Drs S.K. Dey and S.S. Saini, Assistant Professors (PAU), respectively for helpful discussions and critically going through the different parts of the final manuscript. Professor P.P.S. Chauhan, Mathura and Mr. S.K. Mukherjee, the then Assistant Librarian, Indian Veterinary Research Institute, Izatnagar, have provided numerous inaccessible literature during preparation of this Chapter for which I am indebted to them.

REFERENCES

1. Abdussalam, M. and Sarwar, M.M. Schistosomiasis in Pakistan. Proc. 15th Int. Vet. Congr., Stockholm. Part I, pp. 18–23 (1953).
2. Abraham, J., Pillai, K.M. and Iyer, R.P. Incidence of hydatidosis in animals slaughtered in Kerala. Kerala J. Vet. Sci. **11**: 247–251 (1980).

3. Agrawal, M.C. and Shah, H.L. Stephanofilarial dermatitis in India. Vet. Res. Commun. **8**: 93–102 (1984).
4. Agrawal, M.C. and Shah, H.L. A review of *Schistosoma incognitum* Chandler, 1926, Helm. Abstr. **58**: 239–251 (1989).
5. Ahluwalia, S.S. Studies on the host-parasite relationship and control of *Schistosoma incognitum* infection of the domestic pig with observations on the zoonotic potentialities of the parasite. Ph.D. thesis, Agra University, Agra (1968).
6. Aikat, B.K., Bhusnurmath, S.R., Cadersa, M., Chuttani, P.N. and Mitra, S.K. *Echinococcus multilocularis* infection in India—first case report proved at autopsy. Trans. R. Soc. Trop. Med. **72**: 619–621 (1978).
7. Ali, A., Akhtar, M. and Hayat, C.S. A survey report of the prevalence of hydatid disease in camels slaughtered at Faisalabad abattoir. Pakistan Vet. J. **10**: 115–116 (1990).
8. Alwar, V.S. Amphistomiasis—A review of the literature. Indian vet. J. **25**: 417–424 (1949).
9. Alwar, V.S. A note on the incidence and pathogenesis of paracooperiasis in buffalo calves. Madras Vet. Coll. Ann. **22**: 46–47 (1964).
10. Alwar, V.S. and Lalitha, C.M. A check-list of helminth parasites in the Department of Parasitology, Madras Veternary College (additions since 1954). Indian vet. J. **38**: 142–148 (1961).
11. Anantaraman, M. The epizootiology of nasal schistosomiasis in animals. Proc. Indian Acad. Sci. (Anim. Sci.) **90**: 659–663 (1981).
12. Anantaraman, M. Key note address: All India Seminar on Bionomics, Host- Parasite Relationships and Control of Animal Parasites,. Department of Veterinary Parasitology, HAU, Hissar. p. 17 (1983).
13. Anantaraman, M. and Anantaraman, S. A synopsis of Acanthocephala described from India. Professor Ganapati Shastry Commem. Vol. pp. 1–7 (1970).
14. Anon. A bibliography of research publications and post-graduate theses (1959–1976). Department of Parasitology, College of Veterinary Science & A.H., Mathura, Uttar Pradesh, India (1976).
15. Anon. FAO/UNEP/WHO *guidelines for surveillance, prevention and control of echinococcosis/hydatidosis* (J. Eckert, M.S. Gemmell and E.J.L. Soulsby eds), (VPH/81.28), Geneva (1981).
16. Anon. *Guidelines for surveillance, prevention and control of taeniasis/cysticercosis* (M. Gemmell, Z. Matyas, Z. Pawlowski and E.J.L. Soulsby eds), (VPH/83.49), W.H.O.: Geneva (1983).
17. Anon. Brochure of Veterinary Parasitology (1924–1986). Division of Parasitology, Indian Veterinary Research Institute, Izatnagar (1986).
18. Anon. Animal Health Yearbook, FAO (1988).
19. Ansari, J.A. A review of the nomenclature of *Setaria cervi*. Zool. Ang. **177**: 305–310 (1966).
20. Ansari, J.A. A survey on the frequency and intensity of *Setaria cervi*. Indian J. Anim. Sci. **47**: 115–119 (1977).
21. Ansari, M.Z. and Singh, K.S. On the incidence of *Gaigeria pachyscelis* Railliet and Henry, 1910 in sheep and goats. Indian J. Anim. Sci. **51**: 459–465 (1981).
22. Arora, R.G. Studies on the pathology of sheep and goats liver showing microscopic lesions collected from the slaughter houses. M.V. Sc. thesis, Agra University, Agra (1967).
23. Banerjee, D.P., Barman Roy, A.K. and Sanyal, P.K. Public health significance of *Neoascaris vitulorum* larvae in buffalo milk samples. J. Parasitol. **69**: 1124: (1983).
24. Banerjee, M. and Gupta, P.P. Incidence and pathology of verminous pneumonia in sheep and goats in the Ludhiana area (Punjab). J. Res. Punjab Agric. Univ. **17**: 100–103 (1980).
25. Baylis, H.A. *Fauna of British India including Ceylon and Burma*. Nematoda. Vol. I & II. Taylor & Francis: London (1936, 1939).
26. Bhalerao, G.D. *Helminth Parasites of the Domesticated Animals in India*. Scientific Monograph No. 6, ICAR: New Delhi (1935).
27. Bhalerao, G.D. Interesting mode of the life cycle of the lungworm, *Varestrongylus pneumonicus* Bhalerao, 1932. Curr. Sci. **14**: 106–107 (1945).
28. Bhat, T.K., Dhar, D.N., Bansal, G.C. and Sharma, R.L. Blood biochemical changes in lambs infected with normal and gamma irradiated third stage larvae of *Dictyocaulus filaria*. J. Nuclear Agric. Biol. **13**: 77–79 (1984).
29. Bhatia, B.B. *Onchocerca armillata* Railliet and Henry 1909—A study of the infection in Indian sheep with remarks on its bovine hosts. Indian vet. J. **37**: 394–397 (1960).

30. Bhatia, B.B. and Chauhan, P.P.S. *Parasitic diseases of buffaloes in India*. Cosmo Publications: New Delhi (1984).
31. Bhatia, B.B. and Pande, B.P. On the common helminthic infections of the large intestine in Indian sheep with remarks on the two of the pathogenic species. Indian J. Helminthol. **13**: 1–22 (1961).
32. Bhatia, B.B., Upadhaya, D.S. and Juyal, P.D. Epidemiology of *Fasciola gigantica* in buffaloes, goats and sheep in Tarai region of Uttar Pradesh. J. Vet. Parasitol. **3**: 25–29 (1989).
33. Bhattacharyya, H.M., Sinha, P.K. and Sarkar, P.B. Studies on incidence of *Metastrongylus salmi* in West Bengal with observations on its life-cycle. Indian vet. J. **48**: 993–996 (1971).
34. Bhattacharjee, M.L. and Halder, B.R. The occurrence of *Fasciola gigantica* in the liver of an Indian rhinoceros (*R. unicornis*). British Vet. J. **127**: vii (1971).
35. Bhuyan, M.A. A survey of helminths parasitising the livers of domesticated ruminants in East Pakistan. M.Sc. (Vet. Sci.) thesis, East Pakistan, Agric. Univ., Mymensingh, Bangladesh (1970).
36. Boev, S.N. *Protostrongylids*. Oxonian Press: New Delhi (1984).
37. Bryant, C. and Flockhart, H.A. Biochemical strain variation in parasitic helminths. Adv. Parasitol. **25**: 275–319 (1986).
38. Cameron, T.W.M. The incidence and diagnosis of hydatid cysts in Canada. *Echinococcus granulosus* var. *canadensis*. Parasitologia **2**: 381–390 (1960).
39. Chakrabarti, A. and Choudhuri, M.N. Studies on canine filariasis in West Bengal. Indian J. Anim. Hlth. **22**:151–155 (1983).
40. Chakraborty, A., Gogoi, A.R. and Dutta, B.M. Occurrence of *Cysticercus ovis* in tunica vaginalis and urinary bladder of a goat in Assam. Indian J. Parasitol. **8**: 215–216 (1984).
41. Chandra, R. On the occurrence of *Fasciola gigantica* infection in man in India. Seventh National (Indian) Congr. Parasitol., Raipur. Abstr. B-18 (1986).
42. Chandramukhi, A., Muralidhar, K., Katti, B.V., Ravi Kumar, Shankar, S.K., Suryanaryana, V., Ravi, V., Rao, T.V. and Gokul, B.N. Neurocysticercosis—A zoonotic disease and its morbidity on man. Proc. 2nd. National (Indian) Congr. Parasitol., Bangalore. Abstr., lead paper. p. 1 (1988).
43. Chauhan, P.P.S. On the taxonomy and biology of the important nematode parasites of water-buffalo (*Bubalus bubalis*) with remarks on the related forms in some of our domestic and wild ruminants. Agra Univ. J. Res. Sci. (Part I). **27**: 13–18 (1978).
44. Chauhan, P.P.S., Bhatia, B.B. and Pande, B.P. Incidence of gastro-intestinal nematodes in buffalo and cow-calves at the four state livestock farms in Uttar Pradesh. Indian J. Anim. Sci. **43**: 216–219 (1973).
45. Chauhan, P.P.S. and Pande, B.P. Onchocercal lesions in aorta and ligamentum nuchae of buffaloes and bullocks: A. histological study. Indian J. Anim Sci. **42**: 809–813 (1972).
46. Chauhan, P.P.S. and Pande, B.P. Studies on paracooperid infection in buffalo- calves. III. Nodules and their histopathology. Indian J. Anim. Sci. **42**: 930–934 (1972).
47. Chauhan P.P.S. and Pande, B.P. Observations on the incidence of *Thelazia* infection in cattle and buffaloes in Uttar Pradesh with remarks on its significance. Indian J. Anim. Sci. **43**: 300–305 (1973).
48. Chowdhury, N. and Chakraborty, R.L. On concurrent and early aortic onchocerciasis and spirocerciasis in domestic goats (*Capra hircus*). Z. Parasitenk unde **42**: 207–212 (1973).
49. Damodaran, S. and Sundararaj, A. A survey of calf mortality in Madras State. Indian vet. J. **51**: 359–364 (1974).
50. Deka, D.K., Srivastava, G.C. and Chhabra, R.C. Fertility of hydatid cysts from ruminants. Indian J. Parasitol. **6**: 243–244 (1982).
51. Deo, P.G. *Roundworms of Poultry*. Animal Husbandry Series No. 3. ICAR: New Delhi (1964).
52. Dhanda, M.R. and Khera, S.S. Mortality in young calves in India. Indian vet. J. **34**: 384–403 (1957).
53. Dhar, D.N., Sharma, R.L. and Bansal, G.C. Ascarid infection in sheep in the Kashmir valley. Indian J. Parasitol. **7**: 97–99 (1983).
54. Dissanaike, A.S. Observations on *Echinococcus* infection in Ceylon. In: *First regional symposium on scientific knowledge of tropical parasites*. University of Singapore, pp. 223–240 (1962).
55. Dissanaike, A.S. Zoonotic aspects of filarial infections in man. Bull. Wld Hlth. Org. **57**: 349–357 (1979).
56. D'souza, P., Jagannath, M.S. and Rahman, S.A. Studies on the prevalence of subclinical gastrointestinal

parasitism in dairy cattle and buffaloes. First National (Indian) Congr. Vet. Parasitol., Izatnagar. Abstr. (Sect. B) pp. 13–28 (1987).

57. Dutt, S.C. *Paramphistomes and paramphistomiasis of domestic ruminants in India*. pp. 162. Joint Director, Communication Centre, PAU, Ludhiana (1980).
58. Dutt. S.C. and Srivastava, H.D. Studies on *Schistosoma nasale* Rao 1933. II. Molluscan and mammalian hosts of the blood fluke. Indian J. Vet. Sci. **38**: 210–216 (1968).
59. Dutt, S.C. and Srivastava, H.D. The life history of *Gastrodiscoides hominis* (Lewis and McConnell, 1876) Leiper, 1913—The amphistome parasite of man and pig. J. Helminthol. **46**: 35–46 (1972).
60. Farooqi, H.K. Ahmad, K. and Shameem, M. Biocoenotically-linked varying patterns of livestock hydatidosis in the Indian region. Proc. 7th. National (Indian) Congr. Parasitol, Raipur. Abstr. p. 88 (1986).
61. Fernando, S.T. The life cycle of *Mecistocirrus digitatus*, a trichostrongyloid parasite of ruminants. J. Parasitol. **51**: 156–163 (1965).
62. Flisser, A., Espinoza, B., Tovar, A., Plancarte, A. and Correa, D. Host-Parasite relationship in cysticercosis: immunologic study in different compartments of the host. Vet. Parasitol. **20**: 95–102 (1986).
63. Flisser, A., Pe'rez-Monfort, R. and Larralde, C.: The immunology of human and animal cysticercosis: a review. Bull. Wld. Hlth. Org. **57**: 839–856 (1979).
64. Flisser, A., Willms, K., Laclette, J.P., Larralde, C., Ridaura, C., Beltran, F. and Vogt. M.W. (eds). *Cysticercosis: present state of knowledge and perspectives*. Academic Press : New York (1982).
65. Fransen, J.L.A., Vercruysse, J., De Bont, J., Southgate, V.R. and Rollinson, D. Pathology of *Schistosoma nasale* and *Schistosoma spindale* in cattle. Proc. Joint Spring Meeting. British, Netherlands, and Belgian Societies of Parasitology with Belgian Society of Protozoology. University of Southampton, UK. Abstr. 39 (1989).
66. Gaur, S.N.S. and Dutt, S.C. Tissue responses to *Mecistocirrus digitatus* infection in cattle. Phillipine J. Vet. Med. **12**: 64–68 (1973).
67. Gaur, S.N.S., Sethi, M.S., Tewari, H.C. and Om Prakash. A note on the prevalence of helminth parasites in wild and zoo animals in Uttar Pradesh. Indian J. Anim. Sci. **49**: 159–161 (1979).
68. Gill, H.S. and Rao, B.V. On the biology and morphology of *Echinococcus granulosus* (Batsch, 1786) of buffalo-dog origin. Parasitology **57**: 695–704 (1967).
69. Gill, H.S. and Rao, B.V. Incidence and fertility rate of hydatid cysts in Indian Buffalo (*Bos bubalis*). Bull. Off. Int. Epiz. **67**: 989–997 (1967).
70. Gill J.S. Singh, N. and Kwatra, M.S. An outbreak of fascioliasis in goats in the Punjab State and its treatment. Indian J. Parasitol. **7**: 215–216 (1983).
71. Gupta, N.K. *Amphistomes—Systematics & Biology*. Oxford & IBH: New Delhi (1990).
72. Gupta, P.P. and Chowdhury, N. Cerebral coenuriasis in a buffalo. Indian vet. J. **62**: 613–614 (1985).
73. Gupta, R.P. and Paul, J.C. An unusual outbreak of fascioliasis (*Fasciola gigantica*) on a farm. Indian J. Anim. Res. **21**: 41–42 (1987).
74. Gupta, V.P. *Musca domestica*, an intermediate host of two spiruroid nematode parasites of livestock. Orissa Vet. J. **5**: 147–150 (1970).
75. Hoeppli, R. The knowledge of parasites and parasitic infections from ancient times to the 17th century. Exp. Parasitol. **5**: 398–419 (1956).
76. Horak, I.G. Paramphistomiasis of domestic ruminants. Adv. Parasitol. **9**: 33–72 (1971).
77. Irsadullah, M. and Nizami, W.A. *In vivo* developmental pattern of *Echinococcus granulosus* protoscolices from buffalo liver and lung origin in experimental puppies. Proc. 9th. National (Indian) Congr. Parasitol., Ujjain. Abstr. No. 56 (1989).
78. Irshadullah M., Nizami, W.A. and Macpherson, C.N.L. Observations on the suitability and importance of the domestic intermediate hosts of *Echinococcus granulosus* in Uttar Pradesh. J. Helminthol. **63**: 39–45 (1989).
79. Iqbal, Zafar. Studies on epidemiology and control of hydatid disease. Ph.D. thesis. University of Faisalabad, Pakistan (1991) (unpublished).
80. Jain, S.K. and Bandyopadhyay, A.C. A note on the histopathological observations in the lungs of buffalo-calves associated with lungworms (*Dictyocaulus viviparus*) infestation. Indian vet. J. **44**: 661–664 (1967).
81. Katiyar, J.C. *Dictyocaulus arnfieldi* in *Equus asinus*. A histological study. Indian vet. J. **41**: 24–25 (1964).

82. Katiyar, R.D., Srivastava, V.K., Khera, R.C. and Lepcha, N.T. Occurrence of fascioliasis in the Livestock in Sikkim. Livestock Adviser **8**: 37–39 (1983).
83. Kaul, P.L. and Prasad, M.C. Onchocercal aortopathy in goats. Indian, J. Vet. Path. **7**: 40–44 (1983).
84. Kendall, S.B. Fascioliasis in Pakistan. Ann. Trop. Med. Parasitol. **48**: 307–313 (1954).
85. Kendall, S.B. and Parfitt, J.W. Studies on the susceptibility of some species of *Lymnaea* to infection with *Fasciola gigantica* and *F. hepatica*. Ann. Trop. Med. Parasitol. **53**: 220–227 (1959).
86. Khuroo, M.S., Datta, D.V., Koshy, A., Mitra, S.K. and Chuttani, P.N. Alveolar hydatid disease of the liver with Budd-chiari syndrome. Postgrad. Med. J. **56**: 197–201 (1980).
87. Kumar, M., Maru, A. and Pachauri, S.P. Changes in blood cellular components, serum protein concentrations and serum enzyme activities in buffaloes infested with *Fasciola gigantica*. Res. Vet. Sci. **33**: 260–261 (1982).
88. Kumar, M., Pathak, K.M.L. and Pachauri, S.P. Clinicopathological studies on naturally-occurring bovine fascioliasis in India. British Vet. J. **138**: 241–246 (1982).
89. Kumar, V. and Burbure, G. de. Schistosomiasis of animals and man in Asia. Helm. Abstr. **55**: 469–480 (1986).
90. Levine, N.D. Weather, climate and bionomics of nematode larvae. Adv. Vet. Sci. **8**: 215–261 (1963).
91. Levine, N.D. *Nematode parasites of domestic animals and of man*. Burgess Pub. Co., Minneapolis: U.S.A. (1968).
92. Mathur, P.B. and Uppal, R.K. Studies on the economic importance of gastrointestinal helminthiasis in sheep in semi-arid areas of Rajasthan. First Asian Congr. Parasitol., Bombay Abstr. p. 157 (1978).
93. McManus, D.P. and Smyth, J.D. Hydatidosis: changing concepts in epidemiology and speciation. Parasitol. Today **2**: 163–168 (1986).
94. Mia, S., Dewan, M.L. and Chowdhury, M.H.A. The route of infection of buffalo calves by *Toxocara* (Neoascaris) *vitulorum*. Trop. Anim. Hlth. Prod. **7**: 153–156 (1975).
95. Minnett, F..C. Mortality in sheep and goats in India. Indian J. Vet. Sci. **20**: 69–103 (1950).
96. Misra, P.K., Patnaik, G.M. and Misra, S.C. Endemic form of fascioleasis caused by *Fasciola gigantica* in the costal belt of Orissa. Vet. J. **10**: 38–41 (1975).
97. Misra, S.C. and Ruprah, N.S.: Incidence of Helminths in goats at Hissar. J. Res. PAU. **5**: 279–284 (1968).
98. Moghe, M.A. Results of a survey of the nature and incidence of helminth infection in cattle, goats and sheep in the central provinces and Berar and Central India. Indian J. Vet. Sci. **12**: 220–230 (1945).
99. Mohan, R.N. Diseases and parasites of buffaloes. Vet. Bull. **38**: 735–756 (1968).
100. Mohiyuddeen, S. Enzootic bovine paraplegia in some maland tracts (Hilly and heavy rainfall region) of Mysore State with particular reference to cerebrospinal nematodiasis as its probable cause. Indian J. Vet. Sci. **26**: 1–19 (1956).
101. Murleedharan, K. Paracooperiasis in buffalo calves. Livestock Adviser **9**: 55–58 (1984).
102. Murleedharan, K., Kumar, S.P., Hegde, K.S. and Alwar, V.S. Studies on epizootiology of nasal schistosomiasis of bovines. I. Prevalence and incidence of infection. Mysore. J. Agric. Sci. **10**: 105–117 (1976).
103. Murleedharan, K., Ziauddin, S. and Seshadri, S.J. A rare case of fatal colitis in a calf infected with *Oesophagostomum radiatum* Indian J. Parasitol. **8**: 287–288 (1984).
104. Nasseri, A.A., Sastry, G.N.S. and Rajsekharan, D. Incidence and Pathology of bovine onchocerciasis. Indian J. Vet. Path. **71**: 35–39 (1983).
105. Nelson, G.S. The pathology of filarial infections. Helm. Abstr. **35**: 311–336 (1968).
106. Nooruddin, M., Baki, M.A. and Mondal M.M.H.: Pathology of *Fasciola gigantica* infection in the liver of goats. Livestock Adviser **12**: 43–45 (1987).
107. Nooruddin, M., Hoque, M.F. and Islam, K.S. Epidemiological study of bovine stephanofilariasis in Bangladesh. Int. J. Trop. Agri. **4**: 96–98 (1986).
108. Nooruddin, M., Samad, M.A. and Rahman, A. A note on certain haematological and biochemical changes of Black Bengal goats infected with *Fasciola gigantica*. Haryana Vet. **21**: 133–136 (1982).
109. Pachalag, S.V., Mathur, P.B. and Patnaik, B. Intussusception in adult exotic sheep due to nodules of *Oesophagostomum columbianum* Curtice (1890). Indian J. Sci. Indust. **3**: 155–158 (1969).
110. Pande, B.P. and Ahluwalia, S.S. Observations on *Onchocerca* infection of the ligamentum nuchae in Indian buffalo. British Vet. J. **120**: 132–137 (1964).
111. Pande, B.P. and Chauhan, P.P.S. Abomasal lesion in a trichostrongyloid infection of buffalo-calves (below two years) and the morphology and taxonomy of the nematode. Indian J. Anim. Sci. **39**: 79–88 (1969).

112. Pande, B.P., Chauhan, P.P.S., Bhatia, B.B. and Arora, G.S. Studies on bubaline eye-worms with reference to the species composition of *Thelazia* and their pathogenic significance. Indian J. Vet. Sci. **40**: 330–345 (1970).
113. Pande, B.P., Rai, P. and Bhatia, B.B. Nematode affecting the aorta in Indian caprine and equine hosts. J. Parasitol. **47**: 951–952 (1961).
114. Pandey, V.S. Observations on the morphology and biology of *Echinococcus granulosus* (Batsch, 1786) of goat-dog origin. J. Helminthol. **46**: 219–233 (1972).
115. Pandit, B.A., Mir, A.S., Nasreen, S. and Khan, A.A. The epidemiology of ovine fascioliasis in Jammu and Kashmir State. J. Vet. Parasitol. **3**: 13–15 (1989).
116. Paramanathan, D.C. Some observations on brood capsules of hydatid cysts from local animals. Ceylon J. med. Sci. **10**: 57–59 (1961).
117. Parija, S.C. (ed). *Review of Parasitic Zoonoses*. AITBS Publishers: New Delhi (1990).
118. Parry, K.M.A. and Sellors, H.T. *Chest Diseases*. Butterworths: London (1963).
119. Pathak, K.M.L. and Gaur, S.N.S. Surveillance of human taeniasis and porcine cysticercosis in certain parts of Uttar Pradesh. Proc. 2nd. National (Indian) Congr. Vet. Parasitol., Bangalore. Abstr. p. 5 (1988).
120. Patnaik, B. Onchocerciasis due to *O. armillata* in cattle in Orissa. J. Helminthol. **36**: 313–326 (1962).
121. Patnaik, B., Mathur, P.B. and Pachalag, S.V. Gastrointestinal helminthiasis in sheep in semi-arid zone with particular reference to Rambouillet breed. Guj. Vet. **7**: 38–43 (1973).
122. Patnaik, M.M. Studies on certain aspects of filarial infections of livestock and poultry including their wild counterparts in Orissa with special reference to their prevalence, histapathology, diagnosis and chemotherapy. Ph.D. thesis, Agra University, Agra (1975) (unpublished).
123. Patnaik, M.M. and Pande, B.P. A note on parafilariasis in buffalo, *Bos (Bubalus) bubalis*. J. Helminthol. **37**: 343–348 (1963).
124. Patnaik, M.M. and Pande, B.P. Notes on the helminthic infestations encountered in one month old buffalo calves. Indian vet. J. **40**: 128–133 (1963).
125. Patnaik, M.M. and Ray, S.K. Studies on geographical distribution and ecology of *Lymnaea auricularia* var *rufescens*, the intermediate host of *Fasciola gigantica* in Orissa. Indian J. Vet. Sci. **38**: 484–508 (1968).
126. Pattanayak, G.M. and Gupta, P.P. Parasitic infections in pigs. J. Res. PAU. **18**: 457–461 (1981).
127. Pawlowski, Z. and Schultz, M.G. Taeniasis and cysticercosis (*Taenia saginata*). Adv. Parasitol. **10**: 269–343 (1972).
128. Peter, C.T. Studies on the life history of *Gastrodiscus secundus*, Looss, 1907 an amphistomatous parasite of equines in India. Indian J. Helminthol. **12**: 18–50 (1960).
129. Peter, C.T. and Srivastava, H.D. Studies on the life history of *Pseudodiscus collinsi* (Cobbold, 1875) Sonsino, 1895, an amphistomatous parasite of equines and elephants in India. Indian J. Helminthol. **12**: 1–17 (1960).
130. Poynter, D. and Selway, S. Diseases caused by lungworms. Helm. Abstr. **35**: 105–127 (1966).
131. Poynter, D. Parasitic bronchitis. Adv. Parasitol. **6**: 349–359 (1968).
132. Prasad, M.C. and Bhalla, N.P. Pathology of aortic elaeophoriasis in Indian water buffaloes (*Bubalus bublis*). Indian vet. J. **54**: 97–101 (1977).
133. Prasad, M.C., Yadav, R.N. and Gowal, K.N. Bovine dictyocaulosis in Arunachal Pradesh. Indian J. Vet. Path. **3**: 48–49 (1979).
134. Prater, S.H. *The book of Indian animals*. Third (revised) edn., Bombay Natural History Society. Leaders Press: Bombay (1971).
135. Rahman, A., Ahmed, M.U. and Mia, A.S. Diseases of goats diagnosed in slughter houses in Bangladesh. Trop. Anim. Hlth. Prod. **7**: 164: (1975).
136. Rahman, A., Ahmed, M.U. and Mia, A.S. Studies on diseases of goats in Bangladesh: Mortality of goats under farm and rural conditions. Trop. Anim. Hlth. Prod. **8**: 90: (1976).
137. Rahman, M.H., Huq, S. and Dewan, M.L. A study of incidence of onchocercal parasites in cattle in Bangladesh. Bangladesh Vet. J. **11**: 63–65 (1977).
138. Rahman, M.H. and Mandal, M.M.H. (eds). Abstracts of research in parasitology in Bangladesh (1963–1985). Bangladesh Livestock Research Institute, Savar, Dhaka (1986).
139. Rai, M.K. Studies on helminth parasites of yak (*Bos poephagus*) in Sikkim. M. V. Sc. thesis, Birsa Agric. Univ., Ranchi. Abstr.: J. Vet. Parasitol. **2**: 77–78 (1988).

140. Raja, E.E. Parasitic infection in rodents especially those communicable to man and animals. Ph.D. thesis, Madras University. Abstr.: Cheiron **3**: 173–174 (1974).
141. Rajamohan, K., Kalyansundram, R. and Peter, C.T. A preliminary note on ascariasis in goats. Kerala J. Vet. Sci. **1**: 1 : (1970).
142. Ramanujachari, G. and Alwar, V.S. Indian pig (*Sus scrofa domestica*) as a host for *Onchocerca* sp. and *Enterobius vermicularis*. Indian vet. J. **29**: 329: (1993).
143. Ramanujachari, G. and Alwar, V.S. A check-list of parasites (Classes: Trematoda, Cestoda and Nematoda) in the Department of Parasitology, Madras Veterinary College (additions since 1947). Indian vet. J. **31**: 46–56 (1954).
144. Ramaswamy, K., Singh, B.P. and Arora, B.M. Occurrence of *Muellerius* sp. in spotted deer, *Axis* in India. Indian J. Parasitol. **9**: 187: (1985).
145. Ramgopal, Kulkarni, H.G. and Arya, S.N. Mortality in bovine in rural areas of Gujarat. Indian J. Anim. Res. **16**: 80–84 (1982).
146. Ranatunge, P. Common parasitic. problems in sheep and goat husbandry. Ceylon Vet. J. **19**: 126–127 (1971).
147. Rao, A.T. and Acharjyo, L.N. Histopathological changes in the liver of a spotted deer (*Axis axis*) infected with *Fasciola gigantica*. Indian vet. J. **46**: 36–38 (1969).
148. Rao, A.T. and Acharjyo, L.N. Histopathological changes in some of the organs in heart worm infection in an Indian fox (*Vulpes bengalensis*). Indian vet. J. **48**: 342–344 (1971).
149. Rao, A.T. and Acharjyo, L.N. Diagnosis and classification of common diseases of captive animals at Nandan Kanan Zoo in Orissa (India). Indian J. Anim. Hlth **23**: 147–152 (1984).
150. Rao, A.T., Nayak, B.C. and Acharjyo, L.N. Histopathology of intestinal lesions due to *Echinococcus granulosus* in an Indian wolf (*Canis lupus*). Indian vet. J. **50**: 199–200 (1973).
151. Rao, B.V. Studies on helminth parasites of carnivorous mammals. M.Sc. thesis, Faculty of Veterinary Science, Madras University (1958).
152. Rao, B.V. and Anantaraman, M. Observations on the biology and development of some taeniid cestodes. Indian J. Helminthol. **18**: 161–171 (1966).
153. Rao, J.R., Sikdar, A., Sharma Deorani, V.P. and Jha, S.K. A report on the outbreak of fascioliasis in buffaloes in Great Nicobar Islands. Cheiron **14**: 162–163 (1985).
154. Rao, P.V.R. and Indiradevi, T. Nasal schistosomiasis in buffaloes. Indian J. Anim. Hlth. **10**: 185–188 (1971).
155. Rao, S.R. A short note on *Musca viscina* as the possible vector of *Thelazia* sp. Indian vet. J. **25**: 399–401 (1949).
156. Rausch, R.L. Life-cycle pattern and geographic distribution of *Echinococcus species*. In: *Biology of Echinococcus and hydatid disease*. (R.C.A. Thompson ed), pp. 44–48. George Allen & Unwin: London (1986).
157. Reddy, C.R.R.M., Narasiah, I.L., Parvathi, G. and Rao, M.S. Epidemiology of hydatid disease in Kurnool. Indian J. Med. Res. **56**: 1205–1220 (1968).
158. Rickard, M.D. and Williams, J.F. Hydatidosis/Cysticercosis: Immune mechanisms and immunization against infection. Adv. Parasitol. **21**: 229–296 (1982).
159. Roy, B.C., Chakraborty, M., Das, M.M. and Chatterjee, B.P. World incidence of hydatid disease in general and pulmonary hydatid disease in particular with reference to India. J. Indian Med. Assoc. **55**: 212–217 (1970).
160. Roy, B. and Tandon, V. Metacestodiasis in North East India: A study on the prevalence of hydatidosis and cysticercosis in Mizoram, Nagaland and Assam. Indian J. Anim. Hlth. 28: 5–10 (1989).
161. Sabri, M.A., Khan, M.Z., Chaudhry, A.H. and Samad, H.A. Comparative pathology of natural fascioliasis in buffalo and cattle. Pakistan Vet. J. **2**: 27–28 (1982).
162. Sahai, B.N., Singh, S.P. and Verma, A.K. A new microfilaria from a buffalo-bull, tentatively assigned to *Parafilaria sahaii*, Srivastava and Dutt, 1959 (a preliminary report). Zool. Anz. **191**: 205–209 (1973).
163. Saleque, A. On experimental taeniasis-cysticercosis due to *Taenia solium* and *T. saginata*. M.V.Sc. thesis, Assam Agric. Univ., Guwahati (1978) (unpublished).
164. Saleque, A., Chowdhury, N., Iyer, P.K.R. and Baruah, G. Induced *Taenia solium* cysticercosis in rhesus monkeys (*Macaca mulatta*): a clinico-pathological study. Ann. Trop. Med. Parasitol. **82**: 103–105 (1988).
165. Sanyal, P.K. Preliminary studies on hypobiosis—a diapause phenomenon of *Haemonchus contortus*. First National (Indian) Congr. Vet. Parasitol., Izatnagar. Abstr. B-10 (1987).

166. Sarkar, P., Basak, D.K. and Bhattacharjee, H.M. Pathology of *Dirofilaria immitis* infection in dogs. Indian vet. J. **53**: 55–57 (1976).
167. Sarwar, M.M. A new lungworm, *Pneumostrongylus ovis*. n. sp. from sheep in Pakistan. Proc. Pak Sci. Conf. 5th Med. Vet. Sci. Abstr. Sect. 15 (1953).
168. Sarwar, M.M. Studies on some trichostrongylids of ruminants from Indo-Pakistan subcontinent. Biologia **2**: 145–215 (1956).
169. Sarwar, M.M. Cerebrospinal setariasis. British Vet. J. **113**: 295–296 (1957).
170. Sastry, G.J.S., Rao, V.P., Narasimham, M.V. V.L. and Rao, P.K. Animal filariasis in East Godavary district, Andhra Pradesh. Livestock Adviser **10**: 56–60 (1985).
171. Sathianesan, V. and Peter, C.T. Studies on the incidence of gastro-intestinal nematodes of goats in Kerala. Kerala J. Vet. Sci. **2**: 119–122 (1971).
172. Schwabe, C.W. Current status of hydatid disease: a zoonosis of increasing importance. In: *The Biology of Echinococcus and hydatid disease*. (R.C.A. Thompson ed), George Allen & Unwin: London (1986).
173. Seneviratna, P. Helminth infections of some domesticated animals in Ceylon. Ceylon Vet. Ser. Rep. **2**: 161–166 (1954).
174. Seneviratna, P. A check list of helminths in the Department of Veterinary Pathology, University of Ceylon, Peradeniya. Ceylon Vet. J. **3**: 32–37 (1955).
175. Seneviratna, R.D. and Kulasiri, C. Arteriosclerosis of the aorta and pulmonary artery of an ox due to *Cysticercus bovis*. British Vet. J. **111**: 387–390 (1955).
176. Sewell, R.B.S. Cercariae indicae, Indian J. Med. Res. **10**: (Suppl.) 1–370(1922).
177. Shah, H.L. and Agrawal, M.C. Schistosomiasis. In: *Review of Parasitic Zoonoses*. (S.C. Parija ed), AITBS Publishers: New Delhi (1991).
178. Shanmugalingam, K. Neoascariasis in buffalo calves in Ceylon. Ceylon Vet. J. **4**: 46–50 (1956).
179. Shanmugalingam, K. and Seneviratna, P.: A note on the preliminary investigation into the incidence of parasitism in buffaloes in the dry zone. Ceylon Vet. Ser. Rep. **2**: 151–156 (1954).
180. Sharma, R.L., Bhat, T.K. and Dhar, D.N. Control of sheep lungworm in India. Parasitol. Today **4**: 33–36 (1988).
181. Sharma, T.D. and Chitkara, N.L. Hydatid disease in Amritsar: A study of potential risk. Indian J. Med. Res. **51**: 1015–1018 (1963).
182. Shastri, U.V. A preliminary note on filarial infections in cattle (*Bos indicus*) and buffaloes (*Bubalus bubalis*). Indian J. Anim. Hlth. **12**: 193–195 (1973).
183. Sinclair, K.B. Pathogenesis of *Fasciola* and other liver flukes. Helm. Abstr. **36**: 115–134 (1967).
184. Singh, B.P., Sharma Deorani, V.P. and Srivastava, V.K. Exceptionally high hydatid infection in liver of an Indian buffalo and its epidemiological significance. Seventh National (Indian) Congr. Parasitol., Raipur. Abstr. p. 62 (1986).
185. Singh, K.S. and Srivastava, H.D. *Diagnosis and treatmnent of helminthic infections*. ICAR: New Delhi (1977).
186. Singh, N. Chemotherapy against experimental *Fasciola gigantica* infection in sheep. M.V.Sc. thesis, Haryana Agric. Univ.,Hissar (1989) (unpublished).
187. Smyth, J.D. The biology of the hydatid organism: literature review 1964–1968. Adv. Parasitol. **7**: 327–347 (1969).
188. Smyth, J.D. and McManus, D.P. *The physiology and Biochemistry of Cestodes*. Cambridge University Press: Cambridge (1989).
189. Sokolic, A., Deorani, V.P.S., Tewari, H.C. and Rao, B.V. *Dictyocaulus filaria* in India. Present prospects for using radiation-attenuated vaccine in its control. In: *Isotopes and Radiation in Parasitology*. pp. 125–128. IAEA, Vienna (1970).
190. Sood, M.L. Haemonchus in India. Parasitology **83**: 639–650 (1981).
191. Southwell, T. *Fauna of British India including Ceylon and Burma*. Cestoda: Vol. I & II. Taylor & Francis: London (1930).
192. Srivastava, A.K. and Sharma, D.N. Studies on the occurrence, clinical features and patho-morphological aspects of ascariasis in buffalo calves. Vet. Res. J. **4**: 160–162 (1981).
193. Srivastava, H.D. The occurrence of an unrecorded filarid nematode, *Onchocerca cervicalis* Raillliet and Henry, 1910, in the ligamentum nuche of horses in India. Proc. 24th. Indian Sci. Congr., Hyderabad (Sect. VIII). Abstr. p. 399 (1937).

194. Srivastava, H.D. A survey of the incidence of helminth infection in India at the Imperial Veterinary Research Institute, Izatnagar. Indian J. Vet. Sci. **15**: 146–148 (1945).
195. Srivastava, H.D. Helminth parasites of domestic animals. Presidential address 41st Annual Session. Indian Nat. Acad. Sci. (Biol. Sci.). pp. 35–44 (1971).
196. Srivastava, H.D. and Dutt, S.C. Parafilariasis in buffaloes caused by *Parafilaria Sahaii* n. sp. Curr. Sci. **28**: 128–129 (1959).
197. Srivastava, S.C. *Neoascaris vitulorum* (Goeze, 1782) Travassos 1907, in intestinal perforation with its localisation in liver of buffalo calves. Indian vet. J. **40**: 758–762 (1963).
198. Srivastava, S.C. and Pande, B.P. Occurrence of aortic onchocerciasis and spirocerciasis in buffalo-calves with a note on the reports in other domestic animals. Indian J. Vet. Sci. **34**: 222–231 (1964).
199. Subramaniam, T., D'souza, B.A. and Victor, D.A. Bronchopneumonia in baby pigs due to *Metastrongylus apri*. Indian vet. J. **44**: 121–127 (1967).
200. Tewari, H.C., Pandey, M.N. and Basnet, B.M. A note on the incidence of lungworm infections in Nepal. Indian J. Anim. Sci. **45**: 694–695 (1975).
201. Thapar, G.S. Systematic survey of helminth parasites of domesticated animals in India. Indian J. Vet. Sci. **26**: 211–271 (1956).
202. Thapar, G.S. and Tandon, R.S. On the life-history of liver-fluke, *Fasciola gigantica* Cobbold, 1855. Indian J. Helminthol. **4**: 1–36 (1952).
203. Thomas, P.C. and Peter, C.T. Studies on the common nematodes in pigs (*Sus scrofa domestica*). Indian vet. J. **52**: 668–669 (1975).
204. Thompson, R.C.A. (ed). *The biology of Echinococcus and hydatid disease*. George Allen & Unwin: London (1988).
205. Thompson, R.C.A. and Lymbery, A.J. The nature, extent and significance of variation within the genus *Echinococcus*. Adv. Parasitol. **27**: 209–258 (1988).
206. Tripathi, H.N. and Srivastva, H.D. Amphistomes of ruminants. 2. Life history of two species of *Palamphistomum* Srivastava and Tripathi (1980) of sheep, goats and buffaloes. Indian J. Anim. Sci. **57**: 1403–1056 (1987).
207. Tripathi, J.C. Seasonal variations in egg output of gastro-intestinal nematode of goats. I. Total egg counts. Indian J. Vet. Sci. **36**: 203–210 (1966).
208. Tripathi, J.C. Seasonal incidence and bionomics of some common gastro-intestinal nematode larvae of sheep and goats. Agra Univ. J. Res. Sci. **16**: 179–181 (1967).
209. Tripathi, J.C. Seasonal variations in egg output of gastro-intestinal nematodes of goats. III. Recovery of infective larvae. Indian J. Vet. Sci. **40**: 46–60 (1970).
210. Van Gorp, K., Mangelschots, M. and Brandt, J. *Toxocara vitulorum*: A possible agent of larva migrants in humans? In: *Helminth Zoonoses* (S. Geerts, V. Kumar, and J. Brandt eds), Martinus Nijhoff Publishers: Dordrencht. pp. 159–166 (1987).
211. Varma, A.K. On *Fasciola indica* n. sp. with some observations on *Fasciola hepatica*. and *Fasciola gigantica* J. Helminthol. **27**: 185–198 (1953).
212. Varma, T.K. and Ahluwalia, S.S. Some observations on the biology and development of *Cysticercus cellulosae* in lamb, kid and laboratory animals. Livestock Adviser **11**: 47–52 (1986).
213. Varma, T.K. and Malviya, H.C. Unusal occurrence of secondary echinococcosis in goat. Indian J. Parasitol. **10**: 235–236 (1986).
214. Vercruysse, J., De Bont, J., Van Aken, D., De Sit, I., Southgate, V.R. and Rollingson, D. Cattle schistosomiasis in Sri Lanka. Proc. Joint Spring Meeting. British, Netherlands and Belgian Societies of Parasitology with Belgian Society of Protozoology. University of Southampton, UK. Abstr. No. 71 (1989).
215. Wagle, S.M. and Wagle, M.B. Hydatid disease. Indian J. Surg. **25**: 610–613 (1963).

Appendix: Some Important Helminths of Indian Subcontinent

Parasite	Geographical distribution	Hosts	Location	Pathology
(1)	(2)	(3)	(4)	(5)
TREMATODA				
Dicrocoelium dendriticum	Pakistan, India, Sri Lanka	Sheep, goat, cattle, yak	Bile duct	Cholangitis, cholangiectiasis, portal cirrhosis, pre-carcinomatous changes
Eurytrema pancreaticum	India, Bangladesh, Burma	Cattle, sheep, goat	Pancreas, liver, small intestine	Haematuria?
Heterophyes heterophyes	India	Dog, cat	Small intestine	Not reported
Haplorchis taichui	India	Dog, cat, fox	Small intestine	Mild enteritis, atrophy of goblet cells and crypts of Lieberkühn
H. yokogawai	India	Dog, cat, crow-pheasant	Small intestine	Subacute haemorrhagic enteritis
H. pumilio	India	Dog, cat	Small intestine	Necrosis of epithelium
Centrocestus formosanus	India	Fowl, pigeon	Small intestine	Mild enteritis
Prosthogonimus ovatus *P. indicus (=P. putschkowskii* *P. pellucidus*	India	Domestic fowl, pigeon, duck, grey partridges, common quails, guinea fowls, common teal, pin tail	Bursa fabricii, cloaca, oviduct	Salpingitis, submucosal oedema, haemorrhage, hyperplastic changes in cloaca and uterus, granuloma in abdominal muscles
Opisthorchis caninus	Pakistan, India, Bangladesh	Dog, domestic cat, pig, Indian hyaena, wild cat	Liver, gall bladder, bile duct	Cloudy swelling, necrosis periportal hyperplasia, cirrhosis
O. noverca	Pakistan, India	Dog, domestic pig, wild boar, man?	Bile duct, gall bladder, pancreatic duct	Non-suppurative cholangitis, hyperplasia of bile duct, cirrhosis, chronic pancreatitis
O. tenuicollis	India, Bangladesh	Cat, dog, pig, man?	Liver, bile duct, gall bladder, small intestine	Multilobular cirrhosis, parenchymatous degeneration
Nanophyetus salmincola	India	Dog	Small intestine	Bradycardia, glossoplegia? polypnoea
Isoparochis hypselobagri	India	Pig, man	Bile duct	Not reported
Fasciola jacksoni	India	Elephant	Liver, bile duct, duodenum	Clinically resembles ovine fascioliasis

(Contd.)

(1)	(2)	(3)	(4)	(5)
Fasciolopsis buski	India, Bangladesh, Burma	Man, domestic pig, wild boar, rabbit	Small intestine	Epithelial desquamation, chronic greenish diarrhoea, anaemia, oedema
Echinostoma revolutum	India, Bangladesh, Nepal	Fowl, duck, pigeon	large intestine	Haemorrhagic enteritis, proctitis
Echinoparyphium flexum	India	Fowl, duck, pigeon	Small intestine	Catarrhal enteritis
Echinochasmus perfoliatus	India	Dog, cat	Small intestine	Enteritis?
Artyfechinostomum sufrartyfex	India	Man, pig, dog, cat, wild boar, honeybadger, rat, spiny-tailed lizard, palm civet?	Small intestine	Ulceration, haemorrhagic enteritis, hypertrophy of goblet cells, destruction of Bruner's gland
Ogmocotyle indica	India	Sheep, goat, cattle	Intestine, bile duct, abomasum	Haemorrhagic enteritis, oedema, necrosis
Philophthalmus gralli	India, Sri Lanka, Bangladesh	Fowl, duck, man	Eye	Blepharitis, keratitis
Paragonimus westermanii	India, Sri Lanka, Bangladesh	Man, dog, domestic cat, pig, tiger, leopard, clouded leopard, bear cat, rusty spotted cat, golden cat, civet cat, leopard cat, common mongoose, Himalayan pin-marten	Lung	Pneumonia, hyperplasia of bronchial epithelium, peribronchial fibrosis, pseudotubercle around eggs
P. compactus	India, Sri Lanka	Civet cat, fishing cat, Indian mongoose	Lung	Not reported
P. Kellicotti? *P. edwardsii?*	India	Leopard cat, golden cat, common palm civet	Lung	Not reported
CESTODA				
Moniezia expansa	widespread	Sheep, goat, cattle, buffalo, camel, barking deer, black buck, wild boar, giraffe, hippopotamus	small intestine, gall bladder	Catarrhal enteritis, anaemia, stunted growth
M. benedeni	India, Sri Lanka, Nepal	Sheep, goat, buffalo, wild boar, spotted deer	Small intestine	Diarrhoea
M. denticulata	India	Sheep, goat	Small intestine	Not reported
Bertiella studeri	India	Man, crab-eating macacaque, capped langur	Small intestine	Malabsorption of carbohydrate and fat, diarrhoea
B. satyri	India	Man, gibbon	Small intestine	Not reported
Killigrewia sp.	India	Pigeon, dove, man	Small intestine	Colicky pain, nausea, anorexia

(Contd.)

(1)	(2)	(3)	(4)	(5)
Avitellina centripunctata	India, Bangladesh?	Sheep, goat, cattle, camel, antilope	Small intestine	Nodulation, hyperplasia of glands
A. lahorea	Pakistan, India	Sheep, goat, buffalo	Small intestine	Enteritis?
Stilesia globipunctata	Pakistan, India	Sheep, goat, camel, *chital*	Small intestine	Degeneration of villi, hypertrophy of mucosa, granuloma formation
Railietina tetragona	India, Burma, probably at other places	Domestic fowl, pea fowl	Small intestine	Chronic catarrhal enteritis, ulceration, fusion of villi, hypoproteinaemia, lymphocytopaenia
R. echinobothrida	India, Bangladesh, Burma	Domestic fowl, peacock	Small intestine	Extensive chronic catarrhal hyperplastic enteritis, granuloma formation, necrosis
R. cesticellus	India, Burma, probably at other places	Domestic fowl	Small intestine	Catarrhal enteritis, coagulative necrosis of villi, hypoglycaemia, lower egg production
Cotugnia digonopora	India, Burma	Domestic fowl, painted spur fowl, duck	Small intestine	Coagulative necrosis around scolex, hyperplasia of payers' patches
Amoebotaenia sphenoides	India, Burma, probably at other places	Domestic fowl, duck	Small intestine	Subacute/acute enteritis, extensive mucoid degeneration, formation of lymphoid follicles
Choanotaenia infundibulum	India, Burma	Domestic fowl, duck	Small intestine	Subacute catarrhal enteritis, coagulative necrosis
Dipylidium caninum	Widespread	Dog, cat, man, Indian fox, jackal, Indian hyaena	Small intestine	Connective tissue hyperplasia/hypertrophy of cells
Joyeuxiella pasquale	India	Cat	Small intestine	Haemorrhage, destruction of glands
Hymenolepis carioca	India	Domestic fowl	Small intestine	Chronic catarrhal enteritis, hyperplasia of payers' patches, depletion of polysaccharide complex
H. contaniana	India	Domestic fowl	Small intestine	Catarrhal enteritis, vacuolar degeneration, hyperactivity of payers' patches
H. diminuta	Widespread	Man, mice, rat, bandicoot	Small intestine	Abdominal discomfort
H. nana	Widespread	Man, rat	Small intestine	Abdominal pain, diarrhoea
H. collaris, *H. coronula*, *H. gracilis*	India, Burma	Domestic duck, spotted bill duck	Small intestine	Haemorrhagic enteritis
Mesocestoides lineatus	India, Sri Lanka	Dog, cat, jackal, Indian fox, leopard	Small intestine	Not reported

(Contd.)

(1)	(2)	(3)	(4)	(5)
Diphyllobothrium latum	India, Sri Lanka, Bangladesh	Dog, domestic cat, jungle cat, jackal, clouded leopard	Small intestine	Abdominal discomfort, intestinal obstruction, diarrhoea/ constipation
Spirometra ranarum, *S. erinacei*, *S. erinaceaeuropaei*, *S. mansoni*, *S. mansonoides*	India, Bhutan, Bangladesh, Burma	Dog, domestic cat, fishing cat, pig, man, African lion, clouded leopard, Indian python	Small intestine	Not reported
NEMATODA				
Ascaris suum	Widespread	Pig, sheep, goat, wild boar, man?	Small intestine	Degenerative and necrotic changes in liver; pulmonary oedema and consolidation
A. lumbricoides	Widespread	Man, monkey, sheep, buffalo, Malayan giant squirrel	Small intestine	Haemorrhage, congestion necrosis of liver and lung parenchyma
Parascaris equorum	Pakistan, India, Sri Lanka	Horse, ass	Small intestine	Degenerative changes in liver; haemorrhage, oedema, congestion in lung
Toxocara canis	Widespread	Dog, jackal (DH); man, monkey, chicken, several experimental animals (AH)	Small intestine, (DH), elsewhere (AH)	Haemorrhagic/ulcerative enteritis; neuronal degeneration, meningitis
Heterakis gallinarum	Widespread	Domestic fowl, grey jungle fowl, duck, peacock	Large intestine	Haemorrhage, desquamation, necrosis, nodulation
Ascaridia galli	Widespread	Domestic fowl, duck, japanese quail, alexandrine parakeet	Small intestine, gizzard	Anaemia, haemorrhage, desquamative changes in villi, lower egg production
Subulura minetti	India	Domestic fowl, pea fowl, quail, patridge	Large intestine	Epithelial desquamation, necrosis
S. brumpti	India	Domestic fowl, duck, turkey	Large intestine	Degenerative changes
Strongylus vulgaris	India	Horse, donkey	Caecum, colon	Thrombosis, aneurysm, nodulation
S. edentatus	Pakistan? India	Horse, donkey	Caecum, colon	Arteritis, granuloma, caecal nodulation
S. equinus	Pakistan? India	Horse, donkey	Caecum, colon	Arteritis, nodulation
Tridontophorus, *Periostomum* and *Trichonema* spp.	India	Horse, donkey	Caecum, colon	Proliferative changes, endarteritis, mucosal hypertrophy
Stephanurus dentatus	India, Bangladesh, Nepal	Domestic pig, wild boar	Pevis of kidney, erratically liver, lungs, pancreas or elsewhere	Acute phlebitis, eosinophilic granuloma, biliary hyperplasia, periportal fibrosis/ perilobular cirrhosis
Syngamus trachea	India, Sri Lanka	Domestic fowl	Trachea	Coughing and gaping
Mammomonogamus laryngeus	India, Sri Lanka	Cattle, buffalo	Larynx, pharynx	Mucosal oedema, erosion, ulcers

(Contd.)

(1)	(2)	(3)	(4)	(5)
Ancylostoma caninum	Widespread	Dog, domestic cat, jungle cat, leopard cat, civet cat, clouded leopard, jackal, Indian hyaena, lion, tiger, polar bear, sloth bear, man	Small intestine	Anaemia, fusion of villi, increase in RNA content, depletion of calcium in hepatic cells, increase in β-globulin and decrease in γ-globulin
A. braziliense	India, widespread in Sri Lanka	Domestic cat, dog, man, civet cat, leopard cat, wolf	Small intestine	Anaemia, hacmorrhagic enteritis
A. ceylanicum	India, Sri Lanka	Dog, cat, man	Small intestine	Anaemia, gonadal dysfunction
A. duodenale	Widespread	Man; dog and jackal (rarely)	Small intestine	Similar to *A. caninum* but severe sloughing of epithelium
Necator americanus	India, common in Sri Lanka	Man, rhinoceros?	Small intestine	Anaemia, eosinophilia; increase in α-2, β and γ-globulin
Uncinaria stenocephala	India	Pig, Indian fox	Small intestine	Anaemia, enteritis
Grammocephalus hybridatus	India	Elephant	Liver, bile duct, stomach	Biliary hyperplasia, portal fibrosis, gastric nodulation
Galonchus perniciosus	India	Indian lion, leopard, *cheetah*	Small intestine	Eosinophilic granuloma, nodulation, sclerosis, perivascular cuffing
Ashworthius perrilli	India	Nilgai, black buck	Abomasum	Catarrhal inflammation, oedema
Filaroides osleri	India	Dog, cat, leopard cat	Trachea, bronchi, lung	Interstitial pneumonia, hyperplasia of bronchial mucosa, nodulation
Anafilaroides rostratus	Sri Lanka	Domestic cat	Lung	Tracheobronchitis
Angiostrongylus cantonensis	India	Bandicoot, rat, man?	Lung	Not reported
Habronema muscae	Pakistan, India, Sri Lanka	Horse, donkey; cattle, goat (larval form in skin)	Stomach, heart	Chronic catarrhal gastritis
H. majus	Pakistan, India	Horse, donkey	Stomach	Chronic gastritis
Draschia megastoma	Pakistan, India, Sri Lanka, Burma	Horse, donkey, wild ass	Stomach	Lymphofollicular reaction around vessels, carcinomatous growth, submucosal granuloma
Spirocerca lupi	Widespread in Sri Lanka, common elsewhere	Dog, domestic cat (rarely), jungle cat, bear cat (DH); domestic fowl, Indian bran owl, garden lizard, grey shrew etc. (PH); domestic and hill goat, buffalo, monkey and other species (AH)	Oesophagus, stomach mainly; erratically numerous sites	Fibrochondromasarcomatous changes with metastasis in lungs, liver, kidney; gastric ulcers, aneurysm, rupture of aorta, sondylitis, exostosis, paraplegia

(Contd.)

(1)	(2)	(3)	(4)	(5)
Cyclospiura subequalis	India	Domestic cat, jungle cat, tiger	Stomach	Gastric tumour, nodulation
Ascarops strongylina	India	Pig	Stomach	Destruction of glands
A. dentata	India	Pig, wild boar	Stomach	Gastritis
Physocephalus sexalatus	India	Pig, wild boar	Stomach	Gastritis
Simondsia paradoxa	India	Pig	Stomach	Subacute/chronic gastritis suppuration, calcification
Tetrameres mohtedi	India	Fowl, duck	Proventriculus	Acute proventriculitis, lymphoid hyperplasia
Physaloptera praeputualis	India, Sri Lanka, Bangladesh	Cat, dog, clouded leopard	Stomach	Gastritis
P. brevispiculum (=P. fülleborni?)	India	Domestic cat, wild cat, honey-badger	Stomach	Gastritis
Gnathostoma spinigerum	India, Bangladesh, Burma	Man, domestic cat, civet cat, golden cat, leopard, tiger, dog, jackal	Stomach, liver, body cavity, mesentery, diaphragm or elsewhere	Gastric tumour, granulomatous reaction, coagulation necrosis, sclerosis, perivascular cuffing
Dracunculus medinensis	Pakistan, India	Man, dog, cat, cattle, horse, jackal, leopard, chinkara deer, monkey, cobra	Subcutaneous tissue, eye, heart, scrotum, vertebral column	Generalised articaria, dermatitis, ulcer, nodulation
Trichinella spiralis	India, Bangladesh	Pig, cat, civet cat, rat, bandicoot, mole, squirrel, bat	Small intestine, skeletal muscle	Granular degeneration, loss of striation in muscles, myocardial haemorrhage, pneumonia, cloudy swelling, interstitial nephritis, haemorrhage in brain
Capillaria hepatica	India	Dog, rat, bandicoot, man	Liver	Granuloma around egg/parasite, necrobiotic/cirrhotic changes
PENTASTOMIDA				
Linguatula serrata	India, Bangladesh	Man, dog (DH); sheep, goat, cattle, buffalo, pig, camel, rabbit, Indian palm squirrel, bandicoot, Indian python (IH)	*Adult*: Respiratory passage Nymph: Liver, lung, lymphnode or elsewhere	Haemorrhagic nodulation, lymphopoiesis, coagulative necrosis, oedema, hypertrophy

DH: Definitive host; PH: Paratenic host; AH: Abnormal host.
?*Species inquirenda*/occurrence doubtful.

4. Nematode–Cestode Zoonoses

John H. Cross
Department of Preventive Medicine and Biometrics
Uniformed Services University of the Health Sciences
Bethesda, USA

INTRODUCTION

There are more helminth parasites of lower animals infecting humans than any other group of animal parasites. Nematode, cestode, and trematode infections abound in nature and are found in every known species of vertebrate. Most parasites usually exist in harmony in their natural hosts, but when they become zoonotic parasites, they usually cause serious disease in humans.

In this chapter a number of zoonotic helminths of worldwide distribution which often have an impact on human and veterinary public health in tropical countries has been reviewed. Comprehensive additional reviews of zoonotic trematodiases are available [24, 30] which readers may consult for further reading (see also Chapters 12 and 14).

NEMATODE ZOONOSES

There are a large number of roundworms of animals that may cause disease in humans. In the following sections, the nematodes will be discussed under the groups viz., trichurids, ascarids, strongylids, rhabditids, spirurids, and acanthocephala.

Trichurids

Trichinellosis

Trichinellosis or trichinosis, while considered a disease of the Temperate and Arctic regions of the world, has also been reported from some Tropical countries in Africa, Latin America, and Asia. Earlier, all trichinae were considered as *Trichinella spiralis*, but variants of the parasite have been described, first as strains and subspecies and now as separate species: *T. spiralis*, *T. nelsoni*, *T. nativa*, *T. pseudospiralis* and *T. britovi* [31]. Classical *T. Spiralis* is a synanthropic zoonotic from infecting pigs, dogs, cats, and man, while *T. nelsoni*, *T. britovi* and *T. nativa* are sylvatic species infecting wild carnivores. *T. pseudospiralis* is maintained in carnivorous mammals and raptorial birds [29].

Trichina are small nematodes; the males measure 1.4–1.6 mm by 40–60 μm, and the females measure 2.2–3.6 μm by 60–100 μm. The male has two pseudobursae, and the vulva in the female is located on the ventral side of the body in the oesophageal region. In both sexes the anterior end contains gland-like cells or stichocytes associated with the oesophagus. Secretions from the stichocytes are antigenic and antibodies produced are important in the host's response to the parasite.

The complete life cycle occurs in a single host. When raw muscle containing cysts of *Trichinella* larvae is eaten, the cysts are digested out of the muscle and the larvae pass to the small intestine. The larvae enter the intestinal epithelium, molt into adults, and the male dies after fertilisation. The females deposit larvae for several weeks to a few months, until their death and expulsion from the tissue. The larva (100 μm by 6 μm) migrate throughout the body via the blood, and invade nearly any tissue, but survive and develop further in skeletal muscle cells. The invaded muscle cell is known as a "nurse cell" and it forms a cyst around the larva. The larvae reach the muscle in about 5 days and become encapsulated within 3 weeks. The size of the cyst in man averages 400 μm by 250 μm and the larvae within the cyst may reach length of 1 mm and are considered infective. The greatest densities of larvae are found in the diaphragm, tongue, and masseter muscles which are highly active and possess the richest supply of blood.

The intestinal phase spans 10 days to several weeks, and is determined by the immune response of the host to the adult worm, viz intestinal inflammation leading to expulsion of the worms. The encapsulated larvae survive for several years in muscle tissue. The capsule eventually thickens and begins to calcify. The larvae are resistant to chemical and physical factors. They have been known to live for 4 months in rotting tissue and are also resistant to desiccation, salting and smoking.

Most human infections are asymptomatic or subclinical with the severity depending on the number of larvae/g of infected meat eaten. Infections with less than one larvae/g of muscle tissue are usually asymptomatic. Those with more than 100 larvae/g and 1000 larvae/g of muscle tissue, respectively produce clinical symptoms and serious disease. Race, age, sex, general health status, and individual resistance factors influence the severity of the disease. Initial symptoms are associated with the intestinal phase of the life cycle; loss of appetite, abdominal pain, nausea, weight loss, and diarrhoea. These symptoms are often confused with other gastrointestinal infections, and trichinosis at this stage is often diagnosed wrongly. The worm-induced intestinal inflammation may last for 1–2 months. In 7–9 days newborn larvae begin to penetrate muscle cells and an inflammatory response results in muscles of the tongue, larynx, masseter, diaphragm, neck, intercostal muscles, and muscles attached to tendons and joints. The muscle fibers become oedematous and enlarged due to infiltration of neutrophils, eosinophils, lymphocytes, and tissue histocytes. Inflammation is maximum at 5–6 weeks and begins to diminish when the larva is encapsulated. At this stage the clinical symptoms include anorexia, emaciation, fever, dyspnoea, periorbital oedema, eosinophilia, myalgia and pain in joints. In severe cases, myocarditis, endocarditis, encephalitis, or meningitis are also noted. The convalescent phase of infection begins in a few months, but may continue for several years until the larvae and cysts have been completely calcified.

When suspected, the diagnosis of trichinosis is not difficult; it is the mild or asymptomatic infection that is missed. Trichinosis should be suspected when people present similar symptoms after sharing a common meat source. If the meat is available for examination, trichina larvae can be found by direct microscopic examination of "pressed" muscle or after digestion in artificial gastric juice (pepsin-HCl). Physical signs of joint pain, periorbital oedema, fever, myalgia, and malaise are pathognomonic. Eosinophilia is common with splinter haemorrhages on the nail beds and conjunctiva. Serological tests are sensitive and specific, especially the enzyme-linked immunosorbent assay (ELISA) [28] (see Chapter 12). Serology only provides a presumptive diagnosis which is confirmed by demonstrating larvae after by microscopical examination of muscle biopsies.

Treatment of trichinellosis depends on the intensity and duration of infection. Anthelmintics are effective against the intestinal phase of the life cycle, but are of little value against larvae in the musculature. Thiabendazole, mebendazole, albendazole are effective against worms in the intestines. The latter two drugs are contraindicated during the first trimester of pregnancy. Anti-inflammatory and anti-allergic drugs, such as corticosteroids, are used to relieve symptoms associated with the muscle phase of the disease. Prednisolone (40–60 mg/d) can be given until the fever and allergic signs subside.

Most cases of trichinosis in the United States, as well as elsewhere, are associated with common food (except pork) source. Most are attributed to meat of wild animals. The infection and disease occur in Egypt, Kenya and Tanzania [3]. The spotted hyena is the major reservoir host of *Trichinella* in Africa, and the species has been designated *T. nelsoni* because of certain biological characteristics, especially low infectivity to rats and domestic pigs. There are sporadic reports of trichinosis from Iran, Afghanistan, India and Indonesia. In recent years several epidemics with over 10% infection in pigs of hills of Northern Thailand. A local fermented pork dish "Nham" is a major source of infection. The species in Thailand is considered *T. spiralis* based upon experimental studies in pigs and rodents. Enzootic cases of trichinosis occur throughout China and recently outbreaks occurred in Hong Kong also. One epidemic involved Gurkha soldiers in the British Army in Hong Kong after eating raw pork. An outbreak occurred in Japanese hunters who ate a wild bear. Epidemics have been reported from Chile, Uruguay, Argentina, and Mexico. Occasionally cases have been reported from Canada with most associated with consumption of inadequately cooked meat of bear as well as polar bear and walrus whereas epidemics in the United States are associated with the consumption of pork, pork products and meat of wild animals. In European countries the outbreaks are on the decrease and associated with consumption of meat of wild animals.

Related to the problems in sylvatic trichinosis is the uncertainty of the systematics of *Trichinella*. The isolates from hyena in Kenya have been designated as *T. nelsoni* and those from Arctic bears and walruses as *T. nàtiva*. Unlike other species, which die after exposure to freezing temperatures, the latter species is resistant. The isolate *T. pseudospiralis* from a Russian raccoon and raptorial birds, does not induce the formation of a muscle capsule. Isoenzyme and DNA profiles of these species are different (see Chapter 5).

Trichuriasis

The human whipworm, *Trichuris trichiura*, is not considered a zoonosis; however, on rare occasions, animal trichurids, such as *T. vulpis* and *T. suis* have been reported in large intestine of man. The trichurid eggs are characteristically round with polar plugs. *T. Trichiura* eggs are smaller (49–61 μm by 23–29 μm) than those of *T. suis* (47–71 μm by 27–35 μm) and *T. vulpis* (70–90 μm by 32–40 μm). The life cycle is direct with eggs passed into the environment in faeces, embryonate in the soil, and ingested by a host. The larvae escape from the egg in the intestines and mature in the large intestines. Clinical trichuriasis in man with animal parasites, is similar to infection with the species occurring in man. The diagnosis of organism of zoonotic importance is usually based upon egg size. It is possible that human infections, with trichurids of animal origin is not uncommon, but escape detection as the microcopists rarely determine the size of the eggs.

Capillariasis

All capillarid infections in humans are of zoonotic importance. Out of the 200 species of the genus *Capillaria* only three—*C. hepatica*. *C. aerophilia* and *C. philippinensis* have been reported in humans.

Capillaria hepatica is a parasite found worldwide in the livers of rodents. There are probably a number of species infecting various rodents, but *C. hepatica* is the only one reported in humans. The parasitosis in humans is rare, but is reported from diverse parts of the world. Adult worms live, reproduce, and die in the liver. The eggs remain in the liver until the rodent dies and are released from the tissue only after the death of the host and the carcass decomposes or the host is eaten and the eggs in the liver pass in the soil after digestion. The eggs embryonate only in the presence of oxygen. When the egg is eaten by a rodent or human, the egg hatches and the larva migrates to the liver and matures. Human infections cause fever, hepatomegaly, hyperglobulinaemia, leucocytosis and eosinophilia. Most cases are diagnosed at autopsy by the presence of typical thick-shelled trichurid-like eggs in liver tissue.

Capillaria aerophilia, a parasite of carnivores, especially of cats, is probably widely distributed throughout the world, but rarely reported. It is found in the epithelial lining of the pulmonary tree. Not more than 10 infections have been reported in humans, mostly from Russia.

Capillaria philippinensis is an intestinal parasite that was originally described in humans in the Philippines. Freshwater fish serve as the intermediate host of the parasite, and possibly fish-eating birds are the natural definitive hosts [7]. Nearly 2,000 human cases have been reported from the Philippines, a somewhat lesser number from Thailand, 1–3 cases each from Japan, Egypt, Iran and Taiwan. Eggs, larvae, and adult forms of the parasite are excreted in the faeces. After ingestion by small fish, the eggs hatch and larvae (150 μm) are released. The larva grows to about 300 μm in the intestine of fish. When ingested, the larva matures and the first generation female worms produce larvae in two weeks. These larvae are retained and develop into adults in two more weeks and produce eggs. Therefore, autoinfection is part of the life cycle. Many thousands of worms have been recovered from the intestine of humans at autopsy. Infected persons experience borborygmy, diarrhoea and abdominal pain. They develop a protein losing enteropathy, suffer a loss of serum electrolytes, and become cachexic. Untreated patients usually die as a result of the irreversible effects of the infection. Diagnosis is made by finding the parasite or eggs in faeces. Mebendazole (400 mg/d for 20 d) and albendazole (400 mg/d for 10 d) are the drugs of choice for treatment.

Anatrichosoma sp. is a capillarid-like nematode of the skin and nasal mucosa of animals, especially monkeys in which it causes subcutaneous lesions or creeping eruption and blisters. On two occasions, parasites considered to be *A. cutanea* were reported in humans. Little is known regarding the life cycle, although worms and eggs are present in the subcutaneous lesions.

Dioctophyma renale, the giant kidney worm of mustelids, jackals, dogs, and other carnivores is found in many parts of the world. The adult worms are large (females are 20–100 cm by 5–12 mm) and usually one female resides in one kidney; it may destroy the entire kidney parenchyma. In dogs, the worms may lodge in the peritoneal cavity. Eggs are passed in the urine, embryonate after months in water and are ingested by oligochaetes, hatch and eventually develop into a third-stage larvae. Frogs and fish eat the oligochaetes and the larva get lodged in mesenteries or liver. When a carnivore eats the oligochaete or fish or frog, the larvae migrate into the gastric mucosa, liver, peritoneal cavity, and finally into the kidney. Human infections are rare, since oligocheates and paratenic hosts are usually not part of a human diet. Less than 20 human cases of infection shedding eggs in urine and showing symptoms of haematuria and renal colic are on record.

Ascarids

Ascarid infections are among the most common nematode infections in man and lower animals. Over one billion humans are infected at any one time with *Ascaris lumbricoides*, and *A. suum* is equally common among swine populations. Dogs are usually born with *Toxocara canis*, and *T. cati* infections in cats are acquired by egg ingestion or larvae through mother's milk. Other *Toxocara*, viz. *T. vitulorum*, may be acquired similarly.

A. lumbricoides is not considered a parasite of lower animals transmitted to man, but a question remains whether *A. suum* is of zoonotic importance. The two species are closely related and show slight morphological and physiological differences. Experimentally both species infect the heterologous host, but *A. lumbricoides* usually does not mature in swine. Although there are reports of *A. suum* maturing in humans (see Chapter 3), the parasite does not remain in the intestines for long. Among the ascarids, *Toxocara* species are important zoonoses.

Toxocariasis

The major cause of the visceral larval migrans syndrome is *T. canis*. *T. cati* may be moderately involved, and *T. vitulorum* of cattle is only a suspected cause of the syndrome [35].

Adult *T. canis* are found in small intestines of canine; females measure 6–18 cm by 2.5–3 mm, and the males are 4–10 cm by 2.0–2.5 mm. It is larger than *T. cati*. Unembryonated eggs which pass through the faeces, are thick-walled with pitted shells. When eggs with fully developed larvae are ingested by canines, the larvae migrate from the intestine to the liver, then to the lungs. After a period of time in the lungs, they migrate up the pulmonary passages, are swallowed, and return to the intestines. The prepatent period is 4–5 weeks and the life span is 4–6 weeks. Puppies are highly susceptible to infection. In older dogs and other animals eggs may hatch, but the larvae remain in tissues, but the latter serve as paratenic hosts. Dogs may, therefore, acquire infections from eggs, by eating paratenic hosts, by receiving larvae from mother's milk or via the placenta, or by ingesting larvae passed in faeces of suckling pups. Transplacental transmission is quite common. The larvae may reside in the tissues of the bitch and migrate through the placenta to the livers of the foetuses during its pregnancy. When the pups are born the larvae migrate to the lungs and then to the intestines where they mautre and produce eggs in 3–4 weeks.

T. cati is smaller than *T. canis* and has a similar life cycle, except that there is no transplacental transmission and cats of all ages can be infected.

When humans or lower animals ingest embryonated eggs, the released larvae migrate to the liver, then to other organs and tissues. The larvae can remain alive for extended periods and may cause disease, especially in children. Adults may also become infected and manifest the symptoms. In most cases, infections are mild and asymptomatic, except for eosinophilia. In some cases there is hepatomegaly, hyperglobulinaemia and pneumonitis. Infected children may show rashes, asthmatic attacks, abdominal and neurological disturbances. In older children, endophthalmitis may result from worms entering the eye. The retinal lesion resembles retinoblastoma and at times unnecessary enucleation occurs.

The parasitologic diagnosis in humans is difficult unless a larva is found in a biopsy, usually from the liver or from a surgically removed eye. The diagnosis is usually presumptive, based on symptoms, association with dogs, and a history of pica. Among the serologic tests ELISA is considered the most sensitive and specific using antigens prepared from larvae. The diagnosis in the natural definitive host is done by the detection of eggs, larvae or adult parasites in the animal's faeces. Recent research efforts have been directed toward the seroepidemiology of visceral larval migrans using various antigens and immunological tests, such as "dot" ELISA and monoclonal antibodies.

Symptomatic treatment with anti-inflammatory drugs is recommended. Anthelmintics may be of value in acute infections in children with high eosinophilia. Anthelminthics are effective in treating intestinal infections in dogs. Thiabendazole may be of some value, but recent studies recommend albendazole for the treatment of visceral and ocular larval migrans [37].

Throughout the world cats and dogs are infected with *Toxocara* and human disease has been recognised in most countries [16]. Infections often go unrecognised, however, and prevalence data is, for the most part from serosurveys. Black children are infected more often than children of Anglo-saxans. The prevalence rates are higher in people of low socioeconomic status and residing in rural areas.

T. vitulorum is found in bovids, with cows and calves acquiring infection from eggs in wallow water, soil and pastures. Calves may also acquire infections through milk, suckling contaminated udder and teats of the cows. It has been suggested that *T. vitulorum* may be a cause of human visceral larval migrans [33].

Anisakiasis

Anisakiasis is caused by the larval stages of marine anisakid nematodes [2]. The adult worms are parasites of marine mammals and when humans eat marine fish infected with larval stages, the parasite penetrates the intestinal tissue. Europeans and Japanese have most of the infections involving *Anisakis simplex*, the herringworm, and *Pseudoterranova decipiens*, the codfish worm. The larval stages are difficult to speciate; however, the division of the oesophagus into a muscular portion and the glandular ventriculum separate the worms into various types.

Adult worms reside in the intestines of whales, seals, dolphins, and other fish-eating marine mammals which release eggs through the faeces. After hatching, the larvae are eaten by marine invertebrates, such as small crustaceans, which in turn are eaten by larger invertebrates or fish and the larvae develop into the infective stage. The definitive host becomes infected when the intermediate hosts are eaten.

In humans, larvae are acquired by eating raw fish, inadequately cooked or preserved squid or octopus. Infections can cause mild to severe disease, from "tickle throat" to eosinophilic tumors in the stomach and intestines. Treatment of severe infection is by surgery, but in some cases worms can be removed by enteroscopy. Diagnosis is difficult and most of the cases are confirmed as anisakiasis at the time of surgery or enteroscopy. In tissue sections, worms can be identified by the presence of "Y"-shaped lateral cords, muscles, and a thick cuticle.

Infections usually occur in humans living in costal areas and in people who enjoy eating raw marine fish, such as the Dutch who like "green herring" and the Japanese who relish raw fish prepared as "shashimi" or "sushi". Freezing of fish will kill the worms which are usually in the flesh of fish.

Lagochilascariasis

Lagochilascaris minor is a rare ascarid which has been reported 19 times in humans; all in Latin America. Its natural host(s) and life cycle are unknown. Colonies of worms have been found in abscesses, subcutaneous tissue of the neck, mastoid processes, and nose. The source of infection and means of transmission are also unknown, but it is assumed to be through the eggs.

Strongylids

The order Strongylida contains many nematode species that infect man and animals. Male strongylids possess a hand shaped bursa which has finger-like rays that hold the female worms during copulation. The males have a pair of equal spicules and the females a complex ovijector with well developed sphincters. Most members of the order lay thin-shelled eggs. Important superfamilies are the Ancylostomatoidea, Strongyloidea, Trichostrongyloidae and the Metostrongyloidea.

Ancylostomiasis

Human hookworms *Ancylostoma duodenale* and *Necator americanus* are the most important nematodes in the superfamily Ancylostomatoidea. They are not considered zoonotic parasites as they occur only in humans. On the other hand, *A. ceylanicum*, an intestinal parasite of the civet cat, originally reported from Sri Lanka has been reported in Asians. Many hookworms of animals are unable to complete develpment in humans, but are able to invade the skin and are a cause of cutaneous larval migrans or creeping eruption.

Ancylostoma ceylanicum is now known to be distributed widely in Asia and is also reported from parts of Africa and Latin America. For many years it was confused with *A. braziliensis*. Cats and wild felids and dogs are natural hosts, but patent human infections are reported throughout Asia. Infected persons treated for hookworn infection in Calcutta, India, were found to have multiple infection with *N. americanus*, *A. duodenale* and *A. ceylanicum*. Human infections with this parasite are considered rare, and occur sporadically with involvement of a few worms. Disease associated with human infection is minimal. *A. ceylanicum* is not associated with cutaneous larval migrans in humans, while other species of *Ancylostoma* may invade the skin and produce serpiginous lesions.

Ancylostoma braziliensis is found in canines and felines throughout the tropics. It was reported from humans, but it is now thought that the worms were confused with *A. ceylanicum*. This species is an important cause of cutaneous larval migrans in tropical and subtropical parts of the world.

Ancylostoma caninum is also found in dogs and cats, especially in the northern hemisphere and may also cause creeping eruption. Human eosinophilic enteritis may be caused by *A. caninum*. Other hookworms of animals that may cause disease in humans, are *A. tubaeforme*, *A. malayanum*, *Cyclodontostomum purvisi*, *Uncinaria stenocephala* and *Bunostomum phlebotomum*.

The life cycles of hookworm species of animals are similar to that of species infecting human. Adult parasites reside in the small intestine and following copulation, the female worms pass eggs in the faeces. In the soil, the eggs mature producing a rhabditiform larva which after a few days develops into a filariform larva. The infective third-stage larvae enter the skin, migrate through the body to the heart and lungs, and finally to the intestines where the worms develop into adults. The adult worms attach to the intestinal mucosa and feed on blood.

Creeping eruption or cutaneous larval migrans results from penetration into the skin of humans by infective larval stages of hookworms of animals and other helminthic parasites. The major etiologic agent, however, is the third-stage larva of *A. braziliensis*. The parasite cannot complete development in humans and migrates in the germinal layer of the skin.

Following penetration, the third-stage larva produces an itchy papule which eventually develops into serpiginous tracks in the epidermis. The larva moves several millimeters without any evidence, but the lesions eventually become itchy and the dried track is evident. Pruritus intensifies and scratching leads to secondary bacterial infections. The migration may continue for months resulting in extensive lesions and serious illness. Some lesions heal spontaneously without treatment. Treatment consists of oral administration of thiabendazole (50 mg/kg/d for 3 days) or topical applications of thiabendazole ointment. Freezing of the skin may also be done. The diagnosis is based upon symptoms and history of exposure.

Cutaneous larval migrans is reported from areas with a warm and humid climate: Latin America, southern United States, Europe, Africa, the Philippines, Taiwan, Australia and India.

Mammomonogamiasis

Mammomonogamus laryngeus (formerly *Syngamus laryngeus*) is a parasite of respiratory passages of ruminants and has been reported from humans on several occasions. It is reported from animals in tropical America, India and Southeast Asia. Little is known about its life cycle, but it is believed that transmission occurs through the ingestion of larvae in eggs or from paratenic hosts. In man, the worms "in copula" have been found attached to the tracheal wall. Symptoms consist of tracheolaryngeal irritation, cough, and occasional haemoptysis. Detection of eggs in sputum or faeces may be useful for the diagnosis. No specific anthelminthic therapy is reported and most cases have been treated surgically or by extraction of the worms by bronchoscopy.

Oesophagostomiasis

Oesophagostomum species are natural parasites of nonhuman primates and other animals in Africa, Asia, and South America. The larval stages cause nodules in the intestinal wall; the common name is "nodular worm". The life cycle is simple in that females attach to the intestinal mucosa and pass eggs into the faeces. The larvae hatch for the eggs and develop into infective third-stage larvae that are eventually swallowed. The larvae penetrate the gut mucosa and provoke nodules and eventually abscesses. The worms emerge from the mucosa and become sexually active. In most infections of humans, however, immature worms were found. Abdominal pain, diarrhoea, and intestinal obstruction are reported in humans. Nodules have been found mostly in the ileocecal region of the intestine. The eggs are similar to those of hookworm, making it difficult to diagnose. The specific diagnosis usually follows surgical removal of the nodules. The infection is self-limiting, but surgical intervention will relieve symptoms.

Trichostrongylosis

The trichostrongyles are important intestinal parasites of ruminants and although several species are reported in humans, the most important ones belong to the genus *Trichostrongylus*. These parasites are common among sheep, goats, bovines and other domestic and wild animals all over the world. The life cycle of these slender bursated nematodes is direct. Thin-shelled eggs, passed by female worms into the faecal stream, embryonate and hatch in the faeces. The rhabditiform larvae feed on microflora and develop into third-stage larvae. They leave the faeces and collect in the moisture on vegetation, waiting for ingestion by the definitive host. Once ingested larvae develop directly into adults in the digestive tract.

A number of species of *Trichostrongylus* are reported to cause disease in humans; *T. orientalis*, *T. axei*, and *T. colubriformis* are the most important. Infections occur sporadically in South America, Europe, the Middle East, Africa, Korea, and Japan. *T. orientalis* appears to be a parasite of humans and is, transmitted from man to man, while the major host for the other species are sheep. In Iran, seven species of *Trichostrongylus* are common in man and animals throughout the country and in one area, 67% of the population was reported to be infected with *T. orientalis* and *T. colubriformis* [15].

Human infections occur in areas utilizing human and animal feces as fertilizer. The infections are acquired by the oral ingestion of infective larvae on vegetation and other food stuffs. Most infections are asymptomatic, but with infections involving several hundred worms, there may be diarrhoea, abdominal pain, weight loss, anaemia, and eosinophilia. Disease is caused by damage to the doudenal and jejunal mocosa where desquamation and haemorrhage may occur. Infections may last for several years. Thiabendazole and pyrental pamoate have been found effective for treatment.

Angiostrongyliasis

There are approximately 20 species of *Angiostrongylus* reported from the blood vessels and lungs of various species of animals, of which only two are documented pathogens of man and two are considered potential pathogens. *A. cantonensis* causes eosinophilic meningitis upon invasion of the brain and central nervous system (CNS) of man and *A. costaricensis* is a cause of abdominal angiostrongyliasis. *A. mackerrasse* from rats in Australia and *A. malaysiensis* from rats in Southeast Asia may also be capable of causing eosinophilic meningitis in man [8].

A. cantonensis was first reported in 1935 in China as a parasite of rat in and the first human infection was documented in Taiwan in 1945. Later, it was also reported from a human in Hawaii and from many parts of Asia, the Pacific Islands, Australia, India, the Indian Ocean, Africa, Cuba, and Puerto Rico. Infection in rat (*Rattus* and *Bandicota* species) are equally widespread.

The adult worms are located in the pulmonary arteries. The males are 16–19 mm long and 0.26 mm wide. The bursa is characteristic of the species as is the size of the spicule. The females are 21–25 mm long and 0.30–0.36 mm wide and daily lay more than 10,000 embryonated eggs, which hatch in the lungs. The larvae migrating up the pulmonary tree, are swallowed and pass into the faeces. The first-stage larvae enter a molluscan intermediate host and develop into infective thrid-stage larvae. When the molluscan host is eaten by a rat, the larvae migrate from the intestines to the brain, grow into young adults and migrate to the pulmonary artery where they reproduce. The prepatent period is approximately 40 days and they may live for the lifetime of the rat. Infections with large numbers of worms will often lead to the death of the rats, but one infection will confer immunity to further infections [6].

When humans eat an infected molluscan host, the third-stage larvae migrate to the CNS and provoke an eosinophilic meningitis accompanied with fever, headache, nausea, vomiting, stiff neck, paresthesia, paralysis of the eye muscles, and somnolence. Larvae and pre-adult worms are sometimes recovered from the cerebrospinal fluid (CSF) and occasionally from the eye. The incubation period is variable, with symptoms appearing as early as on first day, but usually 5–10 days after ingestion of an infected mollusc. The spinal fluid is usually cloudy with leucocytes, predominantly eosinophils. Granulomatous lesions and necrotic debris develop around dead worms. Deaths have occurred as a result of infection with a large number of worms.

The disease is usually self-limiting, lasting only for 4–6 weeks and symptomatic treatment is recommended. Anthelminthic drugs are not recommended which may increase the inflammatory reaction and enhance the disease. Recently, however, albendazole has been successfully used in treating pediatric cases in Taiwan [20].

The diagnosis is presumptive based on symptoms of eosinophilia and a history of eating snails. The definitive diagnosis is made when worms are recovered from spinal fluid or the eye. Serologic tests especially the ELISA have been used. Monoclonal antibodies have also been used in an ELISA. Mitochondrial DNA has been studied, but a DNA probe has yet to be developed as a diagnostic tool [21].

Humans acquire angiostrongyliasis by eating an infected mollusc or a paratenic host infected with third-stage larve. Some infections occur accidentally when edible vegetation is contaminated with mucus from a slug containing shed larvae. Vegetables may also contain tiny slugs or planaria infected with the parasite. Shrimp and prawn serve as paratenic hosts and sauces prepared from juices from these crustaceans may contain larvae. Intermediate and transport hosts are also used for medicinal purposes by some societies, and when eaten raw, infections may develop.

The giant African snail (*Achatica fulica*) is an important intermediate host in many areas of Southeast Asia. This and other land snails are excellent hosts, often infected with large numbers of larvae. This snail is the main source of infection in Taiwan. *Pila* species of snails, often referred to as apple snails, are the major sources of infection in Thailand. This snail, however, is a poor host as it harbours only a few parasites in its tissues. Thailand has more cases of angiostrongyliasis than Taiwan, but the cases in Taiwan are often more severe. This is probably due to the fact that the infection in Thailand involve fewer worms.

A. cantonensis seems to be spreading throughout the world. It first appears in the rodent and molluscan populations, and then most assuredly will be found in humans. The parasite was recently found in the United States in rats and also in zoo monkeys. It was also found in a zoo monkey in the Bahamas [13]. It should not be too long before human infections are reported from these areas.

Angiostrongylus costaricensis was first found at autopsy in Costa Ricans in Central America and later in the natural host, the cotton rat *Sigmodon hispidus*, and the black rat, *Rattus rattus*. The intermediate host is the common slug, *Vaginulus plebeius*. Subsequently, rats throughout Central and South America have been found infected. It is also reported from Texas in the United States.

The adult worms are similar to *A. cantonensis*, but are located in the arteries in the ileocecal region of the intestine. Females deposit unembryonated eggs that are carried through the arterioles to the wall of the intestine. They embryonate and the first-stage larvae hatch from the egg and migrate through the tissue, enter the intestinal lumen, and are passed with the faeces. The larvae are ingested by a slug intermediate host and migrate to the mantle and foot and develop to the third-stage. When the slug is eaten by the rat, the larvae are released and enter the rat lymphatics, develop into young adults and move to the ileocecal arteries.

Abdominal angiostrongyliasis is well documented in children from Costa Rica and recently in Brazil [14]. Eggs deposited in the tissue cause a granulomatous eosinophilic inflammatory reaction. There is oedema and thickening of the intestinal wall and narrowing of the intestinal lumen. There may be fever, abdominal pain on the right side, anorexia, nausea, vomiting, diarrhoea, leucocytosis with eosinophilia. Palpation reveals a tumorous mass and rectal examination is painful. The symptoms are similar to appendicitis and upon surgery and sectioning of the tissue, the presence of worms and eggs confirms diagnosis.

In humans, larvae are not passed in faeces, therefore, a stool examination will not help in diagnosis. Serological tests are becoming available, but are not widely used. Most diagnoses are made following surgery. Thiabendazole is recommended for treatment. Albendazole may also be of value.

Most people in endemic areas do not eat slugs; therefore, infection with *A. costaricensis* is considered to be accidental. Larvae released from slugs may be innocently ingested from the mucus deposited on fruits and vegetables or water. Infections occur more often in male children.

A. siamensis is a parasite found in the mesenteric artery of rats in Thailand [23]. Its natural intermediate host is unknown, but the parasite will develop in the aquatic snail, *Biomphalaria glabrata*. The parasite causes intestinal lesions similar to those caused by *A. costaricensis*. Infections in humans have not been reported from Southeast Asia, but a monkey imported to Japan from Malaysia was found with intestinal lesions similar to those reported in rats and the first-stage larvae recovered were similar to first-stage larvae of *A. simianensis*. In the future, human infections are likely to be reported from Thailand and elsewhere in Asia.

Rhabditids

The Rhabditita are characterized by having a prominet muscular oesophagus, sometime in their development. Most are free-living nematodes, some of which may accidently infect man and some are recorded as spurious infections.

Micronemiasis

Micronema delitrix is a saprophytic nematode that has been reported from horses and on three occasions from humans who died of meningoencephalitis. Eggs, larvae, and adult females have been found in tissue, but male worms were not found. Male worms, as well as all other stages of the parasite, have been found, however, in soil rich in decomposing matter. Infections in equines have been found in the brain as well as in other locations, such as the kidney, nose, jaw, and maxillary sinuses. Human infections are reported only from Canada and the United States, and equine infections from the United States and Egypt.

Strongyloidiasis

Two species of *Strongyloides* have been reported in humans, *S. stercoralis* which has been reported worldwide, especially in tropical countries, and *S. fuelleborni*, reported only from Africa and Papua New Guinea [17].

S. stercoralis is a parasite of humans as well as dogs and cats. It is one of the smallest nematodes, infecting man. The delicate parasitic females are 2.2–2.7 mm long and 30–40 μm wide. Parasitic males are unknown. The parasite is unique in having a free-living cycle also, which produces rhabditoid males and females.

Eggs from parasitic females in the small intestinal mucosa hatch and yield a rhabdiform larva with a characteristic muscular oesophagus with a posterior bulbus. The first-stage larvae usually pass with the feces and in the soil transform into free-living males and females or third-stage filariform larvae. The first-stage larvae may also transform into filariform or infective stage larvae in the bowel and penetrate the intestinal tissue. This leads to an autoinfection and in the immunosuppressed cases, hyperinfections occur. Free-living adult worms also reproduce, releasing eggs that develop into first-stage larvae, and eventually, infective thrid-stage larvae. Infective larvae penetrate skin of man and animals and migrate via the blood to the lungs, into the alveoli, up the respiratory tree to the pharynx and on to the intestines.

Although infections may continue for years, the sypmtoms are usually mild and well tolerated. Initial symptoms upon penetration of larvae into the skin may cause a pruritus. In repeated infections, there may be an allergic response. Larvae passing out of the anus may enter the skin and cause urticarial lesions on the buttocks and thigh. This has been termed " larval currens". The parasite may cause lesions in the intestinal mucosa provoking epigastric pain, diarrhoea, dyspnoea, nausea, and vomiting. Leucocytosis and eosinophilia are common. In the immunosuppressed autoinfection (not with HIV infections) may lead to severe symptoms with the parasite disseminated throughout the body. The lungs are especially affected, with larvae found in the sputum and with symptoms of asthma. There may also be intestinal symptoms of pain and diarrhoea, malabsorption, dehydration, hypokalemia and hypoproteinaemia, leading to death.

The diagnosis is made by detecting larvae in the faeces or sputum. Stool cultures using the filter paper mehod of Harada and Mori or charcoal cultures followed by Baermannization are useful. The entero-test or string capsule may also be used to isolate larvae from the intestinal mucous. ELISA, and skin-testing may also be helpful. In patients with gastroenteritis, a high eosinophilia, and a history of risiding in the tropics, should be suspected of having strongyloidiasis until the infecion is ruled out.

Treatment is effective with thiabendazole or albendazole.

S. stercoralis is considered a parasite of man, but occasionally it also infects dogs and cats. Man is the main source of infection via faecally contaminated soil. The reported rates of prevalence are exceptionally low, probably due to the use of less sensitive diagnostic methods. Even with culture methods the rates can be low. In the Philippines, the parasite was detected in only 1% of the 30,000 stools, many of which were examined by culture methods [9].

There are a number of *Strongyloides* species of other animals that can cause cutaneous larval migrans in humans and sometimes transient intestinal infections. *S. procyonis* from the raccoon and *S. myopotami* from the nutria as well as *S. ransomi*, *S. papillosus* and *S. mesteri* of swine, sheep, cattle and horses have been incriminated as causes of creeping eruption.

S. fuelleborni, once considered a parasite primarily of non-human primates, has been shown to be transmitted from man-to-man in some endemic areas. The parasite has been reported in urban areas and from monkey and ape populations. The parasite was first reported among populations in several parts of Africa and, more recently, a similar parasite has been found in Papua New Guinea.

The female worms pass eggs rather than larvae and the eggs mature in the soil releasing rhabditiform larvae. There is also a free-living cycle and which most infections are via skin penetration, there is also evidence suggesting transmammary transmission. In Africa the larvae have been found in women's milk. It has been detected more in infants than in children and adults in Papua New Guinea [1]. In one area of Papua New Guinea, children often suffer from the "swollen belly syndrome" and die as early as 3 weeks of age.

Spirurids

The spirurids are nematode parasites of the upper digestive tract, tissues, and tissue spaces of vertebrates. The oesophagus of the worms is divided into anterior muscular and posterior glandular parts. The worms are slender. The females have vulva near the middle of the body and most males have two spicules of unequal length. An intermediate host, usually an arthropod, is required for completion of the life cycle. There are a number of important zoonotic parasites in this group.

Dracunculosis

Dracunculus medinensis, the Guinea worm of humans and a variety of animals, is one of the longest nematodes. It is also known as the "Medina Worm" or "Fiery Serpent of Medina". The adults occur in subcutaneous tissue, with females measuring up to 120 cm in length by 1 mm in width; the males are rarely seen and measure 2–3 cm by 400 μm. An estimated 40 million people were infected with the parasite prior to an international campaign to eradicate it was initiated in the 1980s in Africa and other endemic areas of India and other parts of Asia.

The adult females usually migrate to one of the extremities and the cephalic end of the worm forms a papule in the skin which develops into a blister and ruptures when the limb comes in contact with water. A loop of the uterus of the worm prolapses releasing rhabditoid larvae into the water. First a large number of larvae are released. Then several batches over a period of weeks are released whenever the infected area is exposed to water.

The third-stage larvae are ingested by copepods of the genus *Cyclops* and the larvae migrate from the stomach to the coelomic cavity of the arthropod. Here the larvae moults twice to the second- and third-stage, becoming infective in three weeks. Copepods infected with more than 5 or 6 larvae usually die. Definitive hosts become infected by ingesting copepods in water. The copepod is digested in the host intestine with the release of the larvae, which penetrate the intestinal tissue and travel to the connective tissue via the lymphatics. The females continue to magrate and the uterus becomes distended with rhabditoid larvae. The prepatent period is approximately one year before the females cause the skin lesion and release the larvae. After 4–5 months, the worms apparently die.

The presence of the worm in deep tissue is usually asymptomatic. When the parasite migrates to the skin, systemic and local symptoms appear. The worms release a histamine like toxin which causes urticaria, pruritus, dyspnoea, vomiting, fever, giddiness and syncope. A local lesion develops in a few hours, which is initially a reddish papule with vesicular center and an indurated margin. Once the vesicle is formed and before the parasite emerges, there is a strong burning sensation. Once the vesicle breaks and the parasite emerges, the symptoms disappear apparently from the massive release of toxins during the discharge of the third-stage larvae. Lesions appear on areas of the skin such as the feet, legs, hands, and arms which commonly come in contact with water. Water carriers often have lesions on their backs because of the constant splashing of water on to their backs when they carry water buckets on their shoulders. The lesion may become secondarily infected with the infection extending into the tunnel where the female lies. Fibrosis may occur in the skin, muscle, and tendons leading to interference in locomotion. Possible sequelae may be arthritis, synovitis, ankylosis, and contractures involving limbs, and chronic abscesses. Infected individuals may be incapacitated for three months or more.

A postitive diagnosis of dracunculosis is made only after the appearance of symptoms and lesions and the emergence of larvae when the affected area is placed into water. In endemic areas, the diagnosis can easily be made even by the appearance of the blister-like lesion. No immunodiagnostic techniques are currently available.

The time-honoured treatment, once the worm has emerged, is to wind the worm slowly around a stick and gradually pull it out of the tunnel. Care must be taken to prevent breaking the worm since it will only withdraw into the tissue and cause severe inflammation. Anthelminthics, such as nindazole, thiabendazole and metronidazole reduce inflammation. Aspirin relieves some pain and antibiotics control the secondary bacterial infections.

It is estimated that 140 millon people are at risk of infection in the endemic countries of Africa, the Middle East, and Asia, and probably 10 million people become infected each year. Infection is obtained from the ingestion of water containing infected copepods in areas with low rainfall or areas with distinct rainy and dry seasons. The occurrence of infection is most common in rural areas and during the dry season when water levels are low and drinking water is most polluted. In India, infections are highest in people who drink from step-wells than those who obtain their water from curb wells. Males and females are both infected, with those between the ages of 5 and 50 years are heavily infected. It is usually a rural disease among the farmers; women who collect the water are the sources of contamination. Infected animals are also responsible for contaminating water sources.

Extensive campaigns are uderway from India, Pakistan, Ghana, and Nigeria, to eradicate dracunculosis [18]. A great deal of success has been made by providing safe water, health education, and chemical control of the copepod vectors. Boiling of water is a practical means of control, but it is impractical in areas where wood or fuel is scarce. Abate is an effective insecticide that is used to kill the copepods in contaminated water.

Gnathostomiasis

There are at least 12 species of *Gnathostoma* reported from canines and felines [10]. The adult worms cause gastric tumors in definitive hosts. The larval stages of a few species have been recorded as larval migrans in humans. Most human infections occur in Thailand and Japan. Of the six species of parasites (*G. spinigerum*, *G. hispidium*, *G. doloresi*, *G. vietnamicum*, *G. malaysiensis*, and *G. nipponicum*) endemic to Asia, *G. spinigerum* is the most important. Several hundred cases of gnathostomiasis spinigerum are seen each year in hospitals in Thailand.

G. spinigerum is a short stout nematode with an enlarged, more or less subglobose, spiny head with spines extending halfway down the body. The females and males are 25–54 mm and 11–25 mm long, respectively. One or more worms live in tumors found in the stomach of the definitive hosts. The females exerete the eggs through the host's faeces, reach water and hatch after a week or more, and release a first-stage larva. The larva enters a species of *Cyclops* and migrates to the coelomic cavity where it develops into a second-stage larva. When eaten by a second-intermediate host (fish, frogs, snakes), the larva migrates from the stomach to the musculature and develops into an infective third-stage larva. Many species of animals can serve as paratenic or transport hosts, but the parasite will only reach sexual maturity in the natural canine or feline definitive host. Man becomes infected with the third-stage larvae and serves as a paratenic host. Most human infections are acquired by eating raw freshwater fish.

In humans, the larva passes from the intestines and migrates through the tissue. First symptoms occurring a few days after ingestion of an infected intermediate host include nausea, vomiting, abdominal cramps, and diarrhoea. Few days later, allergic reactions such as urticaria, pruritus, cough, and chest discomfort may occur. Leucocytosis with eosinophilia will develop with transient or migratory swelling. The parasite may cause pain by migrating into any organ. The larva eventually enters subcutaneous tissue, causes cutaneous larval migrans and intermittent subcutaneous oedema. The swelling with redness lasts a week or more and then disappears, only to reappear later at a new location. This may continue for even for years. The migration of the parasite into the CNS, may cause paralysis, coma, bloody xanthochromic cerebrospinal fluid, and death. The parasite also has been recovered from the eye.

The diagnosis of patent infections in the natural feline and canine host is done by faecal detection of ovoid (65 μm by 40 μm), transparent, pitted, unembryonated eggs, with a single mucoid plug. Eggs are usually few in number. In the abnormal human host, the definitive diagnosis is identification of larvae in surgically obtained specimens. In endemic areas, recurrent transient subcutaneous swellings with leucocytosis and eosinophilia are suggestive of infection. A positive skin test reaction using antigens from larvae or adults is of some value, but more recently highly purified antigens have been used in an ELISA to detect antibodies in patients in Thailand. Specific immunologic tests are of special value in differentiating *G. spinigerum* from *A. contonensis* infections in the CNS [38].

Specific treatment is surgical removal of the larva. This is difficult because of the transient nature of the infection. Mebendazole and albendazole have been used, but under the dosages given, the parasite was stimulated and moved but was not eliminated.

Although infections may be acquired by ingesting infected copepods in drinking water, most humans acquire it by eating uncooked freshwater fish. Similarly, the definitive host acquires infections by eating fish or chickens, ducks, frogs, and water snakes, that can serve as paratenic hosts. In Japan and Thailand, species of fish, especially eels, are the major sources of human infection, with the fish harbouring hundreds of third-stage larvae. Infections could be prevented if the intermediate and paratenic host were consumed after properly cooking. It is difficult, however, to change eating habits that have been in practice for generations.

Gongylonemiasis

Gongylonema species are spirurid nematodes that live in tunnels in the mucosa and submucosa of the oesophagus, tongue and oral cavity of sheep, swine, goats, cattle, monkeys, and or rare occasion man. The nematodes are widely distributed geographically and reported mainly from animals and less than 50 infections in human have been documented [22].

Female worms may be as long as 15 cm, and the males 6 cm. The adult worms are covered with scutes or bosses arranged in eight longitudinal series, two each in four submedian fields. The eggs are embryonated when deposited, are swallowed and passed in the feces. The eggs must be eaten by various species of dung beetles or cockroaches. The eggs hatch in the beetles' intestine and the larvae migrate to the haemocoel where in one month, they develop into thrid-stage larvae. The definitive host becomes infected by eating the intermediate arthropod host. The larvae escape from the insects, penetrate the stomach or duodenal mucosa and migrate up to the oesophagus, tongue, and oral mucosa and develop into adults in two months. In humans, the parasite has been recovered from the submucosa of the lips, gums, hard palate, soft palate, and tonsils. There is often pharyngitis and stomatitis, erosive patches in the oesophageal mucosa with blood and sialorrhoea. Recovery of the worms from the mucosa confirms the diagnosis and the worms are removed by the fingers or surgical instruments.

G. pulchrum is the species associated with human infections. *G. neoplasticum* is a species found in the stomach of rats and causes neoplasms. Man acquires infection accidently by ingesting an infected insect host. Animals acquire infection by eating infected beetles on grass. In endemic areas many species of beetles serve as hosts and swine and cattle are usually heavily infected, especially in Russia.

Thelaziasis

Thelazia callipaeda and *T. californiensis* are two of several spirurid species that lodge in the conjunctiva of wild and domestic animals, but rarely in humans. More than 20 human infections have been reported from Asian countries and involve *T. callipaeda*, the oriental eye worm [36]. *T. californiensis* is found in animals in North America and a number of human infections have been reported from California, U.S.A.

The adult worms are approximately 5 mm long and lie on the surface of the eyeball. Eggs are embryonated and mucoid flies feeding on lacrimal secretions ingest the larvae. The larvae develop to the infective stage, migrate to the proboscis and are released by the fly when it feeds on ocular secretions. The worms mature within six weeks and produce eggs.

The worms irritate the conjunctival sac and cause lacrimation, conjunctivitis, scratching of the cornea and fibrous opacity. There may be pain with the sensation of the presence of a foreign body. The diagnosis is made by recovery and identification of the worm. A topical anesthetic is applied to the eye prior to removal with forceps.

Transmission is usually seasonal, when the fly population is large.

Filariasis

Humans have a number of filarial parasites that inhabit the lymphatics, body cavities, and tissues, but there are few that are considered zoonotic parasites. *Brugia malayi*, a subperiodic strain, is of zoonotic importance, but *B. malayi*, periodic strain, is not. Other human filarial species, such as *Wuchereria bancrofti*, *Onchocera volvulus*, *Loa loa*, *Mansonella perstans*, *M. ozzardi*, and *M. streptocerca*, may infect some animals, but they are not considered zoonotic. Filarids of animals, such as the dog heartworm *Dirofilaria immitis*, will infect humans and cause disease as will the tissue filariae *D. ripens* and *D. tenuis*. There are also reports of unidentified species of *Dipetalonema* infecting man.

All filariae require an arthropod intermediate host in their life cycles. Female filarids retain their eggs and release microfilariae. The microfilariae are ingested by arthropods feeding on blood or tissue and the larvae develop into infective thrid-stage larvae in these vectors. When the vector feeds, the third-stage larvae are introduced into the definitive host. In some species of filarids, the microfilariae may appear in the blood in greatest numbers during certain times of the day or night. This is termed periodicity: nocturnal, diurnal or, in some cases, subperiodic. In the latter, the microfilariae are in the blood in early evening, e.g. subperiodic *B. malayi*. This species (strain) is also unique in that the parasite will complete development in some animals as well as in humans. Mongolian gerbils and cats are also excellent hosts for subperiodic *B. malayi* and even some strains (China and Korea) of periodic *B. malayi*.

Malayan filariasis is endemic in Southest Asia, Korea, China, and India. Extensive campaigns to eliminate the parasite has had success in Korea and China, but becuase of the zoonotic nature of the parasite, it has been difficult to control it in Malaysia and Indonesia. Monkeys and felines, especially cats are natural hosts for subperiodic *B. malayi* and the major arthropod vectors are *Mansonia* spp. of mosquitoes. *B. pahangi* is similar to *B. malayi* and is a parasite of canines, felines, and monkeys. The microfilariae of these two species are difficult to distinguish and some parasitologist suspect without conclusive evidence that *B. pahangi* will also infect humans. There are also a few reports of brugian infections in humans, from North and South America but the species of parasites is unknown.

Malayan filariasis in humans is a disease which results in invasion of lymphatic channels causing fever, lymphoedema, lymphadenitis, lymphangitis, and abscess formation. Many infections lead to elephantiasis of the lower extremities, mostly the legs. This microfilariasis is easily diagnosed by its identification from the blood smear stained with Giemsa. A number of workers have developed monoclonal antibodies to detect antigen (see Chapter 12) and some DNA probes are being evaluated. Diethylcarbamazine (6 mg/kg/d for 10 d) along with aspirin and a steroid to suppress side effects is used for treatment.

The dog heartworm, *D. immitis*, is found throughout the world. In canines, the large adult worms (12–30 cm) reside in the right heart and pulmonary vessels and microfilariae circulate in the blood. Many species of mosquito serve as vectors. Occasionally, humans (less than 100 reported cases) become infected with the parasite in which the worms cause lesions in the lungs [32]. Nodules (1–3 cm) are usually noted by radiography and have been characterized as "coin lesions". The diagnosis of dirofilariasis is made after the lesion has been removed and examined. Serologic and antigen detection tests are available to determine infection in canine and could be used for suspected human infections [25]. Most infections in human are amicrofilaremic and have been reported from the United States, Japan and Australia (see review in Parija [30]).

Subcutaneous and intraocular dirofilariasis involving *D. tenuis*, a parasite of raccoon, *D. ursi from* bears, and *D. sepeus* from dogs have been reported.

ACANTHOCEPHALA ZOONOSES

Acanthocephaliasis

The spiny-headed worms are natural parasites of swine, rodents, and other animals, and on rare occasions infect humans. The worms were once considered nematodes but have been reclassified. They are large (10–35 cm) and are characterized by the presence of a proboscis with 5–6 rows of curved spines. The proboscis is usually inserted into the intestinal tissue. Two species *Macracanthorhynchus hirudinaceus* and *Moniliformis moniliformis* have been reported from humans. Beetles and cockroaches serve as intermediate hosts. Animal infections are widespread, but human infections are limited; many have been recorded from China. Humans suffer from abdominal colic, diarrhoea, and intestinal perforations. Secondary bacterial infections leading to peritonitis and death are also reported. Infection occurs from accidental, or sometimes intentional, ingestion of beetles.

CESTODE ZOONOSES

Cestodes, or tapeworms, as adults are parasites of the intestinal tract of vertebrates, and larval stages occur in both in vertebrate and invertebrate hosts. The adult worms are usually host-specific, while the larval stages exhibit less specificity. There are about 21 orders of tapeworms, but only two, Pseudophyllidea and Cyclophyllidea, have species that are of zoonotic importance.

The tapeworms are flattened dorsoventrally, have a head or scolex followed by a neck, and a chain of segments or proglottids, each an individual reproductive unit. There are immature, mature and gravid proglottids. They are hermaphroditic and copulation occurs between mature proglottids, within a mature proglottid, or with proglottids of other worms. Gravid proglottids are filled with eggs, which may pass out of the worm individually with the faecal stream, pass in faeces as packets, or the ripe or gravid proglottid may detach from the strobila and pass out with the faeces or crawl out of the anus.

Except for *Hymenolepis nana* an intermediate host is required for further development. The embryonated egg contains an oncosphere or hexacanth (3 paris of hooklets) which develop into one of several different larval stages depending on the species of tapeworm. The infective stage in the intermediate host is ingested by the definitive host and the adult worm develops in the intestines. Some of the tapeworms are restricted only to humans, with a few also occurring in animals, while in others man is an incidental host and essentially serves as an intermediate host. In one species, *Taenia solium*, man may be infected both with adult and larval stages of the parasite.

Human cestodiasis occurs throughout the world and the prevalence and distribution are generally related to eating habits of population groups.

Diphyllobothriids

Diphyllobothriasis

Diphyllobothrium species are pseudophyllidean tapeworms, characterized by a scolex bearing two bothria or sucking groves. There are a number of species reported in humans, but the most important is *D. latum*. This is one of the longest parasites of man, measuring up to 12 meters. The width of proglottids is more than their length and the uterus in the gravid proglottids are rosette-shaped. The eggs are discharged from the proglottids through a uterine pore. The gravid proglottids are retained and eventually disintegrate. Thin-shelled operculated eggs are passed in the faeces; and must reach water for further development. A ciliated embryo (coracidium) develops and hatches from the egg, swims in the water and is eaten by a specific copepod. The larva penetrates into the haemocoel and develops further into a procercoid larva. When eaten by a fish, the procercoid is released, migrates to the fish muscle and transforms into a pleurocercoid larva or sparganum. Larger fish may ingest smaller infected fish and serve as a paratenic host. When the fish muscle is eaten, the sparganum develops into an adult worm in the intestines.

The parasitosis is endemic to lakes, rivers, and reservoirs of the sub-arctic and temperate regions of Eurasia with highest prevalences in Finland, Sweden and Norway. It has been introduced into the Americas by European immigrants.

Most infections are without symptoms, but when there are many worms entangled in the intestines, diarrhoea, abdominal pain and anaemia may be noted. When the strobila is lodged in the jejunum, the tapeworm competes for the available supply of vitamin B_{12} and its deficiency causes pernicious anaemia in the host. The diagnosis is made by finding eggs (which are numerous) in the faeces. Proglottids may also be found. Praziquantel or niclosamide are effective in treating *D. latum* infections.

Human and animal infections are acquired by eating raw infected fish; therefore, the parasitosis is known only in areas where this eating habit is popular. Dogs, cats, bears, and other carnivorous animals are also hosts for the parasite, but man is usually the major source of infection when faeces are permitted to enter freshwater, where there are susceptible copepods and fish.

There are other species of *Diphyllobothrium* that occasionally infect humans. In Peru, where cerviche (raw marine fish marinated in lime, hot peppers, and salt) is eaten, *D. pacificum* is found in humans and pinnipeds. In Alaska, *D. dendriticum* of fish-eating birds and mammals is often reported in humans. Several unspecified species of *Diphyllobothrium* are reported from Canadian fish. DNA probes are being developed to aid in speciation of the larval stages.

Sparganosis

Sparganosis is infection with the sparganum or pleurocercoid stage of pseudophyllidean tapeworms. The adults are parasites of dogs, cats, and other carnivores and are species of *Spirometra*: *S. mansonoides*, *S. mansoni*, and *S. proleferuns*. The spargana of these species have been reported in the tissues of humans.

The life cycle is similar to that of *D. latum*; coracidium from egg, and procercoid in copepods, but the plerocercoids are found in amphibians, reptiles, birds, and mammals. Dogs and cats acquire the infection from eating these second-intermediate hosts. Humans essentially serve as transport hosts, acquiring the infection either by ingesting procercoid infected copepods or a raw second-intermediate host, or by using animal poultices on their skin, eyes, or vaginal opening. The larva migrates out of the animal and into the human tissue; the sparganum wanders aimlessly throughout the tissue, causing transient swellings. The adult worms in the definitive hosts are of variable size, measuring up to 25 cm and the sparganum is white, ribbon-like and up to 10 cm in length. Although sparganosis is reported from all over the world, the disease is seen mostly in Asians. In North America, the species is *Spirometra mansonoides*, and in Asia, *S. mansoni*. Infections in animals are variable for both the definitive and intermediate hosts. *Spirometra* species has recently been found in a cat in India, and an infection in humans has also been reported [34].

Larvae migrating through tissue usually provoke an inflammatory response with swelling, erythema, pruritus, and pain. Many of the lesions are subcutaneous and transient. Ocular sparganosis resulting from the use of poultices occurs most often in Southeast Asia. There is oedema of the eyelids, lacrimation, pruritus and pain. Eosinophilia is usually present. In recent years, there have been increasing reports of cerebral and spinal cord sparganosis in Asians as well as in Americans [39]. The diagnosis is usually made when the sparganum is removed surgically from a lesion. Species identification is made only after feeding the larva to an animal and recovering the adult worm. Most cases are treated surgically since chemotherapy is not effective.

Taeniasis

There are a large number of *Taenia* species of tapeworms, but only two *T. saginata* and *T. solium* are important as parasites of human. The adult worms are found only in the intestines of humans, while the larval stages, *Cysticercus bovis* and *C. cellulosae* develop in the muscle of their intermediate hosts, bovids and porcines, respectively. The latter larvae will also develop in the tissue of humans and cause a serious disease known as cysticercosis.

The morphology of both species is similar. The scolex of *T. solium* possesses rostellum containing hooklets and has four suckers which is followed by a neck, immature, mature and gravid proglottids. *T. saginata* is longer (4–10 m) than *T. solium* (2–4 m). The worms reside in the small intestine and the terminal gravid proglottids detach from the strobila and are passed with the faeces or crawl out of the anus. Each proglottid may contain as many as 100,000 eggs, The eggs of all *Taenia* species are indistinguishable; they are round, 30–40 µm in size, with a thick striated shell containing a hexacanth embryo. Entire proglottids or individual eggs are ingested by the intermediate host and the oncospheres escape and migrate to the muscles where they develop into fluid-filled translucent cysts with an invaginated scolex. Man acquires the infection after eating raw or incompletely cooked beef or pork. The cysts rupture in the intestine, the scolex everts and attaches to the intestinal mucosa, and produces eggs after 10–12 weeks.

Both species of *Taenia* are found worldwide and are most common in areas with poor sanitation. The prevalence is higher for *T. saginata* than for *T. solium,* but because of the ability of the larval stage to develop in man, *T. solium* infection is a more serious parasitosis. *Taenia solium* cysticercosis are important health problems in Latin America, Africa, Southeast Asia, India, and China.

Intestinal infections with the adult tapeworms are usually asymptomatic and individuals realize they are infected by the passing of proglottids through the anus. In symptomatic cases, there may be nausea, abdominal pain, flatulence, diarrhoea or constipation, anorexia, and weight loss. Some psychological problem may develop after anal passage of segments.

Symptoms associated with cysticercosis vary with the location and number of parasites. Cysts in subcutaneous tissue and muscle are usually asymptomatic, while those lodged in the eyes and CNS can result in a number of clinical manifestations. The symptoms may resemble cerebral tumors, meningitis, encephalitis, intracranial hypertension and hysteria. Patients usually experience headache, epileptiform seizures. Ocular lesions appear like uveitis, iritis, and conjunctivitis. Motor muscles of the eye may also be affected.

The intestinal infection is diagnosed by demonstration of the proglottids or eggs. Species identification of the proglottids is made by enumerating the uterine branches (*T. solium* 7–12, *T. saginata* 16–30) after staining. Eggs can be found by direct faecal examination or on perianal swabs. The diagnosis of cysticercosis is more difficult. Subcutaneous cysts can be removed surgically and examined, or tissue biopsies will reveal typical cysticercus characteristics. X-rays and especially CT scans and magnetic resonance imaging (MRI) are used where deep tissue and CNS are involved. ELISA is widely used for diagnosis of neurocysticercosis. An ELISA to detect specific antibodies in saliva has been developed and HPLC and ELISA have been used to detect antigens in CSF of patients [4, 12].

Cysticercosis in animals is not serious. With heavy doses of eggs, bovids may experience fever, anorexia, debilitation and muscle stiffness. Pigs do not live long enough to develop serious disease, but some infected animals may experience epileptiform convulsions.

Single doses of praziquantel have been found effective in treating intestinal taeniasis in humans and cysticercosis in cattle and pigs. Niclosamide is also used for treating intestinal infections. Praziquantel treatment for 14 days has replaced surgical treatment of cysticercosis in humans. It is of little value, however, when the cysts have become calcified. Albendazole has also been found effective in the treatment of neurocysticercosis [19].

Taeniasis and cysticercosis occur in populations that eat raw or partially cooked beef and pork (without killing the eysticerci). Most of these population groups also have poor sanitary habits, exposing cattle and pigs to human faeces. Bovids are the only source of human infection with *T. saginata.* However, it has been shown recently that the strain of *T. saginata* from Taiwan, and possibly from other Asian areas, may be obtained from pig liver. This strain or species develops in certain strains of pigs in rural areas of Taiwan but develops poorly or not at all in bovids [11]. The population does not eat beef and often eats uncooked pork. Comparative DNA studies between Taiwan *T. saginata* and specimens from other parts of the world show some differences between the parasites.

Taeniasis and cysticercosis can be controlled by improving sanitation and by eating only the well-cooked beef and pork.

Echinococcosis

Hydatid disease is one of the most serious helminthic zoonoses. It is caused by infection with the larval stages of tapeworms of the genus *Echinococcus*. There are four species: *E. granulosus*, *E. multilocularis*, *E. vogeli*, and *E. oligarthrus*. The first three cause disease in humans, and the most important is *E. granulosus*.

Dogs and wild canines are the definitive hosts for *E. granulosus*, with the tiny (3–6 mm) adult tapeworm residing in the small intestine. The worm has a scolex with four suckers and a rostellum with hooklets, and a total of three proglottids: immature, mature, and gravid. The eggs are similar to those of other taenid species and are discharged from the gravid proglottid into the faeces. After ingestion by an intermediate host, usually a herbivore or omnivore, the egg hatches, releases a hexacanth embryo which migrates from the intestines to various organs. The liver of sheep is a common location where the parasite forms a cyst. The cyst for *E. granulosus* is unilocular with a wall consisting of cuticular, laminated outer layers, and a germinal inner layer. Brood capsules containing protoscolices bud off from the germinative laryer and float in the fluid filled cyst. The fluid containing the protoscolices and brood capsules is called **hydatid sand**, and the cyst is a **hydatid cyst**. When the infected organ is eaten by canines, the protoscolices are released and attach to the intestinal wall and grow into adults. In sheep, a hydatid cyst develops in 9 months after ingestion of eggs, and the prepatent period in canines is approximately 2 months. Infection in dogs may last for a year, and the infection is usually heavy since a single hydatid cyst can give rise to thousands of worms.

E. granulosus is enzootic wherever sheep are raised, especially in parts of South America, Mediterranean countries, Russia, North China, India, Australia and New Zealand. *E. multilocularis* is limited to the Northern Hemisphere, Middle East, Northern India, Russia and Eastern Europe, and *E. vogeli* to Latin America. *E. granulosus* and *E. multilocularis* may co-exist in areas of the Northern Hemispheres. Hydatid disease in humans is highest in sheep-raising countries.

Adult *Echinococcus* infections in canines are asymptomatic, but heavy infections may cause enteritis. Natural intermediate hosts suffer very little from the hydatid cysts. Farmers may suffer an economic loss as a result of condemnation and confiscation of infected viscera of sheep and beef cattle.

E. granulosus hydatid cysts most frequently occur in the right lobe of the liver, followed by the lung, kidney, heart, bones, and CNS of humans. Symptoms depend on location and size of the cyst; in some cases, location may depend on the strain of the parasite. Strains from sylvatic areas, involving cervids and wolves, tend to localize in the lungs and are more benign, while strains from domestic cycles between sheep and goats, localize in the liver. The majority of infections in humans conssist of a single cyst, but multiple cysts and multiple organ involvement also occurs. It takes years for cysts to become large enough to cause organ dysfunction and pain. The cyst may enlarge to contain several liters of fluid. Rupture of the cyst may cause an anaphylactic shock and death, due to release of antigens. Release of the protoscolices may set up many new cysts in the abdominal cavity, lungs and other organs. There may be signs and symptoms of hepatic enlargement, a palpable mass in the right upper quadrant of the liver, epigastric pain, nausea, and vomitting. Cysts in the lung are usually in the right lower lobe, without any symptoms, but leakage may cause pain, dry cough, dyspnoea, haemoptysis. The cyst may be coughed up and recovery follows. Many patients have hepatic and pulmonary cysts. Hydatidosis of the bone is rare, but when it occurs, it is symptomless, causes erosion of the marrow, destruction of the trabeculae, necrosis and fracture. Renal cysts cause pain and haematuria, and cerebral cysts lead to epilepsy and usually death.

Alveolar hydatidosis due to *E. multilocularis* usually occurs in the liver where growth is progressive, forming vesicles in all directions, producing a multilocular appearance. As the cyst grows, there is initial necrosis and the cyst becomes a spongy gelatinous mass. Few protoscolices are seen in human infection and there may be metastases to other organs, with secondary alveolar cyst development in the lung and brain. The symptoms are similar to those seen in mucinoid carcinoma of the liver. There may be hepatomegaly and splenomegaly, ascites and jaundice. The disease progresses slowly and the diagnosis is usually not made early enough to prevent death.

Polycystic hydatid disease is caused by *E. vogeli* and the cysts are intermediate between cystic and hydatid disease. The fluid-filled cysts are large, and contain brood capsules and protoscolices. The polycystic structure extends throughout the affected area. The liver is the primary site with cysts extending into the mesenteries, omentum, pericardium lungs, and other organs.

When echinococcosis is suspected, especially in persons from endemic areas, a variety of serologic tests can be used. Autogenic material is readily available from hydatid cysts removed from animals and it can be diagnosed by immunoelectrophoresis, indirect hemagglutination and ELISA (see also Chapter 12). Monoclonal antibodies are also available to assay parasite macromolecules circulating in sera of patients [27]. Clones from specific DNA in an ELISA have also been used for unilocular and alveolar hydatid disease. The Casoni skin test may be initially helpful in indicating disease. In association with serological tests radiography, CT scanning, ultrasonography, and MRI are extremely valuable in making the diagnosis. All *Echinococcus* species share antigens; therefore, the serologic tests can be used for all. Serologic tests for hydatid disease in animals are not reliable, as the antibody production is low and cross-reaction with other parasites are also observed. The confirmed diagnosis is made only after identification of larval stages, recovered from cysts at surgery. Adult worm infections in dogs can be diagnosed by recovery of worms from faeces and followed with treatment. The eggs cannot be differentiated from other taenid species.

Surgical removal of the hydatid cyst is the most expedient treatment, but spilling of the cyst fluid should be prevented. The protoscolices must be removed carefully and be inactivated prior to removal of the cyst. Once the cyst is sterilized, it can be removed. Alveolar hydatid cysts can also be removed surgically, but requires extensive resections, or removal of the affected organ. Little is reported on removal of polycystic hydatids. In cases where surgery would be difficult, anthelmintics have been found to be of some value. Long term use of mebendazole has slowed the growth of some hydatid cysts, but all larvae are not destroyed. Recently, albendazole has been found more effective than mebendazole for cystic hydatid and alveolar hydatid disease.

E. granulosus is maintained primarily by a dog-sheep life cycle, In some parts of the world, animals other than sheep may serve as an intermediate host. In North China, yaks are an important source for canine infections [5]. In the wilds, other canines and herbivores are important (see Table 10; Chapter 3). In pastoral areas, home-butchered sheep viscera is fed to dogs and as a result, exposure to infection is continuous. Eggs of the parasite are dispersed in the fields, predisposing the sheep to infection. Humans acquire infections from dogs, since tapeworm eggs adhere to hair around the anus and children playing with dogs are usually exposed to infection at a young age. Eggs may also contaminate vegetables and water, and can be disseminated by flies and other arthropods.

E. multilocularis is maintained in nature primarily by rodents and foxes. Dogs and cats can become infected by eating infected rodents. Because of the sylvatic nature of the disease, humans rarely become exposed to infection. Most infections are probably from domestic dogs and cats.

E. vogeli is only known from Latin America where wild canids and large rodents are involved in its life cycle. Humans may acquire infections from hunting dogs that have eaten viscera of wild rodent.

Prevalence rates of human infection is limited since most infections go undetected for many years.

Data on sheep infection is often unreliable since sheep farmers are unwilling to provide the information. In some sheep raising countries, 50% of the animals and often 100% of the dogs can be infected.

Echinococcus eggs are very resistant to environmental conditions and can remain viable for long periods under extreme changes in temperature and other environmental conditions.

Dipylidiasis

It would be very unusual for a dog or cat to go through its life, without becoming infected with *Dipylidium caninun.* Fleas are intermediate hosts and usually transmit the parasite to these animals and occasionally to humans. The parasite is reported in animals from all over the world and in humans from most European countries, the Americas and Asia.

The adult worms are 10–70 cm long, have an armed rostellum and a double set of reproductive organs with uterine pores on each lateral margin; hence, the common name, double pore tapeworm. Gravid proglottids are pumpkin-seed shaped; they are motile and pass out with the feces or migrate on their own out of the anus. Each proglottide has packets of eggs enclosed in an embryonic membrane. The proglottids disintegrate with the release of eggs which are then eaten by flea larvae. The eggs hatch in the gut of insects, migrate to the haemocoel and develop into procercoid and then cysticercoid larvae. The larvae survive metamorphosis of the larval fleas to the adults, which are subsequently eaten by the dog or cat host, and the cysticercoids develop into adult tapeworms in 3–4 weeks.

Humans, mostly children, acquire the infection by the accidental ingestion of larvae or adult fleas infected with cysticercoid larvae. Symptoms are usually mild with diarrhoea, irritability, anorexia. Motile proglottids wandering out of the anus are often the only indication of infection and causes a great deal of anxiety to the parents of young children. Diagnosis is made by finding the proglottids and examining "press preparations" under a microscope for double uterine pores and egg packets. The infections are treated with niclosamide or praziquantel.

Animals suffer very little, unless the infections are exceptionally heavy. The diagnosis is easily made by finding proglottids, usually on the surface of the faecal material of the animal or by detecting eggs around the anus of the dog or cat. Animals easily become infected with the parasite, usually repeatedly, and humans undoubtedly are resistent to infection, given the high prevalence of infections in pets.

Hymenolepiasis

Two species of *Hymenolepis* found naturally in rodents will on occasion infect humans. *H. diminuta*, the rat tapeworm, is occasionally found in people, but *H. nana*, the dwarf tapeworm of mice, rats and other rodents, is not uncommon in people. It is considered the most common tapeworm in the United States. Both parasites are found in rodents and human populations all over the world. Only 200 human infections with *H. diminuta* have been reported, but this is probably not a valid figure since all infections have not been included. A number of infections were found during biomedical surveys in rural areas of Indonesia. *H. nana* infections, also detected during surveys, are more common in Indonesia and the Philippines, and the prevalence rates are high especially in infants in Europe, Africa, the Middle East and Latin America. Infections rates in rodents for both the species are usually high.

The strobila of *H. diminuta* varies from 20–60 mm and the proglottids increase in width from 0.5 mm to 3.5 mm towards the distal end and are wider than long. Subspherical eggs (60–80 μm by 70–85 μm), passed by detached gravid proglottids into the faeces are ingested by a larval or adult arthropod. The released oncosphere passes to the haemocoel and develops to a cysticercoid larva. A variety of arthropods, such as beetles, fleas, and lepidopterans serve as intermediate hosts. The cysticercoids will develop into an adult after the arthropod is ingested.

H. nana are smaller (2.5–4 mm by 1 mm), and the eggs are spherical to subspherical (30–45 μm).

The eggs contain an oncosphere which is surrounded by an inner envelop with polar thickenings and filaments. The life cycle is direct, but arthropod intermediate hosts are involved in some rodent strains infecting rodents. Man may act as a definitive as well as an intermediate host. Eggs released by the distal segments pass into the faeces. Upon ingestion, the oncosphere is released and the hexacanth embryos penetrate the intestinal mucosa. After 5 days, the larvae change into a cysticercoid and re-enter the gut lumen where they develop into an adult in 2 weeks and release eggs. Some eggs may be retained to re-initiate the life cycle. The adults survive only for a few weeks and new generations emerge through autoinfection. Resistance to reinfection will eventually develop. In some strains, coprophagous arthropods may serve as intermediate hosts with cysticercoids developing in the haemocoel. There is a question of speciating *H. nana*, of separating them into subspecies, such as *H. nana fraterna* in rodents and *H. nana nana* in humans. There are physiologic differences as well in these tapeworms, but subspecies separation is not accepted readily. DNA analysis of these strains may provide the answers.

Most human infections with *H. nana* and *H. diminuta* are asymptomatic, but symptoms of nausea, vomiting, abdominal pain and diarrhoea have been reproted. Eosinophilia, allergic rections of anal and nasal pruritus have been noted in some patients whereas in neurologic symptoms, such as seizures, dizziness, and irritability have also been reported.

Proglottids may be found in faeces, but the presence of characteristic eggs in the faeces provides the diagnosis. Niclosamide or praziquantel in a single dose are recommended for treatment.

Humans are the natural reservoir for *H. nana*, but there are strains in rodents that readily adapt to humans, with insects serving as intermediate hosts. The reservoir hosts for *H. diminuta*, however, are rodents, with insects serving as vectors of the parasistoses. Beetles are important in the transmission of these tapeworms, especially those that are found in cereal and grains; infected insects are accidently ingested with these products.

Other Cestodiases

There are a number of other cestodiases that are occasionally reported in humans: bertielliasis, coenurosis, raillietiniasis, inermicapsiferiasis, and mesocestodiasis. All are parasites of lower animals and have been reported from various parts of the world.

Bertiella studeri is an anoplocepholid tapeworm of nonhuman primates. Proglottids are passed in groups of about 20 and orbatid mites serve as intermediate hosts. Most human infections are asymptomatic and have been reported in children from Southeast Asia and India. The eggs are irregularly ovoid, have a middle envelop and an inner shell with a bicornate protrusion on one side. The worms may pass spontaneously.

Coenurosis is infection with the larval stage of a *Taenia* or *Multiceps* species of tapeworm. The coenurus is a fluid-filled bladder, the size of a hen's egg filled with protoscolices that bud off either internally or externally. Canines are the definitive hosts and the herbivores intermediate hosts in which the coenurus is found in the CNS and connective tissue. When eggs are ingested with grass or water, the oncosphere migrates from the intestine to the tissue and in 6–8 months, the coenurus develops. When these larvae are eaten by canines, large numbers of adult tapeworms develop in the intestine. Man accidently becomes infected with the eggs leading to the developing of cyst in the brain, eye, or connective tissue. Symptoms are similar to cerebral cysticercosis or hydatid disease and the pre-operative differential diagnosis is difficult. Serologic tests are available. Chemotherapy has not been effective. The parasitosis is reported from sheep in many parts of the world but most infections in humans have been reported from Africa. Disease in sheep can cause death leading to significant economic losses.

Raillietina celebensis is a common parasite of rats and children in Asia. The intermidiate host is unknown, but is assumed to be an arthropod. The infection is usually asymptomatic, with often spontaneous expulision of the worm. Proglottids contain hundreds of egg capsules, with each capsule containing four eggs. Ants may be intermediate hosts.

R. demerariensis is a neotropical species found in Latin America, with infections not unusual in rural areas of Ecuador. Proglottids the size of rice grains contain 200 capsules, each with four eggs. Gastrointestinal disturbances may develop with infection.

Inermicapsifer madagascariensis is related to *Raillietina* sp. and is found in rodents in Africa and parts of Southeast Asia. The segments contain 150 capsules, each with six or more eggs. Infections are asymptomatic and the diagnosis is made by finding the proglottids. The unarmed scolex of the parasite must be examined to differentiate it from *Raillietina*. The life cycle is unknown.

Mesocestoides variabilis and *M. lineatus* have been reported in humans from Asia, Africa and North America. The life cycle and mode of trasmission is unknown. Dogs, cats, and other carnivores are definitive hosts, and the first-intermediate hosts are coprophagous arthropods. Amphibians, birds, and reptiles serve as second-intermediate hosts in which a pleurocercoid larva called tetrathyridium develops in tissue or serous cavities where it multiplies asexually. The tetrathyridium develops into an adult in the intestine of the definitive host. Symptoms in humans consist of digestive disorders, abdominal pain and diarrhoea. The adult worms have a parauterine organ which can be seen in proglottids passed in the faeces. Eggs with double membrane are grouped in the parauterine organ. Infections in human are acquired by eating raw meat of definitive host, usually from birds.

CONCLUSION

Helminthiases of animals that are transmitted to man remain a continuing problem throughout the world. Although no population can be considered free from or immune to the scourges caused by these worms, the predominance of infections occur in the lesser developed parts of the world where poor sanitary practices are rampant. Eating habits are also important in the continued existence of the helminthiases. The eating of raw pork, beef, lamb, fish, and crabs, and the accidental ingestion of arthropods are major sources of these zoonoses. Improvement of sanitation and changes in eating habits would certainly lower the incidence of these diseases, but this is easier said than done. Anyone with experience in dealing with people in impoverished areas is well aware of the difficulty of changing the habit of the people. Helminthic zoonotic diseases have existed for generations and unless changes are made, they will continue into the future.

However, there is hope for the future. Even though people are becoming infected, we are now developing the tools to accurately diagnose these parasitoses. There has been a great improvement in serodiagnosis. Specific antigens for the respective parasites are being purified and the ELISA test is in widespread use. Specific monoclonal antibodies are available to detect antigens from helminthic parasites and DNA probes will be available in the not too distant future. DNA analysis will also aid in the specific identification of those parasites in cases where a taxonomic status is questionable. Unfortunately, it will take some time before these tools of the future become available and are cheap enough to be used in endemic areas where they are really needed.

During the past few decades, several new species of zoonotic parasites have been found to traumatize man. The question arises as to how many new ones are still out there in nature [26]. Great changes are occurring in our environment that jeopardize our ecology. These changes will cause havoc with our wildlife and will thus also change the helminth fauna in these animals. Will these worms adapt to man who is moving into former animal habitats? Future public health workers, physicians, and veterinarians should keep a vigilant eye for the possible emergence of new zoonotic diseases.

REFERENCES

1. Barnish, G. and Ashford, R.W. *Strongyloides* cf *fuelleborni* in Papua New Guinea: epidemiology in an isolated community, and results of an intervention study. Ann. Trop. Med. Parasitol. **83**: 499–506 (1989).
2. Bier, J.W., Deardoff, J.J. Jackson, G.J., Raybourne, R.B. Human anisakiasis. In: *Intestinal Helminthic Infections, Bailliere's Clinical Tropical Medicine and Communicable Diseases* (Z.S. Pawloswki ed), pp. 723–733 (1987).
3. Campbell, W.C. Trichinosis revisited—Another look at modes of transmission. Parasitol. Today **4**: 83–86 (1988).
4. Choromanski, L., Estrada, J.J. and Kuhn, R.E. Detection of antigens of larval *Taenia solium* in the cerebrospinal fluid of patients with the use of HPLC and ELISA. J. Parasitol. **76**: 69–73 (1990).
5. Craig, P.S., Deshan, L. and Zhaoxun, D. Hydatid disease in China. Parasitol. Today **7**: 46–50 (1991).
6. Cross, J.H. Experimental studies on *Angiostrongylus* species and strains in monkeys and laboratory animals. In: *Studies on Angiostrongyliasis in Eastern Asia and Australia* (J.H. Cross ed), pp. 118–137. NAMRU–2 SP–44 (1979).
7. Cross, J.H. Intestinal capillariasis. Parasitol. Today **6**: 26–28 (1990).
8. Cross, J.H. Public health importance of *Angiostrongylus cantonensis* and its relatives. Parasitol. Today **3**: 367–369 (1987).
9. Cross, J.H. and Basaca-Sevilla, V. Biomedical Surveys in the Philippines. NAMRU-2 SP-47 (1984).
10. Daengsvang, S. A monograph on the genus *Gnathostoma* and gnathostomiasis in Thailand. 85 pp., 33 plates. SEAMIC: Tokyo (1980).
11. Fan, P.C. Asian *Taenia saginata*: Species or strain. In: *Proc. 33rd SEAMEO-TROPMED Regional Seminar on Emerging Problems in Food-borne Parasitic Zoonoses*, Nov. 1990, Chiang Mai, Thailand (1991). Southeast Asian J. Trop. Med. Pub. Hlth. **22**: 245–250.
12. Feldman, M., Plancarte, A., Sandoval, M., Wilson, M. and Flisser, A. Comparison of two assays (EIA and EITB) and two samples (saliva and serum) for the diagnosis of neurocysticercosis. Trans. R. Soc. Trop. Med. Hyg. **84**: 559–562 (1990).
13. Gardiner, C.H., Wells, S., Gutter, A.E., Fitzgerald, L., Anderson, D.C., Harris, R.K. and Nichols D.K. Eosinophilic meningoencephalitis due to *Angiostrongylus cantonesis* as the cause of death in captive non-human primates. Am. J. Trop. Med. Hyg. **42**: 70–84 (1990).
14. Gareff-Teixeira, C., de Avila-Pires, F.D., Machado, R.C.C., Camillo-Coura, L. and Lenzi, H.L. Identificacão de roedores silvestres como hospedeiros do *Angiostrongylus costaricensis* no sul do Brasil. Rev. Inst. Med. trop. São Paulo. **32**: 147–150 (1990).
15. Ghadirian, E. and Arfaa, R. Present status of trichostrongyliasis in Iran. Am. J. Trop. Med. Hyg. **24**: 935–941 (1975).
16. Gillespie, S.H. The epidemiology of *Toxocara canis*. Parasitol. Today **4**: 180–182 (1988).
17. Grove, D.I. *Strongyloidiasis: A Major Roundworm Infection of Man.* 336 pp. Taylor & Francis: London (1989).
18. Hopkins, D.R. and Ruiz-Tiben, E. Dracunculiasis eradication: target 1995. Am. J. Trop. Med. Hyg. **43**: 296–300 (1990).
19. Horton, R.J., Cruz, M. and Cruz, J. A companison of the efficacy of albendazole and praziquantel in the treatment of neurocysticercosis. In: *Proc. 33rd SEAMEO-TROPMED Regional seminar on Emerging Problems in Food-borne Parasitic Zoonoses*, Nov. 1990, Chiang Mai, Thailand (1991).
20. Hwang,K.P. and Chen E.R. Clinical studies on angiostrongyliasis cantonensis among children in Taiwan. In: *Proc. 33rd SEAMEO-TROPMED Regional Seminar on Emerging Problems in Food-borne Parasitic Zoonoses*, Nov. 1990, Chiang Mai, Thailand (1991). Southeast Asian J. Trop. Med. Pub. Hlth. **22**: 194–199.
21. Iida, A. A simplified analysis for mitochondrial DNA and its restriction sites of *Angiostrongylus cantonensis*. Trop. Biomed. **4**: 179–187 (1987).
22. Illescas-Gómez, M.P., Osorio, M.R., Garcia, V.G. and Morales, M.A.G. Human *Gongylonema* infection in Spain. Am. J. Trop. Med. Hyg. **38**: 363–365 (1988).
23. Kamiya, M., Oku, Y., Katakura, K., Kamiya, H., Ohbayashi, M., Abe, H. Suzuki, H. and Bhaibulaya, M. Report of the prevalence and experimental infections of *Angiostrongylus siamensis* Ohbayashi, Kamiya et Bhaibulaya, 1979, parasitic in the mesenteric arteries of rodents in Thailand. Jpn. J. Vet. Res. **28**: 114–121 (1980).

24. Kumar, V. Zoonotic trematodiasis in South-East and Far-East Asian countries. In: *Helminth Zoonoses* (S. Geerts, V. Kumar and J. Barandt eds), pp. 106–118. Martinus Nijhoff Publishers: Dordrecht, The Netherlands (1987).
25. Matherne, C.M., Green, S.P., Corwin, R.M. and Kelley, S.T. Detection of circulating *Dirofilaria immitis* antigens in random source laboratory dogs: evaluation of two commercial serodiagnostic tests. Lab Anim. Sci. **38**: 584–587 (1988).
26. Mayr, A. and Hubert, P. Current problems in zoonosis research. Zentralbl Veterinarmed. **36**: 3–9 (1989).
27. McManus, D.P. Molecular technology: improving strategies for controlling hydatid disease and cysticercosis. Southeast Asian J. Trop. Med. Pub. Hlth. **21**: 161–173 (1990).
28. Murrell, K.D., Dubey, J.P., and Fayer, R. Role of biotechnology in the control of food-borne parasites. In: *Biotechnology and Food Safety. Proc. 2nd International Symposium Beltsville, MD* (D.D. Bills and S.D. Kung eds), pp. 215–225. Butterworth-Heinemann: Boston, Massachusetts (1990).
29. Murrell, K.D., Fayer, R. and Dubey J.P. Parasitic organisms. In: *Meat and Poultry Microbiology, Advances in Meat Science* (A.M. Pearson and T.R. Dutsmitir eds), pp. 311–377. Academic Press: New York (1986).
30. Parija, S.C. (ed). *Review of Parasitic Zoonoses.* AITBS Publishers: New Delhi (1990).
31. Pozio, E., La Rosa, G., Murrell, K.D. and Lichtenfels, J.R. Taxonomic revision of the genus *Trichinella*. J. Parasitol. **78**: 654–659 (1992).
32. Risher, W.H., Crocker, E.F. Jr., Beckman, E.N., Blalock J.B. and Ochsner J.L. Pulmonary dirofilariasis. The largest single-institution experience. J. Thorac. Cardiovasc. Surg. **97**: 303–308 (1989).
33. Roberts, J.A. The extraparasitic life cycle of *Toxocara vitulorum* in the village environment of Sri Lanka. Vet. Res. Commun. (Netherlands) **13**: 377–388 (1989).
34. Saleque, A., Juyal, P.D. and Bhatia, B.B. *Spirometra* sp. in a domestic cat in India. Vet. Parasitol. **35**, 273–276 (1990).
35. Schantz, P.M. *Toxocara* larva migrans now. Am. J. Trop. Med. Hyg. **41**: 21–34 (1989).
36. Shi, Y.E., Han, J.J., Yang, W.Y. and Wei, D.X. *Thelazia callipaeda* (Nematoda: Spirurida): transmission by flies from dogs to children in Hebei, China. Trans. R. Soc. Trop. Med. Hyg. **82**: 627: (1988).
37. Sturchler, D., Schubarth, P., Gualzata, M., Gottstein, G., and Oettli A. Thiabendazole vs. albendazole in treatment of toxocariasis: a clinical trial. Ann. Trop. Med. Parasitol. **83**: 473–478 (1989).
38. Tuntipopipat, S., Chawengkiattikul, R., Witoonpanich, R., Chiemchanya, S. and Sirisinha, S. Antigens, antibodies and immune complexes in cerebrospinal fluid of patients with cerebral gnathostomiasis. Southeast Asian J. Trop. Med. Pub. Hlth. **20**: 439–446 (1989).
39. Yamashita, K., Akimura, T., Kawano, K., Wakuta, Y., Aoki, H. and Gondou, T. Cerebral sparganosis mansoni. Report of two cases. Surg. Neurol. **33**: 28–34 (1990).

5. Genetic Variation in Helminths and Its Epidemiological Significance

R.C.A. Thompson and A.J. Lymbery*
Institute for Molecular Genetics and Animal Disease
School of Veterinary Studies, Murdoch University, Western Australia

*Western Australian Department of Agriculture
South Perth, Western Australia

INTRODUCTION

Over the last 20 years, the application of biochemical and molecular techniques has uncovered an enormous amount of genetic variation within and between natural populations of organisms. Parasites have proved to be no exception. The genetic variation which has recently been found within species of both protozoan [96] and helminth [60] parasites is much greater than was anticipated.

What is the epidemiological significance of this genetic variation? Do the genetic differences detected by highly discriminatory biochemical and molecular criteria affect phenotypic traits such as host preference, virulence and drug resistance which influence our attempts to treat and control parasitic disease? If not, can we ignore the evidence for intraspecific variation in our formulation of treatment regimes and control policies?

In this chapter, we attempt to provide answers to these questions by relating the variation measured by biochemical and molecular genetic techniques in species of parasitic helminths to variation in traits of epidemiological significance.

GENETIC VARIATION AND ITS EPIDEMIOLOGICAL SIGNIFICANCE

The Measurement of Genetic Variation

There are three major conditions that a technique should meet to provide a reliable estimate of intraspecific genetic variation [44]:

1. There should be a one-to-one correspondence between phenotype and genotype;
2. All, or a large proportion of allelic substitutions at a locus should be detectable;
3. The loci studied should be a random sample of genes in the genome.

These conditions were first approximated by the technique of enzyme electrophoresis and over the last 20 years there have been numerous electrophoretic surveys of genetic variation in a very wide range of organisms [reviews in 70, 74, 83]. The major advantages of enzyme electrophoresis are its accessibility and its potential to provide a very large sample of genes known to be variable within species. The major limitations are its relative insensitivity (compromising condition 2, above) and its limitation to a particular class of structural gene loci, which may not provide an unbiased sample of genomic variability [10].

DNA-based techniques have the potential to overcome both of these limitations. There are two levels at which DNA sequence variation can be observed; restriction endonuclease analysis and complete sequencing. Restriction endonuclease analysis of defined genomic regions, isolated by gene probes or the polymerase chain reaction, can be made more sensitive by using four-base, rather than six-base, cutters and by mapping restriction sites. Complete sequencing of DNA undoubtedly provides the most sensitive measure of genetic variation and the recent developments of the polymerase chain reaction coupled with primer-mediated direct sequencing [26a, 104] have made this a much more feasible approach. Restriction analysis and sequencing of DNA provide access to both coding and non-coding regions of the genome. Although this provides the opportunity for an unbiased sampling of the genome, the onus is on the investigator to choose for analysis a sequence of DNA which is appropriate to the question in hand. For example, if one wishes to measure genetic variation within populations of a particular species of helminth, it is futile to examine an evolutionary conserved sequence of DNA, such as the small subunit rRNA gene.

The Structure of Genetic Variation in Helminths

The genetic structure of a species refers to the manner in which the genetic variation occurring in that species is partitioned within and between populations. There is a common perception that genetic variation in species of parasites occurs mostly between, rather than within, populations. This is thought to be a consequence of the major ecological problems faced by parasites in first colonising, and then reproducing within the host. Price [77] predicted that these problems would favour the evolution of host specificity and the tolerance of inbreeding, leading to small, isolated populations that are genetically uniform as a consequence of the breeding system and genetically differentiated from other populations because of genetic drift and adaptation to local conditions.

These predictions have not been supported by empirical studies of genetic variation in species of helminths. Electrophoretic analyses have found levels of genetic diversity within, and genetic differentiation between, natural populations of parasitic helminths similar to those found in free-living invertebrates (Table 1). Although we are aware of only five studies on seven species of helminths, which have attempted a complete hierarchical partitioning of genetic variation within and between populations, these

Table 1. Unweighted mean values of genetic diversity within populations and genetic differentiation between populations for parasitic helminths[a] and free-living invertebrates[b]

Taxa	Genetic diversity			Genetic differentiation	
	No. of species	P[c]	H[d]	No. of species	D[e]
Parasitic helminths	23	0.33	0.10	5	0.024
Free-living invertebrates[e]	57	0.47	0.13	7	0.012

[a]Data from [2, 11, 17, 29, 31, 48, 49, 98, 99].
[b]Data from [8, 9, 10].
[c]Proportion of polymorphic enzyme loci.
[d]Observed or expected [63] heterozygosity at enzyme loci.
[e]Average number of base substitutions per enzyme locus [63, 64].

have all found most genetic variation within populations from the same host or from a number of hosts in the same locality (Table 2). In *Paragonimus iloktsuenensis, Fascioloides magna* and *Onchocerca volvulus*, substantial differentiation occurred only between populations separated by very large distances.

These data do not suggest that most species of endoparasitic helminths exist in small, genetically

Table 2. Proportion of genetic variation at different hierarchical levels in subdivided populations of parasitic helminths [modified from 93]

Species	Proportion of genetic variation[a,b]			
	Within infrapopulations	Between infrapopulations	Between Suprapopulations	Between areas
Paragonimus ohirai	—	0.90	0.10	—
P. iloktsuenensis	—	0.48	—	0.52
Fascioloides magna	—	0.83	0.04	0.13
Onchocerca volvulus	0.49	0.33	0.03	0.15
Toxocara cati	0.84	0.16	—	—
Echinococcus granulosus	—	0.95	0.05	—
Ostertagia ostertagi	—	0.99	0.01	—

[a]Infrapopulations and Suprapopulations as defined by Margolis et al [51]; areas refer to different countries, states or islands.
[b]Data from [2, 13a, 31, 49, 50, 59, 60].

invariant, isolated populations. On the contrary, they point to large effective population sizes, substantial within-population diversity and extensive gene flow, limited only by geographic distance between populations.

Genetic Variation in Epidemiological Traits

Many characteristics of parasites are relevant to the treatment and control of parasitic disease. Thompson and Lymbery [91] listed such traits as host preference, host specificity, pathogenicity, virulence, development rate and resistance to chemotherapy (see Chapters 1, 10 and 14).

The genetic control of these traits in helminths is rarely known. Occasionally, they must be controlled by a single major gene, inherited in a simple Mendelian fashion. For example, Woods et al [100] found that resistance to mebendazole in *Caenorhabditis elegans* resulted from a mutation in a single gene coding for β-tubulin. More often, however, epidemiological traits are likely to be controlled by the interaction of a number of minor genes (polygenes), each of small effect. For example, host preference and host specificity in gastrointestinal helminths may be determined by genes influencing dispersal behaviour, mode of attachment of the gut wall and avoidance of a range of nonspecific and immunological host defences. Similarly, virulence may by determined by genes influencing body size, reproduction rate, migratory movement, avoidance of host defences and production of enzymes or toxins.

Traits which are inherited in a polygenic fashion are usually continuously variable, as a result of both the simultaneous segregation of many different genes and the influence of the environment in which those genes are expressed [27]. The information we urgently require about traits of epidemiological significance is the relative importance of genetic and environmental factors in determining their values. In the language of quantitative genetics, this is broad-sense heritability; the proportion of total phenotypic variance in a trait which is due to genetic, as opposed to environmental, differences between individuals.

Estimates of heritabilities for epidemiological traits will provide us with some insight about how we can change the distribution of their values in natural populations of parasites. For example, if virulence is determined principally by the genotype of the parasite, then we should concentrate on controlling virulent strains. On the other hand, if virulence is determined principally by the genotype of the host, then we should concentrate on protecting susceptible individuals. If we wish to predict the consequences of our intervention, however, then we need to consider not only the heritabilities of epidemiological traits, but also the genetic corrections between them. Traits do not evolve in isolation and genetic correlations, arising from pleiotropy or linkage disequilibrium, can direct the course of evolution [97].For example, virulence and host specificity are likely to be influenced by some of the same genes; selection of virulence, induced by control programmes which target particular strains, may produce a correlated response in host specificity, broadening (or narrowing) the range of hosts utilised (see Chapter 1).

Genetic Structure and Variation in Epidemiological Traits: What is a Strain?

Species of helminths are often assumed to exist as a complex of genetically different strains, isolated in different localities or host transmission cycles [15, 23, 37, 38]. We have previously argued that if, as current data indicate, most genetic variation occurs within local populations of helminths, with differentiation only over large geographic distances, then there is little biological justification for recognising informal taxonomic categories between the levels of population and subspecies [91, 93].

We have suggested, however, that the term "strain" does have value as a practical descriptor to aid in the treatment and control of parasitic diseases. To this end, we defined a strain, in operational terms, as a population, or group of populations, which differs statistically from other populations of the same species in gene frequencies, and in one or more traits of epidemiological significance [91]. The rationale for this definition was that differences in gene frequencies provide evidence for restricted gene flow between populations and that, in the absence of heritability estimates, restricted gene flow provides an indication of a genetic basis to traits of epidemiological significance.

There are a number of points to note about this definition.

1. The term "strain" is used for populations which differ in gene frequencies and epidemiological traits, not for individuals within populations. If most traits of epidemiological significance are polygenic, and most species of helminths exist in large, outcrossing populations, we do not expect to find permanent associations between certain alleles at a marker locus and particular values for an epidemiological trait. Such associations will be rapidly broken up by recombination.
2. If differences in gene frequencies are to be used as indicators of restricted gene flow between populations, then the genes which are analysed should not be subjected to selection, for this can produce genetic differentiation even in the presence of gene flow [84]. Enzyme loci are often assumed to be neutral, or nearly so, but this is not always the case [40, 74a] so a large number of loci should be examined when using enzyme electrophoresis. With restriction analysis or DNA sequencing, the sequence of DNA to be studied should be chosen carefully, if possible using introns, flanking regions or silent codon sites.
3. Samples from putative strains should be sufficiently large to detect differences in both gene frequencies and the means of epidemiological traits. Care in interpretation may be required, however; Lewontin [45, 46] has pointed out that the power of statistical tests to distinguish differences between populations in these two classes of characters are vastly dissimilar.

Bearing our definition of a strain, and its caveats in mind, we now proceed to a case-by-case study of epidemiologically significant variation in selected species of helminths.

Schistosoma

The inherent variability within species of the genus *Schistosoma* has been recognised for many years

[65]. Extensive differences have been demonstrated between isolates of all major species affecting humans and other animals from many geographical areas. Reported differences may frequently have an influence on the epidemiology of schistosomiasis (see Chapter 3).

At the level of the intermediate host, variability in characteristics such as host preference and host specificity, duration of infection and cercarial shedding patterns must be taken into consideration if control efforts are to be directed at interrupting transmission. The susceptibility of snails to local isolates of *Schistosoma* must be determined before transmission patterns can be elucidated. For example, isolates of *S. japonicum* from the Philippines develop normally in the snail vector *Oncomelania hupensis*, whereas other oriental isolates of the parasite fail to establish infections in this snail species [106]. Similar differences in intermediate host preference between isolates of *S. japonicum* from South-East Asia [20, 105], as well as between isolates of *S. mansoni* from Brazil [73] have been reported.

Equally important are differences in the developmental behaviour of the asexual generations in snails. For example, Mouahid and Combes [56] demonstrated differences in both the duration of infection and cercarial productivity between Sudanese and Spanish isolates of *S. bovis* in the same species of snail vector. The influence of variability in cercarial shedding pattern on the epidemiology of schistosomiasis was demonstrated very elegantly by Theron [89], who studied the chronological characteristics of *S. mansoni* cercariae using two isolates from the same endemic region in Guadeloupe; one from an urban domestic focus and the other from a sylvatic focus. He found a five-hour difference in cercarial shedding between the two isolates. Transmission of the urban isolate is primarily dependent upon human infection and this isolate was found to have a peak emergence from the snail in late morning. In contrast, rats are the main definitive host for the sylvatic isolate which was characterised by a peak emergence in the late afternoon coinciding with the crepuscular activities of its definitive host. More recently, Pages and Theron [72] demonstrated differences in the timing of cercarial emergence between different isolates of *S. intercalatum* from North Africa. They concluded that the timing of cercarial emergence rhythms may be an additional marker to estimate the genetic variability of *S. intercalatum*.

In the definitive host, isolates of species of *Schistosoma* differ in infectivity, susceptibility to drugs, antigenicity, development and prepatent period, egg laying capacity and tissue distribution of the eggs [13, 18, 24, 32, 54, 65, 85, 87]. Diagnosis and treatment will be affected by such variability and immunoprophylaxis may be jeopardised by antigenic differences. For example, geographic isolates of both *S. japonicum* and *S. mansoni* differ with respect to the passage of eggs in either the urine or faeces [18, 87]. Differences in sensitivity and resistance to currently available drugs [24, 32] is a severe impediment to control schistosomiasis. Such variable sensitivity between isolates of *S. mansoni* in endemic areas means that drug dosage must be varied accordingly [32] and, thus, the identification of local strains is essential for effective treatment.

The development of immunoprophylactic regimes is seen as the most promising approach to control schistosomiasis. However, the existence of antigenic differences between isolates [e.g. 13] may be a serious limitation. For example, a lack of cross-protection was demonstrated in mice immunised with a Chinese isolate of *S. japonicum* which failed to resist a challenge with an isolate of *S. japonicum* from the Philippines [55]. More research is clearly required to determine whether the extensive genetic heterogeneity within species of *Schistosoma* is also reflected in widespread antigenic diversity and whether the antigens in question are likely to be important in the development of vaccines. In this respect, it has been pointed out by Mitchell [54] that this potential problem is likely to be more serious with *S. japonicum* than with *S. mansoni* since the former appears to be inherently more genetically variable.

It is quite clear that species of *Schistosoma* exhibit considerable heterogeneity and that much of this variation has a genetic basis. In this regard, isoenzyme studies have made the most important contribution in elucidating the extensive genetic heterogeneity that exists [80, 99]. However, it is very important, as emphasised earlier in this chapter, to realise that some variation in characters having an important influence on the epidemiology and control of schistosomiasis may not necessarily be reflected genetically. This has been emphasised recently by Jones and Kusel [36] who examined the wide phenotypic variation in *S. mansoni* and demonstrated some to be non-genetic. It would appear that certain still to be defined host or environmental factors can act on the parasite to shift expression of its genetic material in an appropriate direction [36]. The generation of such phenotypic variability is of great benefit in resisting the effects of drugs and vaccine-induced immunity, as well as enabling the parasite to infect a broad range of hosts. The fact that such variability is non-genetic, however, means that variable isolates may not be identified if the only differential criteria are those designed to detect genetic heterogeneity such as isoenzyme electrophoresis and DNA analysis. As we have emphasised above, parasite isolates should be fully characterised using genetic and phenotypic criteria so that the epidemiological significance of intraspecific variation can be fully determined.

Taeniid cestodes

Echinococcus

A considerable amount of information is now known about genetic variation in *Echinococcus* and the occurrence and characteristics of different strains [91, 93, 94]. Most attention has focussed on *E. granulosus* which has been shown to consist of numerous strains, some of which may warrant species status, adapted to different intermediate hosts. In addition, there are a number of isolates which have been studied from hosts such as buffaloes, goats and wild ungulates which require further characterisation but which may also constitute distinct strains according to the definition given earlier in this chapter. It is also clear that some species of intermediate host may be susceptible to more than one strain [91]. Such variability in *Echinococcus* is very important epidemiologically since strain characteristics may influence local patterns of transmission and thus have to be taken into consideration before implementing control strategies. For example, in Britain, sheep and horses are commonly infected with *E. granulosus*. Research over the last twenty years has shown that *E. granulosus of* sheep and horse origin are genetically different strains which differ in a range of characteristics including their host preference [86, 90, 91, 95]. Control efforts can therefore focus on either the sheep or horse cycles without concern over interaction between the two cycles.

In Australia, a similar situation exists in that two cycles operate on the mainland [91]. One of these is a typical sheep-dog cycle whereas the other is primarily sylvatic involving macropod marsupials, such as kangaroos and wallabies, and dingoes. Although there are morphological differences between *E. granulosus* of domestic and wild animal origin on the Australian mainland [42], these appear to be host induced [34] and probably do not reflect strain differences [49]. However, such intermediate host-induced morphological variation is retained by the adult worm and is thus of great value in determining the origin of infection in epidemiological studies [19a]. The absence of genetically distinct strains on the Australian mainland suggests that interaction between domestic and sylvatic cycles is likely. Evidence has been obtained of spillover from domestic foci into sylvatic hosts, although there has been no indication of spillover from sylvatic into domestic cycles. The latter would be very significant epidemiologically since the contamination of livestock grazing land from sylvatic hosts would make control virtually impossible. Yet it is possible that passage of *E. granulosus* through sylvatic hosts reduces infectivity of the parasite to domestic hosts as indicated by preliminary cross infection experiments.

Developmental differences between strains of *Echinococcus* may also have an important influence on the epidemiology of hydatid disease. For example, the identification of a strain of *E. granulosus* adapted to cattle in Switzerland and Germany was very significant [33, 92]. Not only was this the first time that a strain of *E. granulosus* was described which preferentially utilised cattle as an intermediate host but it was also revealed that the adult stage of the cattle strain developed far quicker than was presumed to be the norm for *E. granulosus*. Whereas the horse and common sheep strains have a prepatent period not earlier than 42 days post infection, *E. granulosus* of Swiss cattle origin produced gravid proglottids as early as 33 days after infection [92, 35]. Such rapid development is very significant since the timing of drug treatment regimes for dogs, which are designed to prevent egg production by *E. granulosus*, must vary between endemic areas. For example, the normal six-weekly interval recommended by the World Health Organisation [4] would not be suitable in Germany and Switzerland and should be reduced to an interval of five weeks. It has also been shown that other forms of *E. granulosus*, including the camel and Tasmanian sheep strains, are capable of producing eggs earlier than 42 days post infection [26, 41].

Forms of *Echinococcus* have also been identified which show differences in clinical manifestations, pathogenicity and infectivity to humans [91]. On the basis of current epidemiological data, it appears that *E. granulosus* of horse and pig origin may not be infective to humans. The form of *E. granulosus* which uses cervids as intermediate hosts in North America causes only a benign infection in humans, whereas in Libya and northern Kenya, there has been speculation that local isolates of *E. granulosus* may be particularly virulent in human hosts. As mentioned above, there are several forms of *E. granulosus* and *E. multilocularis* that should be characterised further in order to determine whether reported variability is correlated with genetic differences. There is mounting evidence that forms of *E. granulosus* which affect goats and buffaloes in Asia may represent distinct strains, as may also be the case with *E. multilocularis* in northern Japan, which utilises pigs as an intermediate host [91]. There are also numerous observations in the literature of unusual characteristics associated with particular isolates of *Echinococcus* that should be substantiated by further studies. For example, in Sardinia, sheep are the most commonly infected intermediate host of *E. granulosus*, yet cysts are rarely fertile [14]. In Jordan, morphological differences reported between isolates of *E. granulosus* from different intermediate hosts were thought to be indicative of strain variation [81]. In Spain, comparative studies on isolates of *E. granulosus* from different intermediate hosts revealed genetic differences using DNA hybridisation analysis [21]. In all these cases, further criteria should be applied to the isolates which have been studied in order to determine whether initial differences may reflect strain variation.

Taenia

Although much is now known about the epidemiological significance of variation in *Echinococcus*, little is known of the situation within species of *Taenia*. There may be fundamental differences between the reproductive biology of *Taenia* and *Echinococcus* that favour the adaptive radiation of the latter. However, Rausch [79] considers that the greater degree of intraspecific variation seen in *E. granulosus* is associated with the adaptation of the parasite to domestic host assemblages and that such diversity probably also occurs in species of *Taenia* with synanthropic hosts. Rausch cited evidence of intraspecific variation in *T. hydatigena*, *T. ovis* and *T. saginata*, and speculated that a range of biological variation comparable to that reported for *E. granulosus* would be found in these three species as well as in *T. solium*. Recent studies have confirmed Rausch's prediction, although much further research is required to determine the degree of strain variation within species of *Taenia*.

Abidi et al [1] found differences in the major biochemical components of cysticerci of *T. hydatigena* from goats and pigs in India and concluded that different strains exist. However, it is possible that such differences could be host induced and these observations should be substantiated by genetic studies. Electrophoretic analysis of different isolates of *T. saginata* revealed distinct differences which were considered to reflect strain variation [103]. Fan et al [28] also concluded that different strains of *T. saginata* may exist in Africa since pigs were found to be susceptible to isolates of the parasite from Ethiopia and

Madagascar. Seesee and Worley [82] reported variation in a sylvatic form of *T. ovis*, *T.o. krabbei* from grizzly bears in North America. Genetic differences have also recently been reported between isolates of *T. solium* from India, Mexico and Zimbabwe using DNA techniques [52], but further studies are required to determine whether the genetic differences are associated with any phenotypic characters of epidemiological significance.

Differences in infectivity between isolates of *T. taeniaeformis* have been reported by Conchedda and Ferretti [19] and Nonaka [71]. Because of the widespread use of this taeniid tapeworm as a laboratory model for economically important species of *Taenia*, it is essential that the potential extent of heterogeneity between laboratories is determined. A similar situation has recently been reported in the most widely used tapeworm model, *Hymenolepis diminuta* (reviewed in [16]). Interestingly, the differences reported between isolates of *H. diminuta* do not appear to have a genetic basis [3] and may reflect the regulation of biochemical processes by environmental factors (e.g. see [12]).

Trichostrongylid nematodes

Trichostrongylids and, in particular, *Haemonchus, Trichostrongylus* and *Ostertagia*, are important causes of economic losses to the livestock industries of many countries. An adequate understanding of the epidemiology of these parasites is essential for the development of integrated control strategies yet it is only over the last decade that attention has focussed on the genetic variability of these nematodes. We now know that trichostrongylids can respond rapidly to new selection pressures and that such variation has an important impact on the epidemiology of trichostrongylid infections. This is best exemplified by the widespread development of anthelmintic resistance, as well as increasing evidence of how rapidly they can adjust to new ecological conditions [43, 53, 88]. From the epidemiological viewpoint, the most important practical aspects of such variability are seen in the development of drug resistant strains and in some countries multiple drug resistant strains have emerged [88]. In addition, differences have been reported between isolates in host specificity, infectivity, egg development, survival and hatching requirements, and arrested development.

Most attention has focussed on *Haemonchus contortus* which is considered to represent a classical example of strain variation [16]. Most information on variability within this parasite species concerns the biochemical basis of drug resistance. The existence of strains of *H. contortus* resistant to different benzimidazole drugs is a major factor impeding control [78] (see Chapter 14), and comparative studies on the mechanisms of action of these drugs in susceptible and drug resistant strains of the parasite is an important focus of current research [47].

Michel [53] has also discussed other epidemiological aspects of intraspecific variability in *Haemonchus*. He emphasised that, as a consequence of intraspecific variability the impact of control measures may be transient, since, apart from the selection of drug resistant strains, the annual alternation of sheep and cattle on different pastures will eventually lead to populations of reduced host specificity; control based on seasonal patterns of herbage infestation will select worms with differing temperature responses or the ability, in some other way, to give rise to herbage infestations at atypical times of the year; and the selection of resistant strains of sheep may give rise to nematode strains which are more successful in parasitising resistant sheep [53].

Trichinella

The pioneering work of Nelson on *Trichinella* in Africa, first revealed that *T. spiralis* was not a uniform species and demonstrated that it is not possible to make generalisations about the epidemiology of a particular species of parasite from one geographical area to another (reviewed in [65, 66, 67, 68, 69]). However, since Nelson's studies in Africa during the 1960s, the situation has been shown to be far more complex. There are at least four distinct groups of *Trichinella* with a widespread geographical distribution, although variants exist within each group [23, 76]: The north temperate domestic "*spiralis*" form is restricted to those parts of the world where pigs are widely used for food, primarily North America and Europe. The so-called northern arctic sylvatic "*nativa*" form, is mainly restricted to circumpolar regions of Asia, North America and Europe. The southern sylvatic "*nelsoni*" form is found in Africa, parts of Europe, middle Asia and parts of the USSR. The geographical distribution of "*pseudospiralis*" has not been fully determined but includes parts of the USSR and possibly also India and North America (see Chapter 4).

Until recently, isolates of *Trichinella* have been characterised principally by behavioural features such as host specificity, infectivity, virulence, tissue distribution of larvae, reproductive capacity, temperature sensitivity etc. There are no clear-cut morphological differences between *Trichinella* isolates and it is only as a result of recent isoenzyme and DNA studies that a genetic explanation has been provided for the extensive phenotypic variability described within *T. spiralis* [16, 22, 30, 39, 58, 75, 76, 102].

The phenotypic variability is of great epidemiological significance. For example, antigenic differences [61, 62] may complicate diagnosis, although some antigenic variability may not reflect differences between isolates but rather variable levels of larval infestation [25]. Variation in other characteristics such as infectivity, virulence, pathogenesis and drug sensitivity will have a profound effect on the epidemiology of trichinosis. However, further studies are essential so that a clearer picture of the patterns of geographical distribution can be obtained, especially of isolates from wild animals [76]. For example, it was assumed that isolates of *T. spiralis* (*T.s. nativa*) in arctic regions were resistant to cold temperatures until comparative studies by Worley et al [102] revealed a disparity in cold-hardiness of *T.s. nativa* isolates from wolves in neighbouring zoogeographic regions of North America. This variablity was reflected in different DNA restriction fragment patterns, thus indicating the existence of dissimilar sylvatic strains at higher latitudes where arctic or freeze-resistant biotypes predominate [102]. These studies demonstrate that generalisations that all isolates in arctic regions will be the "cold-hardy" type should not be made, and illustrate the value of DNA analysis is such comparative investigations.

Studies on the host-specificity characteristics of different isolates of *Trichinella* are also required in some endemic areas to determine whether infections in domestic animals, especially pigs, originate from domestic or sylvatic cycles [57]. Similarly, it is important to determine whether domestic strains exist which are infective to wild animals since such reservoirs would pose serious obstacles to the eradication of the parasite from domestic hosts. For example, Dame et al [22] used DNA analysis to compare a number of isolates of *T. spiralis* from sylvatic hosts in North America and demonstrated that domestic-type isolates of *T.s. spiralis* do occur in wild animals. These authors emphasised that previous discussions on the importance of sylvatic reservoirs have neglected the potential for transmission in the reverse directon, i.e. from domestic to sylvatic hosts, and that isolates should not be classified purely on their host or geographical origin.

The application of enzyme electrophoresis and DNA probes will be particularly useful for the identification and discrimination of *Trichinella* isolates from the flesh of wild and domestic animals. For example, isolates of *Trichinella* from both sylvatic and domestic sources contributed to the 993 cases of human trichinosis reported in 24 provinces of Spain from 1981–1987 [6]. The future application of molecular characterisation procedures in such circumstances would assist in elucidating the number of genetic variants being transmitted in such endemic areas. Similarly, recent outbreaks of human trichinosis in France have been traced to the consumption of horsemeat [5, 7] but little is known about *Trichinella* in horses. Once again, molecular characterisation procedures would be useful, particularly in determining the origin of infection in horses.

CONCLUSIONS

1. Empirical studies of genetic variation within and between populations of parasitic helminths suggest

that most species consist of large, outbred, genetically diverse populations with extensive gene flow limited only by geographical isolation.

2. Within-populations, genetic diversity measured by biochemical and molecular techniques is unlikely to be related to variation in traits of epidemiological significance. These traits are probably polygenic and associations with biochemical or molecular genetic markers will be weak and rapidly broken up by recombination.

3. To measure within-population variation in epidemiological traits, the techniques of quantitative genetics are necessary. We urgently require estimates of heritability and genetic correlations between these traits.

4. This does not mean that studies of genetic variation in parasitic helminths, using biochemical and molecular techniques, have no value for the treatment and control of parasitic disease. Firstly, such studies enable us to answer fundamentally important questions about the evolutionary history, mode of reproduction, nature of the breeding system, influence of selection and extent of gene flow in natural populations. Such knowledge is vital to the design of effective control programmes. Secondly, the identification of genetic differences between populations, when accompanied by differences in epidemiological traits, provides us with an operational measure of strain variation. The identification of strains, differing genetically in such traits as host preference, development rate and virulence, has had a major impact on the control of schistosomiasis, hydatid disease and trichinosis.

REFERENCES

1. Abidi, S.M.A., Nizami, W.A., Khan, P., Ahmad, M. and Irshadullah, M. Biochemical characterisation of *Taenia hydatigena* from goats and pigs. J. Helminthol. **63**: 333–337 (1989).
2. Agatsuma, T. and Habe, S. Genetic variability and differentiation of natural populations in three Japanese lung flukes, *Paragonimus ohirai, Paragonimus iloktsuenensis* and *Paragonimus sadoensis* (Digenea: Troglotrematidae). J. Parasitol. **72**: 417–433 (1986).
3. Andrews, R.H., Adams, M., Beverstock, P.R., Behm, C.A. and Bryant, C. Genetic characterisation of three strains of *Hymenolepis diminuta* at 39 enzyme loci. Int. J. Parasitol. **19**: 515–518 (1989).
4. Anon. Echinococcosis, hydatidosis surveillance, prevention and control: FAO/UNEP/WHO guidelines. Food and Agriculture Organisation, Rome (1982).
5. Anon. Trichinosis: Outbreaks associated with horse-meat. Wkly. Epidem. Rec. **61**: 168–170 (1986).
6. Anon. Trichinosis. Wkly. Epidem. Rec. **64**: 98 (1989).
7. Anon. Trichinosis. Wkly. Epidem. Rec. **66**: 177–178 (1991).
8. Avise, J.C. Genetic differentiation during speciation. In: *Molecular Evolution* (F.J. Ayala ed), pp. 106–122. Sinauer: Sunderland, Massachusetts (1976).
9. Ayala, F.J. Genetic differentiation during the speciation process. Evolut. Biol. **8**: 1–78 (1975).
10. Ayala, F.J. The genetic structure of species. In: *Perspectives on Evolution* (R. Milkman ed), pp. 60–82. Sinauer: Sunderland, Massachusetts (1982).
11. Baverstock, P.R., Adams, M. and Beveridge, I. Biochemical differentiation in bile duct cestodes and their marsupial hosts. Mol. Biol. Evol. **2**: 321–337 (1985).
12. Bennet, E.M., Behm, C.A. and Bryant, C. The role of the host in the regulation of end-product formation in two strains of the rat tapeworm, *Hymenolepis diminuta*. Int. J. Parasitol. **20**: 841–848 (1990).
13. Blanton, R., Horowitz, S., Galvano Dos Reis, M., Rottman, F., Andrade Z. and Mahmoud, A.A.F. Stage and species specificity of antigens encoded by two geographic strains of *Schistosoma mansoni* mRNA. J. Parasitol. **72**: 445–453 (1986).

13a. Blouin, M.S., Dame, J.B., Tarrant, C..A. and Courtney, C.H. Unusual population genetics of a parasitic nematode: mtDNA variation within and among populations. Evolution **46**: 470–476 (1992).

14. Bortoletti, G., Gabriele, F., Seu, V. and Palmas, C. Epidemiology of hydatid disease in Sardinia: a study of fertility of cysts in sheep. J. Helminthol. **64**: 212–216 (1990).

15. Bourns, T.K.R. Protozoa and parasitic helminths. In: *Biochemical and Immunological Taxonomy of Animals*. Academic Press: London (1974).
16. Bryant, C. and Flockhart, H.A. Biochemical strain variation in parasitic helminths. Adv. Parasitol. **25**: 275–319 (1987).
17. Bullini, L., Nascetti, G., Paggi, L., Orecchia, P., Mattiucci, S. and Berland, B. Genetic variation of ascaridoid worms with different life cycles. Evolution **40**: 437–440 (1986).
18. Cheever, A.W. *Schistosoma japonicum*: the pathology of experimental infection. Exp. Parasitol. **59**: 1–11 (1985).
19. Conchedda, M. and Ferretti, G. Vaccination of susceptible hosts with uninfective strains of the same parasite (*Taenia taeniaeformis*, Cestoda) provide protection against an infective strain. J. Parasitol. **69**: 1166–1167 (1983).
19a. Constantine, C.C., Thompson, R.C.A., jenkins, D.J., Hobbs, R.P. and Lymbery A.J. Morphological characterization of adult *Echinococcus granulosus* as a means of determining transimission patterns. J. Parasitol. **79**: 57–61 (9193).
20. Cross, J.H., Zaraspe, G., Lu, S.K., Chiu, K.M. and Hung, H.K. Susceptibility of *Oncomelania hupensis* subspecies to infection with geographic strains of *Schistosoma japonicum*. Southeast Asian J. Trop. Med. Pub. Hlth. **15**: 155–160(1984).
21. Cuesta-Bandera, C., McManus, D.P. and Rishi, A.K. Characterisation of *Echinococcus granulosus* of Spanish origin by DNA restriction endonuclease analysis and Southern blot hybridisation. Int. J. Parasitol. **18**: 137–141 (1988).
22. Dame, J.B., Murrel, K.D., Worley, D.E. and Schad, G.A. *Trichinella spiralis*: genetic evidence for synanthropic subspecies in sylvatic hosts. Exp. Parasitol. **64**: 195–203 (1987).
23. Dick, T.A. Species, and infraspecific variation. In: *Trichinella and Trichinosis* (W.C. Campbell ed), pp. 31–73. Plenum: New York (1983).
24. Doong, Y.C., Wong. L.J.C., Bruce, J.I. and Wong, S.S. Enzymatic differences between hycanthone-resistant and sensitive strains of *Schistosoma mansoni*. Comp. Biochem. Physiol. **B87**: 459–464 (1987).
25. Dupouy-Camet, J., Bougnoux, M.E., Ancelle, T., Fagard, R. and Lapierre, J. Antigenic characteristics of two strains of *Trichinella spiralis* isolated during the horsemeat-related outbreaks of 1985 in France. Parasitol. Res. **75**: 79–80 (1988).
26. Eckert, J., Thompson, R.C.A., Michael, S.A., Kumaratilake, L.M. and El-Sawah, H.M. *Echinococcus granulosus* of camel origin: development in dogs and parasite morphology. Parasitol. Res. **75**: 536–544 (1989).
26a. Erlich, H.A. and Arnheim, N. genetic analysis using the polymerase chain reaction. Annu. Rev. Genet. **26**: 479–506 (1992).
27. Falconer, D.S. *Introduction to Quantitative Genetics*. Longman: New York (1981).
28. Fan, P.C., Chung, W.C., Lo, C.T. and Lin, C.Y. The pig as an experimental host of *Taenia saginata* (Ethiopia and Madagascar strains). Ann. Trop. Med. Parasitol. **84**: 93–95 (1990).
29. Fletcher, M., Lo Verde, P.T. and Woodruff, D.S. Genetic variation in *Schistotoma mansoni*: enzyme polymorphisms in populations from Africa, Southwest Asia, South America and the West Indies. Am. J. Trop. Med. Hyg. **30**: 406–421 (1981).
30. Flockhart, H.A., Harrison, S.E., Donaldson, A.R. and James, E.G. Enzyme polymorphism in *Trichinella*. Trans. R. Soc. Trop. Med. Hyg. **76**: 541–545 (1982).
31. Flockhart, H.A., Cibulskis, R., Karano, M. and Albiez, E.J. *Onchocerca volvulus:* enzyme polymorphism in relation to the differentiation of forest and savannah strains of the parasites. Trans. R. Soc. Trop. Med. Hyg. **80**: 285–292 (1986).
32. Foster, R. A review of clinical experience with oxamniquine. Trans. R. Soc. Trop. Med. Hyg. **81**: 55–59 (1987).
33. Hahn, E., Hörchner, F., Huwer, M. and Sanft, S. Zur Frage der Stammspezifität von *Echinococcus granulosus* (Rind) in Deutschland. Wien. Tierärzt. Monatssch. **75**: 476–479 (1988).
34. Hobbs, R.P., Lymbery, A.J. and Thompson, R.C.A. Rostellar hook morphology of *Echinococcus granulosus* (Batsch, 1786) from natural and experimental Australian hosts and its implications for strain recognition. Parasitology **101**: 273–281 (1990).
35. Hörchner, F. Epizootologie der Echinokokkose in Deutschland. In: *Symposium "Weideparasitosen", Fachgruppe Parasitologie und Parasitaire Krankheiten*, pp. 71–75. Deutsche Veterinarmedizinische Gesellschaft: Giessen/Lahn, Germany (1988).

36. Jones, J.T. and Kusel, J.R. Intraspecific variation in *Schistosoma mansoni*. Parasitol. Today **5**: 37–39 (1989).
37. Kennedy, C.R. *Ecological Animal Parasitology*. Blackwell Scientific Publications: Oxford, UK (1975).
38. Kino, H. and Kennedy, C.R. Differences in biological characteristics of two strains of the cestode *Hymenolepis diminuta*. Parasitology **95**: 97–110 (1987).
39. Klassen, G.R. and Thiessen, J.P. Restriction endonuclease analysis of repetitive sequences in the *Trichinella* genome: three strain-specific patterns. J. Parasitol. **72**: 772–775 (1986).
40. Koehn, R.K. and Hilbish, D.J. The adaptive importance of genetic variation. Am. Sci. **75**: 134–141 (1987).
41. Kumaratilake, L.M., Thompson, R.C.A. and Dunsmore, J.D. Comparative strobilar development of *Echinococcus granulosus* of sheep origin from different geographical areas of Australia *in vivo* and *in vitro*. Int. J. Parasitol. **13**: 151–156 (1983).
42. Kumaratilake, L.M. and Thompson, R.C.A. Morphological characterisation of Australian strains of *Echinococcus granulosus*. Int. J. Parasitol. **14**: 467–477 (1984).
43. Le Jambre, L.F. Genetics and the control of trichostrongylid parasites of ruminants. In: *Biology and Control of Endoparasites* (L.E.A. Symons, A.D. Donald and J.K. Dineen eds), pp. 53–80. Academic Press: Sydney (1982).
44. Lewontin, R.C. *The Genetic Basis of Evolutionary Change*. Columbia University Press: New York (1974).
45. Lewontin, R.C. Detecting population differences in quantitative characters as opposed to gene frequencies. Am. Nat. **123**: 115–124 (1984).
46. Lewontin, R.C., A comment on the comments of Rogers and Felsenstein. Am. Nat. **127**: 733–734 (1986).
47. Lubega, G.W. and Prichard, R.K. Specific interaction of benzimidazole anthelmintics with tubulin: high-affinity binding and benzimidazole resistance in *Haemonchus contortus*. Mol. Biochem. Parasitol. **38**: 221–232 (1990).
48. Lymbery, A.J. and Thompson, R.C.A. Electrophoretic analysis of genetic vatriation in *Echinococcus granulosus* from domestic hosts in Australia. Int. J. Parasitol. **18**: 803–811 (1988).
49. Lymbery, A.J., Thompson, R.C.A. and Hobbs, R.P. Genetic diversity and genetic differentiation in *Echinococcus granulosus* (Batsch, 1786) from domestic and sylvatic hosts on the mainland of Australia. Parasitology **101**: 283–289 (1990).
50. Lydeard, C., Mulvey, M., Aho, J.M. and Kennedy, P.K. Genetic variability among natural populations of the liver fluke *Fascioloides magna* in white-tailed deer, *Odocoileus virginiarus*. Can. J. Zool. **67**: 2021–2025 (1989).
51. Margolis, L., Esch, G.W., Holmes, J.C., Kuris, A.M. and Schad, G.A. The use of ecological terms in parasitology (report to an ad hoc committee of the American Society of Parasitologists). J. Parasitol. **68**: 131–133 (1982).
52. McManus, D.P. Characterisation of taeniid cestodes by DNA analysis. Rev. Sci. Tech. Off. Int. Epiz. **9**: 489–510 (1990).
53. Michel, J.F. Epidemiology and control of gastrointestinal helminths in domestic animals. In: *Chemotherapy of Gastrointestinal Helminths* (H. Van den Bossche, D. Thienpont and Janssens, P.F. eds), pp. 67–123. Springer: Berlin (1985).
54. Mitchell, G.F., Rogers, M.V. and Tiu, W.U. Schistosomiasis. In: *Vaccination Strategies of Tropical Diseases* (F.Y. Liew ed), pp. 149–164. CRC Press: New York (1989).
55. Moloney, N.A., Garcia, E.G. and Webbe, G. The srain specificity of vaccination with ultraviolet attenuated cercariae of the Chinese strain of *Schistosoma japonicum*. Trans. R. Soc. Trop. Med. Hyg. **79**: 245 (1985).
56. Mouahid, A. and Combes, C. Genetic variability of *Schistosoma bovis* cercarial production according to miracidial dose. J. Helminthol. **61**: 89–94 (1987).
57. Murrell, K.D., Leiby, D.A., Duffy, C.H. and Schad, G.A. Comparative infectivity of *Trichinella spiralis spiralis* (from domestic swine) and *T. spiralis nativa* (from polar bear) for Arctic foxes and domestic swine. Proc. Helminthol. Soc., Wash. **53**: 286–288 (1986).
58. Mydynski, L.J. and Dick, R.A. The use of enzyme polymorphisms to identify genetic differences in the genus *Trichinella*. J. Parasitol. **71**: 671–677 (1985).
59. Nadler, S.A. Biochemical polymorphism in *Parascaris equorum*, *Toxocara canis* and *Toxocara cati*. Mol. Biochem. Parasitol. **18**: 45–54 (1986).

60. Nadler, S.A. Molecular approaches to studying helminth population genetics and phylogeny. Int. J. Parasitol. **20**: 11–29 (1990).
61. Nakai, Y., Ogimoto, K. and Yamaguchi, T. Detection of human and rabbit antibodies to *Trichinella spiralis* by enzyme linked immunosorbent assay (ELISA). Jpn. J. Parasitol. **36**: 328–333 (1987).
62. Navarrete, I., Bolas-Fernandez, F., Serrane, F. and Habela, M. Serodiagnosis of swine trichinellosis by the ELISA test: antigen efficacy of some *Trichinella* strain isolates. In: *Trichinellosis* (C.E. Tanner, A.R. Martinez-Fernandez and F. Bolas-Fernandez, eds), pp. 231–236. Cientificas Press: Madrid (1989).
63. Nei, M. *Molecular Population Genetics and Evolution*. North-Holland: Amsterdam (1975).
64. Nei, M. Estimation of average heterozygosity and genetic distance from a small number of individuals. Genetics **89**: 583–590 (1978).
65. Nelson, G.S. The epidemiological significance of intraspecific variations in helminths of medical importance; with particular reference to *Trichinella* and *Schistosoma*. In: *H.D. Srivastava Comemoration Volume* (Singh, K.S. and Tandon, B.K. eds), pp. 19–25. Indian Veterinary Research Institute, Izatnagar (1970).
66. Nelson, G.S. Carrion-feeding cannibalistic carnivores and human disease in Africa with special reference to trichinosis and hydatid disease in Kenya. Symp. Zool. Soc. Lond. **50**: 181–198 (1982).
67. Nelson, G.S. Wild animals as reservoir hosts of parasitic diseases of man in Kenya. In: *Tropical Parasitoses and Parasitic Zoonoses* (J.D. Dunsmore ed), pp. 59–72. Tenth meeting, WAAVP, Perth (1983).
68. Nelson, G.S. Parasitic zoonoses. In: *The Biology of Parasitism. A Molecular and Immunological Approach.* (P.T. Englund and A. Sher eds), pp. 13–41. Alan R. Liss: New York (1988).
69. Nelson, G.S. Human behaviour and the epidemiology of helminth infections: cultural practices and microepidemiology. In: *Parasitism and Host Behaviour* (C.J. Barnard and J.M. Behnke eds), pp. 234–263. Taylor & Francis: London (1990).
70. Nevo, E. Genertic variation in natural populations: patterns and theory. Theoret. Populn. Biol. **13**: 121–177 (1978).
71. Nonaka, N. Analysis of innate resistance of rodents to infection by various strains of *Taenia taeniaeformis*. Jpn. J. Vet Res. **35**: 142–145 (1989).
72. Pages, J.R. and Theron, A. *Schistosoma intercalatum* from Cameroon and Zaire: chronobiological differentiation of cercarial emergence. J. Parasitol. **76**: 743–745 (1990).
73. Parense, W.L. and Correa, L.R. Observations on two biological races of *Schistosoma mansoni*. Mem. Inst. Oswaldo Cruz **76**: 287–291 (1981).
74. Powell, J.R. Protein variation in natural populations of animals. Evol. Biol. **8**: 79–119 (1975).
74a. Powers, D.A., Laverman, T., Crawford, D. and Di Michele, L. Genetic mechanisms for adapting to a changing environment. Annu. Rev. Genet. **25**: 629–659 (1991).
75. Pozio, E. Isoenzymatic typing of 23 *Trichinella* isolates. Trop. Med. Parasitol. **38**: 111–116 (1987).
76. Pozio, E., Doby, J.M., LaRosa, G., Rossi, P. and Bruschi, F. Identification biochimique de l'agent etiologique d'une trichinellose asymptomatique observee chez us sujet immunodeprime en France. Bull. Soc. Francaise Parasitol. **7**: 201–208 (1989).
77. Price, P.W. *Evolutionary Biology of Parasites*. Princeton University Press: Princeton (1980).
78. Prichard, R.K., Hall. C.A., Kelly, J.D., Martin, I.C.A. and Donald, A.D. The problem of anthelmintic resistance in nematodes. Aust. Vet. J. **56**: 239–251 (1980).
79. Rausch, R.L. Parasitology: retrospect and prospect. J. Parasitol. **71**: 139–151 (1985).
80. Rollinson, D. Biochemical genetics in the study of schistosomes and the intermediate hosts. Parassitologia **27**: 123–139 (1985).
81. Said, I.M., Abdel-Hafez, S.K. and Al-Yaman, F.M. Morphological variation of *Echinococcus granulosus* protoscoleces from hydatid cysts of human and various domestic animals in Jordan. Int. J. Parasitol. **18**: 1111–1114 (1988).
82. Seesee, F.M. and Worley, D.E. *Taenia ovis krabbei* from grizzly bears, *Ursus arctos*, in Montana and adjacent areas. Proc. Helm. Soc. Wash. **53**: 298–300 (1986).
83. Selander, R.K. Genic variation in natural populations. In: *Molecular Evolution* (F.J. Ayala ed), pp. 21–45. Sinauer: Sunderland, Massachusetts (1976).
84. Slatkin, M. Gene flow in natural populations: Ann. Rev. Ecol. Syst. **16**: 393–430 (1985).

85. Smith, M.A. and Clegg, J.A. Different levels of immunity to *Schistosoma mansoni* in the mouse: the role of variant cercariae. Parasitology **78**: 311 (1979).
86. Smyth, J.D. Parasitological serendipity: from *Schistocephalus* to *Echinococcus*. Int. J. Parasitol. **20**: 411–423 (1990).
87. Soliman, G.N., El Assal, F.M., Monsour, N.S. and Garo, K. Comparison of two Egyptian strains of *Schistosoma mansoni* in hamsters. Z. Parasitenkunde **72**: 353–363 (1986).
88. Taylor, M.A. and Hunt, K.R. Anthelmintic drug resistance in the U.K. Vet. Rec. **125**: 143–147 (1989).
89. Theron, A. Early and late shedding patterns of *Schistosoma mansoni* cercariae: ecological significance in transmission to human and murine hosts. J. Parasitol. **70**: 652–655 (1984).
90. Thompson, R.C.A. and Smyth, J.D. Equine hydatidosis: a review of the current status in Great Britain and the results of an epidemiological survey. Vet. Parasitol. **1**: 107–127 (1975).
91. Thompson, R.C.A. and Lymbery, A.J. The nature, extent and significance of variation within the genus *Echinococcus*. Adv. Parasitol. **27**: 210–258 (1988).
92. Thompson, R.C.A., Kumaratilake, L.M. and Eckert, J. Observations on *Echinococcus granulosus* of cattle origin in Switzerland. Int. J. Parasitol. **14**: 283–291 (1984).
93. Thompson, R.C.A. and Lymbery A.J. Intraspecific variation in parasites—what is a strain? Parasitol. Today **6**: 345–348 (1990).
94. Thompson, R.C.A. and Lymbery, A.J. *Echinococcus*: Biology and strain variation. Int. J. Parasitol **20**: 457–470 (1991).
95. Thompson, R.C.A. *Echinococcus* and Giardia: variation on a theme. Int. J. Parasitol. **21**: 291–297 (1991).
96. Tibayrenc, M., Kjellberg, F. and Ayala, F.J. A clonal theory of parasitic protozoa: the population structures of *Entamoeba, Giardia, Leishmania, Naelgeria, Plasmodium, Trichomonas* and *Trypanosoma* and their medical and taxonomic consequences. Proc. Nat. Acad. Sci., U.S.A. **87**: 2414–2418 (1990).
97. Via, S. Ecological genetics and host adaptation in herbivorous insects: the experimental study of evolution in natural and agricultural systems. Ann. Rev. Entomol. **35**: 421–446 (1990).
98. Vrijenhoek, R.C. Genetic differentiation among larval nematodes infecting fishes. J. Parasitol. **64**: 790–798 (1978).
99. Woodruff, D.S., Merenlender, A.M., Upatham, E.S. and Viyanant, V. Genetic variation and differentiation of three *Schistosoma* species from the Philippines, Laos and Peninsular Malaysia. Am. J. Trop. Med. Hyg. **36**: 345–354 (1987).
100. Woods, R.A. Malone, K.M., Spence, A.M., Sigurdson, W.D. and Byard, E.H. The genetics, ultrastructure and tubulin polypeptides of mebendazole-resistant mutants of *Caenorhabditis elegans*. Can. J. Zool. **67**: 2422–2431 (1989).
101. Worbes, H., Thompson, R.C.A. and Eckert, J. Occurrence of the cattle strain of *Echinococcus granulosus* in the German Democratic Republic. Parasitol. Res. **75**: 495–497 (1989).
102. Worley, D.E., Zarlenga, D.S. and Seesee, F.M. Freezing resistance of a *Trichinella spiralis nativa* isolate from a grey wolf, *Canis lupus*, in Montana, with observations on genetic and biological characteristics of the biotype. J. Helm. Soc. Wash. **57**: 57–60 (1990).
103. Wouters, G., Brandt, J.and Geerts, S. Observations on possible strain differences in *Taenia saginata*. In: *Helminth Zoonoses* (S. Geerts, V. Kumar and J. Brandt eds), pp. 76–80. Martinus Nijhoff: Dordrecht, Netherlands (1987).
104. Wrischnik, L.A., Higuchi, R.G., Stoneking, M., Erlich, H.A., Arnheim, N. and Wilson, A.C. Length mutations in human mitochondrial DNA: direct sequencing of enzymatically amplified DNA. Nucl. Acids Res. **15**: 529–542 (1987).
105. Yuan, H.C., Upatham, E.S., Maleeya, K. and Khunbarivan, V. Susceptibility of snail vectors to oriental anthropohilic *Schistosoma*. Southeast Asian J. Trop. Med. Pub. Hlth. **15**: 86–94 (1984).
106. Yuan, H.C., Upatham, E.S., Kruatrachne, M. and Khunbarivan, V. Susceptibility of snail vectors to oriental anthropophilic *Schistosoma*. Acta Zool. Sinica **35**: 189–195 (1989) [in Chinese].

6. Ultrastructure of Helminths

R.E.B. Hanna
Department of Biology, Queen's University of Belfast
Belfast, Northern Ireland

INTRODUCTION

The last thirty years has seen a revolution in our knowledge of the functional morphology of parasitic helminths, particularly trematodes and cestodes. This has largely resulted from the application of increasingly sophisticated electron microscope technology, coupled with developments in histo- and cytochemistry, biochemistry, immunology and molecular biology. The literature documenting these advances is extensive and broadly scattered, encompassing a wide range of species and developmental stages, some of which have economic and medical significance and many which have mainly scientific interest. This review is not intended to be comprehensive, but rather to provide a general introduction to the ultrastructure of those organs and tissues of platyhelminth and nematode parasites which have particular relevance in the host-parasite relationship. Thus most attention is devoted to parasite surfaces, at which many physiological interactions with the host tissues occur, and to the process of nutrition which also may have a direct bearing on the outcome of parasitic infection. Certain aspects of reproduction, particularly gametogenesis and vitellogenesis are examined since their increased capacity and functional efficiency have particular relevance in parasitic life cycles. On the other hand, reference to neuromuscular systems has been omitted since these bear close morphological and physiological affinities with their equivalents in free-living invertebrates.

Emphasis, throughout, is placed on functional correlates rather than on anatomical detail, so the application of techniques which allow integration of morphological features with biochemical or physiological activity received particular attention.

In the interests of clarity and continuity, attention is focussed on only a few of the many species of parasitic helminths which have been studied to date. Among the platyhelminths, *Fasciola hepatica* has come to be regarded almost as a "model" species in cytological investigations. This is largely a result of its ready availability to electron microscopists, its cosmopolitan distribution and its considerable economic importance. Much that has been revealed regarding the ultrastructure and functional morphology of its various tissues appears relevant to platyhelminths generally. In contrast *Schistosoma mansoni*, arguably the most prevalent and damaging of the medically important helminths, presents many highly specialised facets of ultrastructure and life cycle, which render it less servicable in the elucidation of general principles of platyhelminth organisation.

Amongst the cestodes *Hymenolepis diminuta* has received most attention in cytological and physiological investigations. This is largely due to the ease with which its life cycle may be maintained under controlled conditions. Economically important cestodes such as *Taenia* spp. and *Echinococcus* spp. are less amenable for experimental manipulation. Similarly, among nematodes the "models" on which our understanding of the functional morphology of the phylum is based have tended to be those species which are readily available to electron microscopists, of a convenient size, and easily maintained in the laboratory (e.g. *Nippostrongylus, Ascaris*) rather than those of paramount medical and economic significance (e.g. *Wuchereria, Ancylostoma*).

Bearing in mind the range and diversity of parasitic helminths and the variety of life cycle stages exhibited even within a single group, it has been necessary to restrict the discussion to those taxa having particular relevance in human affairs, and to the life cycle stages (usually adults) which impinge directly on mammalian hosts. Thus, no reference is made to groups such as the monogeneans and acanthocephalans which have considerable scientific interest but little economic or medical significance, or to larval stages which are free-living or parasitise invertebrate intermediate hosts. Information on these organisms may be found in the up-to-date general texts and specialised review articles referred to herein.

TREMATODES

Tegument

The discovery, in the early 1960's, that the surface layer of parasitic platyhelminths comprises a living cytoplasmic syncytium proved to be the starting point for a modern reassessment of the nature of the host-parasite relationship, and concentrated the attention of physiologists and immunologists on the complex biochemical exchanges occurring at this highly important interface. The ultrastructure of the adult digenean tegument is exemplified by that of *F. hepatica* and consists of two regions, an anucleate surface syncytium and a deeper nucleated zone, which are connected by cytoplasmic bridges (Fig. 1). The syncytial layer of cytoplasm covers the entire surface of the fluke. It contains mitochondria and spines composed of crystalline protein and is bounded apically and basally by plasma membranes. The apical membrane is highly folded and is continuous over the spines while the basal one invaginates into the syncytium forming lamellae (Fig. 2). The surface syncytium is continuous, via cytoplasmic tubes (trabecula) through the underlying basal lamina and muscle layers with the nucleated zone. The latter comprises discrete uninucleate cell bodies (tegumental cells, cytons or perikarya) which lie embedded in the parenchymal tissue [118]. This basic organisation is universal among trematodes and cestodes [71, 73, 115, 123] but considerable differences in ultrastructure occur between taxonomic groups and even from species to species. For example, many trematodes have aspinous teguments, while none of the cestodes possess spines as such. In addition to the nucleus, each tegumental cell contains granular endoplasmic reticulum (GER), Golgi complexes, secretory granules and numerous mitochondria (Fig. 3). These features are consistent with a role in protein secretion.

Two types of tegumental cell have been recognised in *F. hepatica* [119]. These vary in the nature of the secretory granules which they manufacture. The so-called "T1" type produce membrane-bound spherical bodies which are homogenous and very electron-dense, while the "T2" cells produce biconcave disc-shaped bodies which are electron-lucid. Both T1 and T2 types of secretory bodies abound in the surface syncytium, having been transported to this zone through the connecting trabecula.

Experimental verification for synthesis of secretory granules at the GER and Golgi apparatus of the tegumental perikarya followed by trans-trabecular transport into the surface syncytium was achieved using electron microscope autoradiography [57, 111]. The amino acid precursor ^{3}H-leucine was found to be incorporated at the GER of T1 perikarya in *F. hepatica*, and radioactively labelled T1 bodies emerging from the maturing face of Golgi bodies were traced during a 3-hour migration through the trabecula and across the cytoplasmic surface syncytium to their points of exocrine discharge at the apical plasma membrane. Their contents contributed to the glycocalyx, a fuzzy filamentous layer of glycoprotein molecules up to 40 nm in thickness which completely invests the outer aspect of surface membrane. Threadgold

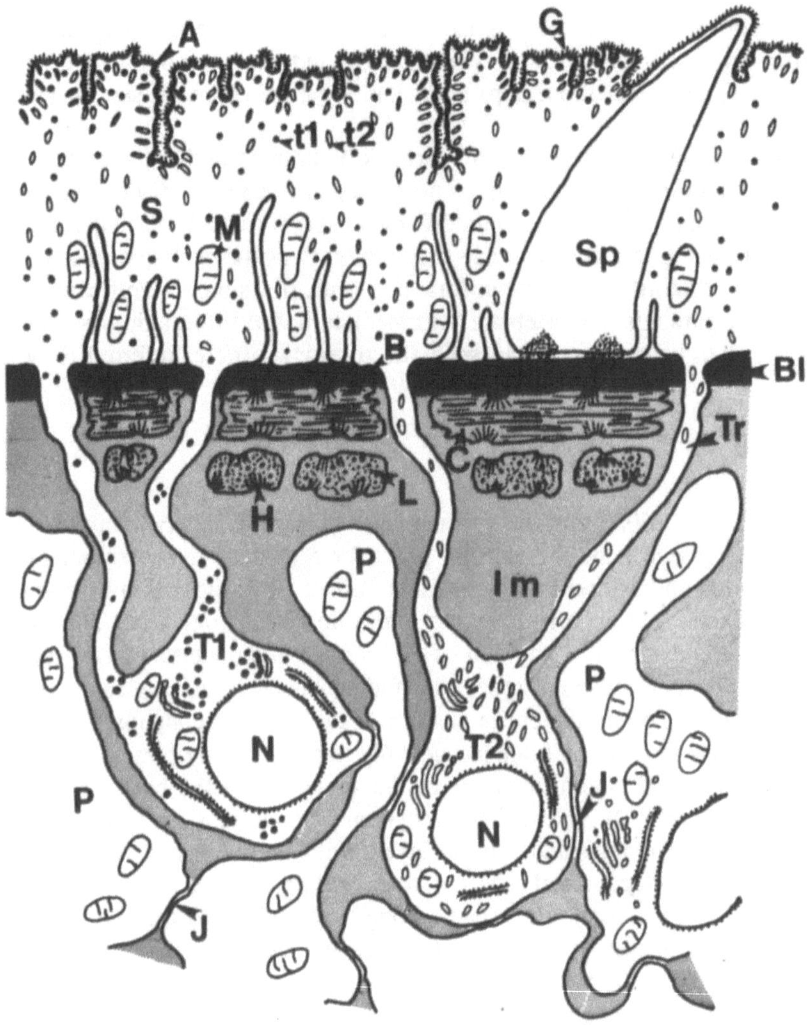

Fig. 1. Ultrastructure of the tegument of *Fasciola hepatica*. The surface syncytium (S) is connected with underlying perikarya (T1 and T2) which contain nuclei (N) and manufacture proteinaceous secretory bodies (t1 and t2) by a GER-Golgi mechanism. These bodies pass to the surface through trabecula (Tr) which traverse the superficial layers of circular (C) and longitudinal (L) muscle. The syncytium is bounded by an apical plasma membrane (A) which bears an external glycocalyx (G), and by an infolded basal plasma membrane (B). Mitochondria (M) are abundant and spines (Sp) of crystalline protein are firmly anchored to the basal lamina (B1). The muscle fibres are also attached to the basal lamina and interstitial material (Im) by hemidesmosomes (H). Ramifications of the parenchymal tissue (P) form gap junctions (J) with each other and with the tegumental perikarya (after Threadgold, 89, 90).

[121] has shown that the glycocalyx in *F. hepatica* comprises an inner layer of predominantly protein molecules closely applied to the outer aspect of the apical bilayer. From these, side chains of oligosaccharides and gangliosides bearing terminal sialic acid residues project distally to form an outer labile zone. The glycoprotein nature of the secretory material enclosed by T1 bodies has been verified by the finding that

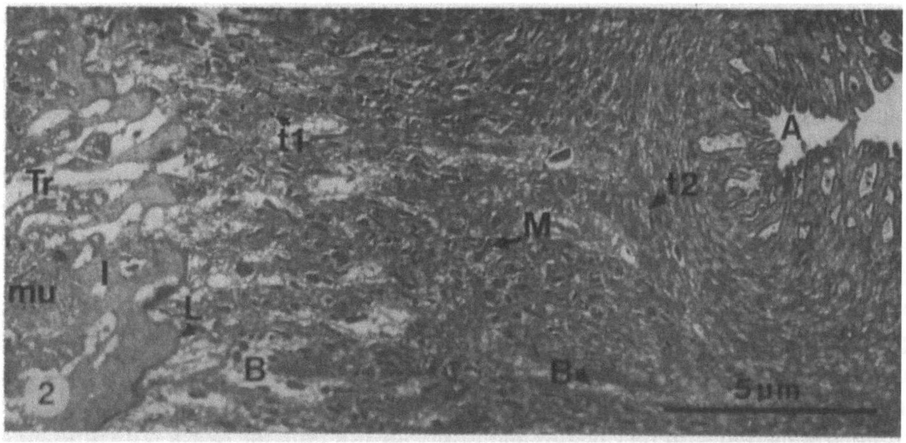

Fig. 2. *Fasciola hepatica*: Surface syncytium of the tegument. The apical surface (A) is highly folded while the basal membrane sends vacuolated invaginations (B) deep into the cytoplasm. Mitochondria (M), t1 and t2 secretory bodies are abundant, and vacuolated trabecula (Tr) connect the surface syncytium with the underlying perikarya. The basal membrane is supported by a basal lamina (L), beneath which muscle fibres (Mu) are embedded in interstitial material (I).

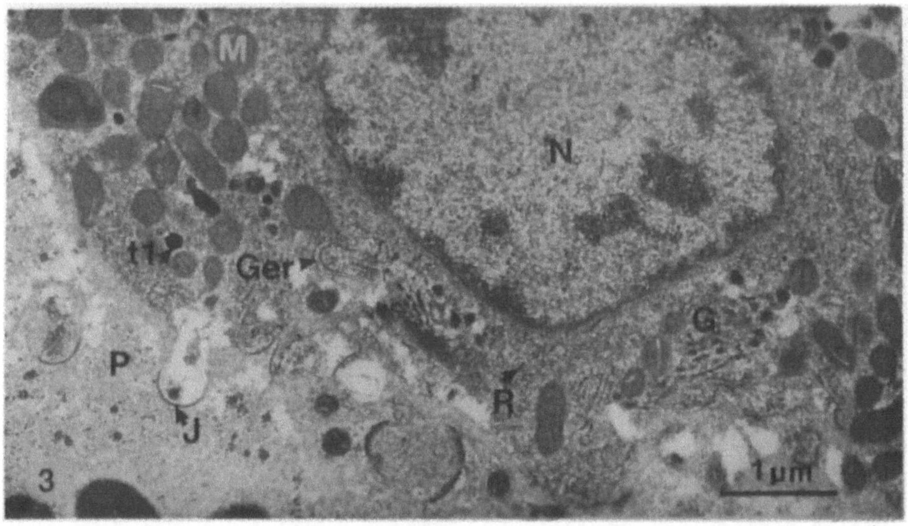

Fig. 3. *Fasciola hepatica*: A tegumental perikaryon. Secretory bodies (t1) are produced at the granular endoplasmic reticulum (Ger) and Golgi fields (G). Ribosomes (R) and mitochondria (M) are also abundant in the cytoplasm. Junctional complexes (J) are formed with the surrounding parenchymal tissue (P). N = nucleus.

^{3}H-galactose and ^{3}H-glucose were readily incorporated into forming T1 bodies at the Golgi fields of tegumental perikarya [54]. In *Megalodiscus* also, ^{3}H-glucose and ^{3}H-galactose were found to be incorporated into secretory bodies and subsequently transported to the surface glycocalyx over a 5 hr period [111]. In some trematodes such as *Schistosoma mansoni* acid and alkaline phosphatases have been demonstrated in the glycocalyx [90], but in *F. hepatica* enzyme activity is associated only with the inner aspect of the apical membrane [120]. Lumsden [77] provided a comprehensive review of the ultrastructure and cytochemistry of platyhelminth surface membranes.

While active production of glycoproteinaceous secretory bodies and their release at the apical membrane of the tegumental syncytium appears to occur in all trematodes and cestodes studied to date, little information is available as to the exact physiological role of the process. Recently, however, evidence has accumulated that active turnover of the glycocalyx, in *F. hepatica* at least, serves to protect the migrating juvenile from the effects of host immunity[55, 56]. It appears that the earliest and most vigorous humoral immune response in sheep, cattle and rats to *F. hepatica* infection is directed against antigenic determinants in the glycocalyx of the juvenile fluke. This is not surprising in view of the fact that the glycocalyx totally envelops the parasite during its tissue migration phase, and represents the only source of antigens available for host recognition. High levels of glycocalyx-specific antibodies build up in the host circulation, and attach to the fluke surface forming immune complexes which are potentially damaging by virtue of their activity in opsonization and complement fixation. Continuous sloughing of the antigen- antibody complex accompanied by renewal of the glycocalyx from beneath by exocrine discharge of secretory bodies allows the surface membrane to escape immunologically mediated attack by the host [55].

The role of the tegumental secretory bodies and glycocalyx in immune evasion by migrating *F. hepatica* was first investigated by indirect fluorescent antibody labelling techniques at light microscope level [55, 56]. Subsequently, the involvement of the tegumental perikarya and secretory bodies in replacement of the surface glycocalyx was confirmed when monoclonal antibodies developed against specific glycocalyx epitopes were found to attach to T1 type secretory bodies throughout the tegumental cytoplasm as well as to the GER and Golgi bodies in the perikarya [58] (Fig. 4). Such immunocytochemical labelling at electron microscope level was achieved by the use of Protein A or second antibody molecules (goat anti-mouse IgG) linked to colloidal gold. This technique of colloidal gold labelling is gaining widespread popularity in the investigation of parasite immunogenicity, and has immense potential for correlating ultrastructral features with the fundamentally important phenomenon of immune evasion.

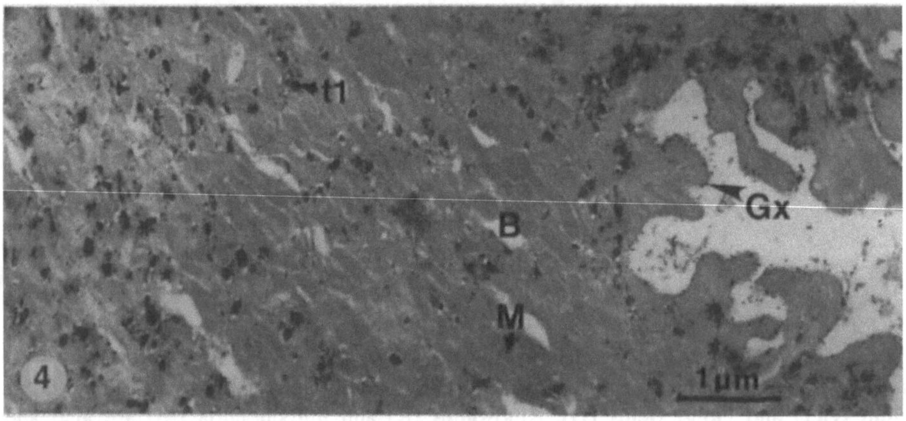

Fig. 4. *Fasciola hepatica*: Surface syncytium of the tegument treated with monoclonal antibody against an antigenic epitope in the glycocalyx and subsequently labelled with a second antibody linked to colloidal gold. The gold particles (small dense spheres) are attached mainly to secretory bodies (t1) and to the surface glycocalyx (Gx). Mitochondria (M) are unlabelled. B = vacuolated invagination of the basal membrane.

In schistosomes, the phenomenon of surface turnover takes a different form to that described in *F. hepatica*. While the invading cercariae of *S. mansoni* have a typical trilaminate surface membrane with associated glycocalyx, this is lost within 3 h of penetration and is replaced by a heptalaminate surface, comprising two closely opposed trilaminate membranes [62]. Multilaminate bodies, produced in the tegumental perikarya, migrate through the surface syncytium and fuse with the apical membrane, allowing continuous sloughing and replacement in the adult worms. Experiments in which the outer membrane was labelled with electron-dense cationic tracer molecules such as ferritin have indicated a membrane "half-life" of 2-3 h under experimental conditions [37].

Elegant studies using the technique of freeze-fracture microtomy have been carried out on the surface membrane of *S. mansoni* [63]. In this method, surfaces of frozen blood flukes are fractured by a glancing blow from a swinging microtome blade, producing "fracture-faces" which may subsequently be replicated and shadowed for electron microscope examination. The fracture planes pass preferentially through the lines of weakness in the surface layers, namely the hydrophobic interiors of each bilayer. An important feature of these freeze-fractured membranes is the appearance of numerous intramembranous particles on the fracture faces. These are interpreted as representing protein or glycoprotein molecules inserted into the membranes. They are particularly numerous on the external face of the outer bilayer, and increase in size as the worms age. Possibly these intramembranous particles revealed by freeze-fracturing represent host antigen glycoproteins attached to parasite antigens or receptors. This interpretation correlates with immunological and biochemical findings that host RBC components are adsorbed unto the tegumental surface of schistosomes, providing a molecular "disguise" which enables the parasite to evade detection by the host's immune system [85].

While elements of the tegumental secretory mechanisms in *F. hepatica* and *Schistosoma* spp. have been correlated with immune evasion phenomena, the situation is by no means fully understood. The role of T2 bodies in *F. hepatica* for instance remains enigmatic. Many species of trematode have only one type of tegumental secretion. In *Schistosoma* two distinct types of tegumental secretory bodies, the multilaminate type and a second discoid type with granular contents are produced in the same perikarya but at separate Golgi fields [136].

In addition to protein and glycoprotein synthetic activities, the tegument in trematodes has been shown to have a role in uptake of nutrient molecules, ions and water, and in exchange of these with the underlying parenchymal tissue. While molecular and ionic flux across the tegument can best be demonstrated by physiological techniques, the apical and basal membranes of the tegumental syncytium in most species examined show morphological features consistent with such phenomena. In general, the apical membrane is highly folded, increasing the potential for absorption by as much as 10-fold in *Schistosoma* [61]. In *F. hepatica*, uptake of monosaccharides, amino acids, etc. from the environment is mainly by simple or facilitated diffusion, whereas in *Schistosoma* active transport of a variety of nutrient molecules is known to occur, the necessary ATP being supplied by energy transduction in the numerous tegumental mitochondria. The situation has been reviewed by Pappas and Read [97]. There is no convincing evidence for uptake of particles or macromolecular entities by endocytosis (i.e. pinocytosis and phagocytosis) in the tegument of any adult trematode.

On the basis of morphological similarity with water-transporting epithelia in other organisms, it has been proposed that the tegument of trematodes (and in particular the infolds of the basal membrane into the cytoplasm of the surface syncytium) is involved with water and ion transport. Threadgold and Brennan [126] demonstrated the existence of Na^+/K^+-dependant ATPase activity associated with the apical membrane and with the basal tegumental infolds in *F. hepatica* and showed that the degree of swelling exhibited by these invaginations correlated with the osmolarity of the external environment. They postulated that a standing ion-concentration gradient was established through the depth of the tegumental syncytium by the activity of ionic pumps situated at the apical membrane and the basal infolds. Water would be drawn osmotically down such a gradient, perhaps bringing small organic molecules into the syncytium by entrainment. Masses of mucopolysaccharide surrounding the infolds probably act as a water sink, accommodating sudden fluctuations in the water content of the environment. The active process of osmoregulation in *F. hepatica* is presumably powered by energy transduction in the tegumental mitochondria, which lie in close proximity to the basal infolds (Fig. 5). It is also likely that a standing ion gradient osmotic flow exists between the base of the tegument and the excretory system by way of the parenchyma, assisting in the extrusion of metabolic waste.

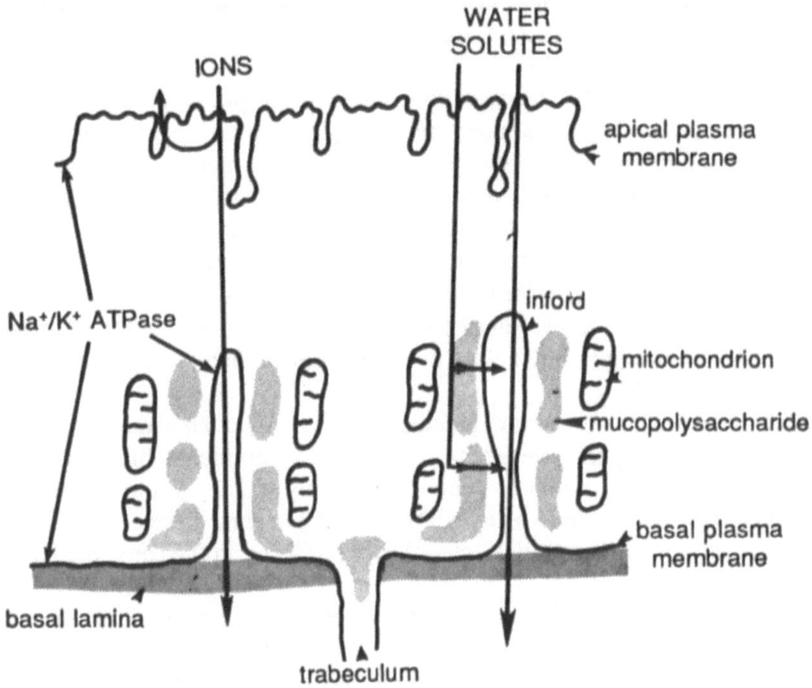

Fig. 5. Diargam illustrating the movement of water and small molecules through the tegument of *Fasciola hepatica*. A standing ion-concentration gradient is established by the activity of ionic pumps at the basal infolds. Water enters osmotically bringing in small molecules by entrainment. Mucopolysaccharide masses absorb the water, holding it in the vicinity of the membranous infolds prior to its translocation through the basal lamina. Energy transduction takes place in the mitochondria which are abundant in the vicinity of the basal infolds (after Threadgold & Brennan, [97]).

In the paramphistomes, many species of which inhabit the rumens of domestic stock worldwide, the tegument lacks mitochondria and its surface is relatively flat, bearing only small tubercles. Presumably in such cases uptake of organic molecules, ions and water is limited, with the gut providing the main route of ingress. In *Srivastavaia indica* and *Gastrothylax crumenifer* from the rumen of buffaloes, the tegumental syncytium is relatively electron dense and may serve mainly as a tough barrier protecting the parasites from mechanical damage in their turbulent environment [27].

In *G. crumenifer*, a further interesting tegumental specialisation has been described in the ventral pouch, a pocket-like invagination of the surface which runs the length of the parasite. Here the thin tegumental syncytium is intimately underlain by numerous large parenchymal cell extensions which are packed with mitochondria and which give a positive histochemical reaction for haemoglobin. Dunn et al [28] postulated that the ventral pouch in this paramphistome facilitates uptake and storage of gaseous oxygen when it is available in the environment (as for example when the host ingests water or fodder).

Finally, in considering the ultrastructure of the tegument in trematodes, mention must be made of the numerous scanning electron microscope (SEM) studies which have revealed striking intraspecific and regional variations in surface topography. SEM has clear potential in taxonomic studies, particularly with regard to mapping of sensory papillae and spine distribution. Systematic investigations on a number of paramphistomes of ruminants have been carried out by Eduardo [29, 30, 31] who emphasised the taxonomic significance of the arrangement of sensory papillae, especially around the anterior end of the worm (Fig. 6). The sense organs themselves represent the terminations of nerve processes which extend to the

Fig. 6. *Gastrodiscoides hominis*: Scanning electron micrograph of nonciliated domed sensory papillae (Ds) which are arranged on concentric tegumental ridges around the oral aperture (0) (Courtesy of Mr G.P. Brennan).

base or into the cytoplasm of the surface syncytium. They have been described in a number of species such as *F. hepatica* (6) and *Megalodiscus temperatus* [93]. Sensory endings may be uniciliated, each bearing a single "9 + 2" cilium which projects from a surface pit (Fig. 7) or is partially surrounded by tegumental tissue. Alternatively, they may appear as non-ciliated domed papillae overlying a neuronal bulb in the syncytium. While functions such as tangoreception, rheoreception and chemoreception have been ascribed to the sensory processes, no physiological evidence has been forwarded to clarify their exact roles in perception of stimuli.

SEM is being used increasingly to investigate the effects of drugs and anthelmintics on parasites. Often the earliest evidence of damage, which may lead to destruction of the parasite, is to be found on examination of the surface topography of treated worms [36]. In view of the relative ease of preparation of specimens for SEM, the method is likely to be useful for *in vitro* screening of compounds with anthelmintic potential.

Gut

The anterior region of the alimentary canal in trematodes is specialised for the intake of food and is lined with tegument continuous with that of the external surface. Generally referred to as the foregut, it

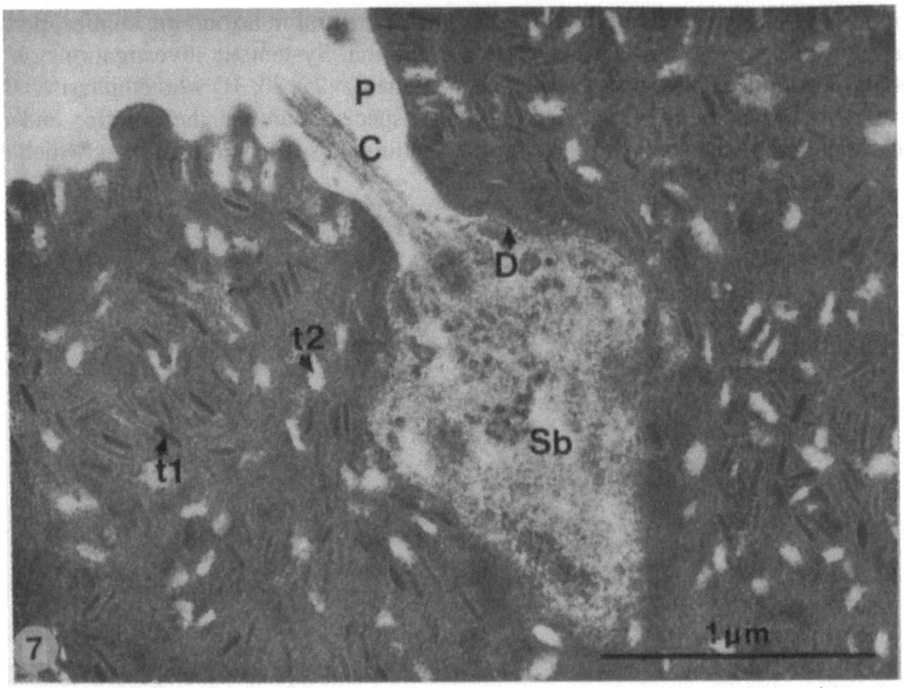

Fig. 7. *Gastrodiscoides hominis*: Uniciliated sensory ending in the tegumental syncytium lining the buccal pouches. The cilium (C) projects from a pit (P) in the surface of the tegument. The sensory bulb (Sb) in which the cilium is rooted is anchored to the tegumental membrane by a septate desmosome (D) and is continuous with a nerve branch. Tegumental secretory bodies (t1 and t2) are evident (Courtesy of Mr G.P. Brennan).

comprises the buccal cavity, which is surrounded by the musculature of the oral sucker, the pharynx, which is usually a muscular pumping organ but which is reduced or absent in many species, and the oesophagus which leads posteriorly to connect with the gut caeca. While the ultrastructure of the tegumental lining of the foregut often closely resembles that of the surface tegument, novel types of secretory body sometimes appear. These are manufactured at the GER-Golgi system in specialised tegumental perikarya lying beneath. In schistosomes for instance, two types of secretory body are found in the lining of the oesophagus which appear nowhere else in the tegument. One type appears as dense membrane-bound bodies, each containing an array of tubules, while the second type contains acid phosphatase. The specialised perikarya which produce these bodies collectively form the oesophagal gland. The foregut secretory bodies in schistosomes are released by an exocrine mechanism into the oesophageal lumen and appear to initiate lysis and digestion of erythrocytes [11, 12, 35].

The tegumental lining of the oesophagus is firmly anchored to the epithelial lining of the paired gut caeca by septate desmosomes. Typically, each caecum is an unbranched, blind-ending tube which runs laterally for at least part of the body length and is lined with a specialised glandular and absorptive gastrodermis. In *Fasciola* and *Fascioloides* each caecum is subdivided into many lateral diverticulae, which enormously increase the internal digestive and absorptive surface. In many species the gastrodermis comprises a nucleated syncytium in which the basal membrane is infolded to partly partition the cytoplasm and separate the nuclei (Fig. 8). This arrangement is found in *Schistosoma* spp. [89, 116]. On the other hand, the gastrodermis is frequently organised on a cellular basis, featuring a single layer of cells which are fairly uniform in morphology and function and which are anchored to one another apically by septate

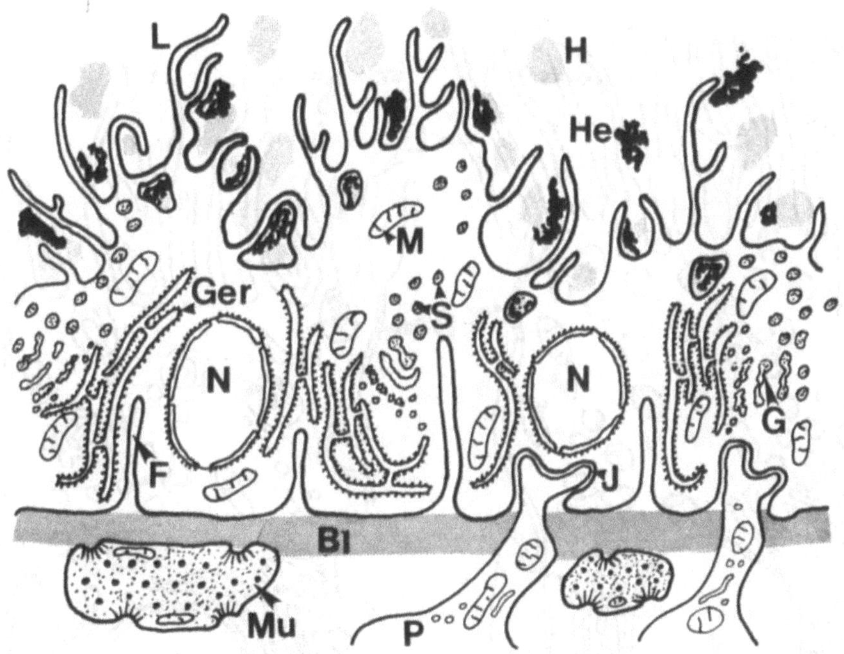

Fig. 8. Ultrastructure of the gastrodermis of *Schistosoma*. The gut is lined by a syncytial epithelium in which infolds (F) of the basal membrane separate the nuclei (N). Secretory bodies (S) which contain hydrolytic enzymes are manufactured at the granular endoplasmic reticulum (Ger) and Golgi fields (G) for exocrine discharge into the gut lumen. These secreted enzymes aid in the extracellular breakdown of haemoglobin (H), a process which was initiated by enzymes from the oesophagal gland. Digestion occurs in close association with the apical lamellae (L) of the gastrodermis, across the membrane of which micromolecular products are transported into the cytoplasm. Undigested haematin (He) remains as residual material in the gut lumen. Inpushings of parenchymal tissue (P) penetrate the basal lamina (B1) and form gap junctions (J) with the basal membrane of the gastrodermis. They probably represent sites of transfer of nutrients to internal tissues. Mu = muscle fibre; M = mitochondrion (after Morris, [67]).

desmosomes (Fig. 9). Such is the case in the caeca of *F. hepatica* [102], and certain ruminant paramphistomes [26].

Whether the gastrodermis is syncytial or cellular, ultrastructural features indicate involvement in firstly, production and release of hydrolytic enzymes for extracellular digestion in the caecal lumen and secondly, absorption of the micromolecular products (amino acids, monosaccharides, etc.) of lumenal digestion. In the digenetic trematodes, heterophagy of macromolecular nutrients and particulate material followed by intracellular digestion has been recorded only rarely, as for example in *F. hepatica*, and is probably not of primary importance in nutrition. This is in direct contrast with the situation in monogenetic trematodes, where intracellular digestion is the rule [34, 115].

The process of synthesis and exocrine secretion of protein-containing material (presumably digestive enzymes) was studied using *in vitro* maintained slices of *F. hepatica* pulse-labelled with tritiated amino acids [53]. Electron microscope autoradiography demonstrated that ^{3}H-leucine was first incorporated into protein at the GER, which occupies most of the cytoplasmic volume towards the base of the gastrodermal cells. During subsequent "chase" incubations up to 45 min in duration, the labelled proteins accumulated first in the Golgi apparatus, then appeared in secretory bodies, and finally moved to the cell apex whence they were discharged by exocytosis (Fig. 10). The limiting membrane of each secretory body became

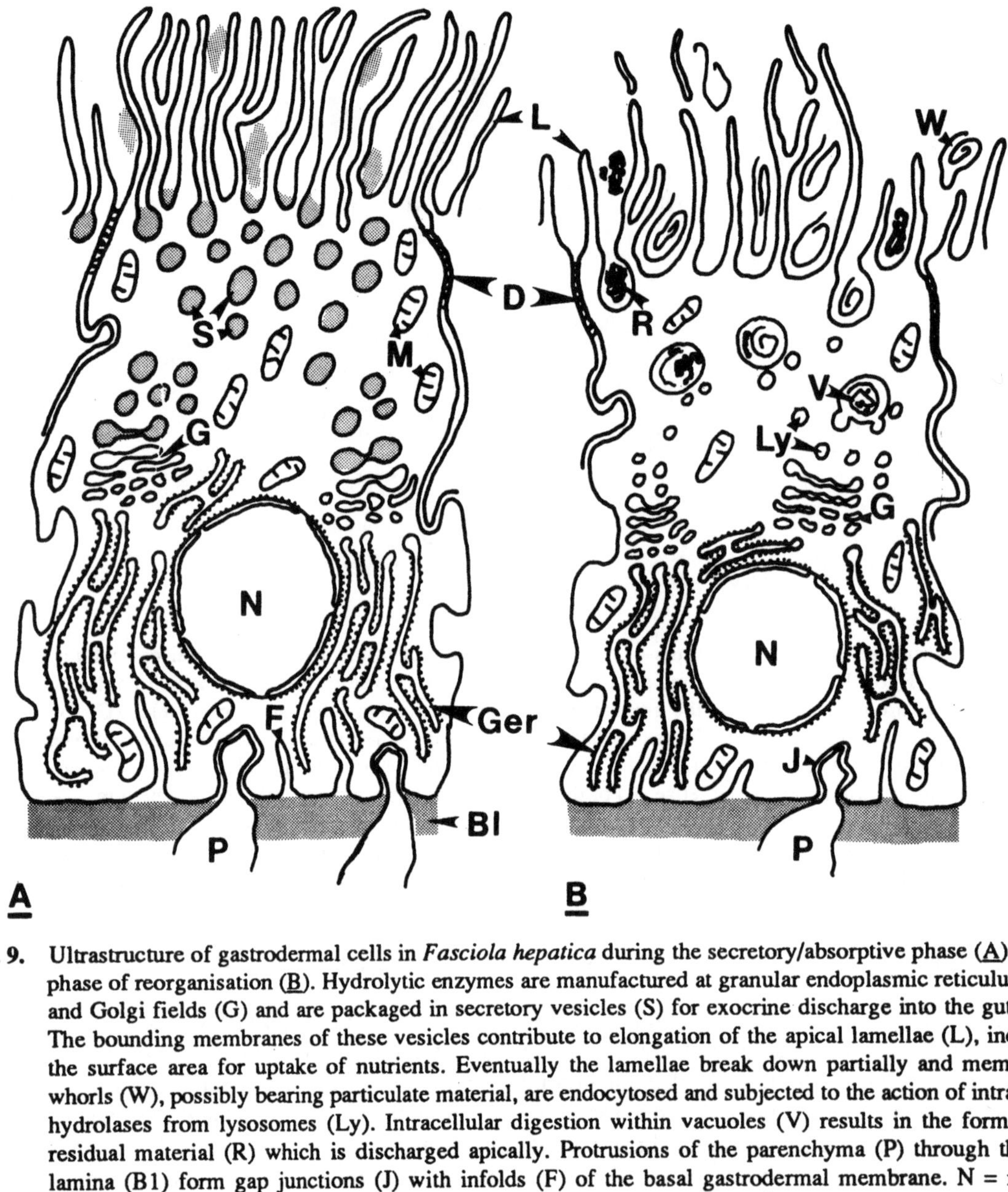

Fig. 9. Ultrastructure of gastrodermal cells in *Fasciola hepatica* during the secretory/absorptive phase (A) and the phase of reorganisation (B). Hydrolytic enzymes are manufactured at granular endoplasmic reticulum (Ger) and Golgi fields (G) and are packaged in secretory vesicles (S) for exocrine discharge into the gut lumen. The bounding membranes of these vesicles contribute to elongation of the apical lamellae (L), increasing the surface area for uptake of nutrients. Eventually the lamellae break down partially and membranous whorls (W), possibly bearing particulate material, are endocytosed and subjected to the action of intracellular hydrolases from lysosomes (Ly). Intracellular digestion within vacuoles (V) results in the formation of residual material (R) which is discharged apically. Protrusions of the parenchyma (P) through the basal lamina (B1) form gap junctions (J) with infolds (F) of the basal gastrodermal membrane. N = nucleus, M = mitochondrion, D = septate desmosomes (after Robinson & Threadgold [80]).

incorporated in the apical lamellae. Thus protein synthesis, segregation and transport in the trematode gastrodermis resembles the process occurring in the tegument, and is comparable with the mechanism which has been established for vertebrate secretory tissues. In addition to proteins, the secretory bodies of the gastrodermis in trematodes contain carbohydrate, probably in the form of glycoprotein. This material is probably concerned with maintenance of the glycocalyx which invests the lumenal surface of the apical lamellae [138].

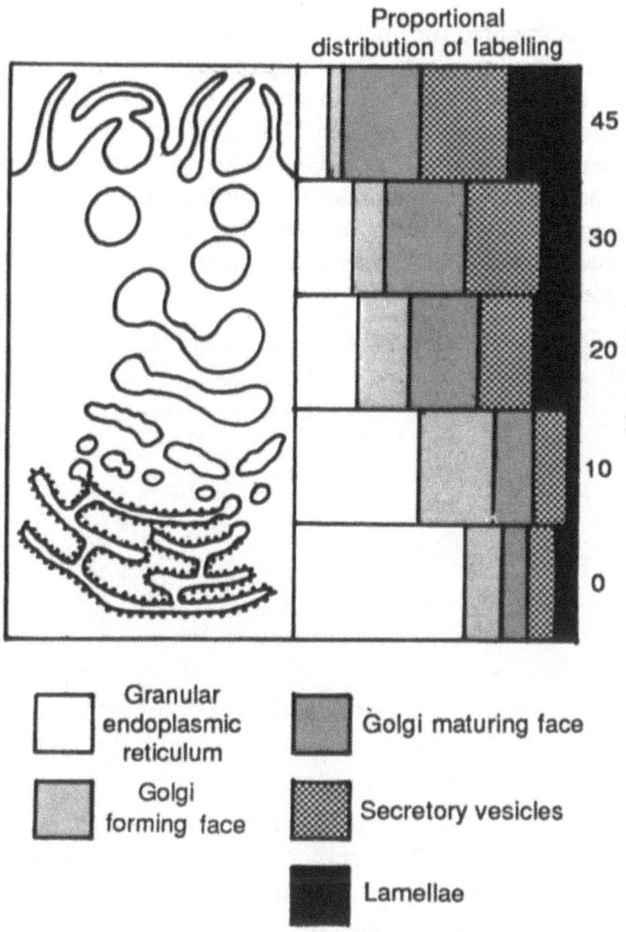

Fig. 10. Distribution of autoradiographic labelling in gut cells of *Fasciola hepatica* after various *in vitro* chase incubation periods following administration of a pulse of ^{3}H-leucine, a radioactive precursor for protein synthesis. Immediately after pulse labelling the largest proportion of radioactivity appeared in the granular endoplasmic reticulum where the amino acid molecules were incorporated into protein, Subsequently the labelled molecules moved through the Golgi fields and accumulated in secretory vesicles at the cell apex prior to exocrine discharge (after Smyth, J.D. & Halton, D.W.: *The Physiology of Trematodes*. Second edition. Cambridge University Press: reference 115, based on data from Hanna [53]).

A variety of digestive enzymes has been demonstrated by histochemical and biochemical methods in the gut of digenetic trematodes. Of these, the most important are probably proteases, acid phosphatase, esterases and leucine aminopeptidase. Information on the occurrence and ultrastructural localization of these enzymes has been collated by Erasmus [34] and Smyth & Halton [115]. In the case of acid phosphatase, activity has been demonstrated at ultrastructural level in association with the plasma membrane of the apical lamellae in *Fasciola* [120] and *Schistosoma* [35]. Activity has also been detected on the infolds of the basal membrane. Possibly the breakdown of phosphate esters at these membrane sites yields a kinetic advantage in the subsequent absorption of digestion products or their transport into the underlying parenchymal cells.

A common feature of the gastrodermis in all trematodes is the extensive development of the apical membrane providing a vast surface area for uptake of the micromolecular products of digestion in the gut lumen. The surface extensions may take the form of digitiform microvilli, as in the ruminant paramphistomes [26] (Fig. 11), or they may appear as pleomorphic lamellae, as in *S. mansoni* [89] and *F. hepatica* [102] (Figs. 8 and 9). Lamellae may branch, anastomose or rejoin the surface to form loops, thus trapping lumenal contents for digestion in close contact with the surface membrane. The basal plasma membrane of the gastrodermis is elaborated by infolds which often accommodate finger-like extensions from the underlying parenchymal cells. The apposing membranes in these regions form junctional complexes across which small molecules, up to a mass of approximately 1000 Daltons, may be passed into and out of the gastrodermis [98].

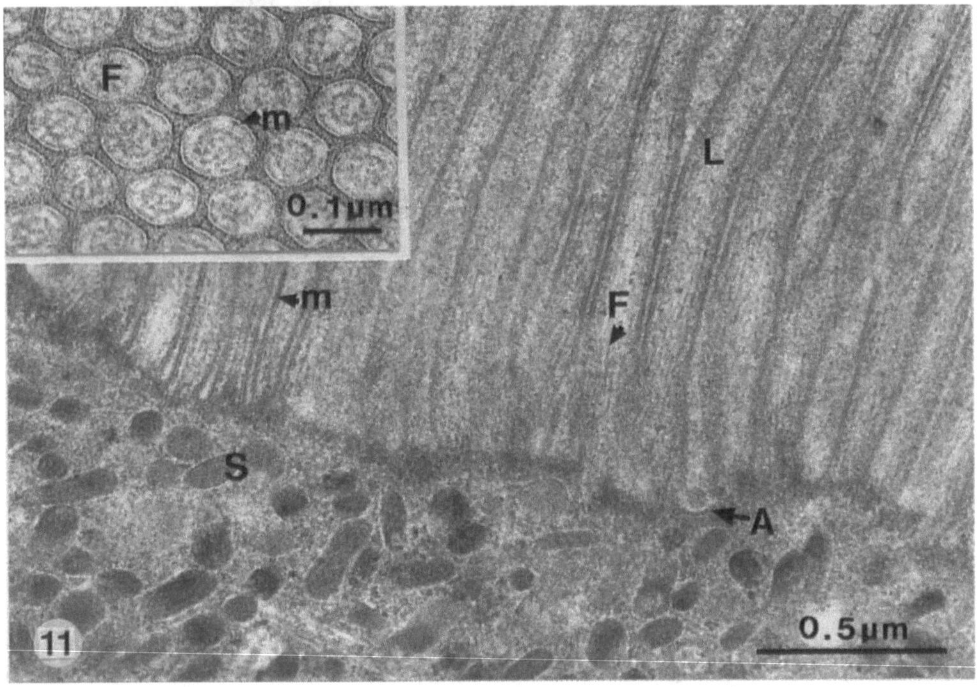

Fig. 11. *Orthocoelium scoliocoelium*: Apex of a gastrodermal cell in a paramphistome parasite of ruminants. Numerous secretory bodies (S) occur below the apical membrane (A). The surface is extended into numerous closely-packed digitiform microvilli (L) which are supported by core fibrils (F) and covered by membrane (m) continuous with the apical cell membrane. The inset shows a transverse section of the microvilli.

Secretion and absorption seem to operate concurrently in the gastrodermis of most trematodes, although in any particular cell or region of the syncytium one or other process may predominate at a particular time. Locally, the ultrastructural appearance probably depends on the principal activity ongoing at the time of fixation, although this is more evident in species having a cellular gastrodermis than in those displaying a syncytial organisation [34]. In metacercariae and early migrating juveniles of *F. hepatica*, the gut appears primarily as a secretory organ which facilitates penetration of the host gut wall and lubricates the passage of the invading fluke. Only later it does assume the highly folded apical surface characteristic of absorptive function [5].

In *F. hepatica*, each gastrodermal cell undergoes a cycle of secretion during which numerous secretory bodies are elaborated at the GER-Golgi system and migrate to the cell apex. Following discharge of the secretory bodies and addition of their bounding membranes to the apical surface, the lamellae appear particularly numerous and closely packed as the cell enters a phase of absorption [102]. At the end of a cycle of secretion, many organelles are sequestered in autophagosomes and broken down by lysosomal hydrolases such as acid phosphatase, aryl sulphatase and cathepsin, presumably to allow reorganisation and reconstitution of the synthetic machinery before the next cycle of secretion. The lamellae too become disrupted after the phase of active absorption, and whorls of membrane bearing particulate material attached to the glycocalyx are shed and endocytosed at the cell apex. These membrane whorls are degraded within the cells yielding residual bodies, which are later extruded (Fig. 9).

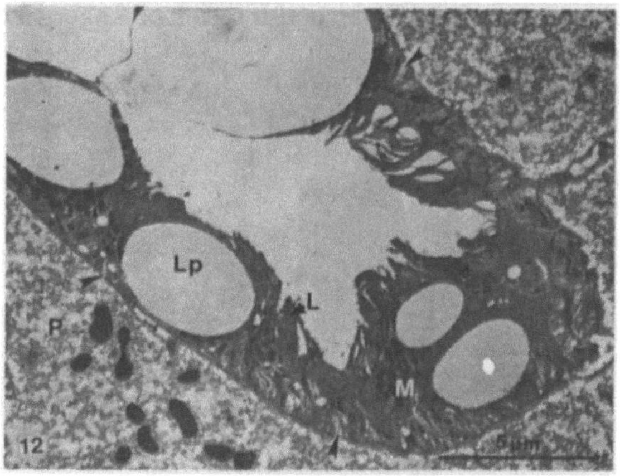

Fig. 12. *Fasciola hepatica*: Syncytial lining of an excretory tubule. The lumenal surface of the epithelium is extended into numerous lamellae (L). Lipid droplets (Lp) are abundant in the cytoplasm of the epithelium, and are discharged directly to the lumen of the duct. Mitochondria (M) are also evident, and protrusions (arrowed) of the parenchymal cells (P) are inserted into the basal cytoplasm.

Bogitsh [10] studied the process of autophagy in *S. mansoni* and demonstrated that starvation of the fluke leads to an enormous increase in hydrolase rich autophagic vacuoles in the gastrodermis. He proposed that autophagosomes are formed by sequestration of areas of gastrodermal cytoplasm (including organelles like mitochondria and GER) in double-membrane vacuoles derived from the infolds of the basal membrane. Lysosomes containing hydrolases which were synthesised by the GER-Golgi system are fused with the outer membrane of each autophagosome, and digestion of the contents proceeds as the inner membrane breaks down, allowing the hydrolases to mix with the enclosed material. The undigested membranous remains are discharged across the apical membrane into the caecal lumen.

Lysosomes, autophagosomes and residual bodies are commonly seen in the gastrodermis of digeneans, but it seems likely that autophagy is more involved with reorganisation of the cytoplasmic machinery to facilitate alternating cycles of secretion and absorption than it is with provision of nutrients.

Excretory System

The protonephridial system in trematodes comprises the flame cells, the efferent and primary collecting ducts, the main collecting ducts and the excretory bladder. While the lay-out of the system varies somewhat from species to species, and from stage to stage during a single life cycle, its main ultrastructural features and functional attributes are fairly uniform throughout the group and indeed for platyhelminths in general. Wilson and Webster [139] comprehensively reviewed the structure and functions of invertebrate protonephridial systems, including those of platyhelminths. The fundamental motive unit of the protonephridial system is the flame cell. Each of these forms a bulbous termination at the distal end of an efferent collecting duct, and comprises a flask-like cap cell from the lumenal surface of which projects a bunch of 50–100 hexagonally packed flagella, each exhibiting the typical "9 + 2" axial filament complex. The flagella are firmly rooted in the cytoplasm of the cap cell by rootlet fibres which are anchored to the terminal web, while their free ends project some distance into the lumen of the adjacent efferent duct (Fig. 13). Successive rows of flagella within the "flame" undulate in opposing directions, setting up a "standing wave" which propels fluid away from the flame bulb and along the collecting duct. It is thought that this forward-directed fluid flow sets up a hydrostatic pressure gradient around the neck of the flame cell which tends to draw fluid into the collecting duct from the surrounding parenchymal tissue [135]. The junction between flame cell and efferent duct is apparently organised as an ultrafiltration unit. From the antero-lateral margins of the flame cell a series of finger-like microvilli or "ribs" stretch

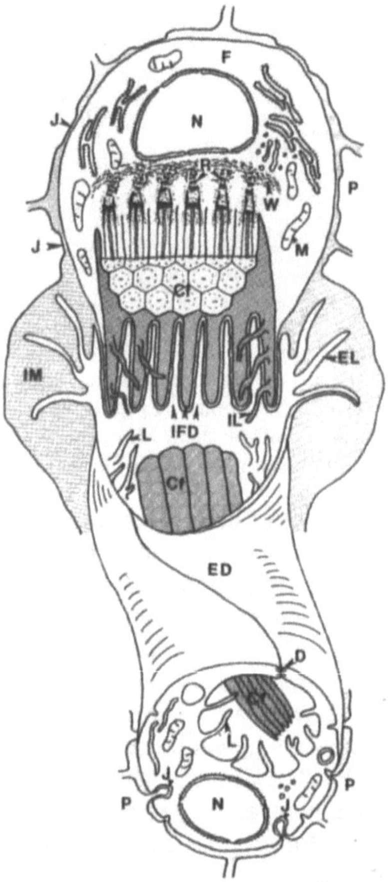

Fig. 13. Diagram illustrating interdigitation (IFD) of the margins of the flame cell (F) and the efferent duct (ED) in the trematode protonephridial system. The ciliary flame (Cf) has been sectioned longitudinally at the proximal end to show anchorage of the ciliary rootlets (R) in the terminal web (W) of the flame cell. The flame has also been cut transversely and removed from the middle region to show the filtration zone. The cavity of the flame cell and the intact surface of the ciliary flame are densely shaded to contrast with the lumenal surface of the efferent duct. The structure is supported internally by the internal leptotriches (IL) and is anchored in the surrounding interstitial material (IM) by the external leptotriches (EL). The efferent duct is a tubular cell, the lumen of which is closed by a longitudinal desmosome (D). The lumen is elevated in lamellae (L) which serve in resorption of nutrient molecules. Note the gap junctions (J) between surrounding parenchymal tissue (P) and the limiting membranes of the flame cell and efferent duct. N = nucleus, M = mitochondrion.

forward, surrounding the flagellar flame. These microvilli interdigitate with others from the margin of the adjacent tube-like efferent duct, and the apposing cytoplasmic "ribs" are tenuously interconnected by fibrous extracellular material forming slit-like filters (Fig. 13). The whole structure is stabilised by further microvilli projecting into the lumen from the cap cell (internal leptotriches) and into the surrounding interstitial material from the margin of the efferent duct (external leptotriches) [96]. It is likely that movement of fluid into the excretory tubules is assisted by the "standing gradient" osmotic flow which is initiated at the tegument (see earlier). The protonephridial network and the surface tegument may be considered as cooperating elements in the maintenance of osmoregulation. In some species such as *F. hepatica* the system may be able to compensate for minor variations in the external environment [126], but most species appear to have limited ability to compensate for osmotic stress [113]. Probably the flow of protonephridial fluid serves mainly to facilitate removal of metabolic waste.

The efferent excretory tubules and the smaller collecting ducts of the excretory system are composed of single tube-like cells placed end-to-end. Each cell bears cytoplasmic extensions which embrace the lumen and which are united along their apposing margins by a desmosome to yield a stable conducting pipe. The nucleus is offset beneath the cytoplasmic ductal wall (Fig. 13). The lumenal surface is amplified by numerous membranous lamellae which may aid in resorption of potentially useful small molecules and electrolytes from the protonephridial fluid. The basal membrane is supported by an underlying layer of interstitial material, which is interrupted at frequent junctional complexes with parenchymal membranes where transfer of metabolites probably occurs. Bennett [7] reported the occurrence of pinocytosis at the lumenal surface of flame cells, efferent ducts and primary collecting ducts in juvenile *F. hepatica*, which may aid in resorption of metabolisable molecules at this rapidly growing stage of development. A further feature which is limited to the juvenile stages in some species is the occurrence of groups of cilia at intervals along the minor ducts. Probably these assist fluid movement against capillary drag [7].

In *F. hepatica* the epithelium of the main excretory ducts and bladder is a nucleated syncytium which is continuous, without appreciable variation throughout the system [96]. Its lumenal plasma membrane is folded into many sheet like lamellae while the basal membrane is frequently infolded to form septa which penetrate deeply into the epithelium. Inpushings of the parenchyma often appear in section as large vesicles in the syncytium (Fig. 12). The lumenal and basal membranes are both reactive for phosphatases, indicating that resorption and transport of metabolites and water is enzyme mediated [120]. Within the excretory epithelium, lipid droplets are abundant, and these are released into the lumen, probably by an apocrine mechanism, for ultimate expulsion through the excretory pore [17, 49]. Since lipid rarely appears in the parenchyma or other tissues adjacent to the excretory tubules, it is likely that fatty acids resulting from anaerobic metabolism are collected in the excretory epithelium and these are condensed with glycerol to yield neutral lipids. The presence of ribosomes, GER, Golgi bodies and mitochondria within the cytoplasm of the epithelium indicates its ability to incorporate such precursors into macromolecules (Fig. 12) [138]. Nitrogenous compounds are also eliminated through the excretory system in *F. hepatica* [92], while large calcareous "excretory corpuscles" of uncertain origin and function (see Chapter 1) abound in the excretory bladder and major ducts of many cercariae and metacercariae [82], but are eliminated in the early stages of establishment in the definitive host.

Parenchyma

Much of the body space in trematodes and indeed all other platyhelminths is taken up by the parenchyma, which is a loose, cellular "packing" tissue surrounding and making intimate contact with all the other organs. Its ultrastructure in *F. hepatica* was studied by Threadgold & Gallagher [128] and Gallagher & Threadgold [51]. Each cell is large, irregular, and mainly occupied by stores of α- and β-glycogen. The nucleus occupies an eccentric position and around it are concentrated most of the other organelles. These include numerous large dense mitochondria, some GER and a few Golgi bodies. The latter have associated vesicles and lysosomes. Groups of mitochondria are congregated in the remaining cytoplasmic areas which are interspersed through the glycogen stores. The parenchymal glycogen is rapidly mobilised during periods of starvation, and is subsequently replenished when the glucose supply is restored. Threadgold & Arme [124] showed in a series of cytochemical tests that glycogenolysis is carried out by an autophagic process in *F. hepatica*. When worms are maintained in a glucose-free medium, areas of glycogen in the parenchymal cells are sequestered by smooth membranes which originate from the mitochondria. Lysosomes from the GER-Golgi system subsequently fuse with each autophagosome and release hydrolytic enzymes to the interior. Mobilized glucose molecules diffuse into the parenchymal cytoplasm leaving undigested membranous material in the form of residual bodies (Fig. 14).

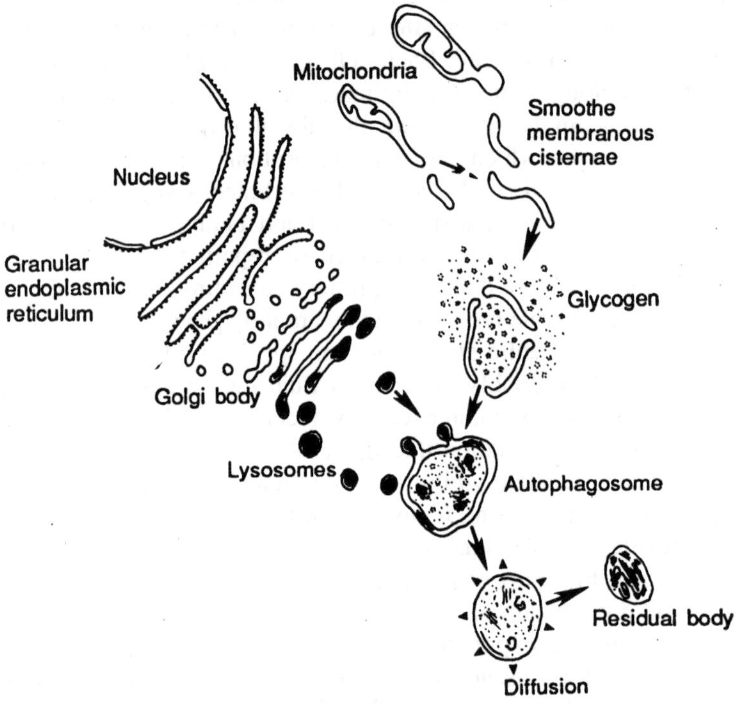

Fig. 14. Diagram illustrating the cytological events associated with glycogenolysis in the parenchyma of *Fasciola hepatica*. Stored glycogen is sequestered by smooth membranous cisternae originating from mitochondria. Addition of hydrolic enzymes from lysosomes leads to the formation of autophagosomes. Mobilised monosaccharide molecules diffuse into the cytoplasm and undigested material remains as a residual body (after Threadgold & Arme [95]).

In addition to its roles in glycogen synthesis and mobilisation, the parenchyma has fundamental importance as a transport tissue, in the absence of any clearly defined circulatory system. The cells form numerous junctional complexes (gap junctions) with each other and with the tegument, gut, muscle, excretory and reproductive systems into which parenchymal protrusions are often extended (Figs. 3, 8, 12). These junctions are sites of active molecular and ionic flux and intercellular communication. In vertebrate tissues, gap junctions have been shown to permit intercellular movement of ions and small molecules up to a mass of approximately 1000 Daltons through highly ordered "permeability channels" from the cytoplasm of one cell to that of an adjacent cell [98].

Ramifying between the parenchymal cells and forming thick layers beneath or around all the other organ systems is the fibrous, non-cellular interstitial material. This is composed of collagen-like fibres in a granular carbohydrate matrix and serves a skeletal function, providing anchorage for the muscle fibres which are attached to it by hemidesmosomes. It is interrupted in the region of junctional complexes but elsewhere supports and stabilises the tissues. In the highly muscular ruminant paramphistomes, the interstitial tissue is particularly well developed beneath the tegument and around the superficial musculature, providing structural stability and flexibility [27].

Lymph System

In a few families of trematodes, notably the Paramphistomatidae, a system of interconnected, blind-ended, fluid-filled canals ramifies throughout the parenchymal tissue. The contents are mobile and "cells", "granules" and "particles" have variously been reported within the fluid matrix. Recent ultrastructural and histochemical studies on three species of paramphistome from Indian water buffaloes have shown that the lymph fluid comprises a proteinaceous, lipid-rich cytoplasmic matrix within which occur naked nuclei, GER complexes and autophagic granules, but no distinct Golgi bodies [25]. Surprisingly, mitochondria were not detected at EM level, although histochemical studies on *Ceylonocotyle* indicated intense succinic dehydrogenase activity in the lymph system [112]. The lymph vessels are delimited by a unit plasma membrane which is supported and separated from adjacent tissues by interstitial material. A few muscle fibres surround the channels, imparting endogenous peristaltic movement to the system (Fig. 15).

The lymph channels display intimate relationships with the parenchymal tissue, in that frequent junctional complexes appear between their delimiting membranes. In addition, membrane-bound protrusions of

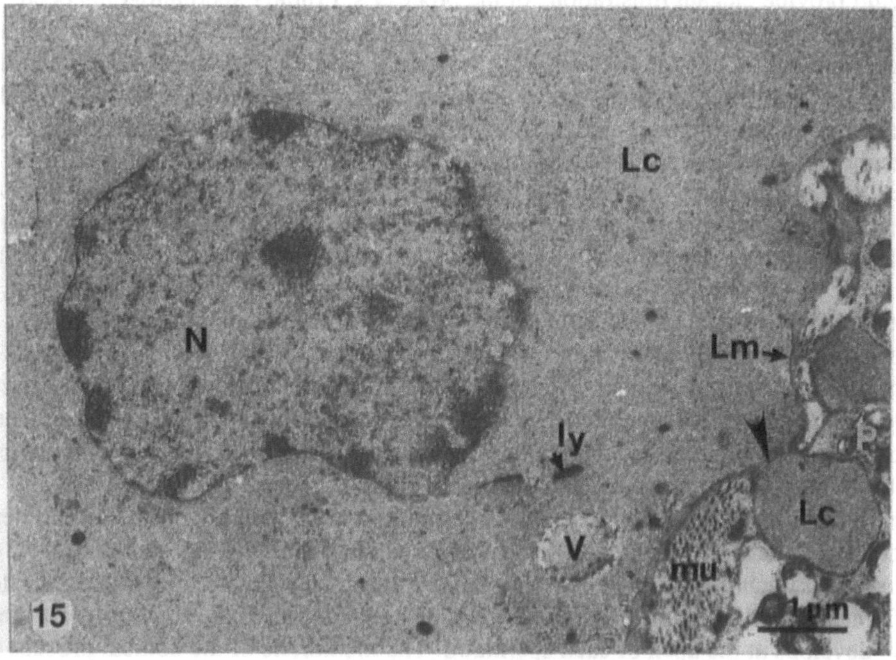

Fig. 15. *Gastrodiscoides hominis*: Part of a lymph channel showing the homogenous syncytial cytoplasmic matrix (Lc) which contains a free-floating nucleus (N), lysosomes (1y) and autophagic vacuoles (V). The channel is delimited by a plasma membrane (Lm) which is underlain by interstitial material and muscle fibres (mu). In certain regions the lymph protrudes (arrow) into surrounding parenchymal cells which show evidence of intense autophagic activity (Courtesy of Mr G.P. Brennan).

lymph are embedded in the cytoplasm of juxtaposing parenchymal cells in some regions, particularly in the area around the pharynx. Within these specialised parenchymal cells, autophagic degradation of sequestered lymph cytoplasm seems to occur [25]. Intercellular connections have not been observed between the lymph and any tissue apart from the parenchyma.

On the evidence to date it seems likely that the lymph system serves to store amino acids in the form of protein and possibly lipoprotein manufactured at the GER complexes and ribosomes. According to local synthetic requirements, the stored precursor molecules may be mobilised by autophagic processes and transported through the extensive network of channels, crossing to sites of reassembly (for example, in the reproductive system and gastrodermis) via the parenchyma. The histochemical finding of haemoglobin in the lymph of some species indicates a role in oxygen storage and transport [25].

Reproductive System

The biology of egg formation in trematodes has been reviewed by a number of authors [32, 115]. In this account, space limitations dictate that consideration be given to the ultrastructural features of only a few organs namely testis, vitellaria and Mehlis' gland, sections of which are frequently encountered in routine electron microscope examination of the parasites.

Testis

Detailed ultrastructural studies on spermatogenesis and spermiogenesis in platyhelminth parasites are few, but the sequence of events and appearance of the various stages seems fairly uniform throughout the phylum. The studies by Burton [18] on *Haematoloechus medioplexus* and Robinson & Halton [104] on *Corrigia vitta* provide elegant illustrations of the processes. Primary spermatogonia occur peripherally in the testis, each characterised by a large nucleus with dense clumps of chromatin and a prominant nucleolus. The scanty cytoplasm is packed with ribosomes and polyribosomes, and in addition contains a few strands of GER and mitochondria. These morphological features are consistent with an undifferentiated or embryonic stage of development. Spermatogenesis is initiated by two successive mitotic divisions producing secondary, then tertiary spermatogonia. Each nuclear division is succeeded by incomplete cytokinesis so that secondary spermatogonia and subsequent stages remain connected by cytoplasmic bridges (Fig. 16). Thus, each group of cells derived from the division of a single primary spermatogonium develops as a syncytium, a mechanism which may ensure synchrony of divisions and differentiation. Primary spermatocytes result from mitosis of each of the tertiary spermatogonia in a tetrad, and exhibit a much increased cytoplasmic to nuclear volume ratio. The nuclei contain synaptonemal complexes, reflecting the pairing of homologous chromosomes at this stage. In the cytoplasm, there is an increase in the amount of GER and in the number of mitochondria while Golgi bodies and associated vacuoles become evident. The subsequent meiotic divisions result in the appearance, first of sixteen secondary spermatocytes which resemble their immediate progenitors, then of thirty-two spermatids connected to one another by thin cytoplasmic stalks. This is the spermatid "rosette".

Differentiation of the spermatids into mature spermatozoa constitutes the process of spermiogenesis which has been documented for a number of species including *F. hepatica*, *Dicrocoelium dendriticum* and *S. mansoni* [88]. Each nucleus decreases in volume as the chromatin condenses first into clumps or short dense filaments, and later into twisted lamellae. The nucleus elongates and migrates into a conical process which develops at the apex of each spermatid. The membrane of the conical process is supported by microtubules, and as development proceeds a circular groove delimiting the process from the spermatid cytoplasm progressively deepens. A centre body and associated basal bodies situated in the conical process give rise to two axial filaments which project laterally (Fig. 17). Each axial filament exhibits a "9 + 1" pattern of microtubules. The axial filaments are anchored in the cytoplasm by associated rootlet

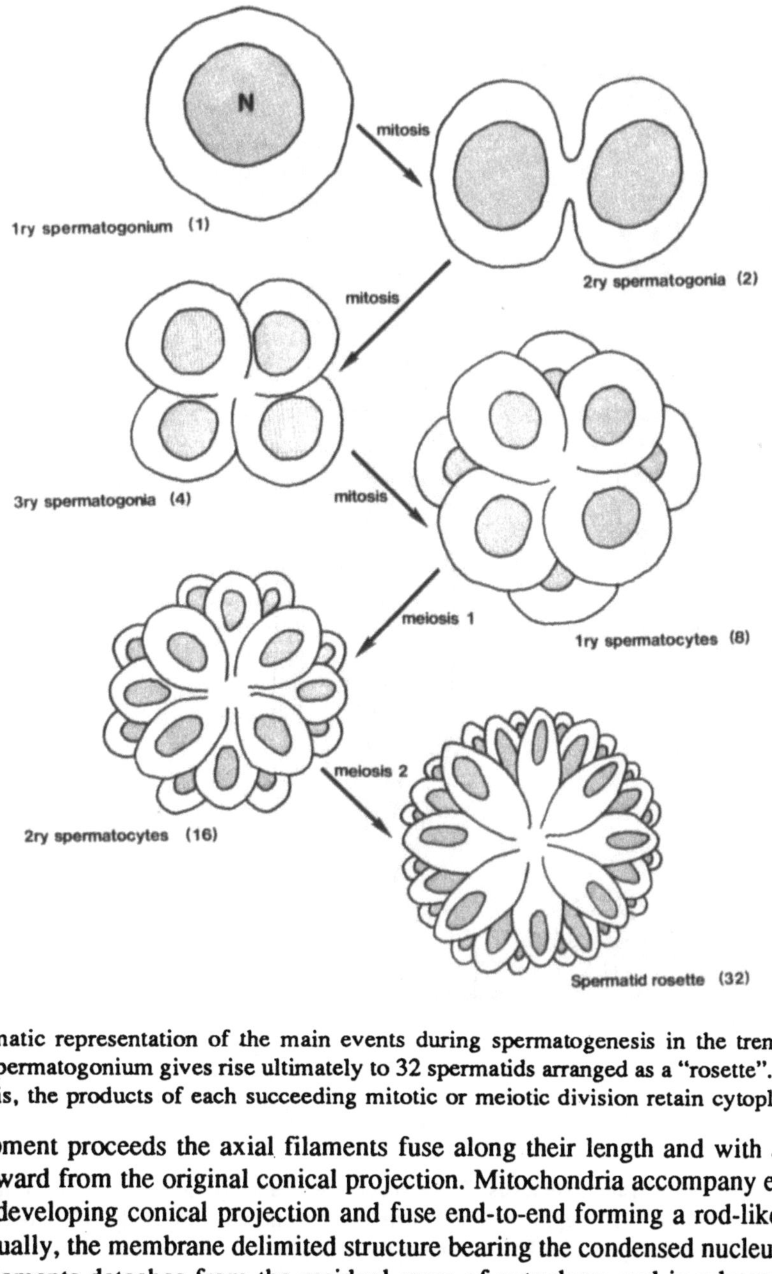

Fig. 16. Diagrammatic representation of the main events during spermatogenesis in the trematode testis. Each primary spermatogonium gives rise ultimately to 32 spermatids arranged as a "rosette". Due to incomplete cytokinesis, the products of each succeeding mitotic or meiotic division retain cytoplasmic continuity.

fibres. As development proceeds the axial filaments fuse along their length and with a central process which extends forward from the original conical projection. Mitochondria accompany each nucleus as it migrates into the developing conical projection and fuse end-to-end forming a rod-like body alongside the nucleus. Eventually, the membrane delimited structure bearing the condensed nucleus, mitochondrion and fused axial filaments detaches from the residual mass of cytoplasm and is released into the lumen of the testis.

Each mature spermatozoon is filamentous, lacking external differentiation into head, middle and tail regions, and contains two axial filaments. Numerous β-glycogen granules are present and presumably serve as an energy source.

In a number of species including *C. vitta* [104] and *S. mansoni* [95] additional non-germinal tissue occurs within the testis. These sustentacular cells appear to provide mechanical support and protection for the spermatogenic cells. They lie on the basal lamina of the testis and send out ramifying cytoplasmic

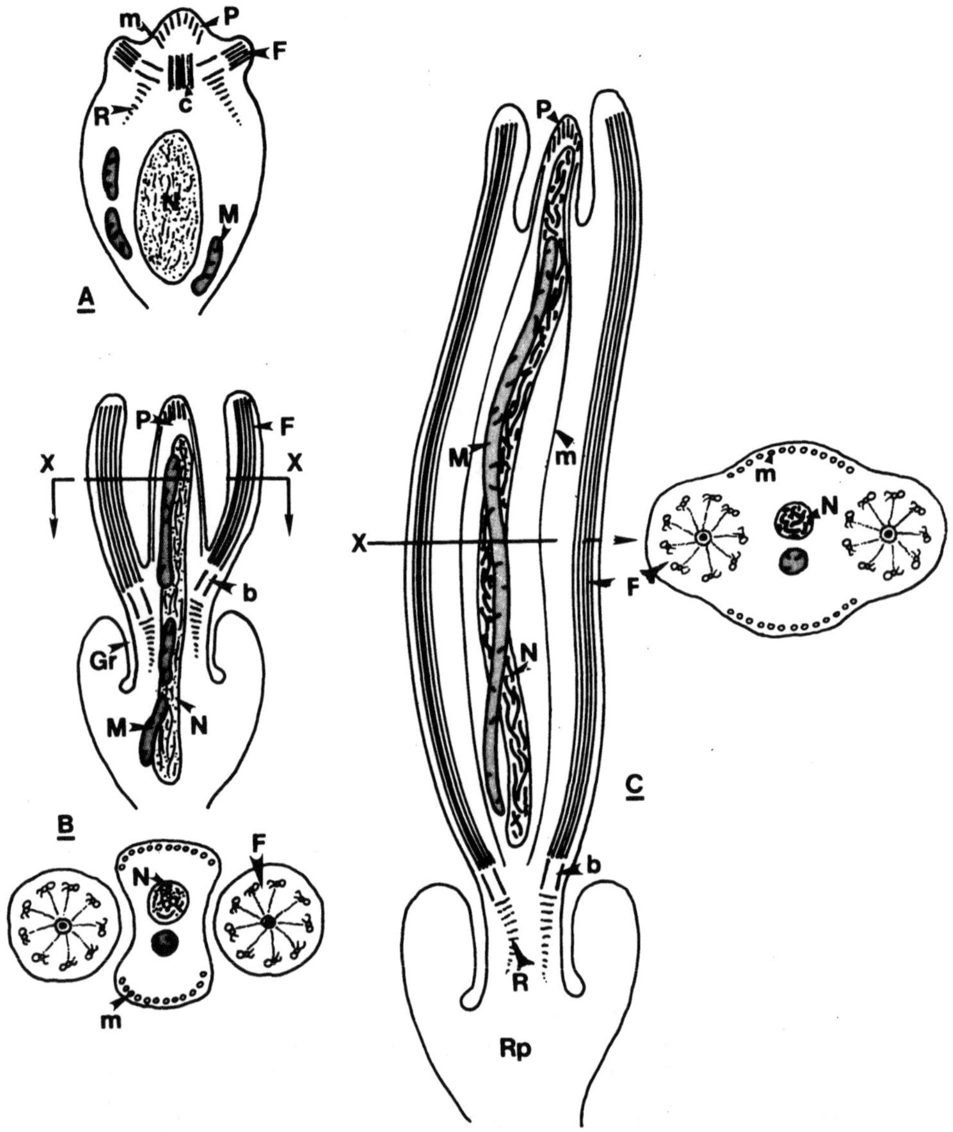

Fig. 17. Spermiogenesis in the trematode testis. Three successive stages A, B and C are illustrated. In the spermatid A, the axial filaments (F) begin to develop as buds from a centre body (c). Rootlets (R) are already evident, and a central process (P) supported by microtubules (m) appears apically. As differentiation proceeds (B) the axial filaments reorientate to run parallel with the developing central process. The basal bodies (b) and rootlets remain near the zone of differentiation which is demarcated by an encircling groove (Gr) at the spermatid apex. As the chromatin condenses, the nucleus (N) elongates and moves into the central process accompanied by mitochondria (M). The section (X–X) illustrates that the axial filaments, which display a "9 + 1" conformation of microtubules, lie to either side of the central process. In the most advanced stage C, the axial filaments have fused with the central process for most of the length of the spermatozoon. In section (X) both axial filament complexes, the nucleus, mitochondrion and microtubules of the central process are seen to lie within a single bounding membrane. The mitochondria have fused into a single cylindrical rod and the nucleus is filamentous. Eventually the mature spermatozoon will detach from the spermatid rosette, leaving the basal bodies and rootlets in the residual cytoplasm (Rp).

processes which partially enclose the developing germinal units. The presence of vesicles and vacuoles containing cellular debris in various stages of degeneration within the cytoplasm of the sustentacular cells indicates that they may have a role in degradation of residual cytoplasm of the developing germ cells, and therefore, possibly in recycling potential nutrient molecules. They may, in addition, provide a route for transfer of amino acids and monosaccharides from the parenchyma to the spermatogenic tissues.

Vitellaria

Numerous studies on egg formation in trematodes have revealed that the main bulk of the egg-shell material originates from proteinaceous secretory bodies produced in the vitelline cells [32, 115]. Detailed studies involving electron microscopy, cytochemistry, autoradiography, stereology and X-ray microanalysis have been carried out with a wide variety of digenean species including *S. mansoni*, *F. hepatica*, *Microphallus* sp., etc. [115].

The vitelline follicles in all species studied contain cells in various stages of cytomorphosis. In a rigorous stereological analysis of vitelline cell development in *F. hepatica*, Threadgold [122] recognised four cell types viz. stem cells, intermediate type 1 cells, intermediate type 2 cells and mature cells. In stem cells the nucleus is large relative to cell volume and the cytoplasm contains many ribosomes and polyribosomes, some mitochondria but little or no GER, a morphological situation characteristic of embryonic or undifferentiated cells. As each cell progresses towards the intermediate stage, the nucleolus and later the nucleus increase in size corresponding with gene activation, ribosome production and protein synthesis for cell growth and manufacture of secretory granules. Mitochondria become more numerous correlating with increased energy demands for growth and synthesis. The GER shows the greatest development, with a 25-fold increase in total surface area between the stem cell and the mature stages. This corresponds with active incorporation of precursor molecules into shell protein which is packaged into membrane bound globules at developing Golgi bodies. Eventually, at the end of the phase of active protein synthesis, numerous tightly packed clusters of mature shell protein globules are located towards the periphery of the cell (Fig. 18).

The termination of the phase of active protein synthesis (i.e. intermediate stage 2) is marked by the onset of a dramatic change in vitelline cell function, with corresponding radical modifications in ultrastructural appearance. As the cells become mature much of the cytoplasm, including GER, ribosomes and mitochondria is sequestered in autophagosomes ("yolk globules") and progressively broken down. The remaining GER and ribosomes recede to the perinuclear region and to the cell periphery as large amounts of α- and β-glycogen begin to accumulate within the cell (Fig. 18). In the mature stage, 20% of the cytoplasmic volume is occupied by glycogen, 21% by the peripheral shell protein globules, 14% by autophagosomes and 40% by residual cytoplasmic material. The nucleus is dense and pyknotic [122].

During maturation from stem cell stage, to mature stage, vitelline cells are displaced from the periphery of the vitelline follicle to the medullary region, whence they detach and pass into the vitelline duct for transport to the vitelline reservoir and ootype, as subsequent generations develop in the more peripheral regions. The supply of stem cells is maintained by mitosis in the peripheral zone of each follicle.

In *F. hepatica* nurse cells are present in the vitelline follicles and cytoplasmic processes from these completely envelop the growing vitelline cells [66]. Junctional complexes exist between parenchymal cell membranes and nurse cell membranes at the periphery of each vitelline follicle, and the nurse cell processes in turn, have junctional complexes with the developing vitelline cells. It seems likely that the nurse cells obtain precursor molecules from the parenchyma and pass these to the vitelline cells for incorporation into proteins and carbohydrates. Experimental evidence for such activity has been obtained in autoradiographic studies [54].

In *S. mansoni* a sequence of events similar to that in *F. hepatica* occurs during maturation of the

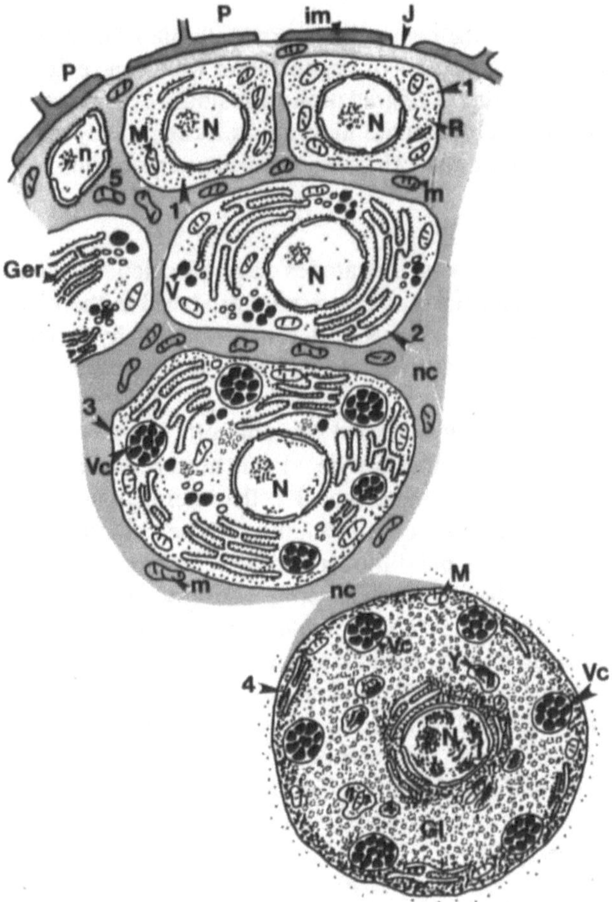

Fig. 18. Stages in the development of vitelline cells in *Fasciola hepatica*. 1 = stem cell, 2 = intermediate type 1 cell, 3 = intermediate type 2 cell, 4 = mature cell, 5 = nurse cell. Globules of shell protein (V) are manufactured at granular endoplasmic reticulum (Ger) and Golgi fields in intermediate cells, and accumulate peripherally in clusters (Vc). In mature cells, organelles are degraded in autophagosomes or "yolk bodies" (Y) while large amounts of glycogen (Gl) accumulate. The developing cells are surrounded by extensions of nurse cell cytoplasm (nc). The vitelline follicle is enveloped by a layer of interstitial material (im) except where junctional complexes (J) occur between nurse cells and parenchymal tissue (P). N = vitelline cell nucleus, n = nurse cell nucleus, M = vitelline cell mitochondrion, m = nurse cell mitochondrion, R = ribosomes (after Irwin & Threadgold [49]).

vitelline cells [33]. However, in the final stages, there is no build-up of glycogen stores in the cytoplasm. This correlates with the fact that embryonic development in *S. mansoni* occurs while the egg is still within the definitive host. Presumably the necessary energy substrates which are available in the internal environment gain access to the developing miracidium across the thin egg shell. *F. hepatica* eggs, on the other hand, undergo embryonic development in the external environment and must be self-contained with regard to energy supply.

A further aspect of vitelline cell activity was revealed in *S. mansoni* by the application of X-ray microanalysis combined with transmission electron microscopy [33]. It appears that treatment of the worms with the anthelmintic Astiban resulted in sequestration of high concentrations of the drug within autophagosomes in the maturing vitelline cells. This conclusion was based on the identification of the characteristic X-ray emission peaks for antimony, the active constituant of Astiban, when " cytosegresomes" containing ribosomes GER and other cytoplasmic components were probed. These bodies also displayed intense acid phosphatase activity, indicating the presence of lysosomal hydrolases. It is possible that drug sequestration in the vitelline cells represents a detoxification effort in the parasites, and may explain the apparent inactivation of the anthelmintic during treatment regimes.

Mehlis' Gland

In those trematodes examined to date, Mehlis' gland consists of two types of secretory cell which are arranged radially around the ootype and lie at different distances from it, supported by parenchymal cells. Each Mehlis' gland cell connects with the lining of the ootype by an individual unbranched cytoplasmic extension along which its secretory products are channeled [130]. The exact role of these secretions in egg-shell formation is still uncertain, but it seems likely that they are concerned with laying down a membrane in the ootype against which the precursor egg-shell protein from the vitelline cells is moulded. This view is supported by the finding that Mehlis' secretion is predominantly lipoprotein in nature [20]. Roles in lubrication of the uterus for passage of the egg and activation of phenolase enzymes for egg-shell tanning have also been proposed [115]. In *F. hepatica*, two ultrastructurally distinct types of Mehlis' gland cell, S_1 and S_2, are recognised [130]. The S_1 cell bodies possess large quantities of GER and many Golgi complexes which give rise to secretory bodies. The latter are conducted to the ootype through long microtubule-lined cytoplasmic extensions which are anchored in the ootype epithelium by septate desmosomes (Fig. 19). The cell bodies are held firmly in place in the parenchyma by invaginations of interstitial material which penetrate deeply into the cytoplasm. These surface invaginations may also aid in transferring metabolites and precursor molecules from the parenchyma to the Mehlis' cells. The S_2 cells are essentially similar to the S_1 type, but lie closer to the ootype, hence have shorter cytoplasmic extensions. Their GER is peculiarly distended, and the secretory bodies they produce are denser than the S_1 bodies and exhibit crystalline contents. The S_1 and S_2 secretory bodies rapidly dissociate in the lumen of the ootype, whereto they are discharged by an exocrine mechanism.

CESTODES

Tegument

Adults

Since tapeworms possess no gut, all nutritive and gaseous exchange must occur across the body tegument which, not surprisingly, exhibits extensive modification of the apical plasma membrane. Otherwise, the organisation of the surface resembles that in trematodes, featuring a syncytial outer zone connected by cytoplasmic processes to perikarya lying amongst the parenchymal cells (Fig. 20). In view of the immense importance of the tegument in the physiology of cestodes, a large number of species from all the major groups within the class have been studied. There is of course some degree of variation in surface topography and morphology between species, and even along the length of an individual tapeworm, but the ultrastructure of the most widely studied species, *Hymenolepis diminuta* seems to be representative of the class as a whole. Specific and regional variations are detailed in several recent comprehensive reviews [73, 78, 80] while the roles of scanning electron microscopy in the elucidation of scolex function and in taxonomic discrimination between closely related species are well illustrated by the studies of Andersen [2] on *Diphyllobothrium* spp.

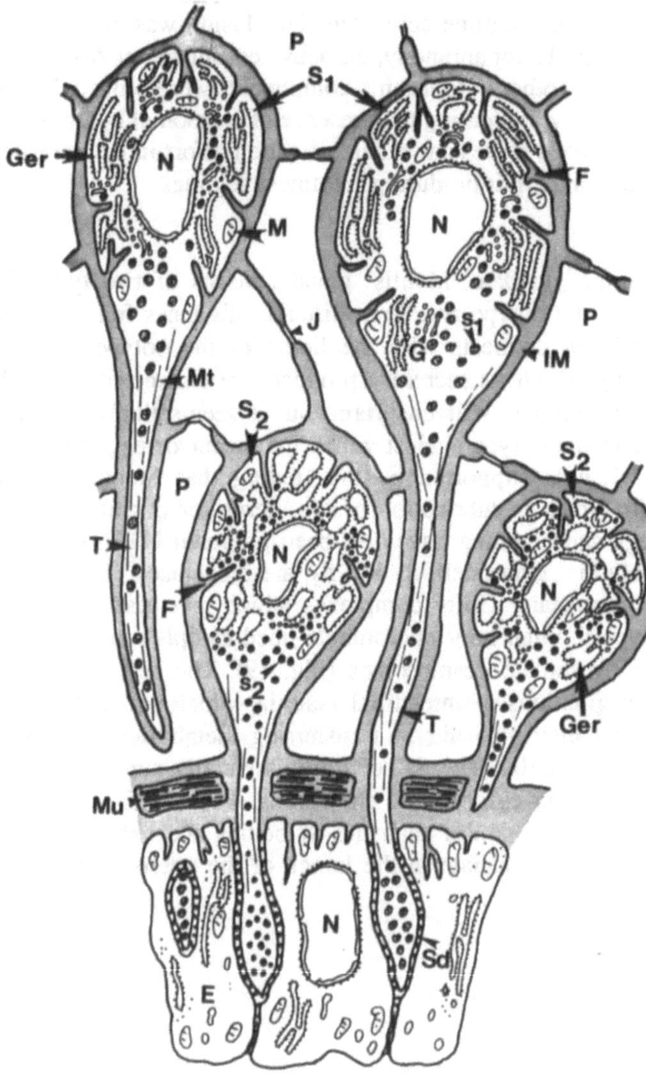

Fig. 19. Mehlis' gland of *Fasciola hepatica*. Two types of secretory cell are recognised (S_1 and S_2) which produce distinctive secretory bodies (s_1 and s_2 respectively) for transport through microtubule (Mt) lined ducts (T) to the epithelium of the ootype (E). Granular endoplasmic reticulum (Ger) and Golgi fields (G) feature in both cell types, and infoldings (F) of interstitial material (IM) anchor the cells in the parenchymal tissue (P). Muscle fibres (Mu) surround the ootype, and the Mehlis' cell ducts are secured between epithelial cells by septate desmosomes (Sd). N = nucleus, M = mitochondrion, J = gap junction (after Threadgold & Irwin [101]).

The apical plasma membrane of the tegument in cestodes is elevated in a regular array of microvillus-like projections termed "microtriches" (singular microthrix). These comprise a "brush-border" which is comparable with that of the vertebrate intestinal epithelium. The microtriches amplify the tegumental surface area by approximately 10 fold, significantly enhancing its absorptive potential [123] (Fig. 21). In *Diphyllobothrium latum*, the rapid increase in the rate of metabolism occasioned by transfer of the pleurocercoid from the cold-blooded intermediate host (fish) to the warm-blooded definitive host (man)

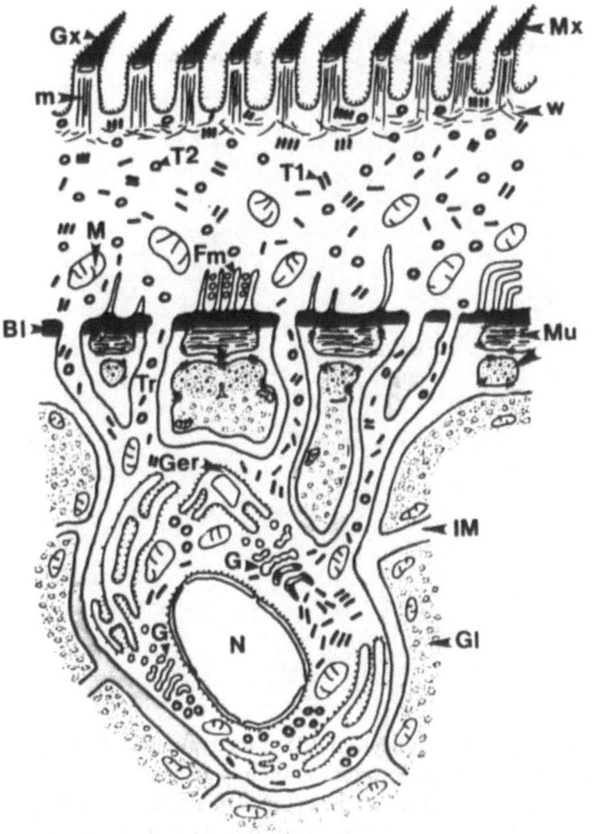

Fig. 20. Ultrastructure of the tegument of *Hymenolepis diminuta*. The surface syncytium bears microtriches (Mx) apically which feature electron dense tips and proximal shafts supported by core fibres (m) extending to the cell web (w). The apical plasma membrane which covers the microtriches completely, bears an external glycocalyx (Gx). Within the cytoplasm lie tegumental secretory bodies (T1 and T2) which are manufactured at granular endoplasmic reticulum (Ger) and Golgi fields (G) in the perikarya and transported to the surface syncytium through trabecula (Tr) which penetrate the basal lamina (B1). The basal membrane of the surface syncytium features numerous digitiform infolds, often arranged as multitubular complexes (Fm). Mu = muscle fibres, IM = interstitial material, N = nucleus, M = mitochondrion, G1 = glycogen of myocytons and parenchymal extensions.

is accompanied by multiplication and elongation of the microtriches, resulting in a great increase in surface area for uptake of nutrients [15]. Each microthrix in *H. diminuta* has an electron opaque pointed tip which is composed of densely packed keratinous fibrils. These fibrils end at a multilaminate base plate, demarcating the tip from the cylindrical proximal shaft which has an electron-lucent core with 5 nm thick core fibres orientated longitudinally. The core fibres often extend into the apical cytoplasm of the tegument [79]. The apical plasma membrane bears on its outer surface a glycocalyx composed of glycoprotein with exposed carbohydrate side chains. This is a dynamic zone with a turnover time of about 6 h, as has been demonstrated in autoradiographic studies. The surface components are synthesised in the perikarya by the GER-Golgi system and transported to the apical plasma membrane, as is the case with trematodes [94]. In the surface cytoplasm of the tegument, dense discoidal secretory bodies containing the glycoprotein precursors of the glycocalyx abound.

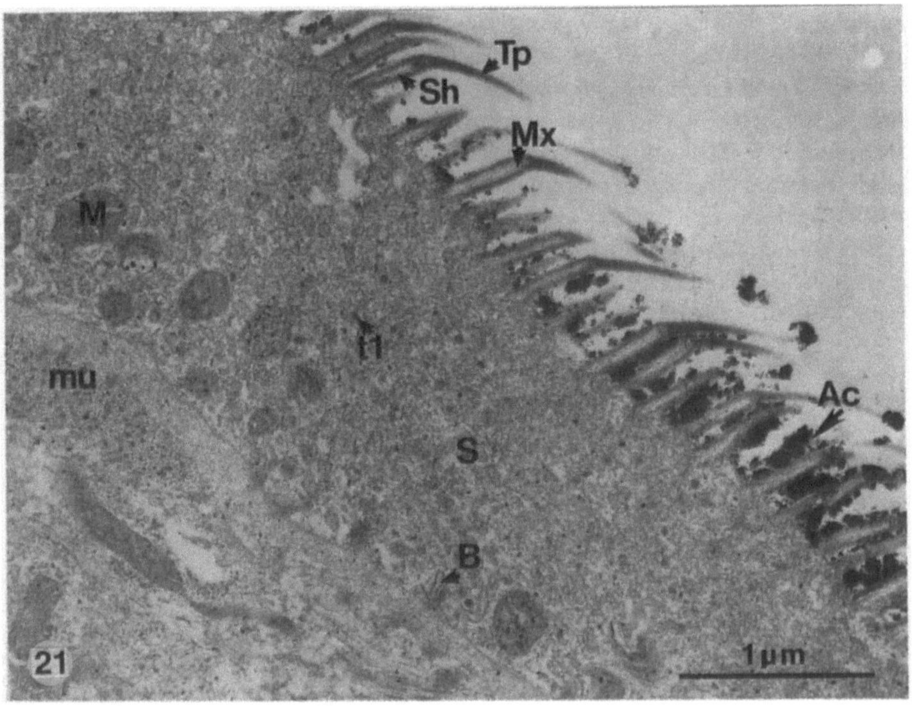

Fig. 21. *Hymenolepis diminuta*: Surface syncytium (S) of the tegument showing the distribution of acid phosphatase (Ac). The cytochemical reaction product is mainly located on the outer aspect of the apical membrane which covers each microthrix (Mx). In the cytoplasm are numerous large mitochondria (M) and tegumental secretory bodies (t1). The basal membrane is infolded (B) and underlain by interstitial material and muscle fibres (mu). Note that each microthrix features an electron dense pointed tip (Tp) and a cylindrical proximal shaft (Sh) supported by core fibrils (Courtesy of Professor L.T. Threadgold).

Mitochondria too are abundant in the surface syncytium especially along the basal membrane. They undoubtedly represent sites of energy transduction for powering the active uptake processes characteristic of the cestode tegument.

The basal membrane of the tegumental syncytium lies on a fibrous basal lamina and is infolded to form septa and digitiform invaginations often arranged as multitubular complexes [131]. The perikarya, which are irregular or flask shaped contain many ribosomes and polyribosomes, GER cisternae which are often swollen, Golgi complexes with associated vesicles, maturing secretory bodies and mitochondria. These features are consistent with a role in synthesis of macromolecular secretory material. The trabecula which connect the perikarya with the surface syncytium have longitudinally orientated microtubules which are involved in transporting secretory bodies towards the surface.

In a few adult diphyllobothriid cestodes there is experimental evidence that endocytosis occurs at the surface of the tegument. In *Schistocephalus*, pinosomes are found at the apical plasma membrane which are capable of engulfing electron-dense tracers such as ruthenium red [129]. These pinosomes quickly move to the basal membrane and discharge their contents by exocytosis into the interstitial material. Some pinosomes pass into the perikarya where they become involved in a heterophagic process. The procercoids of *D. latum* display vesicles suggestive of pinocytotic activity between the bases of microtriches and in the surface syncytium, but there is no evidence of these occurring in the adult stage [14, 15]. To date, endocytosis has not been demonstrated in any adult cyclophyllidean cestode.

The ultrastructural features of the cestode tegument are consistent with roles in protection of the parasite from host digestive enzymes and immune effectors and uptake of nutrient molecules from the environment. The protective functions are closely related to synthetic activities and surface renewal. Areas of the surface which have been damaged by antibodies or by the intestinal secretions of the host are blebbed off and replaced by exocytosis of secretory bodies, whose membranes become inserted into the apical membrane. Endocytosis at the surface of pseudophyllideans may represent a further mechanism of excising damaged areas from the absorptive surface, with concomitant nutritional advantages. In either case, the tapeworm surface may be envisaged as a mosaic of new, fully functional areas of plasma membrane and areas in various stages of functional incapacity [123]. Bearing in mind that growth is an ongoing process in cestodes, continuous addition of new membrane to the apical surface is probably necessary to keep pace with development of the proglottids. A further aspect of the protective role of the surface may be that the massed array of sharply pointed microtriches serves to ward off attack by the immunocompetent cells of the host [78].

The nutritive role of the cestode surface has already been highlighted in considerations of surface membrane amplification. Cytochemical studies have shown that a number of phosphatases and non-specific esterases are associated with the outer aspect of the apical plasma membrane (Fig. 21). These enzymes facilitate the transport of molecules such as sugar phosphate and nucleosides into the apical cytoplasm [79]. In addition the surface of *H. diminuta* exhibits endogenous lipase activity and adsorbs pancreatic amylase. Conversely, tapeworms inhibit certain host enzymes such as pancreatic lipase, trypsin and chymotrypsin, presumably as a protective mechanism. Numerous experimental studies have revealed the existence of specific carrier sites, active transport and facilitated diffusion associated with the apical plasma membrane in a variety of cestodes [97].

Elaboration of the basal plasma membrane, and in particular the appearance of multitubular complexes, suggests that the tegument in cestodes may have a role in regulation of water and ion movement, as seems to be the case with *F. hepatica* [126]. Experimental studies with *H. diminuta* have tended to support this contention [125, 131] although the occurrence of Na^+/K^+-ATPase has not been demonstrated cytochemically.

Metacestodes

In view of the medical and veterinary importance of the cysticercus stage of *Taenia* spp. and the hydatid stage of *Echinococcus* spp., the ultrastructure of the surface layer in these larval forms will be briefly considered. The cysticercus of *Taenia crassiceps* has been studied in detail [127] and while the structure of the tegument generally resembles that of adult worms, it presents several interesting morphological features. The cysticercus exhibits three distinct regions, the bladder, the invagination canal and the scolex. Microtriches vary in form and abundance in the different regions, and certain specialised ones on the scolex develop as the hooklets. Active pinocytosis occurs at the apical surface of the bladder and invagination canal. The pinosomes so formed move into the perikarya where they contribute to the formation of heterolysosomes by fusion with lysosomes from the Golgi complexes. These heterolysosomes and the associated residual bodies have been shown to contain acid phosphatase activity [127]. It is likely that pinocytosis and heterophagy in the cysticercus of *T. crassiceps* is involved with protection, surface turnover and nutrition, as appears to be the case with adult *Schistocephalus* [129]. However, endocytosis in the scolex region of the metacestode results in the formation of "coated vesicles" which may have quite a different role. Such vesicles in other organisms apparently have receptors for proteins and other molecules, so it is possible that specific molecular requirements, hormones or developmental triggers present in the surrounding host fluid are selectively taken up in the actively growing and differentiating scolex region by this process [127]. Endocytosis has not been reported in any other cyclophyllidean metacestode, but Slais et al [114] in a study on *Cysticercus bovis* recorded the occurrence of numerous "vacuolar structures" in the distal cytoplasm of the bladder wall, which may be indicative of such activity. Tegumental secretory bodies, elaborated in the perikarya, are also found commonly in the cytoplasmic syncytium of the bladder wall of cysticerci. Presumably these discharge by an exocrine mechanism at the apical membrane, contributing their contents to the glycocalyx. In the case of *Cysticercus cellulosae*, it has been shown that the glycocalyx investing the surface membrane of the cyst probably embodies the major antigenic component responsible for clinical manifestations in man [21] (see also Chapter 3).

A number of investigations have been carried out on the ultrastructure of the hydatid cyst of *Echinococcus granulosus* [70, 91] and *E. multilocularis* [108, 109, 110]. In both species the wall of the cyst comprises three layers. The outermost is composed of host reaction tissue, mainly fibroblasts and collagen. The intermediate or laminar layer is non-cellular, consisting mainly of carbohydrate and glycoprotein. It is secreted by the underlying germinal layer and may represent a modified and greatly extended glycocalyx [123]. The ultrastructure of the germinal layer resembles that of the adult tegument, featuring a surface syncytium bearing microtriches which are in direct contact with the laminar layer, and underlying perikarya connected with the syncytium by trabecula. The cytoplasm of the surface layer contains numerous electron lucent vesicles which may well be pinosomes, but electron-dense secretory bodies and mitochondria are lacking. In the perikarya, Golgi complexes have a few lysosome-like bodies associated with them, and crystalline structures are also present, together with a few mitochondria. Morseth [91] suggested that in *E. granulosus* the pinosomes, lysosome-like bodies and crystalline residual material represent elements of a heterolysosomal system in the germinal membrane, but as yet experimental evidence from cytochemistry and tracer studies is lacking.

The brood capsule wall is derived by invagination of the germinal membrane into the hydatid fluid. Its microtriches face the cavity of the capsule while its perikarya lie towards the interior of the cyst. Its ultrastructure closely resembles that of the germinal membrane. The tegument of the enclosed protoscoleces contains numerous dense secretory bodies in addition to the electron lucent vesicles. The secretory bodies are produced at Golgi complexes in the perikarya. Mitochondria increase in number in the perikarya and surface syncytium during development and differentiation of the protoscoleces. Each fully formed free protoscolex bears suckers, rostellum and hooks which represent modified microtriches [91, 123].

In addition to the usual functions of absorption of nutrients, protection and secretion, the germinal membrane and brood capsules of the hydatid have a great potential for polyembryony and growth. This indicates that stem cells or anlagen may be present but the mechanism and control of budding have not been investigated.

In species such as *Moniezia expansa* [19], *H. diminuta* [1] and *Raillietina cesticillus* [4] where the cysticercoid larvae occur in arthropod intermediate hosts, the cyst wall is composed of several layers of tough, chemically resistant "fibrous", "collagenous" or "hyaline" material which are apparently produced by various types of secretory cell in the surface zone of the developing larva. In *R. cesticillus*, the "hyaline" layer of the cyst wall has been interpreted as the product of an immune reaction with the insect host [4]. Clearly, the tough inert cyst wall serves to protect the larval scolex during mastication and preliminary digestion in the definitive host.

Excretory System

The ultrastructural organisation of the protonephridial system in tapeworms closely resembles that in trematodes, although the arrangement of the major collecting ducts differs. Most species posses a pair of dorsolateral and a pair of ventrolateral collecting ducts which run the length of the strobila, with the ventral canals connected by a transverse canal near the posterior border of each proglottid. The dorsal

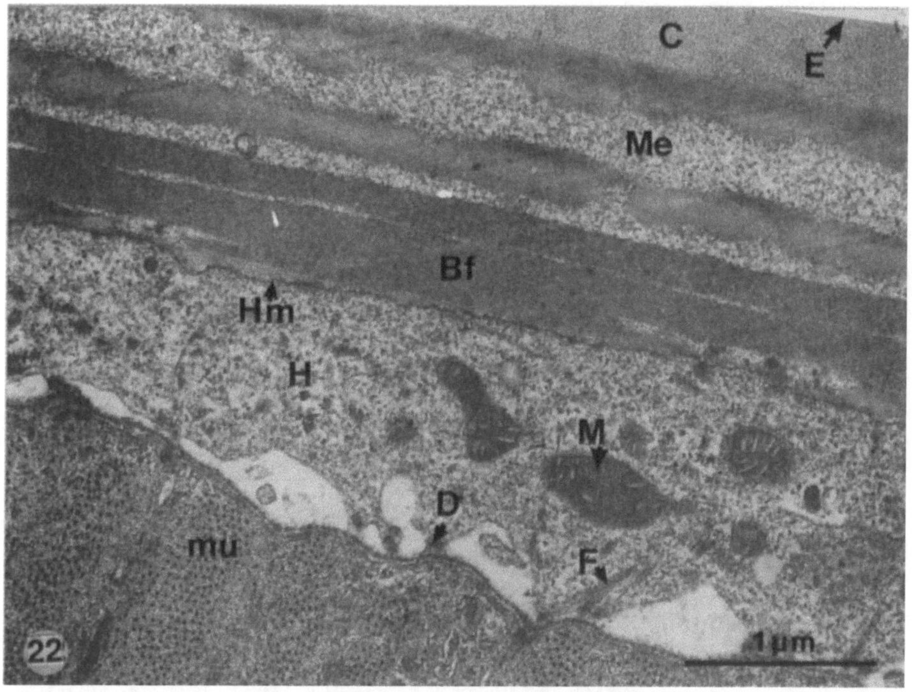

Fig. 22. *Syphacia stroma*: Cuticle and hypodermis of an intestinal nematode of mice. The cuticle is limited apically by a thin trilaminate epicuticle (E) beneath which lie the cortical (C) median (Me) and basal (Bf) zones. The hypodermal membrane (Hm) is attached to the basal region of the cuticle and within the hypodermal cytoplasm (H) occur mitochondria (M) and various other organelles. The basal hypodermal membrane is connected to the musculature (mu) by desmosomes (D), and fibres (F) traverse the cytoplasm to anchor the muscles to the cuticle.

canal often connects the ventral canal on the same side in the scolex region. The finest collecting tubules of the ramifying system are capped by flame cells which closely resemble those in trematodes. The antero-lateral margin of each flame cell interdigitates with "ribs" from the adjoining tubule, forming ultrafiltration zones, and the whole structure is supported by leptotriches [139] (Fig. 23).

The walls of even the finest collecting tubules are syncytial, in contrast with the cellular collecting ducts in trematodes, and the nuclei are offset beneath the lumenal epithelium as "ductal cytons". Microvilli amplify the surface membrane of the ducts, but these vary somewhat in morphology from species to species. In *H. diminuta* and *C. bovis* they are short and bulb-like, but in certain other species they are elongated and extensively developed [78, 114]. The basal membrane of the collecting ducts often extends long cytoplasmic processes into the parenchymal tissue, presumably to facilitate molecular, ionic and osmotic exchange.

The main role of the excretory system in cestodes appears to be removal of nitrogenous waste and end products of energy metabolism. Lipid droplets have not been reported in the epithelium or ductal lumen. It is likely that the microvilli of the lumenal surface serve to resorb organic molecules and electrolytes, a view which was reinforced by the demonstration of phosphatase activity in the ductal epithelium, and by the existence of extensive basal infoldings in some species [78]. The involvement of the excretory system in osmoregulation is open to question since, in general, tapeworms have only a limited capacity to regulate the water content of the body [139].

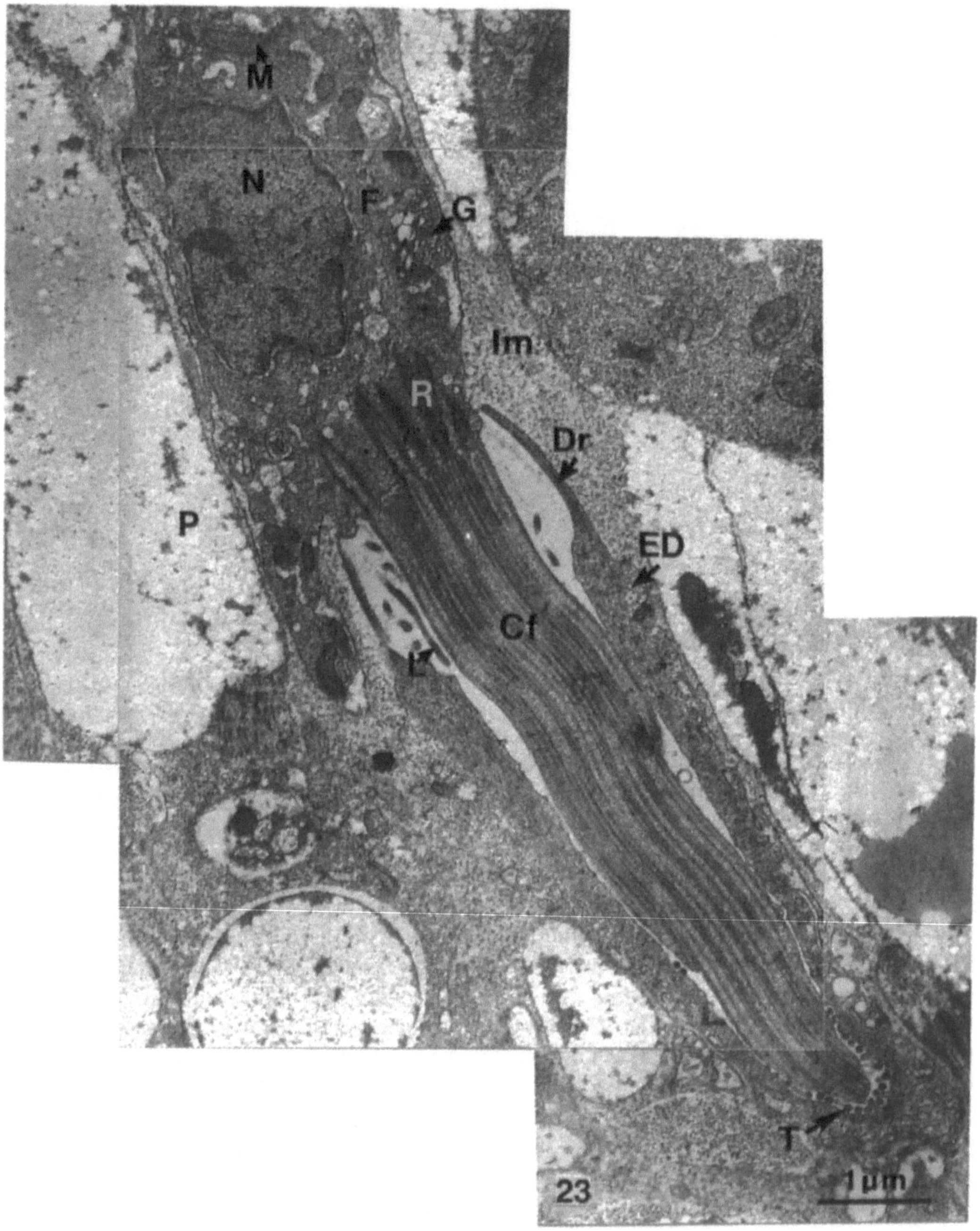

Fig. 23. *Hymenolepis diminuta*: Flame cell and efferent duct of the protonephridial system. The ciliary flame (Cf) is anchored in the cytoplasm of the flame cell (F) by rootlets (R) and extends distally into the lumen of the efferent duct (ED). Cytoplasmic "ribs" (Dr) from the efferent duct interdigitate with corresponding processes from the flame cell forming the filtration zone, which is supported by leptotriches (L). The lining of the efferent duct bears tubercle-like microvilli (T). The flame cell is surrounded by interstitial material (Im) and parenchymal tissue (P) and contains a nucleus (N), mitochondria (M) and Golgi fields (G) (Courtesy of Professor L.T. Threadgold).

Reproductive System

In some pseudophyllideans and most cyclophyllideans, there is a region of cell proliferation just behind the scolex where new proglottids are produced. Stem cells in this region exhibit a typical "embryonic" morphology, with a relatively large nuclear volume and scanty cytoplasm packed with ribosomes. In *Hymenolepis* spp., certain of these stem cells divide to form a cluster, the primary anlage, which subsequently elongates and subdivides to produce several genital anlagen in each developing proglottid. Within these anlagen are stem cells of at least two morphologically distinguishable types. Further differentiation together with migration of new cells to the sites results in the formation of testes, ovary and vitellaria from one stem cell type and associated ducts from the other [13]. In diphyllobothriid cestodes, primary anlagen develop segmentally at defined points along the neck in pleurocercoids and in adults of some species. Growth of each primary anlage is achieved by inward migration of actively dividing germinative cells from elsewhere in the segment as well as by division of cells already present in the anlage. Subsequent cytodifferentiation of the germinal cells gives rise to all the somatic cell types as well as to the genital rudiments [52, 134].

Testis

The process of spermatogenesis is basically similar in all cestode orders studied to date, and in fact closely resembles the pattern in trematodes [24, 106, 107]. A particularly detailed and valuable study of cell divisions and cytodifferentiation in the testis of *Taenia hydatigena* was carried out by Featherston [37]. Primary spermatogonia undergo four mitotic divisions (as opposed to three in trematodes) and the eight quaternary spermatogonia so produced are arranged in a syncytial rosette. Spermatogonia at all stages possess relatively large nuclei and an abundance of ribosomes and mitochondria. GER cisternae increase in frequency as the rosettes develop. Quaternary spermatogonia divide mitotically once more to yield rosettes of 16 primary spermatocytes. The two successive meiotic divisions give rise to clusters of 64 spermatids, in which the nuclei are peripheral and the individual cytons are connected by cytoplasmic bridges to a central cytoplasmic mass [68, 103].

During spermiogenesis, the spermatid nuclei elongate, condense and migrate out into tubular evaginations of the spermatid plasma membrane. These evaginations are supported by microtubules and are further delimited from the remaining spermatid cytoplasm by encircling membranous clefts, a situation which resembles that in trematode spermiogenesis. During migration of the spermatid nuclei, axonemes grow outwards from the surface of the tubular evaginations. In some cestodes, including all the cyclophyllideans, only one axoneme develops from each spermatid. In other groups such as the pseudophyllideans, proteocephalideans and trypanorhynchideans two axonemes emerge and subsequently fuse with each other and with the anterior process of the tubular evagination, although their posterior extremities remain free. Thus two types of spermatozoa are distinguished in cestodes, the "one-axoneme" type and the "two-axoneme" type [24, 117]. The spermatozoon axonemes in cestodes, as in trematodes, exhibit the characteristic platyhelminth "9 + 1" organisation of microtubules which features a single electron-dense central core around which two chains or tubules of differing electron density are helically wound. This arrangement has been described in detail for the spermatozoon of *M. expansa* by Swiderski [117].

Unlike the situation in trematodes, cestode spermatozoa lack mitochondria although most species studied do possess glycogen. This may indicate that the predominant energy pathway in cestode spermatozoa is glycolytic [24].

Vitellaria

In mature vitelline cells of cyclophyllideans such as *H. diminuta,* electron lucent vesicles occur together with clusters of electron-dense shell protein globules. Such vesicles have been reported to contain acid mucopolysaccharide, while the shell protein clusters may contain phenolic protein and phenolase [23].

Development from the undifferentiated stem cells which are located at the periphery of the follicles, closely resembles the process which has been described for trematodes. In the early stages, there is extensive development of GER cisternae, appearance of Golgi complexes and increase in the size and number of mitochondria. Subsequently, clusters of shell protein globules are formed and increase in number while the electron lucent vesicles make their appearance. As the cells approach maturity, protein synthesis ceases and glycogen stores accumulate in the cytoplasm. Meanwhile GER, mitochondria and ribosomes begin to disintegrate [79].

The shell protein globules in the vitelline cells contribute to the formation of the thin capsule in cyclophyllidean eggs and to the relatively thick shell in the eggs of pseudophyllideans. In the cyclophyllideans, only a few (1–3) vitelline cells contribute to each egg whereas in pseudophyllideans the number may reach 20 or more. These differences reflect the differing environments in which the eggs embryonate. The thick shell and ample glycogen supply in pseudophyllidean eggs correlates with embryonation in the external environment, while in the cyclophyllideans, the thinner capsule probably facilitates diffusion of nutrient molecules from the parent worm to the embryo, which develops *in utero* [23].

NEMATODES

Cuticle

Structure

The application of electron microscopy to the study of the nematode cuticle has not brought about any great revolution in our knowledge, in contrast to the case with platyhelminths [71]. Basically, it is a three layered structure comprising an outer cortical zone covered by a membrane-like epicuticle, a median zone composed of fine fibres, and a basal layer containing larger collagenous fibres which have an important skeletal role. Beneath lies the cytoplasmic hypodermis, which is responsible for the secretion of a new cuticle at each moult (Figs. 22 and 24). Each of these zones will be considered in turn.

There has been considerable debate as to whether or not the epicuticle is a highly modified cell membrane [9]. It shows great variability in structure between species ranging from 6 nm to over 40 nm in thickness. In some forms it is multilayered. Freeze-fracture studies on *Nippostrongylus brasiliensis* [75] revealed the presence of intramembranous particles in the epicuticle, a finding which taken together with physiological evidence of protein secretion from the cuticle in *Trichinella spiralis* [99] supported the concept of the epicuticle as a modified plasma, membrane. However, Martinez-Palomo [83] did not encounter intramembranous particles in freeze-fracture replicas of the epicuticle of *Onchocerca volvulus*, and recent tracer studies with a variety of filarial nematodes have provided no conclusive evidence for turnover of the epicuticular material between larval moults and in the adult stage. Such findings have led Howells (64) to propose that the filarial surface at least differs fundamentally from the surface membrane of trematodes and cestodes. The epicuticle of *O. volvulus*, in common with other members of the genus exhibits folding which is independant of the underlying cuticular layers. In the male worm it forms a honeycomb-like pattern on the surface while in the female it is elevated in long protuberances [44, 45, 46].

The cortical zone of the nematode cuticle is often subdivided into outer and inner layers which are usually amorphous and electron-dense, merging with each other and even with the base of the epicuticle. While the principal chemical component of this zone is collagenous protein, perhaps modified to form keratin by the presence of disulphide and sulphydryl groups, there is evidence that in *Ascaris* at least another structural protein called cuticulin is present [50].

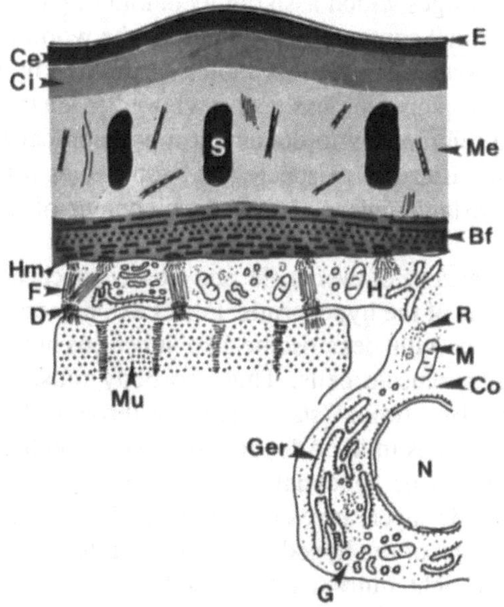

Fig. 24. Ultrastructure of a generalised nematode cuticle. The cortical zone often consists of an outer (Ce) and an inner (Ci) layer and is covered externally by the trilaminate epicuticle (E). The median zone (Me) may contain fibres and struts (S) which support and anchor the cortical zone, while the basal zone (Bf) comprises several layers of collagenous fibres which have skeletal functions. The apical membrane (Hm) of the cytoplasmic hypodermis (H) is attached to the basal fibres by means of desmosomes. Similar structures (D) are involved in attachment of the muscle fibres (Mu) to the basal membrane of the hypodermis, and fibres (F) passing through the cytoplasm ensure anchorage of the musculature to the cuticle. The nuclei of the hypodermis (N) are restricted to the inwardly bulging cords (Co). Ger= granular endoplasmic reticulum, G = Golgi body, M = mitochondrion, R = ribosomes.

The median zone tends to be relatively structureless, comprising a fluid matrix in many species. In this fluid may be suspended globular bodies, fibrous "struts" anchoring the cortical to the basal zone, and fine fibrous material. It is liable to undergo considerable changes during growth and development, increasing four-fold in thickness during growth to the adult stage in *Ascaris lumbricoides* for instance. In *N. brasiliensis*, the median zone is filled with a red fluid containing haemoglobin and esterases. Esterase activity is also present in the median zone of *Ascaris* cuticle. The presence of enzymes in the cuticle together with the finding that ^{14}C-proline can be incorporated during the intermoult stage and after the last moult indicates that the cuticle may have some metabolic capabilities [74].

The basal zone varies in thickness and morphology between species and between life cycle stages. In parasites it is often a thick support layer containing tough, almost crystalline protein with numerous disulphide linkages between the molecules [73]. In some nematodes such as *Ascaris* two or three fibre layers are present. The large collagen fibres comprising each layer run spirally round the body at approximately 75° to the longitudinal axis. The fibres in the middle of the three layers in *Ascaris* cross those of the other two layers at an angle of 135° to form a lattice which alters its angles as the nematode moves, thus allowing for elongation and contraction of the body [60].

Depending on species, the cuticle may bear longitudinal or transverse ridges, annulations, spines, pores or inflated areas. These modifications usually involve only the cortical zone, but may reach down to the median zone also. Lateral extensions of the cuticle termed alae are present in many species. These are fin-like structures or blunt ridges which assist in locomotion, particularly in those gut-parasitic forms which swim freely in the intestinal contents. They stabilise the worm and provide thrust against the fluid medium. The numerous variations in cuticle morphology exhibited by nematodes can all undoubtedly be related to specific functions, a topic which has been reviewed at length by several authors [8, 71, 73, 81]. The use of scanning electron microscopy to document topographical features of larval and adult filarial nematodes has contributed significantly to species and subspecies discrimination within the group [48, 76, 140, 141] and will increasingly influence the development of taxonomic schemes for nematodes generally.

Immediately beneath the basal zone of the cuticle, and closely attached to it by means of hemidesmosomes, lies the apical plasma membrane of the hypodermis. Fibres extend across the cytoplasm of the hypodermis linking these hemidesmosomes with desmosome-like junctions which connect the basal hypodermal membrane with the underlying muscle cells. Thus the body musculature is securely anchored to the "skeletal" zone of the cuticle by fibrous "stays" passing through the cytoplasmic zone (Fig. 24). The hypodermis itself may be cellular, as in the Adenophorea (which includes important animal parasites such as the trichurids and capillarids) or syncytial as in the Secernentea. In either case it forms a thin, membrane delimited cytoplasmic layer, investing the body like a sleeve beneath the cuticle. In filarial nematodes the limiting membrane of the hypodermis incorporates sites for active transport and facilitated diffusion of low molecular weight nutrients such as D-glucose and L-amino acids. Apparently, the overlying cuticle may be considered an inert permeable "unstirred layer" distal to the transport loci [64]. The nuclei of the hypodermis are confined to four inwardly bulging cytoplasmic ridges or cords which run the length of the body in the mid-dorsal, mid-ventral and lateral positions. The so-called excretory canals, when present, are enclosed in the lateral cords. The cytoplasm of the hypodermis contains mitochondria, ribosomes, GER, Golgi bodies and large amounts of lipid and glycogen [74] (Fig. 22).

Moulting

Each of the four moults which occurs during the life cycle of a nematode is probably triggered by an external or internal stimulus, and is mediated by neuroendocrine secretions. The cytological mechanism has been described in detail for *N. brasiliensis* by Lee [72] and for *Brugia pahangi* by Howells & Blainey [65]. The first stage, which is initiated in response to neurosecretory discharge, involves separation of the old cuticle from the apical membrane of the hypodermis. This may be achieved by release of stored leucine aminopeptidase from the apical surface of the hypodermis, beneath the basal zone of the cuticle [105]. Subsequently, a second hypodermal membrane is laid down beneath the original one, resulting temporarily in a double apical plasmalemma. Elaboration of the GER in the hypodermal cytoplasm, followed by active protein synthesis, leads to the accumulation of membrane-bound vesicles containing fluid precursors of the cuticular proteins below the apical membrane. These are discharged by exocytosis into the space between the two apical membranes, driving them apart (Fig. 25).

The new cuticle is laid down in a series of layers which gradually increase in thickness and complexity. It is initially folded into ridges, allowing for rapid elongation of the nematode after ecdysis, a phenomenon which has been studied in detail for *B. pahangi* during and after the third-stage moult [65]. The new epicuticle originates from the former apical membrane of the hypodermis, while the new hypodermal membrane develops attachements with the basal zone of the new cuticle. In some species, during formation of the new cuticle, the inner zones of the old one are partially digested and the resulting breakdown products absorbed for reassembly into new cuticular proteins. This does not appear to be the case in *N. brasiliensis*, however [72]. The final stage in the moulting process is ecdysis or shedding of the old cuticle. In *Phocanema*, a parasite of seals, this is achieved by release of leucine aminopeptidase from the excretory gland into the gap between the old and the new cuticle, resulting in rupture of the old cuticle

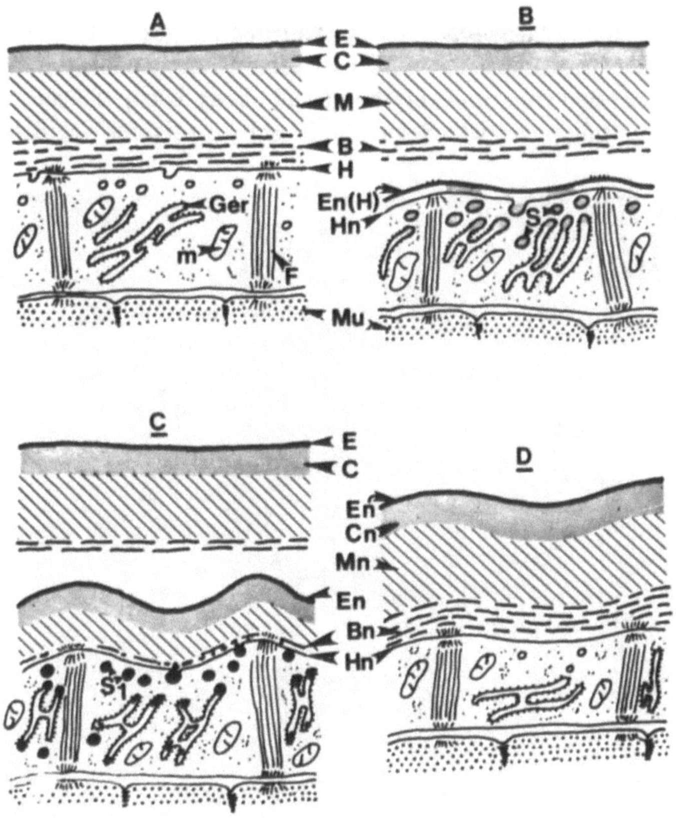

Fig. 25. Diagrams illustrating successive stages (A–D) in moulting of the nematode cuticle. Separation of the old cuticle (comprising the epicuticle [E], cortical [C], median [M] and basal [B] zones) from the apical hypodermal membrane (H) is achieved by exocrine discharge of hydrolytic enzymes into the region between the basal zone and the hypodermal membrane. Following the formation of a new hypodermal membrane (Hn) beneath the original one (which becomes the new epicuticle, En), precursors of the various zones of the new cuticle are secreted sequentially into the intervening space. These precursors are manufactured at the granular endoplasmic reticulum (Ger) and transported to the hypodermal membrane in secretory vesicles (S, S_1). Once the new cortical (Cn), median (Mn) and basal (Bn) zones have been laid down, the old cuticle is ecdysed and the nematode elongates as the folds in the new cuticle flatten out. F = fibres anchoring the musculature (Mu) to the cuticle, m = mitochondrion (after Lee [54]).

and escape of the newly moulted nematode. The synthesis and release of this hydrolytic enzyme are under neurosecretory control [22]. Certain animal parasitic nematodes (e.g. *Trichostrongylus, Haemonchus, Ancylostoma*) produce infective larvae which, having moulted incompletely, are still enclosed within the second-stage cuticle, known as the sheath. Ecdysis (exsheathment) is triggered by specific physiochemical stimuli provided in the definitive host, and is probably mediated by leucine aminopeptidase.

Alimentary Canal

The digestive tract is fundamentally similar in all groups of nematodes. It consists of a straight tube from mouth to anus, and is divided into three parts namely the stomodaeum which includes the lips, mouth, buccal cavity and pharynx, the intestine, and the proctodaeum which consists of the rectum, anus and, in males, the cloaca. The stomodaeum and proctodaeum are lined with cuticle which is shed at each moult.

Considerable diversity is exhibited between species and life cycle stages in the structure of the stomodaeum, and individual modifications are associated with particular feeding habits. In some animal parasitic species such as *Ancylostoma* and *Bunostomum*, the buccal cavity is cup shaped and its cuticular lining is elaborated into teeth or cutting plates which serve to breach host tissues, facilitating access to the food source (Fig. 26).

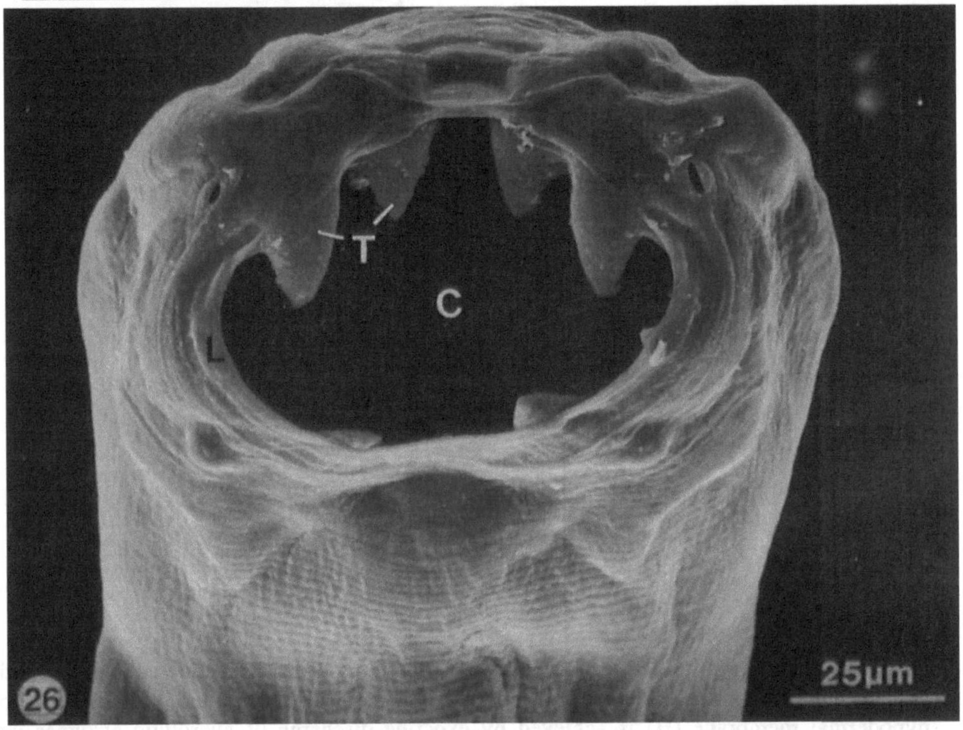

Fig. 26. *Ancylostoma duodenale*: Scanning electron micrograph of the anterior end of an important human hookworm. The buccal cavity (C) is cup-shaped and the cuticular lining is thickened to form hook-like teeth (T) which are employed to breach blood vessels in host intestine. The lips (L) are fused to form a ring around the mouth which acts as a sucker, engulfing a plug of host tissue (Courtesy of Dr. L. Gibbons. After Gibbons, L.M. 1986. SEM guide to the morphology of nematode parasites of vertebrates. C.A.B. International, Slough, England).

The pharynx (oesophagus) is a muscular pumping organ, usually triradiate in section (Fig. 27), which often contains cuticular modifications for processing the food mechanically and passing it into the intestine. In addition, there are usually three uninucleate glands in the pharynx which discharge their secretions into the lumen of the oesophagus or buccal cavity via cuticle lined ducts. These glandular secretions vary in their enzymic nature from species to species and between life cycle stages according to nutritional or functional requirements. In some blood feeders the pharyngeal glands produce anticoagulants (e.g. *Ancylostoma*, *Haemonchus*), while in other species the secretions are voided through the mouth to initiate extracorporeal digestion of host tissues. The partly digested material is then drawn in by the pumping action of the pharynx (e.g. *Anisakis*). The diversity in structure shown by the stomodaeum in nematodes and the mechanical functioning of the various types of pharynx have been reviewed by Lee & Atkinson [74]).

In most nematodes, the intestinal epithelium consists of a single layer of cells supported on a basal

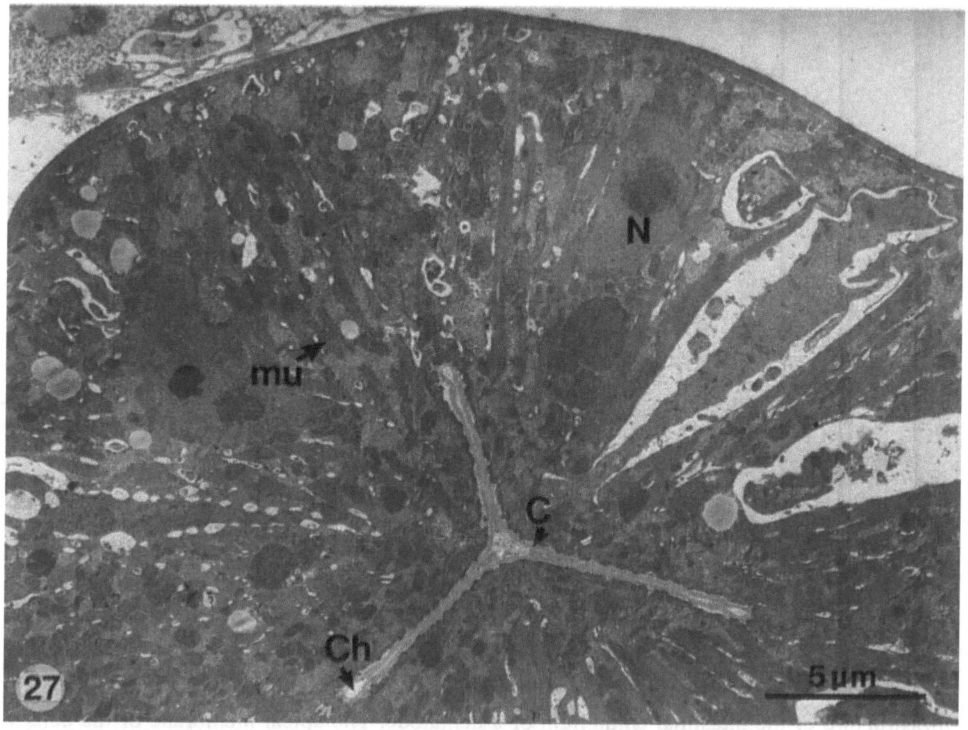

Fig. 27. *Syphacia stroma*: Transverse section of the triradiate pharynx. The lumen is lined by cuticle (C) continuous with that of the body surface. At the extremities of the three arms are longitudinal channels (Ch) which concentrate the food by collecting water. Radial muscle fibres (mu) help provide the pumping action of the organ. N = nucleus of muscle cell.

lamina (Fig. 28). Apically, the cells are linked by belt-like tight junctions which encircle the lumeno-lateral margins. In some species the intestinal lining is syncytial (e.g. *Ancylostoma*). The basal membrane of the cells is often extensively folded and sometimes these infolds are associated with large numbers of mitochondria suggesting involvement with active transport of metabolites from the epithelium into the pseudocoelomic fluid.

The lumenal membrane is extended to form a regular array of microvilli which increase the absorptive surface for active transport of carbohydrates and amino acids into the intestinal cells. The outer aspect of the membrane bears a thick glycocalyx which may absorb enzymes to facilitate surface digestion. However, the main function of the intestinal cells in most species is probably absorption, since digestion seems to be carried out predominantly by enzymes secreted from the pharyngeal glands. In most nematodes, the intestinal microvillar border is the only surface available for uptake of nutrients, the cuticle being impermeable to glucose, amino acids etc. [74, 133]. However, in *Bradynema*, which lives in the haemocoel of insects, areas of the hypodermis are exposed and bear microvilli [101], and in the filarial nematodes transcuticular uptake of nutrient molecules is well established (Howells [64]). In the case of microfilariae and some adult female filariae the alimentary canal is apparently non-functional in the ingestion of food [47, 86].

The cytoplasm of the intestinal epithelium in most nematodes contains GER and Golgi bodies, indicating the occurrence of secretory activity, and there is evidence of pinocytosis and heterolysosomal digestion in some species. The intestinal cells of *Pontonema vulgaris* carry out an absorptive and then a secretory cycle of activity [67], while apocrine and merocrine secretion of esterases occurs in the intestine of

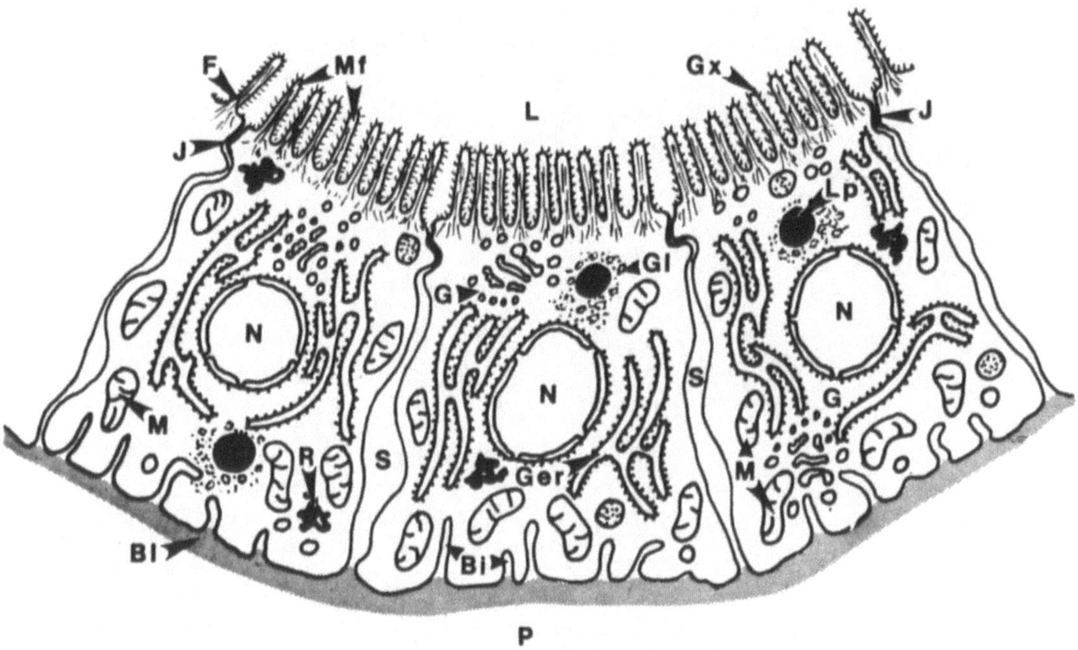

Fig. 28. Ultrastructure of the intestinal epithelium of a nematode. The apical membrane of the cells which faces the gut lumen (L), is extended by numerous regular microvilli (Mf) bearing an external glycocalyx (Gx) and supported by core fibres (F). The cells are linked by apical tight juctions (J), but intercellular spaces (S) expand or contract to accommodate fluid exchange with the pseudocoelom (P), across the basal lamina (Bl). The presence of granular endoplasmic reticulum (Ger) and Golgi fields (G) in the cytoplasm denotes secretory activity, while autophagy and, in some species, heterophagy results in the formation of residual bodies (R). Storage of glycogen (Gl) and lipid (Lp) is sometimes evident in the gastrodermis. N = nucleus, M = mitochondrion, Bi = infolds of basal membrane.

Ascaris and *Ascaridia galli* [100, 133]. Energy reserves such as glycogen and lipid droplets are frequently stored in the cytoplasm of intestinal epithelial cells, particularly those towards the middle and posterior regions of the gut [86].

In *Ascaris*, fluid can move in either direction across the intestine wall in response to osmotic gradients, but when the lumenal contents are isosmotic with the pseudocoelomic fluid, water always moves from the mucosal to the pseudocoelomic side. This presumably helps to maintain internal turgor pressure. Water enters the epidermal cells across the microvillar membrane and is expelled to the intercellular spaces (Fig. 28), which consequently dilate. Since the lumeno-lateral margins of the cells are sealed by tight junctions, fluid cannot leak back into the lumen, but must pass inwards through the basal lamina to the pseudocoelomic space [59]. It is likely that fluid regulation involves expenditure of energy and that ionic pumps are situated at the lateral membranes, as in *F. hepatica* [126], but appropriate cytochemical investigations have yet to be carried out.

Reproductive system

Testis

In male nematodes the testis usually consists of an elongated blind-ending tubular organ which twists back and forth in the pseudocoelom. In most species only one testis is present, but some have a pair.

According to the location of the proliferative germinal zone, nematode testes are classified as either telogonic or hologonic. In the former type the spermatogonia are produced only in the terminal portion of the organ while in the latter, which occurs only in the superfamilies Trichuroidea (e.g. *Trichuris, Trichinella, Capillaria*) and Dioctophymoidea, spermatogonial proliferation takes place along the length of the gonad. In telogonic testes the germinal cells are enclosed by a sheath of elongated multinucleate epitheliod cells which are apparently derived from smooth muscle tissue. External to these lies a basal lamina. By contrast, the proliferative cells in hologonic testes lie directly on a thick basal lamina beneath which occur circular and longitudinal muscle bands.

During spermatogenesis in a telogonic testis the spermatogonia and spermatocytes retain cytoplasmic continuity one with another via central strands of cytoplasm (the rachis). Germ cells originate either from a single relatively undifferentiated terminal cell or a nucleated syncytial mass [3] and development progresses in a linear fashion along the length of the testis, the most mature spermatozoa occurring in the seminal vesicle.

The earliest divisions of each spermatogonium are mitotic, and result in the formation of primary spermatocytes. This stage is accompanied by the appearance of large amounts of GER, numerous Golgi bodies, mitochondria and homogenous GER-derived dense bodies in the cytoplasm. Subsequently, each spermatocyte undergoes the two divisions of meiosis, giving rise to four spermatids, which eventually mature into spermatozoa.

As development proceeds, membranous saccules originating from the Golgi bodies associate with the homogenous dense bodies in the cytoplasm, inducing the formation of paracrystalline material composed of packed microfilaments. These peculiar heterogenous structures containing fibrillar, membranous and electron-dense components increase in number throughout the development period and ultimately become incorporated in the spermatozoa (Fig. 29). Here, the fibrillar material separates to associate with the plasma membrane at one pole of the cell. The microfilaments, which consist of actin [43], are thought to supply the contractile elements which bring about pseudopodial movement in the mature spermatozoon. The remaining membranous and dense material persists in the cytoplasm, and take part in further structural modification of the spermatozoon once it enters the female tract.

Following breakdown of the nuclear envelope at meiotic division it never reforms. However, the DNA condenses, and may appear in the mature spermatozoon as multiple dense clumps (*Brugia*), coiled threads (*Ancylostoma*), a single homogenous mass (*Ascaris*) or as electron opaque material dispersed throughout the cytoplasm (*Dioctophyma*). In *Nippostrongylus* and *Ancylostoma* it extends posteriorly to resemble a tail. Nematode spermatozoa do not possess a flagellum, although microtubules are frequently present.

Following development in the testis, the spermatozoa are stored as spheroidal or elongated immotile cells in the seminal vesicle. They only become active and achieve their final form after stimulation by glandular secretions from the vas deferens, thus maturation is delayed until they arrive in the reproductive tract of the female worm [41]. The morphological transformations undergone by spermatozoa of *Brugia pahangi* in the uterus of the female worm have been documented by Burghardt & Foor [16]. In this situation they acquire the ability to form pseudopodia, and the membranous saccules which once were associated with the microfilament bundles, fuse with the bounding plasma membrane and release their matrix material into the lumen of the uterus. Acid phosphatase activity has been demonstrated in these membranous bodies, and their discharge has been compared with the acrosomal reaction in flagellated sperm [43]. In *Ascaris* spermatozoa, the electron-dense globules remaining from the membrane/microfilament/dense material complexes amalgamate to form a single large refringent body, which, despite its visual prominence, has uncertain significance.

Mature active spermatozoa generally feature a region of clear cytoplasm anteriorly which contains the actin fibrils and sends forward pseudopodial processes. The posterior region is relatively rigid, often

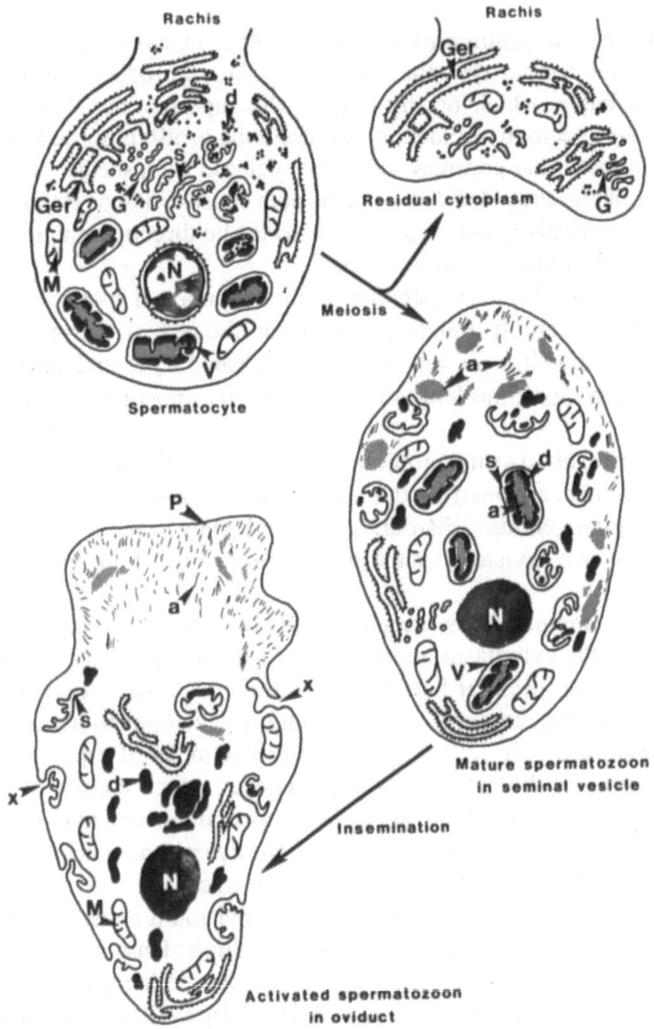

Fig. 29. Spermiogenesis in the nematode testis. During development, spermatocytes retain cytoplasmic continuity one with another via the rachis. Once the nuclear envelope dissolves at meiosis it does not reform, and the nuclear material (N) appears as a highly condensed mass in later stages. Complex heterogenous vacuoles (V) are formed by the association of membranous saccules (s) derived from Golgi fields (G) with dense bodies (d) originating from the granular endoplasmic reticulum (Ger). Actin fibrils (a) exhibiting paracrystalline organisation appear in each vacuole. In maturing spermatozoa the vacuoles dissociate and the actin fibrils migrate to one pole of the cell, supplying contractile elements for activity of the pseudopodia (P). The remaining saccular elements eventually fuse with the bounding plasma membrane (X) when the spermatozoon is activated in the female tract, while the dense material aggregates to form refringent bodies. M = mitochondrion.

conical in shape, and bears the nuclear material, mitochondria, membranous organelles and refringent bodies (Fig. 29). There is wide variation between species in the nature and abundance of cytoplasmic inclusions, the form of the nuclear material, and the general shape of the spermatozoon.

Spermatogenesis in hologonic testes differs from that in telogonic testes mainly in that there is no persistent cytoplasmic connection between the dividing cells. Following division of a peripheral spermatogonium, the cells destined to become spermatocytes undergo meiotic divisions as they progress towards the lumen. Ultimately, spermatids separate from a residual cytoplasmic mass and complete their differentiation on reaching the lumen. Thus, in a single cross-section of a hologonic testis all stages in the development process may be observed. An exception to the usual hologonic pattern is seen in *Trichinella* where spermatozoa develop in cytoplasmic continuity with their neighbours as a "rosette". These groups are transferred intact to the female tract [43].

The numerous intraspecific variations which occur in spermatogenesis, spermiogenesis and spermatozoon morphology are reviewed by Foor [43].

Ovary

Most female nematodes possess a pair of genital tubes, each consisting of a uterus which is continuous with narrow oviduct and a terminal ovary. The two uteri may run parallel or in opposite directions in the body, and they unite distally to form a common vagina which opens at the vulva. As in the male system, ovaries are classified as either telogonic or hologonic depending on whether oogonia are produced only at the terminal end or throughout the gonad. The hologonic type is limited to the superfamilies Trichuroidea and Dioctophymoidea. Telogonic ovaries are invested by a layer of elongated, multinucleated epithelioid cells underlain by a basal lamina. Towards the proximal end of the ovary the epithelial tissue appears as a thin syncytium in cytoplasmic continuity with the terminal cap cell. The epithelial cells contain vast glycogen reserves, numerous mitochondria and bundles of microfilaments which indicate their origin from smooth muscle cells. Hologonic ovaries lack the epithelioid envelope but are encompassed by circular and longitudinal muscle bands which lie external to the basal lamina.

Oogenesis in the telogonic ovary is initiated by mitotic and possibly amitotic division in the relatively undifferentiated oogonia situated at the proximal end of each genital tube. These divisions culminate in the formation of oocytes which are much larger than oogonia and have an increased cytoplasmic to nuclear volume ratio.

The oogonia and oocytes are in cytoplasmic continuity throughout their development with a central cytoplasmic axis, termed the rachis. In the proximal region this is branched, having thread-like connections with clusters of oogonia, but throughout most of the length of the ovary it is large, centrally located, and connected to the oocytes by distinct cytoplasmic bridges which are supported by microtubules and fibres. In *Ascaris* ovaries, as the oocytes develop they assume a characteristic radial arrangement around the central rachis. The cytoplasm of the rachis contains glycogen particles, ribosomes, lipid droplets, microtubules, mitochondria and synaptonemal complexes. All of these organelles appear to be contributed to the rachis by the oogonia and oocytes [39]. Active cytoplasmic movement through the bridges linking the rachis and oocytes probably facilitates distribution of precursor materials and regulatory factors.

As each oocyte matures its connection with the rachis is lost and it moves distally into the oviduct region. In many species the female germ cells remain as primary oocytes throughout their development in the ovary and only undergo maturation divisions following penetration by the spermatozoon [38].

During the movement of oocytes down the ovary, they accumulate large quantities of carbohydrates, lipids and proteins in their cytoplasm. The presence of abundant SER and GER is in accordance with biochemical evidence that active synthesis of triglycerides and proteins occurs. SER is involved in synthesis of lipids, while the GER gives rise to large heterogenous inclusion bodies which contain protein and ascaroside [38].

In addition, the oocytes of *Ascaris* take up proteinaceous material from the pseudocoelomic fluid by phagocytosis [40]. This is accumulated as heterogenous inclusions and granules in the cytoplasm. It appears that phagocytotic and synthetic activities occur sequentially in developing oocytes, with the latter predominating in more mature cells. The rachis may facilitate uniform distribution of precursor substances, ensuring that the oocytes at any given level in the ovary develop synchronously.

The ultrastructure of fertilisation of the oocyte and its subsequent development in the female tract has been studied in most detail in *Ascaris* and *Parascaris* [38, 42]. Elegant illustrations of oogenesis and embryogenesis in *Brugia malayi* were provided by Mehlhorn et al (87) while Franz & Büttner [47] documented the stages of larval development within the female reproductive system of *O. volvulus*. Fertilisation occurs as the oocyte passes through the oviduct, where penetration by the amaeboid spermatozoon triggers a sequence of events leading to formation of the egg shell. Initially the bounding plasma membrane of the spermatozoon fuses with that of the oocyte, releasing the contents of the male gamete into the egg cytoplasm. At this point a second plasma membrane is laid down immediately beneath the original oolemma which in turn becomes visually evident due to slight shrinkage of the cytoplasm (Fig. 30) [3, 84]. This first layer of the developing egg shell is termed the vitelline layer.

Into the space between the vitelline layer and the new plasma membrane, material from the peripheral GER is discharged. As this zone thickens to some 3 μm (in *Ascaris*), its contents transform into the chitinous layer, which contains protein as well as the polysaccharide chitin. The protein clearly originates from the GER, but the site of assembly of chitin molecules from precursor mono- or oligosaccharides in the egg cytoplasm is uncertain. It is likely that the synthetic enzymes involved are located at the plasma membrane [42]. The chitinous layer provides rigidity and physical strength.

The third layer of the egg shell, the ascaroside layer, is formed by discharge of refringent granules stored in the cytoplasm into the region between the plasma membrane and the base of the chitinous layer. This lipid zone influences the permeability of the egg and provides resistance to dessication. As the ascaroside layer forms the first meiotic division occurs in the female pronucleus, and the resultant polar body is extruded into this third shell layer [38].

As the fertilised eggs move through the uterus they acquire a final outer layer which is secreted by the cells lining the uterus wall. The latter have abundant GER and Golgi bodies and exocytose synthesised proteinaceous material into the uterine lumen [38].

In parasitic nematodes, which are ovoviviparous such as *Trichinella, Dipetalonema, Dracunculus* and the filariae, the egg shell is thin and delicate, lacking the chitinous layer. Presumably, in these cases the female provides protection for the egg during embryogenesis [69, 84]. The microfilariae of some filarial nematodes are born enveloped in the thin egg shell (e.g. *Wuchereria bancrofti*) whereas others are already hatched when they are born (e.g. *O. volvulus*) [47, 87]. Further information on the formation and variety of nematode eggs is provided in reviews by Anya [3], Bird [8] and Foor [42].

CONCLUSION

The foregoing is necessarily a superficial and over-generalised account based on the limited range of parasitic helminths which has, so far, been readily available to electron microscopists. Whilst the most important ultrastructural features and functional principles of each organ system are now quite well established and have been found to be relatively uniform within each of the major groups of parasitic helminths, many particular morphological adaptations which are restricted to smaller taxa and individual species remain to be described and elucidated. This is especially true of those parasites which, despite their medical or veterinary significance, occur only in regions where electron microscope facilities are lacking. Many fascinating aspects of functional morphology within groups such as the Filariata, the Camallanata, the Taeniidae, and the Paramphistomatidae remain to be examined in detail and will provide a source of intellectual stimulation for parasitologists and electron microscopists in years to come.

As information on the ultrastructure of parasitic helminths accumulates, it is inevitable that it will exert an increasing influence on the development of taxonomic systems. This is particularly the case for

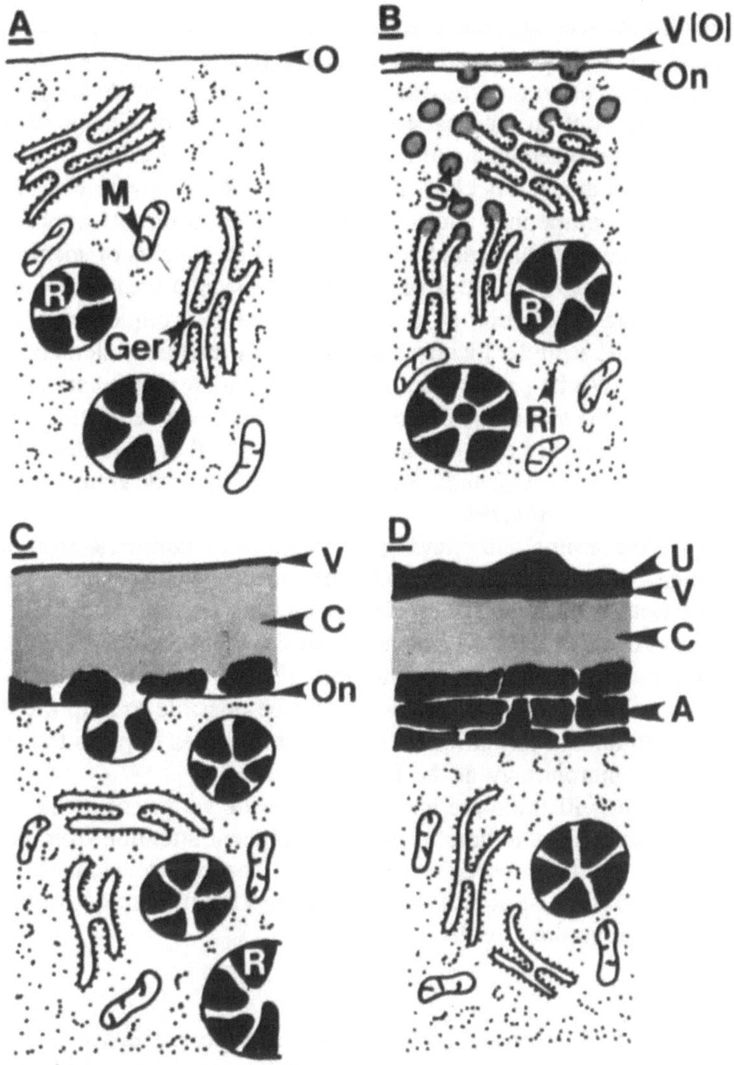

Fig. 30. Successive stages (A-D) in the formation of a nematode egg shell. Immediately after fertilization of the oocyte a new oolemma (On) is laid down beneath the existing one (O). The latter becomes the vitelline layer (V) of the egg shell. Into the space between the vitelline layer and the new oolemma, secretory bodies (S) derived from the peripheral granular endoplasmic reticulum (Ger) discharge precursors of the chitin/protein layer (C). The ascaroside layer (A) is added subsequently by the discharge of lipid-containing refringent granules (R) below the chitin/protein layer. Finally a layer of uterine secretions (U) is added to the outside of the shell. M = mitochondrion, Ri = ribosomes (after Lee & Atkinson [56]).

scanning electron microscope studies, and indeed the ultrastructural features of surface architecture already have a significant role in species discrimination within certain groups of nematodes. Technical developments in recent years have yielded high-resolution "bench-top" scanning electron microscopes which are robust, compact and relatively inexpensive. The availability of such instruments, together with the ease of specimen preparation should bring scanning electron microscopy within the range of taxonomists working in fairly remote areas. It should be noted, however, that functional interpretations generally require the intervention of transmission electron microscopy, and that accurate measurement of structures is impossible in scanning electron micrographs due to the parallax effect. At the more applied level, transmission and scanning electron microscopy have much to offer in the development and screening of new drugs and anthelmintics. Often the earliest effects of chemotherapeutic agents occur at the surfaces of parasites and can readily be detected in ultrastructural studies. Bearing in mind the high cost of conventional drug-screening methods, it is likely that electron microscopy will be increasingly employed in primary selection of active compounds and to monitor concentration effects on target organisms.

The development of ever more sophisticated equipment and techniques of specimen preparation leads to the progressive integration of parasite ultrastructure with molecular biology. Noteworthy among such developments is the increasing application of X-ray microanalysis coupled with electron microscopy. This technique permits exact intracellular localisation and quantification of elements with atomic weights greater than about 20 daltons. Whilst primary accumulations of high molecular weight elements may be uncommon in parasite tissues, the technique offers extremely accurate delineation of secondary deposits of metallic salts precipitated at sites of enzymic activity by classical or modified cytochemical methods. The resulting increase in scope, sensitivity and specificity of such methods greatly facilitates the alignment of ultrastructure with biochemical function.

Potentially the most wide-reaching and exciting innovation in modern electron microscopy has been the application of monoclonal antibody linked electron-dense tracers, which enable ultrastructural localisation of antigenic sites within parasite tissues. The rapidly developing field of parasite immunocytochemistry has revolutionised our approach to the study of host-parasite relationships, and has much to offer for future development of immunodiagnostic tests and vaccines. As an increasingly wide range of commercially prepared monoclonal antibodies becomes available, it will be possible to map the distribution of many biochemically defined structural and physiologically active molecules in parasite cells, vastly expanding our knowledge of functional morphology at the molecular level.

It is hoped that this account will stimulate interest in the wide variety of medically and economically important parasite species which have to date received insufficient attention from electron microscopists but which are accessible to workers in regions where the appropriate technology has recently been introduced. Thus, the data base may be appropriately expanded for future investigations at the molecular level in both fundamental and applied fields.

ACKNOWLEDGEMENTS

My thanks are due to Mr G.P. Brennan and Mr G.W. McCartney for assistance with electron micrography, to Mrs E.J. Purdy and Mrs K.E. Moore for secretarial services, to Dr W.A. Nizami of the Department of Zoology, Aligarh Muslim University, India, for help with provision of material, and to the British Council and the Indian University Grants Commission for travelling and research expenses.

REFERENCES

1. Allison, V.F., Ubelaker, J.E. and Cooper, N.B. The fine structure of the cysticercoid of *Hymenolepis diminuta*. II. The inner wall of the capsule. Z. Parasitenkunde **39**: 137–147 (1972).
2. Andersen, K. The functional morphology of the scolex of *Diphyllobothrium* Cobbold (Cestoda, Pseudophyllidea). A scanning electron and light microscopical study on scoleces of adult *D. dendriticum* (Nitzsch), *D. latum* (L.) and *D. ditremum* (Creplin). Int. J. Parasitol. **5**: 487–493 (1975).
3. Anya, A.O. Physiological aspects of reproduction in nematodes. Adv. Parasitol. **14**: 267–351 (1976).
4. Baron, P.J. On the histology, histochemistry and ultrastructure of the cysticercoid of *Raillietina cesticillus* (Molin, 1858) Fuhrmann, 1920 (Cestoda, Cyclophyllidea). Parasitology **62**: 233–245 (1971).

5. Bennett, C.E. *Fasciola hepatica*: development of caecal epithelium during migration in the mouse. Exp. Parasitol. **37**: 426–441 (1975).
6. Bennett, C.E. Surface features, sensory structures, and movement of the newly excysted juvenile *Fasciola hepatica* L. J. Parasitol. **61**: 886–891 (1975).
7. Baennett, C.E. *Fasciola hepatica*: development of excretory and parenchymal systems during migration in the mouse. Exp. Parasitol. **41**: 43–53 (1977).
8. Bird, A.F. *The structure of nematodes*. Academic Press: New York (1971).
9. Bird, A.F. Chapter 14 Nematoda. In: *Biology of the Integument*. Vol. 1. Invertebrates (J. Bereiter-Hahn, A.G. Matoltsy and K.S. Richards eds), pp. 212–233. Springer Verlag: Berlin (1984).
10. Bogitsh, B.J. Cytochemistry of gastrodermal autophagy following starvation in *Schistosoma mansoni*. J. Parasitol. **61**: 237–248 (1975).
11. Bogitsh, B.J. and Carter, O.S. *Schistosoma mansoni*: ultrastructural studies on the oesophageal secretory granules. J. Parasitol. **63**: 681–686 (1977).
12. Bogitsh, B.J. and Shannon, W.A. Cytochemical and biochemical observations on the digestive tracts of digenetic trematodes. VIII. Acid phosphatase activity in *Schistosoma mansoni* and *Schistosoma douthitti*. Exp. Parasitol. **29**: 337–347 (1971).
13. Bolla, R.I. and Roberts, L.S. Developmental physiology of cestodes. IX. Cytological characteristics of the germinative region of *Hymenolepis diminuta*. J. Parasitol. **57**: 267–277 (1971).
14. Bråten, T. An electron microscope study of the tegument and associated structures of the procercoid of *Diphyllobothrium latum* (L.). Z. Parasitenkunde **30**: 95–103 (1968).
15. Bråten, T. The fine structure of the tegument of *Diphyllobothrium latum* (L.). Z. Parasitenkunde **30**: 104–112 (1968).
16. Burghardt, R.C. and Foor, W.E. Rapid morphological transformations of spermatozoa in the uterus of *Brugia pahangi* (Nematoda: Filaroidea). J. Parasitol. **61**: 343–350 (1975).
17. Burren, C.H., Ehrlich, I. and Johnson, P. Excretion of lipids by the liver fluke (*Fasciola hepatica* L.). Lipids **2**: 353–356 (1967).
18. Burton, P.R. Fine structure of the reproductive system of a frog lung fluke. III. The spermatozoa and its differentiation. J. Parasitol. **58**: 68–83 (1972).
19. Caley, J. Ultrastructural studies of the cysticercoid of *Moniezia expansa* (Anoplocephalidae) with special reference to the developmnet of the cyst. Z. Parasitenkunde **48**: 251–262 (1976).
20. Clegg, J.A. Secretion of lipoprotein by Mehlis' gland in *Fasciola hepatica*. Ann. N.Y. Acad. Sci. **118**: 969–986 (1965).
21. Chandramukhi, A., Katti, M.K., Kumar, B.V.R., Shankar, S.K., Suryanarayana, V., Ravi, V., Rao, T.V. and Gokul, B.N. Neurocysticercosis—a zoonotic disease and its morbidity on man. Second National (Indian) Congr. Parasitol., Bangalore. Lead Paper No. 1 p. i (1988).
22. Davey, K.G. and Kan., S.P. Moulting in a parasitic nematode, *Phocanema decipens*. IV. Ecdysis and its control. Can. J. Zool. **46**: 893–898 (1968).
23. Davis, R.E. and Roberts, L.S. Chapter 4. Platyhelminthes-Eucestoda. In: *Reproductive Biology of Invertebrates*. Vol. I: *Oogenesis, Oviposition and Oosorption* (K.G. Adiyodi and R.G. Adiyodi eds), pp. 109–133. John Wiley & Sons: Chichester, England (1983).
24. Davis, R.E. and Roberts, L.S. Chapter 5. Platyhelminthes-Eucestoda. In: *Reproductive biology of Invertebrates*. Vol. II: *Spermatogenesis and Sperm Function* (K.G. Adiyodi and R.G. Adiyodi eds), pp. 131–149. John Wiley & Sons: Chichester, England (1983).
25. Dunn, T.S., Nizami, W.A. and Hanna, R.E.B. Studies on the ultrastructure and histochemistry of the lymph system in three species of amphistome (Trematoda: Digenea) *Gigantocotyle explanatum, Gastrothylax crumenifer* and *Srivastavaia indica* from the Indian Water Buffalo *Bubalus bubalis*. J. Helminthol. **59**: 1–18 (1985).
26. Dunn, T.S., Hanna, R.E.B. and Nizami, W.A. Ultrastructural and histochemical observations on the foregut and gut caeca of *Gigantocotyle explanatum, Gastrothylax crumenifer* and *Srivastavaia indica* (Trematoda: Paramphistomidae). Int. J. Parasitol. **17**: 1141–1152 (1987).
27. Dunn, T.S., Hanna, R.E.B. and Nizami, W.A. Ultrastructural and cytochemical observations on the tegument of three species of paramphistomes (Platyhelminthes: Digenea) from the Indian Water Buffalo, *Bubalus bubalis*. Int. J. Parasitol. **17**: 1153–1161 (1987).

28. Dunn, T.S., Hanna, R.E.B. and Nizami, W.A. Ultrastructural and cytochemical studies on the ventral pouch of *Gastrothylax crumenifer* (Digenea: Paramphistomidae). Int. J. Parasitol. **17**: 1163–1173 (1987).
29. Eduardo, S.L. *Bilatorchis papillogenitalis* N.G., N. Sp. (Paramphistomidae: Orthocoelinae), a parasite of the red lechwe (*Kobus leche* Gray, 1850) from Zambia. Syst. Parasitol. **1**: 141–149 (1980).
30. Eduardo, S.L. *Orthocoelium indonesiense*, a new species of amphistome from ruminants in Indonesia. Syst. Parasitol.. **1**: 203–210 (1980).
31. Eduardo, S.L. A new genus, *Leiperocotyle*, for *Cotylophoronokapi* Leiper, 1935 and *C. congolense* Baer, 1936 and redescription of *C. okapi*. Syst. Parasitol. **1**: 255–263 (1980).
32. Erasmus, D.A. *The biology of trematodes*. Edward Arnold: London (1972).
33. Erasmus, D.A. *Schistosoma mansoni*: development of the vitelline cell, its role in drug sequestration and changes induced by Astiban. Exp. Parasitol. **38**: 240–256 (1975).
34. Erasmus, D.A. The host-parasite interface of trematodes. Adv. Parasitol. **15**: 201–242 (1977).
35. Ernst, S.C. Biochemical and cytochemical studies of digestive-absorptive functions of esophagus, cecum, and tegument in *Schistosoma mansoni*: acid phosphatase and tracer studies. J. Parasitol. **61**: 633–647 (1975).
36. Fairweather, I., Anderson, H.R. and Baldwin, T.M.A. *Fasciola hepatica*: tegumental surface alterations following treatment *in vitro* with the deacetylated (amine) metabolite of diamphenethide. Parasitol. Res. **73**: 99–106 (1987).
37. Featherston, D.W. *Taenia hydatigena* III. Light and electron microscope study of spermatogenesis. Z. Parasitenkunde **37**: 148–168 (1971).
38. Foor, W.E. Ultrastructural aspects of oocyte development and shell formation in *Ascaris lumbricoides* (Nematoda). J. Parasitol. **53**: 1245–1261 (1967).
39. Foor, W.E. Cytoplasmic bridges in the ovary of *Ascaris lumbricoides*. Bull. Tulane Med. Fac. **27**: 23–30 (1968).
40. Foor, W.E. Origin and possible utilization of small dense granules in the oocytes of *Ascaris suum. J. Parasitol.* **58**: 524–538 (1972).
41. Foor, W.E. Morphological changes of spermatozoa in the uterus and glandular vas deferens. J. Parasitol. **62**: 849–864 (1974).
42. Foor, W.E. Chapter 10. Nematoda. In: *Reproductive Biology of Invertebrates*. Vol. 1: *Oogenesis, oviposition, and oosorption* (K.G. Adiyodi and R.G. Adiyodi eds), pp. 223–256. John Wiley & Sons: Chichester, England (1983).
43. Foor, W.E. Chapter 12. Nematoda. In: *Reproductive Biology of Invertebrates*. Vol. II: *Spermatogenesis and sperm function* (K.G. Adiyodi and R.G. Adiyodi eds), pp. 221–256. John Wiley & Sons: Chichester, England (1983).
44. Franz, M. Electron microscope study of the cuticle of male and female *Onchocerca volvulus* from various geographic areas. Tropenmed. Parasitol. **31**: 149–164 (1980).
45. Franz, M. The fine structure of adult *Onchocerca volvulus*. I. The cuticle, the hypodermis and the muscle cell of the male worm. Tropenmed. Parasitol. **33**: 69–75 (1982).
46. Franz, M. and Büttner, D.W. The fine structure of adult *Onchocerca volvulus*. III. The cuticle, the interchordal hypodermis and the muscle cells of the female worm. Tropenmed. Parasitol. **34**: 61–69 (1983).
47. Franz, M. and Büttner, D.W. The fine structure of adult *Onchocerca volvulus*. V. The digestive tract and the reproductive system of the female worm. Tropenmed. Parasitol. **34**: 155–161 (1983).
48. Franz, M. and Lenze, W. Scanning electron microscope study of adult *Brugia malayi*. Tropenmed. Parasitol. **33**: 17–22 (1982).
49. Fried, B. and Appel, A.J. Excretion of lipids by *Echinostoma revolutum* (Trematoda) adults. J. Parasitol. **63**: 447: (1977).
50. Fujimoto, D. and Kanaya, S. Cuticlin: A non-collagen structural protein from *Ascaris* cuticle. Archs Biochem. Biophys. **157**: 1–6 (1973).
51. Gallagher, S.S.E. and Threadgold, L.T. Electron microscope studies of *Fasciola hepatica*. II. The interrelationship of the parenchyma with other organ systems. Parasitology **57**: 627–632 (1967).
52. Gustafsson, M.K.S. Studies on cytodifferentiation in the neck region of *Diphyllobothrium dendriticum* Nitzsch, 1824 (Cestoda, Pseudophyllidea). Z. Parasitenkunde **50**: 323–329 (1976).

53. Hanna, R.E.B. *Fasciola hepatica*: an electron microscope autoradiographic study of protein synthesis and secretion by gut cells in tissue slices. Exp. Parasitol. **38**: 167–180 (1975).
54. Hanna, R.E.B. *Fasciola hepatica*: a light and electron microscope autoradiographic study of incorporation of monosaccharides into glycogen and glycoprotein. Exp. Parasitol. **39**: 204–213 (1976).
55. Hanna, R.E.B. *Fasciola hepatica*: glycocalyx replacement in the juvenile as a possible mechanism for protection against host immunity. Exp. Parasitol. **50**: 103–114 (1980).
56. Hanna, R.E.B. *Fasciola hepatica*: an immunofluorescent study of antigenic changes in the tegument during development in the rat and the sheep. Exp. Parasitol. **50**: 155–170 (1980).
57. Hanna, R.E.B. Autoradiography of protein synthesis, transport and secretion by the tegument. Exp. Parasitol. **50**: 297–304 (1980).
58. Hanna, R.E.B. and Trudgett, A.G. *Fasciola hepatica*: development of monoclonal antibodies and their use to characterize a glycocalyx antigen in migrating flukes. Parasite Immunol. **5**: 409–425 (1983).
59. Harpur, R.P. and Popkin, J.S. Intestinal fluid transport: studies with the gut of *Ascaris lumbricoides*. Can. J. Physiol. Pharmacol. **51**: 79–90 (1973).
60. Harris, J.E. and Crofton, H.D. Structure and function in the nematodes: Internal pressure and cuticular structure in *Ascaris*. J. Exp. Biol. **34**: 116–130 (1957).
61. Hockley, D.J. Ultrastructure of the tegument of *Schistosoma*. Adv. Parasitol. **11**: 233–305 (1973).
62. Hockley, D.J. and McLaren, D.J. *Schistosoma mansoni*: changes in the outer membrane of the tegument during development from cercaria to adult worms. Int. J. Parasitol. **3**: 13–25 (1973).
63. Hockley, D.J., McLaren, D.J., Ward, B.J. and Nermut, M.V. A freeze-fracture study of the tegumental membrane of *Schistosoma mansoni* (Platyhelminthes: Trematoda). Tissue and Cell **7**: 485–496 (1975).
64. Howells, R.E. Dyamics of the filarial surface. In: *Ciba Foundation Symposium 127 Filariasis* (D. Evered and S. Clark eds), pp. 94–106. John Wiley & Sons: Chichester, England (1987).
65. Howells, R.E. and Blainey, L.J. The moulting process and the phenomenon of intermoult growth in the filarial nematode *Brugia pahangi*. Parasitology **87**: 493–505 (1983).
66. Irwin, S.W.B. and Threadgold, L.T. Electron microscope studies on *Fasciola hepatica*. VIII. The development of the vitelline cells. Exp. Parasitol. **28**: 399–411 (1970).
67. Jennings, J.B and Colam, J.B. Gut structure, digestive physiology and food storage in *Pontonema vulgaris* (Nematoda: Enoplida). J. Zool., Lond. **161**: 211–221 (1970).
68. Kelsoe, G.H., Ubelaker, J.E. and Allison, V.F. The fine structure of spermatogenesis in *Hymenolepis diminuta* (Cestoda) with a description of the mature spermatozoon. Z. Parasitenkunde **54**: 175–187 (1977).
69. Kozek, W.J. The moulting patterns in *Trichinella spiralis*. II. An electron microscopic study. J. Parasitol. **57**: 1029–1038 (1971).
70. Lascano, E.F., Coltorti, E.A. and Varela-Diaz, V.M. Fine structure of the germinal membrane of *Echinococcus granulosus* cysts. J. Parasitol. **61**: 853–860 (1975).
71. Lee, D.L. The structure and composition of the helminth cuticle. Adv. Parasitol. **4**: 187–254 (1966).
72. Lee, D.L. Moulting in nematodes: The formation of the adult cuticle during the final moult of *Nippostrongylus brasiliensis*. Tissue and Cell **2**: 139–153 (1970).
73. Lee, D.L. The structure of the helminth cuticle. Adv. Parsitol **10**: 347–379 (1972).
74. Lee, D.L. and Atkinson, H.J. *Physiology of Nematodes*. 2nd edn. The Macmillan Press: London (1976).
75. Lee, D.L. and Bonner, T.P. Freeze etch studies on nematode body wall. Parasitology **84**: xliv (1982).
76. Lim, K.C., Loh, A.K. and Singh, M. Scanning electron microscopic study of *Brugia malayi* (Filarioidea: Onchocercidae). Southeast Asian J. Trop. Med. Pub. Hlth. **14**: 113–121 (1983).
77. Lumsden, R.D. Surface ultrastructure and cytochemistry of parasitic helminths. Exp. Parasitol. **37**: 267–339 (1975).
78. Lumsden, R.D. and Hildreth, M.B. The fine structure of adult tapeworms. In: *Biology of the Eucestoda*. Vol. 1. (C. Arme and P.W. Pappas, eds), pp. 177–233. Academic Press: London (1983).
79. Lumsden, R.D. and Specian, R. The morphology, histology and fine structure of the adult stage of the Cyclophyllidean tapeworm *Hymenolepis diminuta*. In: *The biology of the rat tapeworm Hymenolepis diminuta* (H.P. Arai ed), pp. 157–280. Academic Press: New York (1980).
80. Lyons, K.M. Epidermal adaptations of parasitic platyhelminths. Symp. Zool. Soc., Lond. **39**: 97–144 (1977).

81. Maggenti, A.R. *General nematology*. Springer: New York (1981).
82. Martin, W.E. and Bils, R.F. Trematode excretory concretions: formation and fine structure. J. Parasitol. **50**: 337–344 (1964).
83. Martinez-Palomo, A. Ultrastructural characterisation of the cuticle of *Onchocerca volvulus* microfilaria. J. Parasitol. **64**: 127–136 (1978).
84. McLaren, D.J. Oogenesis and fertilization in *Dipetalonema viteae* (Nematoda: Filaroidea). Parasitology **66**: 465–472 (1973).
85. McLaren, D.J. *Schistosoma mansoni:* the parasite surface in relation to host immunity. John Wiley & Sons: London (1980).
86. Mehlhorn, H., Dubremetz, J.F., Franz. M., Gustafsson, M., Peters, W., Taraschewski, H., Walldorf, V. and Voigt, W.P. Chapter 3. Morphology. In: *Parasitology in Focus. Facts and Trends* (H. Mehlhorn ed), pp. 161–311. Springer Verlag: Berlin (1988).
87. Mehlhorn, H. Taraschewski, H., Franz, M., Walldorf, V. and Voigt, W.P. Chapter 4. Reproduction. In: *Parasitology in focus. Facts and Trends* (H. Mehlhorn ed), pp. 312–367. Springer Verlag: Berlin (1988).
88. Mohandas, A. Chapter 4. Platyhelminthes-Trematoda. In: *Reproductive Biology of Invertebrates*. Vol. II. *Spermatogenesis and Sperm Function* (K.G. Adiyodi and R.G. Adiyodi eds), pp. 105–129. John Wiley & Sons: Chichester, England (1983).
89. Morris, G.P. Fine structure of the gut epithelium of *Schistosoma mansoni*. Experientia **24**: 480–482 (1968).
90. Morris, G.P. and Threadgold, L.T. Ultrastructure of the tegument of adult *Schistosoma mansoni*. J. Parasitol. **54:** 15–27 (1968).
91 Morseth, D.J. Fine structure of the hydatid cyst and protoscolex of *Echinococcus granulosus*. J. Parasitol. **53**: 312–325 (1967).
92. Moss. G.D. The excretory metabolism of the endoparasitic digenean *Fasciola hepatica* and its relationship to its respiratory metabolism. Parasitology **60**: 1–19 (1970).
93. Nollen, P.M. and Nadakavukaren, M.J. *Megalodiscus temperatus*: scanning electron microscopy of the tegumental surfaces. Exp. Parasitol. **36**: 123–130 (1974).
94. Oaks, J.A. and Lumsden, R.D. Cytological studies on the absorptive surfaces of cestodes. V. Incorporation of carbohydrate-containing macromolecules into tegumental membranes. J. Parasitol. **57**: 1256–1268 (1971).
95. Otubanjo, O.A. *Schistosoma mansoni*: the sustentacular cells of the testes. Parasitology **82**: 125–130 (1981).
96. Pantelouris, E.M. and Threadgold. L.T. The excretory system of adult *Fasciola hepatica*. Cellule **64**: 63–67 (1963).
97. Pappas, P.W. and Read, C.P. Membrane transport in helminth parasites: a review. Exp. Parasitol. **37**: 469–530 (1975).
98. Peracchia, C. Structural correlates of gap junction permeation. Int. Rev. Cytol. **66**: 81–146 (1980).
99. Philipp, M., Parkhouse, R.M.E. and Ogilvie, B.M. Changing proteins on the surface of a parasitic nematode. Nature (Lond.) **287**: 538–540 (1980).
100. Reznik, G.K. The intestinal secretion of the parasitic nematode of domestic fowl *Ascaridia galli*. Dokl. Akad. Nauk SSSR **182**: 735–736 (1968).
101. Riding. I. Microvilli on the outside of a nematode. Nature (Lond.) **226**: 179–180 (1970).
102. Robinson, G. and Threadgold, L.T. Electron microscope studies of *Fasciola hepatica* XII. The fine structure of the gastrodermis. Exp. Parasitol. **37**: 20–36 (1975).
103. Robinson, J.M. and Bogitsh, B.J. A morphological and cytochemical study of sperm development in *Hymenolepis diminuta*. Z. Parasitenkunde **56**: 81–92 (1978).
104. Robinson, R.D. and Halton, D.W. Fine structural observations on spermatogenesis in *Corrigia vitta* (Trematoda: Dicrocoeliidae). Z. Parasitenkunde **68**: 53–72 (1982).
105. Roggen, D.R., Raski, D.J. and Jones, N.O. Further electron microscopic observations of *Xiphinema index*. Nematologica **13**: 1–16 (1967).
106. Rosario, B. An electron microscope study of spermatogenesis in cestodes. J. Ultrastruct. Res. **11**: 412–427 (1964).
107. Rybicka, K., Embryogenesis in cestodes. Adv. Parasitol. **4**: 107–186 (1966).
108. Sakamoto, T. and Sugimura, M. Studies on echinococcosis. XXI. Electron microscopal observations on general structure of larval tissue of multilocular *Echinococcus*. Jpn. J. Vet. Res. **17**: 67–101 (1969).

109. Sakamoto, T. and Sugimura, M. Studies on echinococcosis. XXIII. Electron microscopical observations on histogenesis of larval *Echinococcus multilocularis* Jpn. J. Vet. Res. **18**: 131–144 (1970).
110. Sakamoto, T., Yamashita, J. and Ohbayashi, M. Studies on echinococcosis. XVIII. Observations on tissue cultured germinal cells of larval *Echinococcus multiloculatirs* Jpn. J. Vet. Res. **15**: 75–84 (1967).
111. Shannon, W.A. and Bogitsh, B.J. *Megalodiscus temperatus*: comparative radioautography of glucose-^{3}H and galactose-^{3}H incorporation. Exp. Parasitol. **29**: 309–319 (1971).
112. Sharma, P.N. Histochemical distribution of succinic dehydrogenase in the lymphatic system of a trematode *Ceylonocotyle scoliocoelium*. J. Helminthol. **52**: 159–162 (1978).
113. Siddiqi, A.H., Islam, M.W. and Nizami, W.A. Osmotic and ionic behaviour of some digenetic trematodes. Comp. Biochem. Physiol. **51A**: 929–935 (1975).
114. Slais, J., Serbus. C. and Schramlova, J. The microscopical anatomy of the bladder wall of *Cysticercus bovis* at the electron microscope level. Z. Parasitenkunde **36**: 304–320 (1971).
115. Smyth, J.D. and Halton, D.W. *The physiology of trematodes*. 2nd edn. Cambridge University Press: Cambridge, England (1983).
116. Sodeman, T.M., Sodeman, W.A. and Schnitzer, B. Lamellar structures in the gut of *Schistosoma haematobium*. Ann. Trop. Med. Parasitol. **66**: 475–478 (1972).
117. Swiderski, Z. The fine structure of the spermatozoon of sheep tapeworm, *Moniezia expansa* (Rud., 1810) (Cyclophyllidea, Anoplocephalidae). Zoologica Pol. **18**: 475–486 (1968).
118. Threadgold, L.T. The tegument and associated structures of *Fasciola hepatica*. Q. Jl. microsc. Sci. **104**: 505–512 (1963).
119. Threadgold, L.T. Electron microscope studies of *Fasciola hepatica*. III. Further observations on the tegument and associated structures. Parasitology **75**: 633–637 (1967).
120. Threadgold, L.T. Electron microscope studies of *Fasciola hepatica*. VI. The ultrastructural localization of phosphatases. Exp. Parasitol. **23**: 264–276 (1968).
121. Threadgold, L.T. *Fasciola hepatica*: Ultrastructure and histochemistry of the glycocalyx of the tegument. Exp. Parasitol. **39**: 119–134 (1976).
122. Threadgold, L.T. *Fasciola hepatica*: Stereological analysis of vitelline cell development. Exp. Parasitol. **54**: 352–365 (1982).
123. Threadgold, L.T. Chapter 11. Parasitic Platyhelminths. In: *Biology of the integument*. Vol. 1. *Invertebrates* (J. Bereiter-Hahn, A.G. Matoltsy and K.S. Richards eds), pp. 132–191. Springer Verlag: Berlin (1984).
124. Threadgold, L.T. and Arme, C. Electron microscope studies of *Fasciola hepatica*. XI. Autophagy and parenchymal cell function. Exp. Parasitol. **35**: 389–405 (1974).
125. Threadgold, L.T. and Arme, C. *Hymenolepis diminuta*: An electron microscope study of ion absorption. Exp. Parasitol. **35**: 475–491 (1974).
126. Threadgold, L.T. and Brennan, G. *Fasciola hepatica:* Basal infolds and associated vacuoles of the tegument. Exp. Parasitol. **46**: 300–316 (1978).
127. Threadgold, L.T. and Dunn, J. *Taenia crassiceps*: Regional variations in ultrastructure and evidence of endocytosis in the cysticercus tegument. Exp. Parasitol. **55**: 121–131 (1983).
128. Threadgold, L.T. and Gallagher, S.S.E. Electron microscope studies of *Fasciola hepatica*. I. The ultrastructure and interrelationship of the parenchymal cells. Parasitology **56**: 299–304 (1966).
129. Threadgold, L.T. and Hopkins, C.A. *Schistocephalus solidus* and *Ligula intestinalis*: Pinocytosis by the tegument. Exp. Parasitol. **51**: 444–456 (1981).
130. Threadgold, L.T. and Irwin, S.W.B. Electron microscope studies on *Fasciola hepatica*. IX. The fine structure of Mehlis' gland. Z. Parasitenkunde **35**: 16–30 (1970).
131. Threadgold, L.T. and Read, C.P. *Hymenolepis diminuta*: Ultrastructure of a unique membrane specialization in tegument. Exp. Parasitol. **28**: 246–252 (1970).
132. Ubelaker, J.E., Cooper, N.B. and Allison, V.F. The fine structure of the cysticercoid of *Hymenolepis diminuta*. I. The outer wall of the capsule. Z. Parasitenkunde **34**: 258–270 (1970).
133. Von Brand, T. *Biochemistry of Parasites*. 2nd edn. Academic Press: New York (1973).
134. Wikgren, B.-J.P., Gustafsson, M.K.S. and Knuts, G.M. Primary anlage formation in diphyllobothriid tapeworms. Z. Parasitenkunde **36**: 131–139 (1971).

135. Wilson, R.A. The protonephridial system in the miracidium of the liver fluke, *Fasciola hepatica* L. Comp. Biochem. Physiol. **20A**: 337–342 (1967).
136. Wilson, R.A. and Barnes, P.E. The tegument of *Schistosoma mansoni*: observations on the formation, structure and composition of cytoplasmic inclusions in relation to tegument function. Parasitology **68**: 239–258 (1974).
137. Wilson, R.A. and Barnes, P.E. The formation and turnover of the membranocalyx on the tegument of *Schistosoma mansoni*. Parasitology **74**: 61–71 (1977).
138. Wilson, R.A. and Barnes, P.E. Synthesis of macromolecules by the epithelial surfaces of *Schistosoma mansoni*: an autoradiography study. Parasitology **78**: 295–310 (1979).
139. Wilson, R.A. and Webster, L.A. Protonephridia. Biol. Rev. **49**: 127–160 (1974).
140. Zaman, V. Scanning electron microscopy of *Brugia malayi*. In: *Ciba Foundation Symposium 127 Filariasis* (D. Evered and S. Clark eds), pp. 77–93. John Wiley & Sons: Chichester, England (1987).
141. Zaman, V. and Narayanan, E. Differentiation of infective larvae of *Brugia malayi* and *Wuchereria bancrofti* by scanning electron microscopy. Ann. Trop. Med. Parasitol. **80**: 333–338 (1986).

7. Biochemistry of Helminths

J. Barrett
Department of Biological Sciences, The University College of Wales
Aberystwyth, Dyfed, UK

INTRODUCTION

Biochemistry can be divided into variant and invariant: the invariant features include the use of adenosine phosphates in energy transfer, nucleic acid metabolism and the role of NAD^+ and $NADP^+$ as redox carriers, processes which are similar in all living organisms. The variant features cover metabolic pathways and how they can be modified to meet the environmental needs of the organism. Parasitic helminths come from three different phyla, and within each group parasitism may have arisen on several separate occasions. Parasites occupy a wide variety of different sites within their hosts, each with different physico-chemical characteristics. In addition, during its life cycle a single parasite may pass through a series of different environments ranging from free-living, to the tissues of an invertebrate, to the intestines of a vertebrate. It must, therefore, be expected that parasitic helminths will show a wide range of metabolic adaptations with many examples of parallel and convergent evolution (see Chapter 1).

In looking at the biochemistry of helminths we are interested first, in the way the metabolism of the parasite differs from that of its host and secondly, how the metabolism of parasitic forms differs from that of their free-living relatives. Differences between host and parasite represent possible sites for chemotherapy whilst the differences between parasitic and free-living forms may give some insight into the molecular basis of parasitism.

ENERGY METABOLISM

The main storage carbohydrate of parasitic helminths is glycogen and parasites can contain surprisingly large amounts, up to 50% of the dry wight in some tapeworms, although 25% is probably a fairly average figure for most helminths.

Another carbohydrate which has been found in appreciable quantities in nematodes and acanthocephalans is trehalose. In *Ascaris lumbricoides* it constitutes 2% of the dry weight [15]. Trehalose is a non-reducing disaccharide which is also found in insects and yeasts. The amounts of free glucose in helminths, as in other tissues, is small (less than 0.1 per cent of the dry weight).

Parasitic helminths have an absolute dependency on carbohydrate, either in the form of glycogen or exogenous glucose, as their sole energy source. In adult helminths there is no beta-oxidation of fatty acids and no significant catabolism of amino acids, nor is there any evidence for the cofermentation of amino acids and carbohydrate or of fatty acids and carbohydrate. Characteristically, helminths break down carbohydrate to reduced organic acids but occasionally to alcohols, which are then excreted [8]. The pathways of carbohydrate catabolism are essentially anaerobic, and on the basis of the end-products produced parasites can be divided into two major groups. First, there are those which rely essentially on glycolysis and produce as end-products of carbohydrate breakdown lactate or some other derivative of pyruvate and, secondly, there are those which fix carbon dioxide and have what may be described as an '*Ascaris*' type metabolism. In the '*Ascaris*' type the initial end-products of carbohydrate catabolism are succinate and pyruvate, but these are frequently further metabolised to short chain fatty acids such as acetate, propionate, 2-methylbutyrate and 2-methylvalerate. These two types of metabolism are, however, in no way absolute, they are convenient divisions and in reality there is a continuous spectrum between the two. Most homolactic fermentors fix carbon dioxide to a certain extent and most '*Ascaris*' type parasites excrete variable quantities of lactic acid.

Glycolytic Helminths

A number of parasitic helminths breakdown carbohydrate more or less quantitatively to lactic acid (Tables 1–4). Homolactic fermentation seems more characteristic of tissue or blood dwelling parasites than intestinal parasites. In the filarial worm *Litomosoides carinii* lactate production is markedly influenced by the oxygen tension. Anaerobically 80% of the carbohydrate is metabolised to lactate, the rest to acetate. Under aerobic conditions only 30–40% of the end-products is lactate, acetate accounts for about 35% and the remainder is acetoin. Both acetate and acetoin are probably produced by the pyruvate dehydrogenase complex [6].

A variant of the glycolytic scheme is found in parasites such as the acanthocephalan *Moniliformis moniliformis* which produce ethanol as a major end-product. Ethanol is also an anaerobic end-product in

Table 1. Major end-products of carbohydrate catabolism in nematodes

Species	End-product
Ancylostoma caninum	acetate, propionate
Angiostrongylus cantonensis	acetate, lactate, alanine
Ascaridia galli	acetate, lactate, propionate
Ascaris lumbricoides	2-methylvlaerate, 2-methylbutyrate
Brugia pahangi	lactate
Chandlerella hawkingi	lactate
Dipetalonema (= *Acanthocheilonema*) *viteae*	lactate
Dirofilaria uniformis	lactate
D. immitis	lactate
Dracunculus insignis	lactate
Haemonchus contortus	acetate, ethanol, propanol, propionate
Heterakis gallinae	acetate, propionate
Litomosoides carinii	acetate, lactate, acetoin
Mecistocirrus digitatus	acetate, butyrate, lactate
Nippostrongylus brasiliensis	lactate, succinate
Oesophagostomum radiatum	acetate, lactate, methylbutyrate, propionate
Onchocerca gibsoni	lactate
O. gutturosa	lactate
O. lienalis	lactate
O. volvulus	lactate, succinate
Parascaris equorum	lactate, 2-methylbutyrate, propionate
Setaria cervi	lactate
Trichinella spiralis	acetate, propionate, *n*-valerate
Trichostrongylus colubriformis	acetate, propionate
Trichuris vulpis	lactate, propionate, succinate

free-living and plant parasitic nematodes. Amongst animal parasitic nematodes alcohol dehydrogenase is extremely active in the larvae of *Anisakis* and is also present in *Strongyluris brevicaudata*, so ethanol may also be an end-product in these parasites.

In free-living invertebrates, alanine is often an important anaerobic end-product. However, although most parasitic helminths contain appreciable amounts of alanine, it is not excreted to any significant extent.

Carbon Dioxide Fixation

Carbon dioxide fixation is characteristic of the large intestinal parasites such as *Moniezia expansa, Taenia taeniaeformis, Fasciola hepatica* and *Ascaris lumbricoides* (Tables 1–4). The nematode *A. lumbricoides* has been studied in considerable detail and can be used as a model for the others [4]. In *Ascaris* (Fig. 1) glycogen is broken down by a normal glycolytic sequence, as far as phosphoenolpyruvate. The activity of pyruvate kinase in *Ascaris* is low, and so instead of forming pyruvate, carbon dioxide fixation takes place and phosphoenolpyruvate is converted to oxaloacetate by the action of a cytoplasmic phosphoenolpyruvate carboxykinase (phosphopyruvate carboxylase). Phosphoenolypyruvate carboxykinase also occurs in vertebrates, but the enzyme from *Ascaris* differs from that of vertebrates in being more active with IDP than GDP (ADP is inactive), also the enzyme from vertebrates functions in the direction of phosphoenolpyruvate synthesis. The oxaloacetate formed is reduced, in the cytoplasm, to malate by malate dehydrogenase. This reoxidizes the NADH produced by the glyceraldehyde-3-phosphate dehydrogenase step in glycolysis. Malate is then transported into the mitochondrion via a phosphate dependent translocase. Inside the mitochondrion malate undergoes a dismutation reaction, part is oxidatively decarboxylated to pyruvate via an NAD^+-linked malic enzyme (in vertebrates the malic enzyme is cytoplasmic and $NADP^+$-linked). In the mitochondrion malate is in equilibrium with fumarate via the enzyme fumarase (in *Ascaris*, fumarase occurs both in the cytoplasm and the mitochondrion) and fumarate is reduced to succinate by a fumarate reductase. So in *Ascaris* there is a partial, reversed tricarboxylic acid cycle, the parasite uses the second span of the tricarboxylic acid cycle, from succinate to oxaloacetate, but in the reverse direction.

There are thus two systems for the reoxidation of NADH in *Ascaris*; in the cytoplasm NADH formed in glycolysis is reoxidized by the reduction of oxaloacetate to malate, whilst in the mitochondrion the NADH produced by the malic enzyme is reoxidized in the reduction of fumarate to succinate. The catabolism of carbohydrate to succinate and pyruvate is, therefore, in redox balance and could proceed entirely anaerobically. A variation occurs in the tapeworms *Hymenolepis diminuta* and *H. microstoma* where the malic enzyme is $NADP^+$-linked but the fumarate reductase remains NAD^+-linked. The mitochondria of these cestodes have a non-energy linked transhydrogenase which catalyses the reaction:

$$NADPH + NAD^+ + H^+ \rightleftharpoons NADP^+ + NADH + H^+$$

thus enabling the malic enzyme and fumarate reductase to be coupled [17, 18]. Many of the small gastrointestinal nematodes of mammals may be like *Nippostrongylus brasiliensis*. Aerobically *N. brasiliensis* produces lactate, anaerobically lactate and succinate, with succinate formation involving carbon dioxide fixation and a partial reversal of the tricarboxylic acid cycle [22].

The redox state of the free $[NAD^+]/[NADH]$ couple in the cytoplasm of *Ascaris* muscle is between 725 : 1 and 2214 : 1 [9], a value comparable to the ratio of 1000 : 1 found in normal rat liver (anaerobically the liver ratio falls to 100 : 1). By coupling the reoxidation of cytoplasmic NADH to the reduction of oxaloacetate rather than to pyruvate (as in mammals) the cytoplasmic free $[NAD^+]/[NADH]$ couple in *Ascaris* muscle remains in a relatively oxidized state, despite metabolism being essentially anaerobic. In the mitochondrion of *Ascaris* the free $[NAD^+]/[NADH]$ ratio is considerably lower than that in rat liver

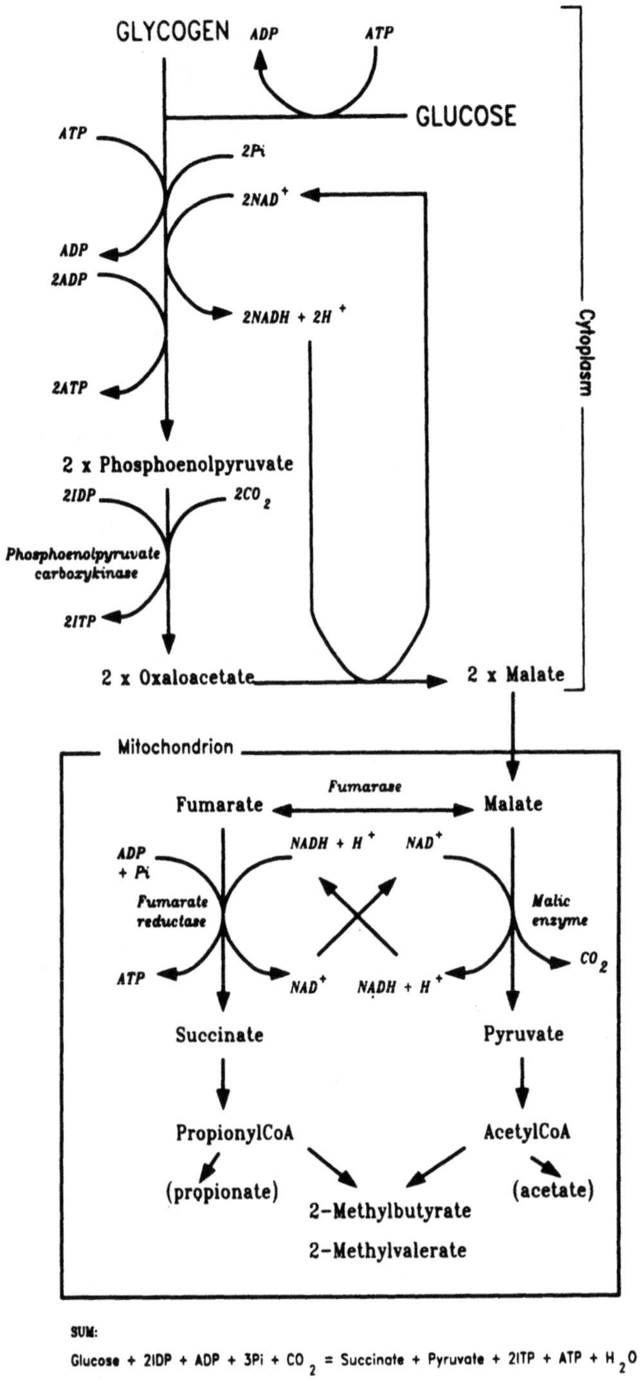

Fig. 1. Carbohydrate catabolism in *Ascaris* [8].

mitochondria (0.1 : 1 as compared to 10 : 1). This may be related to the fact that in the *Ascaris* mitochondrion fumarate is being reduced to succinate, whilst in the mammal, succinate is being oxidized to fumarate.

Table 2. Major end-products of carbohydrate catabolism in cestodes

Species	End-product
Cotugnia digonopora	lactate
Diphyllobothrium dendriticum	lactate, succinate
Echinococcus granulosus	acetate, lactate, succinate
Hymenolepis diminuta	acetate, lactate, succinate
H. microstoma	acetate, lactate, propionate, succinate
Ligula intestinalis	lactate
Mesocestoides corti	acetate, lactate, succinate
Moniezia expansa	lactate, succinate
Schistocephalus solidus	acetate, propionate
Spirometra mansonoides	acetate, propionate

Table 3. Major end-products of carbohydate catabolism in digeneans

Species	End-product
Calicophoron ijimai	acetate, lactate, propionate, succinate
Clinostomum complanatum	lactate
Dicrocoelium dendriticum	lactate
Echinostoma liei	*n*-valerate
Fasciola gigantica	acetate, lactate
Fasciola hepatica	acetate, propionate
Paragonimus westermanii	acetate, propionate
Schistosoma mansoni	lactate
S. margrebowiei	lactate
S. douthitti	lactate

Table 4. Major end-products of carbohydrate catabolism in acanthocephalans

Species	End-product
Echinorhynchus gadi	lactate, succinate
Moniliformis moniliformis	ethanol
Neoechinorhynchus emydis	lactate
N. pseudemydis	lactate
Polymorphus minutus	lactate, succinate

In *Ascaris* the reduction of fumarate to succinate involves part of the cytochrome chain and results in a site 1 phosphorylation of ADP (ATP/malate ratio = 0.5). There is some controversy as to whether the succinate dehydrogenase and fumarate reductase activity of helminths are different functions of the same enzyme or two separate enzymes. The helminth enzyme resembles mammalian succinic dehydrogenase in being membrane bound, and at least in *F. hepatica* and *H. diminuta* existing in active and inactive forms. The succinate oxidase: fumarate reductase ratio of helminth enzymes ranges from 0.3 to 5, midway between the classical aerobic and anaerobic enzymes (mammals, 60; anaerobic bacteria 0.03). Work on *A. lumbricoides* and *A. galli* suggests that they have a single reversible enzyme. However, in *N. brasiliensis* there may be two succinic dehydrogenases, one reversible like the *Ascaris* enzyme, the other similar to the typical mammalian enzyme [21].

The pathway of carbohydrate breakdown in *Ascaris* yields four moles of nucleoside triphosphate for every C_6 unit of glycogen catabolised. Of these, two are produced by the phosphoenolpyruvate carboxykinase reaction and will be in the form of ITP or GTP, and not ATP. *Ascaris* has high levels of the enzyme nucleoside diphosphate kinase, which rapidly catalyses the transfer of high energy phosphate from GTP or ITP to ADP.

$$\text{NTP + ADP} \rightleftharpoons \text{NDP + ATP} \qquad (\text{Keq} = 0.9)$$

The initial end-products of carbohydrate breakdown in helminths which fix carbon dioxide are thus pyruvate and succinate, both produced in the mitochondrion. Succinate is often a major end-product in parasitic helminths, although it can be further metabolised to propionate and short chain fatty acids. Pyruvate is rarely excreted unchanged and is almost always metabolised to some other derivative. However, an interesting variant is found in the cercariae of *S. mansoni* which excrete appreciable amounts of pyruvate, with the excess NADH being reoxidized by the cytochrome chain [49].

The decarboxylation of succinate to propionate in helminths is the same as in mammalian liver, succinate being metabolised by succinic thiokinase, methylmalonyl CoA mutase, methylmalonyl CoA racemase and propionyl CoA carboxylase to propionyl CoA. The conversion of succinate to propionyl CoA does not result in any net production of ATP (one ATP is produced during the decarboxylation of methylmalonyl CoA, but one is required in the initial thiokinase reaction). However, propionyl CoA is a high energy compound and the bond energy may be conserved via CoA transferase reactions [29]:

$$\text{Propionyl CoA + Succinate} \rightleftharpoons \text{Succinyl CoA + Propionate}$$

The pathway from succinate to propionate is reversible, and it is possible that in ruminants, where propionate levels are high, parasites may use this route for glyconeogenesis [51]. The nematode *Haemonchus contortus* excretes propanol as a major end-product. The direct reduction of propionate to propanol is energetically unfavourable and the most likely route is an NAD dependent reduction of propionyl CoA.

In helminths the decarboxylation of pyruvate to acetate is catalysed by the pyruvate dehydrogenase complex. Acetyl CoA produced by the complex is cleaved via a CoA transferase system (as for propionyl CoA) or via deacetylases to yield acetate [43]. The pyruvate dehydrogenase complex is also the source of acetoin, as well as acetaldehyde in those species which excrete ethanol.

In *Ascaris*, the formation of 2-methylvalerate and 2-methylbutyrate from succinate and pyruvate involves a series of mitochondrial condensation and reduction reactions, very similar to the steps of beta-oxidation [30, 31]. The penultimate step in the synthesis of the branched chain fatty acids in the *Ascaris* mitochondrion is the reduction of 2-methylcrotonyl CoA to 2-methylvaleryl CoA. These reductions are catalysed by a flavoprotein enzyme system, and may be linked to the electron transport chain in a reaction analogous to the reduction of fumarate. So the formation of 2-methylbutyrate and 2-methylvalerate in *Ascaris* (Fig. 2), like the formation of succinate may be associated with an electron chain mediated site I phosphorylation of ADP [32].

The anaerobic fermentation of glucose to lactate results in the net formation of 2 moles of ATP mole^{-1} of glucose catabolised. In the *Ascaris* type system the conversion of 1 mole of glucose to 1 mole each of succinate and pyruvate gives a net yield of 3 ATP mole^{-1}, an increase of a third. The formation of acetate and propionate and of the branched chain fatty acids may also provide additional ATP.

Helminths often show marked differences in the proportions of the different end-products produced under aerobic and anaerobic conditions. In general, under anaerobic conditions succinate and lactate production increases, whilst acetate production decreases. The relative proportions of the end-products

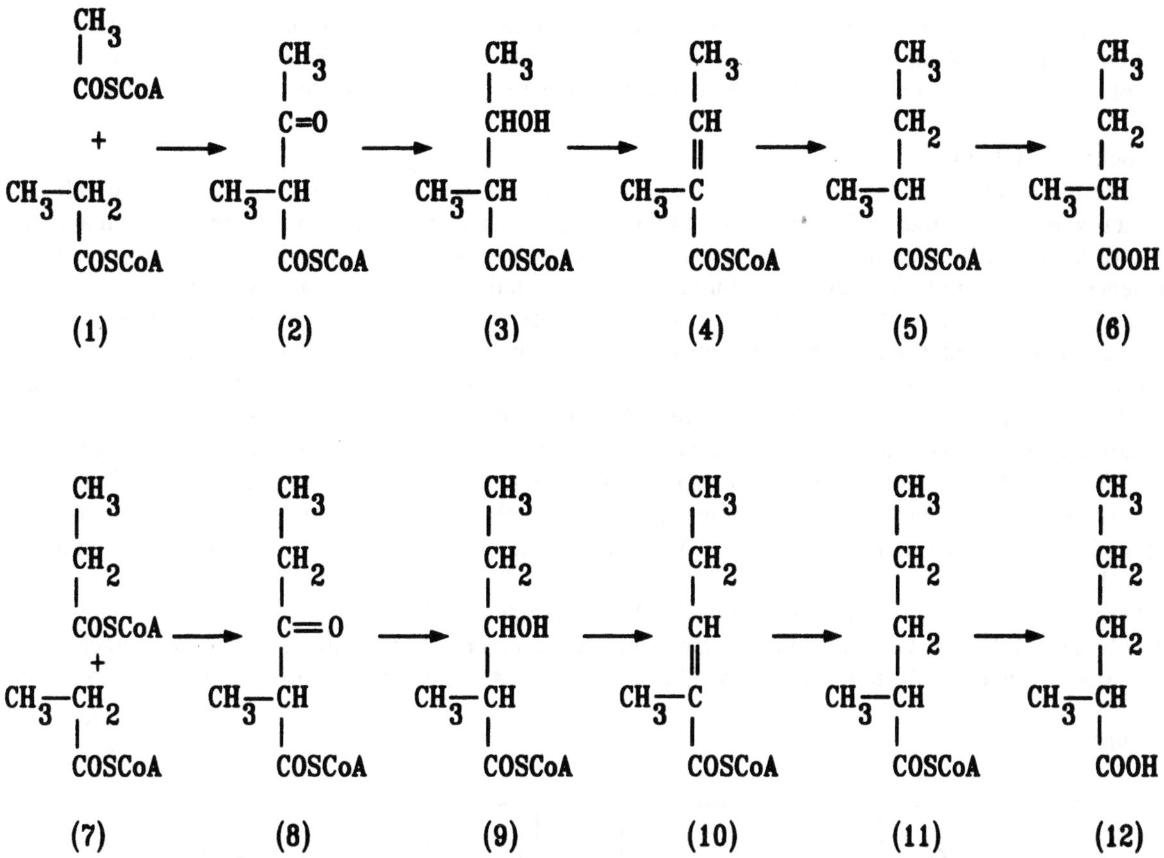

Fig. 2. Pathways of 2-methylbutyrate and 2 methylvalerate formation in *Ascaris*. Acetyl CoA and propionyl CoA (1) condense to give 2-methylacetoacyl CoA (2) which is then reduced by NADH to give 3-hydroxy-2-methylbutyryl CoA (3). 2-methylcrotonyl CoA (tiglyl CoA) (4) is formed by dehydration and then reduced by a flavoprotein dehydrogenase to 2-methylbutyryl CoA (5); this step may be associated with ATP production. 2-methylbutyrate (6) is formed by the removal of CoA either by CoA transferase or a hydrolase. The corresponding steps for 2-methylvalerate(12) formation are the condensation of 2 propionyl CoA (7), 3-keto-2-methylpentanoyl CoA (8), 3-hydroxy-2-methylpentanoyl CoA (9), 2-methyl-2-pentenoyl CoA (10) and 2-methylvaleryl CoA (11) [30–32].

may also be different in males and females, and can vary with age. There can also be tissue differences within the same parasite, and in *Ascaris*, for example, branched chain fatty acid synthesis is restricted to the muscles.

Phosphoenolpyruvate Branchpoint

Phosphoenolpyruvate occupies a central position in helminth energy metabolism (in some way analogous to acetyl CoA in mammals). In helminths phosphoenolpyruvate can either be metabolised by pyruvate kinase to pyruvate and hence to lactate, acetate or ethanol or else by phosphoenolpyruvate carboxykinase to succinate and so to propionate, propanol and short chain fatty acids. There is a general correlation between the major end-products of carbohydrate catabolism in helminths and their pyruvate kinase/phosphoenolpyruvate carboxykinase ratios. Parasites which rely primarily on glycolysis have ratios in the region of 2–10, whilst helminths which fix carbon dioxide have ratios of the order of 0.1–0.05.

Changes in the ratios of the different metabolic end-products are largely brought about by modulation of the branchpoint. A range of activators and inhibitors has been described for helminth pyruvate kinases, but phosphoenolpyruvate carboxykinase does not appear to be physiologically modulated [10].

Cytochrome Chains

Despite the essentially anaerobic nature of carbohydrate catabolism in parasitic helminths they all use oxygen when it is available. That is, at least in air, they have a measurable oxygen consumption and the rates of oxygen utilization are often comparable with those found in free-living organisms. Moreover, the presence of oxygen often results in quantitative and qualitative changes in metabolic end-products.

Despite earlier reports that cytochromes were absent from some parasitic helminths, when investigated in detail, they have all been shown to possess cytochromes and to be capable of oxidative phosphorylation.

Cytochromes corresponding to *b*, *c*, c_1 and a/a_3 have been demonstrated in a wide range of parasites. In addition, a number of helminths, including *A. lumbricoides, A. galli, M. expansa* and *F. hepatica* contain a carbon monoxide reactive haemoprotein, with an absorption spectrum similar to cytochrome *o* [12, 35]. In both *M. expansa* and *A. lumbricoides* the quinone component of the cytochrome chain is not, as in mammals, ubiquinone, but rhodoquinone (an O-demethylated derivative). Rhodoquinone has been isolated from a whole range of helminths, and is found either alone or in conjunction with ubiquinone [1, 19].

In parasites such as *Ascaris* or *Moniezia*, which have *o*-type cytochromes, the cytochrome chain appears to be branched (Fig. 3). One branch is similar to the classical mammalian sequence involving cytochromes *b*, c_1, and *c*, with a/a_3 as the terminal oxidase. Quantitatively, in *Ascaris*, the alternative pathway to cytochrome

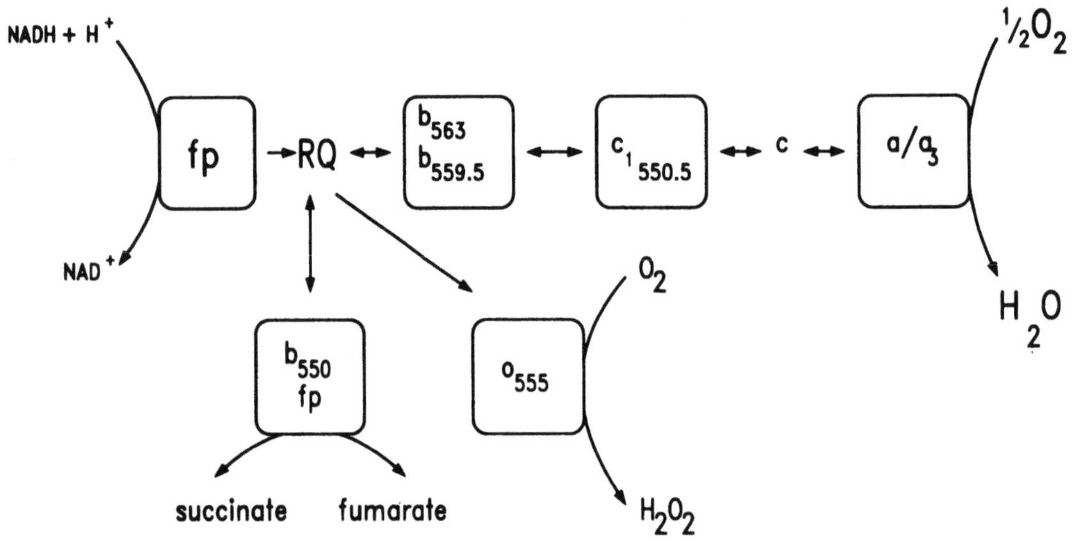

Fig. 3. Proposed cytochrome system in *Ascaris*. The position of non-haem iron components is uncertain. The exact relationship between cytochrome *o* and the classical chain is unclear and there may be a *b* cytochrome associated with the branch [5].

o is the major one, whilst the classical chain only contributes about 30 per cent of the total oxidase activity. The total cytochrome concentration in *Ascaris* mitochondria is about one half of that found in vertebrate muscle mitochondria, and the concentration of cytochrome a/a_3 about one tenth. Nevertheless, the concentration of cytochromes in helminth mitochondria is sufficient to account for the observed rates of oxygen uptake.

The details of the alternative cytochrome chain in helminths are unclear. There is probably at least one *b* cytochrome associated with the branch to cytochrome *o*, but whether the branch connects to the quinone or the *b* cytochrome of the classical chain is unknown. There is no evidence for oxidative phosphorylation associated with the alternative branch and the oxidase is specifically inhibited by radical scavenging agents such as salicylhydroxamic acid. The alternative oxidase is relatively insensitive to inhibition by cyanide and may be the site of cyanide insensitive respiration in helminths. Indeed, in a number of helminths (*A. lumbricoides*, *H. contortus*, *F. hepatica*, *Paramphistomum cervi*) cyanide has been reported to stimulate oxygen uptake.

When mitochondria from helminths which are thought to have a branched cytochrome chain oxidize substrates *in vitro* hydrogen peroxide is formed [25, 36, 38, 39]. It seems likely that hydrogen peroxide is formed by the alternative oxidase, but other sites may be involved [36]. Catalase activity is often very low in parasitic helminths (particularly in cestodes and digeneans). However, most helminths show peroxidase and superoxide dismutase activity.

Different species of helminths show considerable variation in the *in vitro* effects of oxygen on survival, motility, reproduction and development. Some helminths show a Pasteur effect, others do not (Table 5). The ability to accumulate an oxygen debt similarly varies from species to species [5]. Parasitic helminths can all survive anaerobically, often for long periods of time, but most, if not all, probably require at least some oxygen for normal growth and survival.

Table 5. The Pasteur effect in parasitic helminths

Pasteur effect	No Pasteur effect
Nematodes	
Ancylostoma caninum	*Ascaris lumbricoides (adult)*[a]
Ascaridia galli	*Dracunculus insignis*
Ascaris lumbricoides (egg)	*Trichuris vulpis*
Dirofilaria immitis	
Eustrongylides ignotus (juvenile)[a]	
Haemonchus contortus[c]	
Litomosoides carinii[b, d]	
Nippostrongylus brasiliensis[d]	
Digeneans	
Fasciola hepatica (juvenile)	*Dicrocoelium dendriticum*
Schistosoma mansoni (cercariae)	*Fasciola hepatica* (adult)
	S. mansoni (adult)[b,c]
	S. japonicum
Cestodes	
Hymenolepis diminuta[a]	*Echinococcus granulosus*
Schistocephalus solidus (plerocercoid)	
Taenia taeniaeformis	
Acanthocephalans	
Macracanthorhynchus hirudinaceus	

[a]show oxygen debt
[b]no oxygen debt
[c]require oxygen for egg production
[d]require oxygen for motility

Tricarboxylic Acid Cycle

As well as a cytochrome chain, all parasitic helminths seem to have a complete sequence of tricarboxylic acid cycle enzymes. However, the enzymes at the beginning of the cycle often have very low activities, NAD^+-linked isocitrate dehydrogenase is usually absent and aconitase can be difficult to detect. Citrate synthase, however, in contrast to the other enzymes in the first span of the tricarboxylic acid cycle, has a relatively high activity in parasitic helminths. In addition to its role in the tricarboxylic acid cycle, citrate synthase may also be involved as part of a shuttle transporting acetyl CoA across the mitochondrial membrane.

Carbon balance studies suggest that, in the majority of helminths, the tricarboxylic acid cycle accounts for no more than 10% of the total carbohydrate catabolised. However, since the complete aerobic catabolism of 1 mole of glucose yields 36 moles of ATP, compared with 2–6 moles for anaerobic catabolism, the complete oxidation of 10% of the total carbohydrate catabolised will in fact yield more ATP than the anaerobic catabolism of the remaining 90%. So despite its low activity in helminths, the tricarboxylic acid cycle could still be making a significant contribution to overall energy metabolism [4].

Larval Stages

In contrast to adult helminths, the major energy source in the free-living stages of parasites is usually lipid rather than carbohydrate. The free-living and intermediate stages of parasites generally have a normal aerobic metabolism with a functional tricarboxylic acid cycle. The change from free-living to parasitic stages during the life cycle is therefore, accompanied by major metabolic reorganization.

An interesting feature occurs in *Ascaris* eggs, where during development there is a net conversion of lipid to carbohydrate. This is brought about by a glyoxalate cycle, which is only active at this one stage of development. The glyoxalate cycle enzymes have also been found in a number of free-living and plant parasitic nematodes where they are involved in glyconeogenesis from ethanol and acetate [8].

Pentose Pathway

Although it can carry out the complete oxidation of glucose, the pentose pathway is not a major energy source. It is a multifunctional sequence whose main role is to provide NADPH and pentose sugars for synthetic reactions. It also provides a route for the degradation of pentoses. The complete pentose pathway has been demonstrated in several parasites and the first two enzymes of the cycle, glucose-6-phosphate dehydrogenase and 6-phosphogluconate dehydrogenase are known to be widely distributed.

Biosynthesis of Storage Polysaccharides

In parasitic helminths the main substrates for glyconeogenesis are dietary hexoses. Although parasitic helminths have all the enzyme systems necessary to carry out glyconeogenesis from glycolytic or tricarboxylic acid cycle intermediates, there is little evidence that they do so. The synthesis of glycogen from glucose in helminths is similar to that in mammals, starting with glucose-6-phosphate and proceeding via UDP-glucose. The physical and chemical properties of helminth glycogen, such as the lengths of external and internal branches are similar to those of mammalian glycogen.

The synthesis of trehalose in nematodes and acanthocephalans is the same as in insects, again involving UDP glucose. Mammals, however, are unable to synthesize trehalose.

LIPID METABOLISM

Lipids are not a single class of compounds but a heterogenous collection of molecules with similar physical properties, being relatively insoluble in water, but soluble in organic solvents. Lipids have a variety of functions, they are major components of cell membranes and occur in association with proteins and carbohydrates as lipoproteins and glycolipids. Lipid components are involved in the cytochrome chain, as carriers in glycosyltransferase reactions and, as prostaglandins and steroid and prenoid hormones, play an important role in metabolic regulation. Finally, in many organisms triacylglycerols are the major energy reserve.

Composition

Parasites usually contain between 1 and 10% of their dry weight as lipid, although this is often higher in infective stages. The tissue distribution of lipids in parasitic helminths has been extensively studied. In cestodes, lipid is found mainly in the parenchyma, although lipid droplets are often found in or around the excretory canals. When the ripe proglottids are shed, the lipid in the parenchyma is essentially excreted, since it is not used by the adults or the eggs. It is possible that the lipids in the proglottids may stimulate the intermediate host to eat it. There have been reports of a gradient in lipid content along the length of the strobila in cestodes and lipid content may also vary with age [7]. Compared with other helminths digeneans contain relatively little lipid and again lipid droplets are found in and around the excretory canals. In nematodes, lipid occurs in the lateral cords, muscles, intestine and reproductive organs. The intestine is a particularly important site of lipid storage in infective nematode larvae. Finally, in acanthocephalans most of the lipid occurs in the body wall, much of it in association with the lacunar canal system. A feature of lipid distribution in nematodes, cestodes and digeneans is that much of the lipid in the eggs occurs in the space between the embryo and the egg shell. When the eggs hatch, the lipid is discarded.

With the exception of ascarosides and some peculiar triacylglycerols, which are unique to ascarids, the different lipid classes found in parasitic helminths are the same as those found in free-living organisms. The relative amounts of the different lipid fractions are also fairly similar. The proportion of phospholipids in parasitic helminths is usually between 20 and 50% of the total lipids. As in free-living organisms, phosphatidylcholine and phosphatidylethanolamine are the major phospholipids in helminths, with smaller amounts of phosphatidylserine. Even though there are only relatively small amounts of nervous tissue in parasites, sphingolipids are found in significant amounts.

Some very unusual triacylglycerols have been found in ascarids. These lipids contain 2 moles of long chain fatty acids at 2 positions, but the third position is esterified with a short-chain fatty acid [48]. Small amounts of triacylglycerols containing 2 moles of short-chain fatty acid at 2 positions also occur. In *A. lumbricoides* the main short-chain fatty acids in the glycerides are 2-methylvalerate and 2-methylbutyrate with smaller amounts of *n*-valerate, acetate, propionate, isobutyrate and *n*-butyrate. The short-chain fatty acids are synthesized in the muscles, but are only incorporated into lipids in the eggs and ovaries. In the ovaries of *A. lumbricoides*, 80% of the triacylglycerols contain short-chain fatty acids, but no short-chain acids are found in the phospholipids. These peculiar lipids seem to function as an energy source during the early stages of egg embryonation.

The ascarosides (Fig. 4) are a series of unique alpha-glycosides which have been isolated from the eggs and ovaries of ascarid and oxyurid nematodes [47]. The only known function of these compounds is to form the extremely impermeable, inner lipoidal layer of the nematode egg-shell. The ascarosides can be grouped into three related classes: the monol ascarosides, the diol ascarosides and the diol diascarosides. In all cases the glycone (sugar) is the same, 3,6 dideoxy-L-arabinohexose (ascarylose), a sugar not found in any other eukaryote. The aglycones of the monol ascarosides are a mixture of long chain (25-32 carbon atoms) secondary alcohols, with the hydroxyl group on carbon 2. The aglycones of the diol and diol diascarosides are a mixture of long chain (29–33 carbon atoms) diols, with the hydroxyl groups on carbon 2 and (ω–1) giving a symmetrical molecule.

In the ovaries and unfertilized eggs the ascarosides occur almost entirely as esters, the two hydroxyl

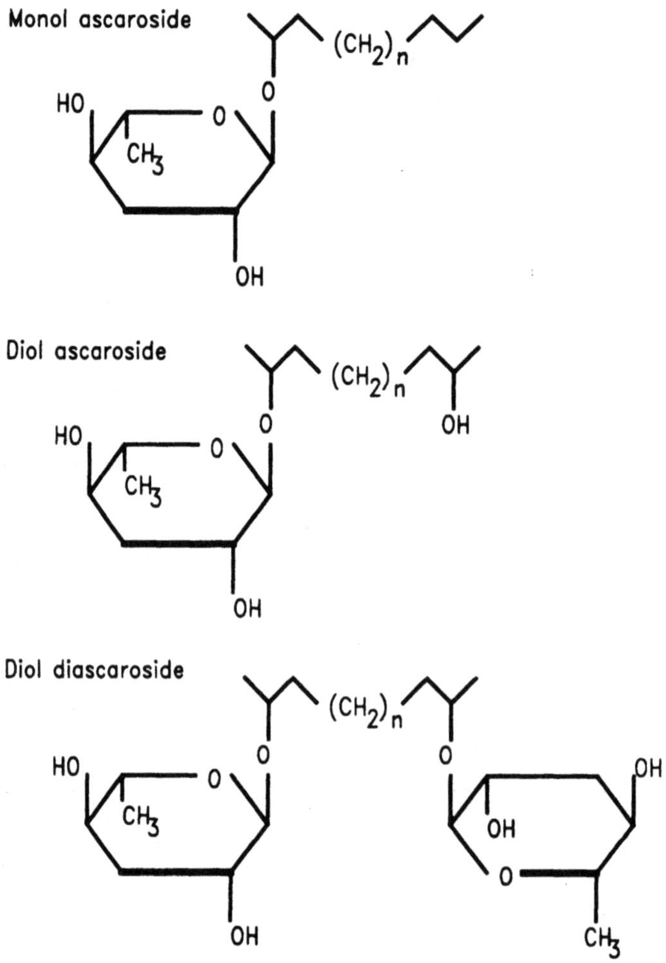

Fig. 4. The structure of ascarosides [5].

groups of ascarylose and the free hydroxyl group of the diol ascaroside aglycone being esterified, primarily with acetate. The ascaroside esters have a melting point of about 30°C, so at the temperature of the host, they are present in the egg as liquid lipid droplets. Following fertilization the ascarosides are de-esterified and the free ascarosides, which have a melting point of about 70°C condense to form an impermeable membrane. Interestingly, when infective ascaris eggs are stimulated to hatch, the first thing that happens is that the ascaroside membrane suddenly becomes permeable. This change in permeability may be enzymatic, but the mechanism is unknown.

Some other unusual features of parasite lipids are the often high levels of lysophosphoglycerides and the high levels of plasmalogens found in some species. Free-fatty acids are also found in unusually large amounts in some species, particularly in the larval stages where they may be involved in the processes which accompany infection. Finally, 80% of the lipids in the cystacanths of the acanthocephalan *Polymorphus minutus* are wax esters (esters of a long chain acid and a long chain alcohol).

There have been numerous studies on the fatty acid composition of whole lipid extracts and of isolated lipid classes from a range of helminths. As in other organisms different lipid classes have different and often characteristic fatty acid compositions. In general a similar range of fatty acids is found in the lipids of parasites and free-living organisms, the major fatty acids in both being C_{16} and C_{18}. In mammals C_{18} acids account for 50–60% of the total, although in some cestodes the percentage of C_{18} acids is 70–80 [7]. The proportion of unsaturated fatty acids in mammals is 50–60% (unsaturation index 0.7–0.8). Most parasitic helminths also fall within this range, but a few have markedly higher levels of unsaturated fatty acids: 80% in *M. moniliformis, M. hirudinaceus* and *Diphyllobothrium dendriticum*, 75% in *A. lumbricoides*.

Helminth lipids, like the lipids from other animals usually contain small amounts of odd numbered or branched chain fatty acids. However, in the adults of *Dirofilaria immitis, Dipetalonema* (= *Acanthocheilonema*) *viteae* and *L. carinii* branched chain fatty acids form 7.4 to 8.6% of the total. Prostaglandins are important derivatives of polyunsaturated C_{20} fatty acids. These potent pharmacological agents have so far been detected in nematodes and cestodes.

There is often a close similarity between an organism's fatty acid composition and the fatty acids present in its diet. This relationship is particularly marked in cestodes where there is usually a highly significant rank correlation coefficient between total fatty acid composition and host intestinal contents. The similarity is easily seen in shark tapeworms, since sharks characteristically contain large amounts of C_{20} and C_{22} polyunsaturated fatty acids. These unusual fatty acids are found in high concentrations in shark tapeworms and are incorporated into all lipid classes. Nevertheless, the incorporation of dietary fatty acids into parasite lipids is a regulated process and the helminths are able to maintain the physical properties of their different lipid components despite variations in diet.

The major sterol found in helminths is cholesterol and in the majority of helminths it is the only sterol present. Usually about one-third of the sterol fraction is present as sterol esters. Traces of plant sterols, presumably derived from the host's diet, have also been found. In many animals steroid derivatives have important hormonal functions. Beta-ecdysone and polar ecdysone derivatives have been isolated from a number of helminths including *A. lumbricoides, D. immitis, S. mansoni, M. expansa* and *E. granulosus* [33]. However, the physiological role of ecdysteroids in helminths has not been established. A number of mammalian steroid hormones such as testosterone and progesterone have been isolated from helminths; these are probably derived from the host and again no physiological role is known for them in the parasite. Steroids may, however, be an important component of the sex pheromones produced by nematodes and digeneans.

Biosynthesis

Parasitic helminths readily incorporate labelled precursors into all of their neutral and phospholipid classes although different lipids become labelled to different extents, reflecting different intrinsic rates of turnover [20]. In the few instances where the pathways of lipid synthesis have been studied in helminths they appear to be similar to those found in mammals, for example the phosphatitidic acid diacylglycerol pathway for triacylglycerol synthesis and the synthesis of phosphatidylcholine from GDP-choline and 1,2-diacylglycerol.

Despite being able to synthesize complex lipids from simple precursors, parasitic helminths all appear unable to synthesize long chain fatty acids or the steroid nucleus *de novo*, nor can they desaturate preformed fatty acids. In mammals fatty acid desaturation requires molecular oxygen and involves the microsomal oxygenase system. In helminths fatty acid synthesis seems to be limited to simple chain lengthening by the sequential addition of acetyl CoA. The rat tapeworm *H. diminuta* can, for example, elongate $C_{16:0}$ and $C_{18:0}$ (but not $C_{18:1}$ or $C_{18:2}$) to C_{20}, C_{22}, C_{24} and C_{26} acids. The mechanism of chain lengthening in helminths in unknown, but may be similar to the mammalian mitochondrial fatty acid elongation system.

Parasitic helminths, therefore, have a nutritional requirement for exogenous long chain unsaturated fatty acids and sterols. They do, however, have the ability to metabolise dietary sterols and nematodes can reduce the double bond at C-5, carry out C-24 dealkylation and methylate the sterol nucleus at C-4 [42]. There are also reports of C-17 hydroxylation. Despite the absence of steroid synthesis, helminths can synthesize a range of polyisoprenoids, including farnesol and dolichols from acetate or mevalonate [28]. Farnesol and related compunds are juvenile hormone mimics and it has been suggested that farnesol might have a similar hormonal function in parasites. The development of parasites in *in vitro* culture is affected by high concentrations of juvenile hormone and parasite extracts show juvenile hormone activity in insects. However, experiments designed to demonstrate a hormonal role for farnesol in parasites have so far been inconclusive.

Steroid synthesis, like fatty acid desaturation, has an absolute requirement for molecular oxygen, the cyclization of squalene via squalene epoxide to give lanosterol being catalysed by a microsomal mono-oxygenase. Parasitic helminths seem to lack microsomal oxidase systems [37] and this may be related to their low oxygen environment. However, in helminths the block to sterol synthesis appears to be in squalene formation rather than in its cyclization. The inability to synthesize sterols may not necessarily be related to the parasitic mode of life, but be a peculiarity of helminths as a whole. The ability to synthesize sterols shows no clear cut phylogenetic pattern and sterol synthesis is lacking in platyhelminths, nematodes, acanthocephalans, arthropods, molluscs, echinoderms and terrestrial annelids. The inability to synthesize long chain unsaturated fatty acids may similarly be a general feature of platyhelminths, both free-living and parasitic. However, free-living nematodes are capable of synthesizing long chain fatty acids *de novo*.

Catabolism

In mammals, complex lipids are broken down by lipases to yield fatty acids, glycerol and the various polar groups from phospholipids. The released fatty acids are then oxidized by the beta-oxidation sequence. Non-specific esterases, capable of hydrolysing lipids have been demonstrated histochemically, in a wide range of parasites. However, there is no good evidence that any adult parasitic helminth can catabolise its lipid stores, nor is there any evidence for an active beta-oxidation sequence.

Despite the absence of a functional beta-oxidation sequence, the beta-oxidation enzymes have been demonstrated in a range of parasitic helminths, including *F. hepatica*, *A. lumbricoides* and *H. diminuta*. The role of the beta-oxidation enzymes in adult parasites is unclear, they may be primarily associated with the eggs or larval stages or they may be involved in chain elongation of fatty acids by condensation with acetyl CoA. Unlike the inability to synthesize sterols, the lack of a functional beta-oxidation sequence is a feature of the adult parasite only. Free-living nematodes and platyhelminths, and the free-living stages of parasitic nematodes and platyhelminths all have active beta-oxidation sequences and catabolise endogenous lipids. The absence of a functional beta-oxidation sequence in adult parasites, despite the presence of the enzymes, may be related either to the relative unimportance of the tricarboxylic acid cycle in parasitic helminths, or to the low environmental pO_2. The absence of an active tricarboxylic acid cycle would restrict the further metabolism of the large amounts of acetyl CoA produced by beta-oxidation. Beta-oxidation also produces large amounts of reducing power (NADH, reduced flavoprotein) which requires reoxidation via an oxidase system. The beta-oxidation system can, therefore, only function aerobically.

Digeneans excrete significant amounts of lipid, mostly via the excretory canal system. In *F. hepatica* lipid excretion can amount to as much as 2% of the wet weight per day. All classes of lipid appear to be excreted, cholesterol, cholesterol esters, triacylglycerols, free fatty acids and phospholipids, suggesting a general loss of lipid rather than specialised secretion [5].

The absence of lipid catabolism in adult parasites poses the problem as to why parasites accumulate lipid when they are unable to catabolise it. Some of the lipid may be destined for incorporation into the eggs, since the free-living stages unlike the adults can catabolise lipids. Alternatively parasites may have to absorb large amounts of lipid in order to get enough of a particular fatty acid or fat soluble vitamin and the excess lipid is stored rather than being excreted again. Changes in lipid content in parasites are often correlated with the onset of the host's immune response and lipid accumulation may be related to the retention of potentially immunogenic compounds.

METABOLISM OF NITROGENOUS COMPOUNDS

Nitrogen compounds, particularly in the amine form, are highly reactive chemically, an essential attribute for the formation of peptide links and other co-valent bonds in metabolic reactions. The principal nitrogen containing compounds in cells are the amino acids and their derivatives (peptides, proteins, porphyrins, amines) and the purines and pyrimidines and their derivatives (nucleosides, nucleotides, nucleic acids).

Amino Acids

There are some twenty common amino acids, in addition there are over one hundred and fifty known non-protein amino acids which occur in cells, compounds such as taurine, ornithine and citrulline. Amino acids occur in the tissues either as free-amino acids or as components of proteins or peptides. *In vivo*, proteins are constantly turning over, with half-lives ranging from minutes to several months.

There have been numerous studies on the total amino acid composition of different helminths. In practically all cases the majority of the standard amino acids have been found and the occasional absence of a particular amino acid in one or two of the analyses may be due to differences in preparation or analytical technique. In general there are usually no qualitative differences between the total amino acid composition of different species and total amino acid composition has not proved particularly useful as an aid to taxonomy. However, the discovery of unusual amino acids during the analysis of tissue extracts can indicate the presence of novel enzymes or pathways.

Compared with the protein amino acids, the free amino acid fraction in tissues is small and is probably present as several independent pools. Parasitic helminths show a wide range of free amino acid levels ranging from 0.1 to 2.6% of the fresh weight, with the highest levels being found in marine monogeneans [3]. The free amino acid pools of parasitic helminths are, in general, similar to those of free-living invertebrates, being dominated by a few non-essential amino acids (usually alanine or proline). Unfortunately, with the exception of *A. lumbricoides* and *A. galli* there are no data on the amino acid pools in different tissues in helminths, nor on how the amino acid pools vary under different physiological conditions. Similarly, nothing is known about the rate of turnover of individual free-amino acids. There is some evidence in parasitic helminths that, as in free-living invertebrates, free-amino acids may have an osmotic role [34].

Animals differ greatly in their ability to synthesize amino acids. The rat can synthesize ten out of the twenty common amino acids, the remaining ten (arginine, histidine, isoleucine, leucine, lysine, methionine, phenylalanine, threonine, tryptophan, valine) are essential amino acids for this species. A survey of the amino acid requirements of animals suggests that they all require, as a minimum, the ten essential amino acids. The few studies which have been carried out with parasitic helminths suggest that they require, in addition to the ten essential amino acids, a dietary source of tyrosine.

The different amino acids are synthesized by multi-enzyme sequences, and as is usual in metabolism the synthetic pathways are separate from the degradative ones. Some of the intermediates in amino acid breakdown are important precursors for other reactions. In general, the non-essential amino acids (alanine, aspartate, asparagine, cysteine, glutamate glutamine, glycine, proline, hydroxyproline, serine, tyrosine) have short synthetic pathways (less than five steps) whilst the essential amino acids have long synthetic pathways (five to ten steps).

Amino acid synthesis, and degradation, in helminths is a relatively neglected area and the information available is often of uneven or uncertain quality. Helminths seem to be able to synthesize the non-essential amino acids by pathways similar to those found in mammals. Reports that parasitic helminths can synthesize essential amino acids have not been substantiated.

Amino acids are not an important energy source in parasitic helminths, only in digenean rediae and sporocysts and in some plant parasitic nematodes, there is evidence for appreciable amino acid catabolism [41]. However, parasitic helminths all appear able to catabolise amino acids by pathways which again appear to be identical with those found in mammals. Despite the absence of microsomal oxidases, parasitic helminths have been shown to be able to carry out a number of oxidative reactions associated with amino acid metabolism, including cysteine dioxygenase, proline hydroxylase, tyrosine hydroxylase and tryptophan hydroxylase.

There are, however, differences in detail between the pathways of amino acid metabolism in helminths and mammals, particularly in the metabolism of the sulphur amino acids and of arginine and proline. Cystathionine β-synthase is a multifunctional enzyme which catalyses the synthesis of cystathionine from homocysteine and serine. The enzyme from nematodes is also able to catalyse the synthesis of hydroxy-thioethers from cysteine and an hydroxyl containing thiol compound. This reaction does not occur with the mammalian enzyme.

The major pathway for the catabolism of arginine is hydrolysis to ornithine (via arginase), followed by transfer of the 5-amino group to a suitable acceptor, usually 2-oxoglutarate, to give glutamic-γ-semialdehyde, which spontaneously cyclizes to Δ^1-pyrroline-5′-carboxylic acid. In mammals the latter is oxidized to glutamate via a dehydrogenase. However, in *F. hepatica* (Fig. 5) the dehydrogenase is absent and Δ^1-pyrroline-5′-carboxylate is instead reduced to proline [26]. Almost all helminths have been found to excrete some proline, but digeneans in particular produce large amounts. *In vitro*, adult *F. hepatica* produce up to 50 mg/100 g wet weight h^{-1} proline and the levels of biliary proline increase up to 10,000 fold in animals with fascioliasis. This increase may not be solely due to excreted proline from the parasite as there are indications that proline catabolism in the host is also impaired during infection. There is, however, considerable evidence that proline production may be responsible for the bile duct hyperplasia found in fascioliasis and may also cause anaemia. The excretion of proline by the adults and eggs of schistosomes has also been implicated in causing bile duct hyperplasia and fibrosis around egg granuloma in infected hosts [27].

Amino Acid Derivatives

Helminths are capable of rapid protein synthesis and the incorporation of labelled amino acids into proteins has been demonstrated in a wide variety of species. Although it has not been investigated in detail, protein synthesis in helminths seems similar to that in mammals and other organisms. Cell free-systems capable of synthesizing proteins have been prepared from helminths and helminth *m*RNA has been translated in a number of heterologous systems (mammalian *m*RNA has, however, not yet been translated in a helminth system).

A range of what appear to be vertebrate peptide hormones have been demonstrated immunocytochemically in helminth tissues. The parasite peptides show cross reactivity with FMRF-amide, substance P, gastrin, gastrin releasing peptide, cholecystokinin, bovine pancreatic polypeptide, neurotensin and several others [24]. However, so far, only one of the peptides, an FMRF-amide like peptide called AF1 from *A. lumbricoides* has actually been isolated and sequenced [14].

A range of physiologically active amines has been found in helminths, including 5-hydroxytryptamine, dihydroxyphenylalanine, dopamine, adrenalin, nor-adrenalin, histamine, octopamine and tyramine. Serotonin is synthesized from tryptophan in two steps, the amino acid is first hydroxylated to give 5-hydroxytryptophan

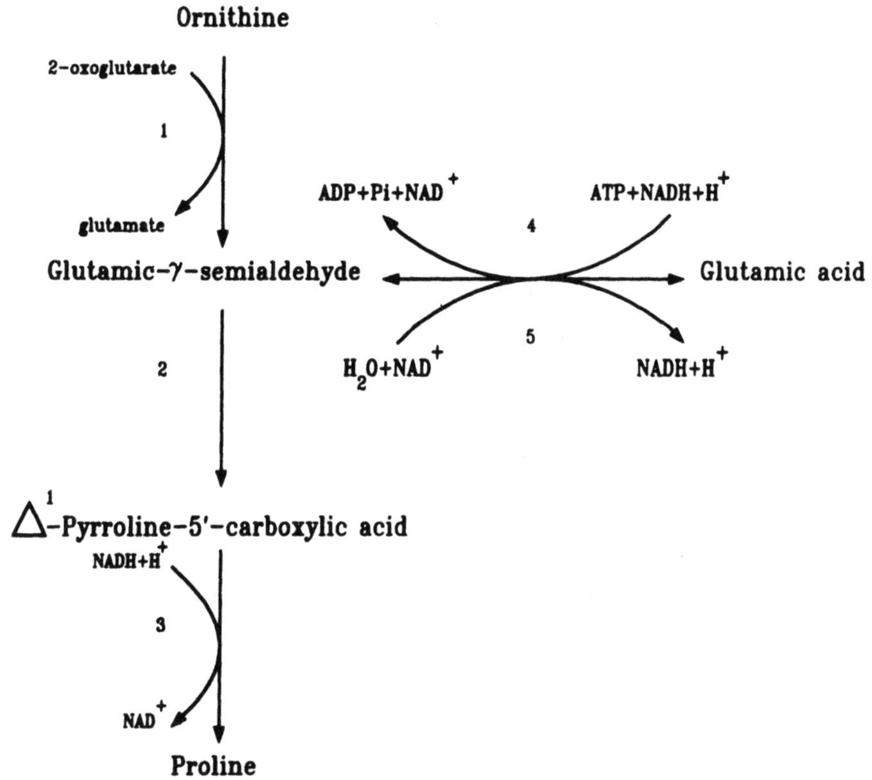

Fig. 5. Proline metabolism in *Fasciola hepatica*: (1) ornithine transaminase: (2) spontaneous reaction; (3) pyrroline-5' carboxylate reductase; (4) glutamate kinase + glutamate dehydrogenase; (5) Δ^1–pyrroline-5'-carboxylate dehydrogenase [25].

which is then decarboxylated. Despite indications that tryptophan hydroxylase might be absent from helminths, it has now been demonstrated in *A. lumbricoides* [11]. However, parasitic platyhelminths have specialised transport mechanisms to enable them to take up 5-hydroxytryptophan from their hosts.

In many cases the physiological amines and the bioactive peptides have been located in the nervous system where the may be functioning as neurotransmitters or neurohormones. So, although the nervous systems of parasitic helminths are structurally simple, biochemically they may be very complex.

The synthesis of porphyrins, for which glycine is a major precursor, is an important pathway in mammals because of the role of the porphyrin nucleus in haem compounds. However, it is doubtful if parasitic helminths can synthesize porphyrins from simple precursors and all seem to require a dietary source of haem. But given a source of protoprophyrin IX, helminths are able to incorporate iron into the ring to synthesize haem and can incorporate the haem nucleus into globin to give haemoglobin. Haemoglobin has been found in nematodes, digeneans and monogeneans, but not in cestodes or acanthocephalans. Where they have been examined in detail, parasite haemoglobins are always distinct from those of their host. The most interesting feature of parasite haemoglobins is their unusually high affinity for oxygen. The perienteric fluid haemoglobin from *A. lumbricoides* has a P_{50} O_2 of 0.0015 mm Hg, the highest oxygen affinity known for any haemoglobin. The extremely high oxygen affinity of *A. lumbricoides* haemoglobin makes it unlikely that it could function in oxygen transport or as an oxygen reserve.

Polyamine Metabolism

Putrescine and the polyamines spermidine and spermine play an important role in the regulation of cellular growth and differentiation. Parasitic helminths seem, however, to lack ornithine decarboxylase, the first enzyme in the polyamine biosynthetic pathway [50]. Putrescine, spermidine and spermine are taken up from the host and interconverted in the parasite. Excess polyamines are excreted as N-acetyl derivatives.

Purines and Pyrimidines

Purines and pyrimidines are precursors of DNA and RNA as well as a wide variety of other compounds including NAD^+, $NADP^+$, ATP, GTP, UTP, CTP and ITP. Six of the nine carbon atoms of purines and four of the six carbon atoms of pyrimidines are derived from amino acids. Purine and pyrimidine nucleotides may be synthesized *de novo* or via salvage pathways, which convert the free-bases back into nucleosides and nucleotides.

Adult *S. mansoni* are unable to synthesize purines *de novo* and they rely entirely on exogenous purines and salvage pathways for their purine nucleotide requirements [13]. In schistosomes hypoxanthine appears to be a major source of purine (Fig. 6). By contrast, nematodes are capable of *de novo* purine synthesis, although salvage pathways may still play an important role [52].

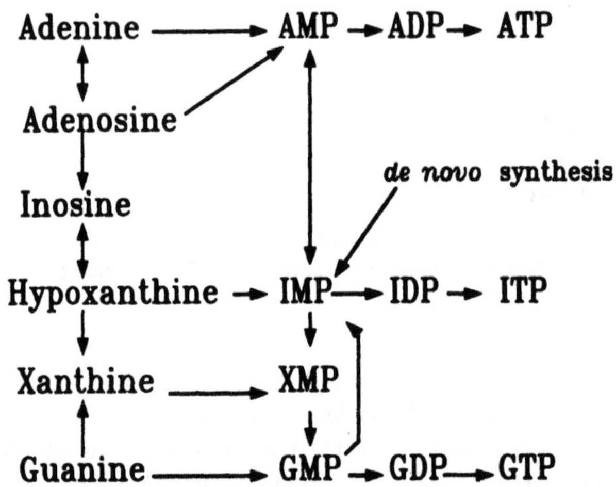

Fig. 6. Purine metabolism in helminths [5].

It was initially thought that *de novo* pyrimidine synthesis was also lacking in parasitic helminths. However, *de novo* pyrimidine synthesis has now been demonstrated in nematodes and digeneans and a pyrimidine specific carbamoyl-phosphate synthetase (CPSase II) has been purified from *A. lumbricoides* and *S. mansoni*. The properties of helminth CPSase II are similar to those of mammals and there is evidence that, as in mammals, the enzymes of pyrimidine synthesis are present as a multi-enzyme complex [2]. In *A. lumbricoides* it has been estimated that the ratio of *de novo* pyrimidine synthesis to salvage pathway is 1 : 10, whilst in *S. mansoni* it is 1 : 2. This can be correlated with the low pyrimidine levels in blood as apposed to tissues.

Deoxyribonucleotides are synthesized by the reduction of the 2′ carbon of the corresponding ribonucleotide. There is evidence that helminths can reduce the 2′ carbon of ribose, but the mechanism is unknown.

Purine and pyrimidine biosynthesis, like other synthetic pathways is under tight metabolic control and activity may be inhibited by exogenous nucleotides. Little is known about the synthetic pathways in the free-living stages of parasites. It is possible that in synthesis, like catabolism, pathways which appear to be absent in the adult may be active at the other stages of the life cycle.

Excretion of Nitrogenous Compounds

About 90% of excretory nitrogen comes from the alpha amino group of amino acids, the remainder from the breakdown of purines and pyrimidines. In parasitic helminths the major excretory product is ammonia, with small quantities of urea, uric acid, amino acids and amines.

The amino groups from the different amino acids are collected via transaminases into glutamate. The enzyme glutamate dehydrogenase then catalyses the oxidative deamination of glutamate to release ammonia:

$$\text{L-glutamate} + H_2O + NAD(P)^+ \rightarrow \text{2-oxoglutarate} + NH_4^+ + NAD(P)H$$

The glutamate dehydrogenase from mammals and parasites is an allosteric enzyme and its activity is modified by a variety of effectors. About 2–10% of the total nitrogenous end-products in helminths is urea, but the mechanism of formation is unclear and it is doubtful if a classical urea cycle is present in any of the parasitic species. Two of the enzymes of the urea cycle, arginase and ornithine carbamoyl transferase are widely distributed in helminths, but they have functions other than in urea production.

The small quantities of uric acid produced by helminths probably come from the breakdown of purines. Uric acid production is higher in tissue parasites, presumably because of the high nucleic acid content of their diet.

Most of the helminths so far studied have been found to excrete significant amounts of nitrogen in the form of amino acids, amino nitrogen constitutes 35% of the nitrogenous end-products of *Nematodirus*, 49% in *T. spiralis* and 28% in *H. diminuta*. Amino acid excretion is widespread in free-living invertebrates as well as parasites, but its significance is uncertain. Much of the amino nitrogen excreted by helminths probably comes from the partial products of digestion, whilst some may not be a normal excretory product, but are the result of leakage from moribound worms. However, in digeneans and cestodes some, at least, of the amino acids are excreted by the protonephridial system and may represent true excretory products. Excreted amino acids may also have a signalling function. Arginine released from the post-acetabular glands of *S. mansoni* cercariae stimulate the attachment response [23], whilst peptide transfer occurs between male and female *S. mansoni*.

In addition to amino acids, some helminths also excrete amines. Both adult and larval *T. spiralis* excrete a range of amines including methylamine, ethylamine, *n*-propylamine, *n*-butylamine, *n*-pentylamine, *n*-hexylamine, *n*-heptylamine, ethylene diamine, 1,5-pentanediamine, *iso*-propylamine, *sec*-butylamine, *iso*-butylamine, 1-amino 2-propanol and ethanolamine. Amine excretion has also been described in *A. lumbricoides*, *N. brasiliensis* and *T. taeniaeformis*. The origin of these amines is obscure, some can be formed by the decarboxylation of the appropriate amino acid, others must involve specialised, and as yet, unknown pathways. A possible function of amine production in helminths may be to help neutralize the acidic end-products from carbohydrate breakdown. Amines also disrupt the keratin layer of the epidermis and so could aid the penetration of infective larvae.

NUCLEIC ACIDS

Helminths like all other living organisms can synthesize DNA and RNA, often at high rates. However, there have been no detailed studies on the mechanisms of nucleic acid synthesis in helminths.

Composition

There is no evidence that the DNA of parasites is in any way peculiar. No novel bases have been discovered, and there is nothing to suggest that the organisation of the genome is unusual. The mitochondrial

DNA from helminths is likewise unremarkable. Interestingly, however, parasitic nematodes and platyhelminths appear to have more complex genomes (in terms of their information content) than their free-living relatives [44]. This is presumably a reflection of the complex life cycle of most parasites.

Chromatin Diminution

During development, the eggs of ascarid nematodes undergo a dramatic chromosomal alteration called chromatin diminution. In the cells destined to form the soma (presomatic blastomeres), the chromosomes disintegrate into small fragments and large pieces of telomeric heterochromatin are shed. The latter fail to move during cell division and are eliminated from the daughter nuclei. Cells in the germ line lineage do not undergo chromatin elimination and maintain their zygote chromosome complement [45]. It has been shown that the eliminated chromatin constitutes about 83% of the total genomic DNA in *Parascaris equorum*, and is made up of higly repeated sequences. Some copies of these sequences are, however, retained in the somatic nuclei, so the coding capacity of the somatic genome is probably the same as in the germ line.

Recombinant DNA Technology

There has been considerable progress in recent years in applying the techniques of molecular biology to helminth parasites. The starting point for recombinant DNA technology is the construction of a 'gene library' and this has been done for several helminth species including *S. mansoni* and *O. volvulus*. Fragments of genomic DNA (containing both exons and introns) can be prepared from extracted DNA by mechanical shearing or by the use of restriction endonucleases. Alternatively specific DNA sequences can be synthesized chemically or synthesized biochemically from extracted messenger RNA (*c*DNA). The DNA fragments are inserted into an appropriate vector which is then introduced into a suitable expression system such as *E. coli*. By screening the progeny, cell lines expressing different parasite genes can be isolated and cloned. Cloning genomic DNA fragments is used to study gene function and control, and for sequence analysis. *c*DNA clones are primarily useful for the production of foreign proteins. Using these techniques with *S. japonicum*, the gene coding for an antigen which appeared to be a specific target for the immune response in resistant mice has been isolated and sequenced [46]. The sequence analysis showed a close homology with the enzyme glutathione transferase and purified parasite glutathione transferases have shown some promise in vaccination experiments (see Chapters 10 and 12).

CONCLUSIONS

Although relatively few species of parasite have been looked at, it is possible to make some general conclusions. The metabolism of helminth parasites is not coupled to that of the host as it is in viruses. However, compared with free-living animals, parasitic helminths show considerable reductions in both catabolic and anabolic pathways.

Adult parasitic helminths appear to have an absolute dependency on carbohydrate for their energy metabolism and produce a bewildering array of end-products. Energy metabolism would apear to be essentially anaerobic, although all helminths have cytochrome chains and appear to be capable of oxidative phosphorylation. The relative contribution of aerobic processes to the overal energy balance in helminths is uncertain, nor is it known why anaerobic metabolism persists in helminths when oxygen is available.

Adult helminths lack what would apear to be a number of basic synthetic pathways, such as the ability to synthesize unsaturated fatty acids, sterols, porphyrins and in some cases purines. This has been compensated for by the development of extremely efficient uptake mechanisms and helminths are capable of extremely high rates of growth and gamete production.

Although adult parasites lack a number of the anabolic and catabolic pathways found in free-living organisms, these pathways are often present in the free-living and intermediate stages of the life cycle. So the genetic information necessary to synthesize, for example the beta-oxidation sequence, is still present in the adult, but not expressed. This has been described as epigenetic loss of function [16].

The biochemical and morphological adaptations found in parasitic helminths are all related to different features of the parasites' environment. The interaction of the parasite with its environment is responsible for tissue migration and the development of host specificity. Few, if any, of the biochemical adaptations found in parasites are unique. Anaerobic pathways of carbohydrate catabolism involving carbon dioxide fixation are found in many free-living invertebrates, including molluscs and annelids and the ability to take up nutrients through the body surface occurs in many marine organisms. Even the extensive metabolic switches which take place during the parasite's life cycle have parallels in the metamorphosis of insects and amphibians.

REFERENCES

1. Allen, P.C. Helminths: comparison of their rhodoquinone. Exp. Parasitol. **34**: 211–219 (1973).
2. Aoki, T., Oya, H., Mori, M. and Tatibana, M. Control of pyrimidine biosynthesis in the *Ascaris* ovary: regulatory properties of glutamine-dependent carbamoylphosphate synthetase and copurification of the enzyme with aspartate carbamoyltransferase and dihydroorotase. Mol. Biochem. Parasitol. **1**: 55–68 (1980).
3. Arme, C. Amino acids in eight species of Monogenea. Z. Parasitenkunde **51**: 261–263 (1977).
4. Barrett, J. Energy metabolism in nematodes. In : *The organization of nematodes* (N.A. Croll ed), pp. 11–70. Academic Press: London (1976).
5. Barrett, J. *Biochemistry of parasitic helminths*. Macmillan: London (1981).
6. Barrett, J.Biochemistry of filarial worms. Helm. Abstr. **52**: 1–18 (1983).
7. Barrett, J. Lipid metabolism. In: *Biology of the eucestoda* (C. Arme and P.W. Pappas eds), Vol. **2**, pp. 391–419. Academic Press: London (1983).
8. Barrett, J. The anaerobic end-products of helminths. Parasitology **88**: 179–198 (1984).
9. Barrett, J. and Beis, I. The redox state of the free nicotinamide adenine dinucleotide couple in the cytoplasm and mitochodria of muscle tissue from *Ascaris lumbricoides* (Nematoda). Comp. Biochem. Physiol. **44A**: 331–340 (1973).
10. Bryant, C. The regulation of respiratory metabolism in parasitic helminths. Adv. Parasitol. **16**: 311–331 (1978).
11. Chaudhuri, J., Martin, R.E. and Donahue, M.J. Tryptophan hydroxylase and aromatic L-amino acid decarboxylase activities in the tissues of adult *Ascaris suum*. Int. J. Parasitol. **18**: 341–346 (1988).
12. Cheah, K.S. Properties of *Ascaris* muscle mitochondria l. Cytochromes. Biochim. Biophys. Acta **387**: 107–114 (1975).
13. Coles, G.C. Recent advances in schistosome biochemistry. Parasitology **89**: 603–637 (1984).
14. Cowden. C., Stretton, A.O.W. and Davis, R.E. AFl, a sequenced bioactive neuropeptide isolated from the nematode *Ascaris suum*. Neuron **2**: 1465–1473 (1989).
15. Fairbairn, D. Trehalose and glucose in helminths and other invertebrates. Can J. Zool. **36**: 787–795 (1958).
16. Fairbairn, D. Biochemical adaptation and loss of genetic capacity in helminth parasites. Biol.Rev. **45**: 29–72 (1970).
17. Fioravanti, C. Mitochondrial malate dehydrogenase, decarboxylating (malic enzyme) and transhydrogenase activities of adult *Hymenolepis microstoma* (Cestoda). J. Parasitol. **68**: 213–220 (1982).
18. Fioravanti, C.F. and Kim, Y. Phospholipid dependence of the *Hymenolepis diminuta* mitochondrial NADPH; NAD transhydrogenase. J. Parasitol. **69**: 1048–1054 (1983).
19. Fioravanti, C.F. and Youngee, K. Rhodoquinone requirement of the *Hymenolepis diminuta* mitochondrial electron transport system. Mol. Biochem. Parasitol. **28**: 129–134 (1988).
20. Frayha, G.J. and Smyth, J.D. Lipid metabolism in parasitic helminths. Adv. Parasitol. **22**: 310–387 (1983).
21. Fry, M. and Brazeley, E.P. NADH-fumarate reductase and succinate dehydrogenase activities in mitochondria of *Ascaridia galli* and *Nippostrongylus brasiliensis*. Comp. Biochem. Physiol. **77B**: 143–150 (1984).

22. Ginger, C.D. The biochemistry of *Nippostrongylus* and other nematodes. Symp. British Soc. Parasitol. **7**: 17–30 (1969).
23. Granzer, M. and Haas, W. The chemical stimuli of human skin surface for the attachment response of *Schistosoma mansoni* cercariae. Int. J. Parasitol. **16**: 575–579 (1986).
24. Gustafson, M.K.S. Cestode neurotransmitters. Parasitol. Today **1**: 72–75 (1985).
25. Hayashi, E. and Terada, M. The influence of oxygen presure on the survival time of *Ascaris lumbricoides suum*. 4. The electron transport system containing cytochrome *o* peroxidase in *Ascaris* muscle. Jpn. J. Parasitol. **22**: 1–12 (1973).
26. Isseroff, H. and Ertel, J.C. Proline in fascioliasis: III. Activities of pyrroline-5-carboxylic acid reductase and pyrroline-5-carboxylic acid dehydrogenase in *Fasciola*. Int. J. Parasitol. **6**: 183–188 (1976).
27. Isseroff, H., Bock, K., Owczarek, A. and Smith, K.R. Schistosomiasis: proline production and release by ova. J. Parasitol. **69**: 285–289 (1983).
28. Johnson, W.J. and Cain, G.D. Biosynthesis of polyisoprenoid lipids in the rat tapeworm *Hymenolepis diminuta*. Comp. Biochem. Physiol. **82B**: 487–495 (1985).
29. Köhler, P., Bryant, C. and Behm, C.A. ATP synthesis in a succinate decarboxylase system from *Fasciola hepatica* mitochondria. Int. J. Parasitol. **8**: 399–404 (1978).
30. Komuniecki, R., Campbell, T. and Rubin, L. Anaerobic metabolism in *Ascaris suum*: acyl CoA intermediates in isolated mitochondria synthesizing 2-methyl branched-chain fatty acids. Mol. Biochem. Parasitol. **24**: 147–154 (1987).
31. Komuniecki, R. Komuniecki, P.R. and Saz. H.J. Pathway of formation of branched-chain volatile fatty acids in *Ascaris* mitochondria. J. Parasitol. **67**: 841–846 (1981).
32. Komuniecki, R., Rioux, A. and Thissen, J. NADH-dependent tiglyl-CoA reduction in disrupted mitochondria of *Ascaris suum*. Mol. Biochem. Parasitol. **10**: 1–10 (1984).
33. Mercer, J.G., Munn, A.E. and Rees, H.H. *Echinococcus granulosus*: occurrence of ecdysteroids in protoscoleces and hydatid cyst fluid. Mol. Biochem. Parasitol. **24**: 203–214 (1987).
34. Ossikovski, E. Comparative studies of the proteins and amino acids and metabolism of helminths. V. Influence of osmotic pressure on the free-amino acid content of adult *Ascaridia galli* (Schrank, 1788) Freeborn, 1923. Khelmintologiya **17**: 41–51 (1984).
35. Paget, T.A., Fry, M. and Lloyd, D. Haemoprotein terminal oxidases in the nematodes *Nippostrongylus brasiliensis* and *Ascaridia galli*. Biochem. J. **256**: 295–298 (1988).
36. Paget, T.A., Fry, M. and Lloyd, D. The O_2-dependence of respiration and H_2O_2 production in the parasitic nematode *Ascaridia galli*. Biochem. J. **256**: 633–639 (1988).
37. Precious, W.Y. and Barrett, J. The possible absence of cytochrome P–450 linked xenobiotic metabolism in helminths. Biochim. Biophys. Acta **992**: 215–222 (1989).
38. Prichard, R.K. and Schofield, P.J. *Fasciola hepatica*: cytochrome c oxidoreductase and effects of oxygen tension and inhibitors. Exp. Parasitol. **29**: 215–222 (1971).
39. Raj. R.K., Puranam, R.S., Kurup, C.K.R. and Ramasarma, T. Oxidative activities in mitochondria-like particles from *Setaria digitata*, a filarial parasite. Biochem. J. **256**: 559–564 (1988).
40. Ramp, Th. and Köhler, P. Glucose and pyruvate catabolism in *Litomososides carinii*. Parasitology **89**: 229–244 (1984).
41 Richards, R.J. The effects of starvation *in vitro* on the free amino acids and sugars in the daughter sporocysts of *Microphallus pygmaeus* (Levinsen, 1881) (Trematoda: Microphallidae). Z. Parasitkunde **35**: 31–39 (1970).
42. Salt, T.A., Chitwood, D.J. and Lusby, W.R. Sterol metabolism in the nematode *Panagrellus redivivus*. Lipids **24**: 325–328 (1989).
43. Saz. H.J. and DeBruyn, B.S. Separation and functions of two acetyl CoA transferases from *Ascaris lumbricoides* mitochondria. J. Exp. Zool. **242**: 241–245 (1987).
44. Searcy, D.G. Measurement by DNA hybridization *in vitro* of the genetic basis of parasite reduction. Evolution **24**: 207–219 (1970).
45. Seidl, C., Bauer, M. and Moritz, K.B. Chromatin diminution and early cleavage in *Parascaris univalens* (Nematoda). Roux's Arch. Dev. Biol. **197**: 307–320 (1988).

46. Smith, D.B., Davern, K.M., Board, P.G. Tiu, W.U., Garcia, E.G. and Mitchell, G.F. Mr 26,000 antigen of *Schistosoma japonicum* recognized by resistant WEHI 129/J mice is a parasite glutathione-S-transferase. Proc. Nat. Acad. Sci. U.S.A. **83**: 8703–8707 (1986).
47. Tarr, G.E. Comparison of ascarosides from four species of ascarid and of one oxyurid (Nematoda). Comp. Biochem. Physiol. **46B**: 167–176 (1973).
48. Tarr, G.E. and Fairbairn, D. Glycerides, waxes and sterols in ovaries of *Ascaris lumbricoides* (Nematoda). Lipids **8**: 303–310 (1973).
49. Van Oordt, B.E.P., Tielens, A.G.M. and Van Den Bergh, S.G. Aerobic to anaerobic transition in the carbohydrate metabolism of *Schistosoma mansoni* cercariae during transformation *in vitro*. Parasitology **98**: 409–417 (1989).
50. Walter, R.D. Polyamine metabolism of filaria and allied parasites. Parasitol. Today **4**: 18–20 (1988).
51. Ward, P.F.V. Aspects of helminth metabolism. Parasitology **84**: 177–194 (1982).
52. Wong. P.C.L. and Yeung, S. B. Pathways of purine ribonucleotide biosynthesis in the adult worm *Metastrongylus apri* (Nematoda: Metastrongyloidea) from pig lung. Mol. Biochem. Parasitol. **2**: 285–293 (1981).

8. Pathophysiology of Helminth Infections

P.H. Holmes
Deaprtment of Veterinary Physiology, University of Glasgow Veterinary School
Bearsden Road, Glasgow, UK

INTRODUCTION

Helminths are major pathogens of both man and animals. The diseases they cause and the associated clinical signs usually reflect the route of infection and in the case of helminth diseases this is predominantly by ingestion into the gastrointestinal tract. This chapter is, therefore, primarily concerned with gastrointestinal helminth infections and nematodes and trematodes in particular, in which there are many shared pathophysiological characteristics. Although helminths affect both man and animals, it is in the latter group that most studies have been conducted on the underlying pathophysiology, and this review focuses on the research conducted in this area over the past two decades.

GASTROINTESTINAL PARASITES

The adverse effects of gastrointestinal parasitism in man are of major importance especially in tropical countries whilst in animals these parasites cause economic losses on a large scale throughout the world. Heavy untreated infections may be fatal and an obvious cause of loss, but non-fatal infections, apart from resulting in poor productivity, also cause affected animals to be culled prematurely and others not to be used as replacement breeding stock.

Much recent work has uncovered deleterious effects on production in cases of parasite infection in which clinical signs of the disease are not evident. Such 'subclinical' losses of production are often difficult to assess but frequently are of major importance [12].

In animals, the major pathogenic parasites of the gastrointestinal tract are numerous nematodes, particularly of the family *Trichostrongylidae,* and trematodes, such as *Fasciola hepatica, F. gigantica, Paramphistomum* spp. and *Schistosoma* spp. In man, the principal pathogenic gastrointestinal nematodes are hookworms and *Ascaris.*

Clinical Effects

Gastrointestinal parasitism is associated with a range of clinical signs, including a failure to gain weight or weight loss, inappetance and usually diarrhoea (see also Chapter 3). In addition to changes in body weight, alterations in body composition also occur and these can be of considerable importance in judging the impact of gastrointestinal parasitism. Changes in body composition as a result of trichostrongylosis have been reported in cattle [20, 27] and in sheep [49, 50]. Most authors have found that in infected animals the percentage of water was higher and the deposition of fat, protein and skeletal calcium and phosphorus lower than in worm-free controls.

There are a number of characteristic changes in blood constituents, the most notable being hypoalbuminaemia and commonly, a depression in the serum total protein concentration. Abomasal parasitism is associated with characteristic increases in the concentration of pepsinogen in the blood.

It has been recognized for some time that gastrointestinal parasites such as *F. hepatica* can adversely affect wool production of sheep [43]. More recently, the inter-relationship between wool production and the level of exposure to trichostrongyles has been examined [48]. There is evidence that milk production may be reduced in ruminants infected with trichostrongyles [9, 55] and *F. hepatica* [41]. Trematode infections can also reduce conception and pregnancy rates, and delay the onset of puberty in cattle [34, 38] and sheep [31].

Many of the pathological findings reflect the clinical signs, for example, poor body condition and weight loss, while others relate to specific changes within the gastrointestinal tract. In the case of abomasal parasitism caused by *Ostertagia* spp., the main damage occurs as the 4th-stage larvae emerge from the gastric glands. As a result of the loss of functional parietal cells, acid secretion is impaired and pH values are elevated [42]. The lesions associated with *Haemonchus contortus* are broadly similar to other trichostrongyle infections of the abomasum, although it is the haematophagic habit of the 4th-stage larvae and adults following their emergence from gastric glands which are most detrimental to the host.

Blood loss into the gastrointestinal tract is also caused by other nematodes, such as hookworms and trematodes, *F. hepatica* and *F. gigantica*.

In trichostrongyle infections of the small intestine, the main damage results from activity of the adult worms. Around 90% of worms are normally present in the proximal third of the small intestine and this is where the greatest damage occurs. Heavy infections are associated with severe enteritis. There is extensive villous atrophy, mucosal thickening and stunting of the microvilli, erosion of the epithelium and infiltration of lymphocytes and neutrophils into the damaged areas. Parasites are located in tunnels in the epithelium of villi and intestinal crypts at all stages of development.

Ascaris spp. may also cause damage to the intestinal epithelium and heavy infections can be associated with intestinal obstruction.

Food Consumption

Reduction in voluntary food intake, or inappetence, is an important feature of gastrointestinal parasitism and is widely recognized as a major factor in the pathogenesis of such infections. The degree of inappetence can vary considerably. In moderate to heavy infections with trichostrongyles, food intake can be reduced by 20% or more, in lighter infections it may be less obvious. Recently, the degree of inappetence has been shown to vary not only with the level and duration of parasitism but also with the level of protein nutrition [3]. In fluke infections, inappetence generally does not occur until patency and there does not appear to be a direct correlation between adult fluke burdens and the degree of inappetence [8]. Depressed weight gains in calves infected with lungworms (*Dictyocaulus viviparus)* have been shown to be primarily due to reduced feed intake [33].

Despite the obvious importance of inappetence in parasitized hosts, it is still not known why it occurs. A number of recent reviews have highlighted the inadequate state of current knowledge on this topic [15, 52].

There is no doubt that many infections are associated with marked pathological changes and it is possible that, in addition to pain or discomfort, alterations in the rate of passage of ingesta even in the absence of diarrhoea could be associated with changes in voluntary feed intake [25]. Others have suggested that the increased levels of gastrointestinal hormones such as gastrin and cholecystokinin observed in parasitized ruminants, may be responsible [23, 53, 56].

Gastrointestinal Function

Although reduced appetite is an important aspect of gastrointestinal parasitism, studies utilizing pair-feeding techniques have demonstrated that inappetance is not the only consequence of infection and such animals frequently show reduced utilization of nutrients relative to parasite-free animals on the same feed intake.

In view of their location and pathogenic effects in infected animals, it is generally assumed that gut parasites must be associated with changes in gastrointestinal function. However, in many infections it has been difficult to relate the two. This is partly because the number of studies has been limited and partly because it is difficult to evaluate the impact which dysfunction in one part of the tract may have on the whole animal. Nevertheless, some recent studies permit a better assessment of the effects of parasitic infections on gastrointestinal motility, digestion and absorption.

Effects of nematode infections on motility have been investigated only in a small number of experiments. In lambs infected with *Trichostrongylus colubriformis*, it has been reported that digesta flow was reduced from the rumen but increased in small and large intestine [44]. More recent studies have shown that subclinical infections of sheep with *T. colubriformis* alter the normal pattern of gastrointestinal motility in the absence of any diarrhoea and cause inhibition of abomasal and proximal small intestinal motility and digesta flow. However, the increased frequency of Migrating Myoelectric Complexes (MMCs) helps to maintain digesta flow through the proximal small intestine [26].

From the limited studies conducted on gastrointestinal motility in parasitized ruminants, it can be concluded that infections of the abomasum, small intestine and possibly the large intestine can seriously disturb the normal pattern of gastrointestinal motility and digesta flow, even in the absence of diarrhoea [25]. Where studied, the rate of flow through the gut was reduced rather than increased, an effect which appears to be partly due to reduced food intake and partly to the influence of the parasites *per se*. With clinical infections, the onset of diarrhoea is preceded by the disruption of MMCs and severe inhibition of the reticulo-rumen and abomasum, at which time there may be an increase in the bacterial populations which could contribute to the occurrence of the diarrhoea.

Intestinal myoelectric activity is also altered during hookworm infection with changes in slow wave patterns and spike activity in infected dogs [57].

Gastrointestinal secretions have been shown to be altered in parasitized ruminants. Infections of the abomasum with *Ostertagia* spp. and *Haemonchus contortus* are associated with elevations of abomasal pH and rises in blood pepsinogen concentrations. However, the impact of these changes on digestion are uncertain.

In humans, hypersecretion of gastric hydrochloric acid has been reported in fish tapeworm (*Diphyllobothrium latum)* infections, and some cases of hookworm infection [11]. Biliary secretion may be affected by direct damage to the liver as in cases of fascioliasis and schistosomiasis or by obstruction to the release of bile into the small intestine as has been reported in some *Ascaris* infections.Very few studies have been conducted to examine pancreatic secretion in parasitized hosts. However, it has been demonstrated that in dogs infected with *T. spiralis* pancreatic secretion was depressed as a result of impairment of gut hormone synthesis or release [11]. Intestinal secretions are also affected by gastrointestinal nematodes with a net flux of fluid into the lumen. Parasitic infections of the small intestine are associated with decreased activity of a variety of brush-border enzymes [32], probably as a result of mucosal damage.

In view of the structural damage and altered motility and secretions, numerous attempts have been made to determine whether impaired digestion and absorption are major causes of poor utilization of food by parasitized animals and man. Although some studies have suggested a reduction in the digestion and absorption of dietary nitrogen or other nutrients, such values are pool indicators of malabsorption. This is largely because of the significant passage of endogenous nitrogen and other materials into the gastrointestinal tract of parasitized animals (*vide infra*).

Fewer difficulties arise with the evaluation of the results of balance studies using pair-fed controls, although this approach still fails to permit direct assessment of digestion and absorption in different parts of the gastrointestinal tract. Ideally, such studies should be conducted in animals with indwelling gastrointestinal cannulae. The results of such studies are discussed later but, in general, they indicate that impaired digestion and absorption are not important causes of the poor utilization of nutrients by parasitized hosts, rather, it is the increased metabolic demands on the host as a result of the parasites' activities. Unfortunately, in many parasitic infections, a diagnosis of malabsorption has been based on indirect evidence such as diarrhoea associated with weight loss. Furthermore, although some studies have clearly shown depressed absorption of radio-isotope labelled substrates in parasitized portions of the intestinal tract, absorption may be enhanced in non-parasitized portions and, as a result, absorption overall may not be affected [11, 51]. However, there are situations in which malabsorption may be significant. These include severe infections with ascariasis [37], mixed trichostrongyle infections in ruminants, and the competitive uptake of vitamin B_{12} by *D. latum* [11].

Gastrointestinal Protein Loss

A distinctive feature of gastrointestinal parasitism is loss of proteins into the gastrointestinal tract. The proteins represent plasma and red cells, exfoliated epithelial cells and mucus. Radioisotopic techniques have permitted measurement of blood protein losses into the tract in man and animals and these techniques have been applied to a variety of gastrointestinal nematode and trematode infections (reviewed in Chapter 11). The results have consistently shown high losses of plasma proteins in infected hosts, and in infections associated with gastrointestinal haemorrhage such as *H. contortus*, *O. radiatum* hookworms, *Fasciola* spp. and *Schistosoma* spp., there is also a massive loss of red cells [10, 18, 28, 29, 30].

Unfortunately, methods to quantify losses of exfoliated epithelial cells and mucus are not yet available. However, there is little doubt that they are increased in many types of gastrointestinal parasitism [6, 13, 35, 45].

Nitrogen balance studies have demonstrated that reduced nitrogen retention is a characteristic effect of helminth and protozoan infections and is associated with depressed growth rates and other effects on productivity. In the majority of experiments in mammals, this has been associated with increased urinary nitrogen loss [39, 44]. However, nitrogen balance studies do not permit direct assessment of the levels of endogenous protein loss or the fate of the proteins within the tract. Such information can only be obtained from animals with indwelling gastrointestinal cannulae. Studies in cannulated sheep infected with *O. circumcincta* or *T. colubriformis* indicated that nitrogen digestion and absorption was influenced by the site of infection and the level of feed intake [46, 47]. Later studies have tended to confirm these findings. A study by Rowe et al [45] in sheep infected with the abomasal parasite, *H. contortus*, showed that blood loss accounted for virtually all the increase in nitrogen flow into the duodenum and all of this was absorbed by the end of the ileum.

Protein Deposition

Attempts to evaluate protein deposition and tissue metabolism in parasitized hosts have been limited. One approach has been to examine food utilization in infected animals. For example, in lambs chronically infected with *O. circumcincta* [14, 50] a significant reduction in the deposition of fat and protein in infected animals compared with pair-fed controls was demonstrated. More recently, carcase evaluation studies have been conducted in cattle infected with mixed trichostrongyles [20] and at slaughter various parameters of protein deposition were found to be reduced in parasitized animals compared with parasite-free controls.

A controlled series of studies by Symons and his co-workers to examine tissue protein synthesis using L-[^{14}C] leucine and L-[^{14}C] tyrosine in parasitized animals is particularly relevant [52]. These showed that protein synthesis was reduced in skeletal muscle of infected sheep, whilst protein synthesis by the liver was increased in such animals.

Measurements of protein synthesis in gastrointestinal tissue have, for technical reasons, so far been restricted to guinea-pigs infected with *T. colubriformis*. In such animals the amount of protein synthesis per day was increased in both the small and large intestine [54]. It is important that this finding is confirmed in ruminants as these sites of increased protein synthesis may be an important cause of reduced nutrient utilization by infected sheep and cattle.

As a result of such studies, Symons [52] concluded that, due to inappetance, gastrointestinal losses of protein and increased rates of gastrointestinal tissue protein metabolism, there is a net movement of amino acid nitrogen from muscle and skin to the liver and gatrointestinal tract which decreases the availability for growth, milk and wool production. However, it is still not possible to construct accurate balance sheets of protein synthesis in parasitized animals nor is there any detailed information on the underlying mechanisms, including hormonal, which may be responsible for the changes in protein synthesis.

Energy Balance

Reduced food intake is undoubtedly a major factor limiting the availability of energy for maintenance and growth. Reductions in apparent energy digestibility have been reported [21, 36, 50] but these have, in general, been very small, even in diarrhoeic animals. Similarly, the few neasurements that have been made of heat production have failed to show any significant differences between infected and control animals [40, 36, Stevenson, Thomas and Holmes, unpublished observations]. However, comparative slaughter experiments in lambs infected with *T. colubriformis* [49] or *O. circumcincta* [14] have shown such animals to have a lower energy retention than parasite-free, pair-fed controls.

The reasons for this have not been accurately determined. However, the most likely cause for the reduction in the efficiency of utilization of digestible energy, which is supported by the findings of Symons and Jones [54], is that there are marked increases in synthetic rates of protein in the liver and gastrointestinal tissue.

Host Factors Modulating Pathophysiological Responses

In addition to the pathogenic mechanisms associated with different species of parasite and different levels of infection, it is now recognized that there are a number of host factors* which can influence the pathophysiology of parasitic infections. These include age and immunological status, the genetic background of the host, and the plane of nutrition. In general, older immune hosts are less susceptible to the pathogenic effects of parasites. However, in some circumstances the performance of older animals may be impaired by parasite challenge. For example, studies have shown that older animals, previously exposed to field [4] or experimental [7, 59] trichostrongyle infections, although showing low or negative faecal egg counts and low worm burdens, show pathophysiological changes when under subsequent larval challenge possibly as a result of an immune hypersensitivity reaction induced by larvae [4, 5].

Genetic resistance to helminth infections has been most clearly identified in some breeds of ruminants which show resistance to certain gastrointestinal nematode infections [19].

---

*An excellent review on this topic entitled "Parasites and Domestic Animals: Host Factors in Pathogenesis" by E.J.L. Soulsby has already been appeared in "Parasites—Their World and Ours", D.F. Mettrick and S.S. Desser (eds). Elsevier Medical Press (Publ.): Amsterdam (1982).

Most studies have been conducted in Australia where it has been clearly established that in Merino sheep, there are marked differences in susceptibility to gastrointestinal helminthiasis between individual animals. This characteristic is heritable and offers the potential for the development and commercial exploitation of helminth resistant sheep in the future. There is also considerable variation in susceptibility to gastrointestinal parasites between different breeds of sheep and this characteristic is the subject of growing interest. Genetic resistance to infection is believed to be immunologically mediated (see Chapter 15) and may be correlated with levels of parasite specific IgA and with eosinophilia, but these possibilities require further investigation.

It is frequently suggested that the nutritional status of the host can influence the pathogenesis of parasitic infections and it is generally accepted that well-nourished animals resist parasitism better than those less adequately fed [24, 58]. Similarly, there is much suggestive evidence that malnutrition influences the pathogenicity of intestinal parasites in man [16, 17]. However, the mechanisms involved have rarely been elucidated. Some studies have been conducted on the role of host nutrition on the pathogenesis of ovine fascioliasis [8] and *Ascaris* infections in young protein-deficient pigs [22]. More recently, a series of studies have been conducted in sheep using isocaloric diets to investigate the influence of dietary protein on the pathogenesis of *H. contortus*. These experiments showed that although the protein content of the diet *per se* did not influence the establishment of single primary infections, lambs given a low protein diet showed more severe clinical signs such as weight loss, anaemia and inappetence, despite similar levels of blood loss. Furthermore, lambs given a high protein diet and subjected to repeated infections were more likely to develop resistance to reinfection than animals given a low protein diet [1, 2, 3, 28].

Similar studies should be conducted on other infections since such effects could have been considerable importance in field situations where suboptimal nutrition commonly occurs. In animals the problem may be further exacerbated by the additional metabolic demands arising from extensive foraging on poor quality pasture.

OTHER PARASITES

This review has concentrated primarily on nematode and trematode infections since they have major pathogenic efferts in man and animals. Cestodes have so far received little mention since the adult forms are rarely thought to have significant pathogenic effects at least in animals.

However, some species of cestodes have considerable public health significance. *Taenia saginata* occurs commonly amongst beef-eating areas of the world and the adult tapeworm can be associated with abdominal discomfort and diarrhoea in humans but in most cases the infection is asymptomatic. Similarly in calves the presence of larval cysticerci in the muscles of cattle are not associated with clinical signs except in very heavy infections. Cysticercosis caused by *Taenia solium* is of greater public health significance. The clinical signs in man depend on the location of the cysts. Most serious are the cases where cysts develop in the central nervous system or eye. Other cysticercoses may also be associated with clinical signs in sheep and cattle by causing lesions in the C.N.S. e.g. *T. multiceps* in sheep or in the liver or lungs e.g. *Echinococcus granulosus*.

In addition to the tissues lesions described above a number of other important helminth parasites of man and animals are associated with clinical signs resulting from obstruction and granuloma formation. These include *Dictyocaulus* spp., *Trichinella spiralis*, *Wuchereria bancrofti*, *Onchocerca volvulus* and *Dicrofilaria immitis*. The wide spectra of pathological changes manifested by many helminth species in their hosts have been discussed elsewhere (see Chapters 10 and 15; see also Chapter 3).

CONCLUSIONS

Helminth parasites are major pathogens of man and animals. They affect various tissues and organs but those of greatest importance are nematodes and trematodes which infect the gastrointestinal tract. The pathophysiological effects of such parasites have been the subject of considerable research over the past two decades. The most advanced studies have been conducted on sheep infected with gastrointestinal nematodes. These have shown that infection is associated with significant effects on feed intake, gastrointestinal function and protein turnover. A central feature of such infections is the loss of large quantities of host protein into the gastrointestinal tract and this results in many of the other changes associated with gastrointestinal helminth infections.

ACKNOWLEDGEMENTS

The assistance of Diana McKechnie and Dorothy McGruer in the preparation and typing of this review is gratefully acknowledged.

REFERENCES

1. Abbott, E.M., Parkins, J.J. and Holmes, P.H. Influence of dietary protein on parasite establishment and pathogenesis in Finn Dorset and Scottish Blackface lambs given a single moderate infection of *Haemonchus contortus*. Res. Vet. Sc. **38**: 6–13 (1985).
2. Abbott, E.M., Parkins, J.J. and Holmes, P.H. Influence of dietary protein on the pathophysiology of ovine haemonchosis in Finn Dorset and Scottish Blackface lambs given a single moderate infection. Res. Vet. Sci. **38**: 54–60 (1985).
3. Abbott, E.M., Parkins, J.J. and Holmes, P.H. The effect of dietary protein on the pathophysiology of acute ovine haemonchosis. Vet. Parasitol. **20**: 275–289, 291–306 (1986).
4. Anderson, N. Trichostrongyloid infections of sheep in a winter rainfall region. I. Epizootiological studies in the western district of Victoria. Aust. J. Agri. Res. **23**: No. 6, 1113–1129 (1972).
5. Armour, J., Bairden, K., Duncan, J., Jennings, F.W. and Parkins, J.J. Observations on ostertagiasis in young cattle over two grazing seasons, with special reference to plasma pepsinogen levels. Vet. Rec. **105**: 500–503 (1979).
6. Armour, J., Jarrett, W.F.H. and Jennings, F.W. Experimental *Ostertagia cicumcincta* in sheep. Development and pathogenesis of a single infection. Am. J. Vet. Res. **27**: 1267–1278 (1966).
7. Barger, I.A. and Southcott, W.H. Trichostrongylosis and wool growth. III. The wool growth response of resistant grazing sheep to larval challenge. Aust. J. Expt. Agri. Anim. Husb. **15**: No. 73, 167–172 (1975).
8. Berry, C.I. and Dargie, J.D. The role of host nutrition on the pathogenesis of ovine fascioliasis. Vet. Parasitol. **2**: 317–332 (1976).
9. Bliss, D.H. and Todd, A.C. Milk production by Vermont dairy cattle after deworming. Vet. Med. Small Anim. Clin. **68**: 1034–1038 (1976).
10. Bremner, K.C. Pathogenic factors in experimantal bovine oesophagostomosis, III. Demonstration of protein-losing enteropathy with ^{51}Cr-albumin. Exp. Parasitol. **24**: 364–374 (1969).
11. Castro, G.A. Physiology of the gastrointestinal tract in the parasitized host. In: *Physiology of the Gastrointestinal Tract* (L.R. Johnson ed), pp 1381–1406. Raven Press: New York (1981).
12. Coop, R.L. The impact of subclinical parasitism in ruminants. In: *Parasites—Their World and Ours* (D.F. Mettrick and S.S. Desser eds), pp. 439–450. Elsevier Medical Press: Amsterdam (1982).
13. Coop, R.L. and Angus, K.W. The effect of continuous doses of *Trichostrongylus colubriformis* larvae on the intestinal mucosa of sheep and on liver Vitamin A concentration. Parasitology **70**: 1–9 (1975).
14. Coop, R.L., Sykes, A.R. and Angus, K.W. The effect of three levels of intake of *Ostertagia circumcincta* larvae on growth rate, food intake and body composition of growing lambs. J. Agri. Sci. **90**: 247–255 (1982).
15. Crompton, D.W.T. Influence of parasitic infection on food intake. Fed. Proc. **43**: 239–245 (1984).

16. Crompton, D.W.T. Chronic ascariasis and malnutrition. Parasitol. Today **1**: 47–52 (1985).
17. Crouch, P.A.R. A review of intestinal parasitism and nutrition. Papua New Guinea Med. J. **25**: 43–49 (1982).
18. Dargie, J.D. Applications of radioisotope techniques to the study of red cell and plasma protein metabolism in helminth diseases of sheep. In: *Pathogenic Processes in Parasitic Infections* (A.E.R. Taylor and R. Muller eds), pp 1–26. Blackwell Scientific Publications: Oxford (1975).
19. Dargie, J.D. The influence of genetic factors on the resistance of ruminants to gastrointestinal nematode and trypanosome infections. In: *Animal Models in Parasitology* (D. Owen ed), pp 17–51. Macmillan Press Ltd.: London (1982).
20. Entrocasso, C., Parkins, J.J., Armour, J., Bairden, K. and McWilliam, P.N. Production, parasitological and carcase evaluation studies in steers exposed to trichostrongyle infection and treated with a morantel bolus of fenbendazole in two consecutive grazing seasons. Res. Vet. Sci. **40**: 76–85 (1986).
21. Entrocasso, C., Parkins, J.J., Armour, J., Bairden, K. and McWilliam, P.N. Metabolism and growth in housed calves given a morantel sustained release bolus and exposed to natural trichostrongyle infection. Res. Vet. Sci. **40**: 65–75 (1986).
22. Forsum, E., Nesheim, M.C. and Crompton, D.W.T. Nutritional aspects of *Ascaris* infection in young protein-deficient pigs. Parasitology **83**: 497–512 (1981).
23. Fox, M.T., Gerrelli, D., Shivalkar, P., and Jacobs, D.E. Effect of omeprazole treatment on feed intake and blood gastrin and pepsinogen levels in the calf. Res. Vet. Sci. **46**: 280–282 (1989).
24. Gibson, T.E. The influence of nutrition on the relationships between gastrointestinal parasites and their hosts. Proc. Nut. Soc. **22**: 15–20 (1963).
25. Gregory, P.C. Parasitic infection and stomach motility: relation to intestinal motility and food intake. In: *Veterinary Research Communication Vol. 1. The Ruminant Stomach* (L.A.A. Ooms, A.D. Degryse and R. Marsboom eds), pp. 267–286. Janssen Research Foundation (1985).
26. Gregory, P.C., Wenham, G., Poppi, D., Coop, R.L., Macrae, J.C. and Miller, S.J. The influence of a chronic subclinical infection of *Trichostrongylus colubriformis* on gastrointestinal motility and digesta flow in sheep. Parasitology **91**: 381–396 (1985).
27. Halliday, G.J., Dalton, R.C., Anderson, N. and Mulligan, W. Total body water and total body solids in cattle affected with ostertagiasis. British Vet. J. **121**: 547–551 (1965).
28. Holmes, P.H., Abbott, E.M. and Parkins, J.J. Use of nuclear techniques to investigate the influence of host nutrition on the pathogenesis of ovine haemonchosis. In: *Nuclear Techniques aimed at improving Meat, Milk and Wool Production from Ruminant Animals*. pp. 185–195. IAEA: Vienna (1986).
29. Holmes, P.H., Dargie, J.D., Maclean, J.M. and Mulligan, W. The anaemia in fascioliasis: studies with ^{51}Cr-labelled red cells. J. Comp. Path. **78**: 415–420 (1968).
30. Holmes, P.H. and Maclean, J.M. The pathophysiology of ovine ostertagiasis: A study of the changes in plasma protein metabolism following single infections. Res. Vet. Sci. **12**: 265–271 (1971).
31. Hope Cawdery, M.J. The effects of fascioliasis on ewe fertility. British Vet. J. **132**: 568–575 (1976).
32. Jones, D.G. Intestinal enzyme activity in lambs chronically infected with *Trichostrongylus colubriformis*: Effect of anthelmintic treatment. Vet. Path. **12**: 79–89 (1983).
33. Kroonen, J.E.G.M., Verstegen, M.W.A., Boon, J.H. and van der Het, W. Effect of infection with lungworms (*Dictyocaulus viviparus*) on energy and nitrogen metabolism in growing calves. British J. Nutr. **55**: 351–360 (1986).
34. McCauley, E.H., Majid, A.A. and Tayeb, A. Economic evaluation of the production impact of bovine schistomiasis and vaccination in the Sudan. Prevent. Vet. Med. **2**: 735–754 (1984).
35. MacDonald, T.T. and Ferguson, A. Small intestinal epithelial cell kinetics and protozoal infection in mice. Gastroenterology **74**: 496–500 (1978).
36. MacRae, J.C., Smith, J.S., Sharman, G.A.M., Corrigall, W. and Coop, R.L. Energy metabolism of lambs infected with *Trichostrongylus colubriformis*. In: *Energy Metabolism in Farm Animals* (A. Ekern and F. Sundstol eds), pp 112–115. EAAP Pub. No. 29, Agricultural University, Norway (1982).
37. Nesheim, M.C. Nutritional aspects of *Ascaris suum* and *A. lumbricoides* infections. In: *Ascariasis and its Public Health Significance* (D.W.T. Crompton, M.C. Nesheim and Z.S. Pawlowski eds), pp. 147–160. Taylor and Francis: London (1985).

38. Oakley, G.A., Owen, B. and Knapp, N.H.H. Production effects of subclinical liver fluke infections in growing dairy heifers. Vet. Rec. **104**: 503–507 (1979).
39. Parkins, J.J., Holmes, P.H. and Bremner, K. The pathophysiology of ovine ostertagiasis: some nitrogen balance and digestibility studies. Res. Vet. Sci. **14**: 21–28 (1973).
40. Randall, R.W. and Gibbs, H.C. Effect of clinical and subclinical gastrointestinal helminthiasis on digestion and energy metabolism in calves. Am. J. Vet. Res. **42**: 1730–1734 (1981).
41. Randell, W.F. and Bradley, R.E. Effects of hexachlorethane on the milk yields of dairy cows in North Florida infected with *Fasciola hepatica*. Am. J. Vet. Res. **41**: 262–264 (1980).
42. Ritchie, J.D.S., Anderson, N., Armour, J., Jarrett, W.F.H., Jennings, F.W. and Urquhart, G.M. Experimental *Ostertagia ostertagi* infections in calves: parasitology and pathogenesis of a single infection. Am. J. Vet. Res. **27**: 659–667 (1966).
43. Roseby, F.B. The effect of fascioliasis on the wool production of merino sheep. Aust. Vet. J. **46**: 361–365 (1970).
44. Roseby, F.B. Effects of *Trichostrongylus colubriformis* (Nematoda) on the nutrition and metabolism of sheep. III. Digesta flow and fermentation. Aust. J. Agri. Res. **28**: 155–164 (1977).
45. Rowe, J.B., Abbott, E.M., Dargie, J.D. and Holmes, P.H. The effect of haemonchosis and blood loss in the abomasum on nitrogen digestion in sheep. Proc. Nutr. Soc. **41**: 74A (1982).
46. Steel, J.W. Pathophysiology of gastrointestinal nematode infections in the ruminant. Proc. Aust. Soc. Anim. Prod. **10**: 139–147 (1974).
47. Steel, J.W. Inter-relationships between gastrointestinal helminth infection, nutrition and impaired productivity. In: *Recent Advances in Animal Nutrition* (D.J. Farre ed), pp 98–110. University of New England, Armidale (1978).
48. Steel, J.W., Jones, W.O. and Symons, L.E.A. Effect of a concurrent infection of *Trichostrongylus colubriformis* on the productivity of physiological and metabolic responses of lambs infected with *Ostertagia circumcincta*. Aust. J. Agri. Res. **33**: 131–140 (1982).
49. Sykes, A.R. and Coop, R.L. Intake and utilisation of food by growing lambs with parasitic damage to the small intestine caused by daily dosing with *Trichostrongylus colubriformis* larvae. J. Agri. Sci. **86**: 507–515 (1976).
50. Sykes, A.R. and Coop, R.L. Intake and utilisation of food by growing sheep with abomasal damage caused by daily dosing with *Ostertagia circumcincta* larvae. J. Agri. Sci. **88**: 671–677 (1977).
51. Symons, L.E.A. Malabsorption. In: *Pathophysiology of Parasitic Infection* (E.J.L. Soulsby ed), pp 11–21. Academic Press: New York (1976).
52. Symons, L.E.A. Anorexia: occurrence, pathophysiology and possible causes in parasitic infections. Adv. Parasitol. **24**: 103–133 (1985).
53. Symons, L.E.A. and Hennessy, D.L. Cholecystokinin and anorexia in sheep infected by the intestinal nematode *Trichostrongylus colubriformis*. Int. J. Parasitol. **11**: 55–58 (1981).
54. Symons, L.E.A. and Jones, W.O. Intestinal protein synthesis in guinea pigs infected with *Trichostrongylus colubriformis*. Int. J. Parasitol. **13**: 309–312 (1983).
55. Thomas, R.J. and Ali, D.A. The effect of *Haemonchus contortus* infection on the pregnant and lactating ewe. Int. J. Parasitol. **13**: 393–398 (1983).
56. Titchen, D.A. The role of hormones in the reactions of the host to enteric parasites. In: *Parasites—Their World and Ours* (D.F. Mettrick and S.S. Desser eds). pp. 244–245. Elsevier Medical Press: Amsterdam (1982).
57. Weisbrodt, N.W. and Castro, G.A. Intestinal myoelectric activity of the dog during hookworm infection. Fed. Proc. **36**: 595 (1977).
58. Whitlock, J.H. The relationship of nutrition to the development of trichostrongylidoses. Cornell Vet. **39**: 146–182 (1949).
59. Yakoob, A., Holmes, P.H. and Armour, J. The pathophysiology of gastrointestinal trichostrongyles in sheep: Plasma losses and changes in plasma pepsinogen levels associated with parasitic challenge of immune animals. Res. Vet. Sci. **34**: 305–309 (1983).

9. Host Responses to Helminths with Emphasis on Eosinophils and Mast Cells

Y. Nawa, T. Abe and M. Owhashi
Department of Parasitology, Miyazaki Medical College, Kiyotake, Japan

INTRODUCTION

Helminth parasites are extraordinarily large as compared to other infectious organisms (Fig. 1). Most of the infectious organisms have a size small enough for engulfment by single phagocytic cell. But, even the smallest helminth parasite is too large to be phagocytosed. Although defence mechanisms operating against various infectious agents are basically the same but some of the immunological events occurring at cellular level are, however, unique in helminth infection.

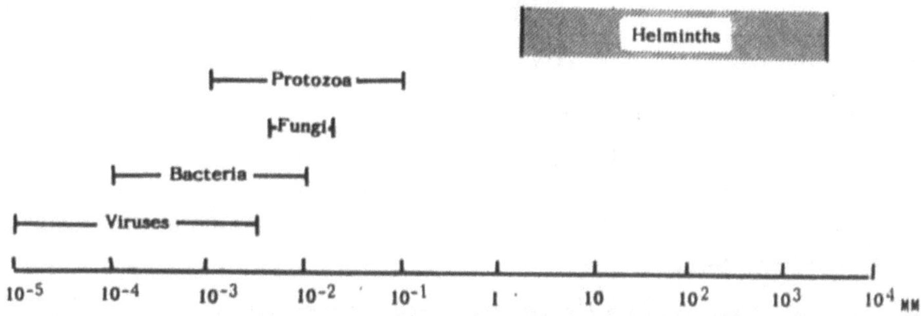

Fig. 1. Relative size distribution of infectious organisms. Note the substantial difference between protozoa and helminth.

To obtain nutrients and a site for lodging, the helminth parasite should invade an appropriate host, generally through percutaneous route or orally (Fig. 2). In certain parasites such as hookworms or schistosomes, which are transmitted through soil and water respectively, the infective stages penetrate directly through the skin of host animals. In filaria, the larvae are, however, inoculated percutaneously by the intermediate hosts such as mosquitoes and black flies. Apart from these percutaneously-infecting parasites, a large number of helminths enter their hosts via the oral route in the form of fertilized eggs or larvae. In certain species, infective larvae lodged in the intermediate host, reach the final host when the intermediate host is ingested by the final host. Such routes of infection are extremely important when we consider the uniqueness of the host's defence mechanism against helminths. In the first section of this review an outline of host's defence systems against helminth infection is provided with details on the route of invasion, whereas in the second section, the uniqueness of certain cellular responses observed in helminth infection is reviewed with emphasis on the role of eosinophils and mast cells.

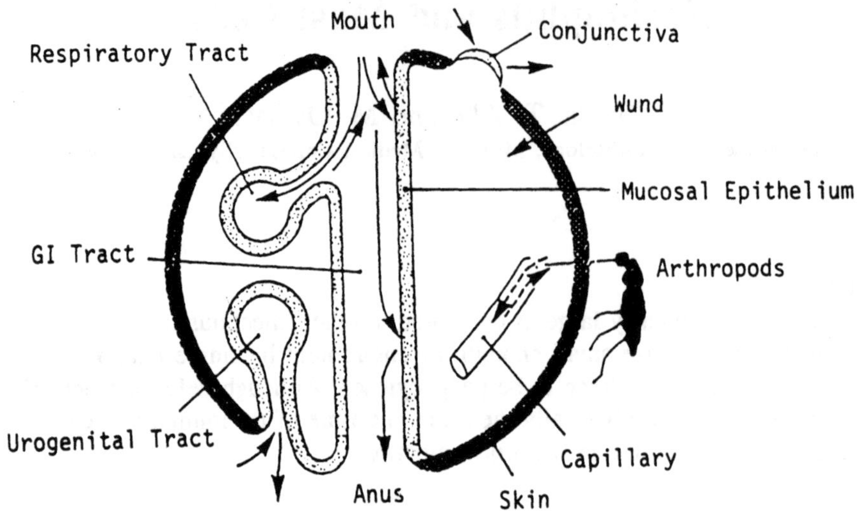

Fig. 2. Routes of infection of the host. (Reproduced from [51] with permission of Dr. Mims).

GENERAL VIEW OF HOST'S DEFENCE SYSTEM

Because of the limitation of the space, the readers who wish to learn general immunology and/or parasite immunology are recommended to refer several excellent text-books [23, 24, 51, 68, 76] for the better understanding of the modern concept in this field. Only very rough sketch of the framework of host's defence mechanisms will be outlined here.

CELLS INVOLVED IN THE DEFENCE MECHANISMS OF HOST

It is obvious from Fig. 2 that the surface epithelium of the skin and of the gastrointestinal mucosa are critically important as the first line of defence. The surface of the skin is covered with hair or is keratinized, so that it can act as a physical barrier to avoid "attachment" by infectious organisms. Similarly, the surface of the intestinal mucosa is covered by a glycocalyx and mucous layer. The rapid turn-over of epithelial cells of the skin or intestinal tract is also helpful in getting rid of the pathogens attached on the surface. The intestinal epithelium is not a simple physical barrier but shows dynamic changes during infection with intestinal nematodes. For example, after infection with *Trichinella spiralis* or *Nippostrongylus brasiliensis*, significant increase in number of goblet cells and a massive discharge of goblet cell mucus is observed at the time of immune-mediated worm expulsion [49]. The worms in the intestine are trapped by the mucus so that they seem to be easily detached and removed by peristaltic movement of the intestine.

Once helminth parasites break through the epithelial barrier and gain access to the inside of the body of host animals, cells of mesenchymal origin act as the second line of defence. Although the simplest means of defence at this stage is phagocytosis and subsequent intracellular digestion by macrophages and neutrophils, helminth parasites are usually too large to be phagocytosed. Against such extraordinary large invaders, not only phagocytic cells but also various other cells, e.g. eosinophils, basophils/mast cells, accumulate around the invaders. These cells, including macrophages and neutrophils, are callled inflammatory cells and they release various potentially helminthotoxic substances such as superoxide anion, hydrogen peroxide, nitric oxide, proteolytic enzymes etc. This is an extracellular form of defence.

The molecular structures that determine "foreignness" (i.e. the sum of their antigenic determinants) of the parasite are processed by macrophages and presented to lymphocytes. Lymphocytes are divided into two major subpopulations, B and T cells. Small B lymphocytes in resting stage transform into large lymphocytes by specific binding with antigens on their surface receptors. Some B cells differentiate into mature plasma cells which can produce and secrete antibodies (with definite specificity), the plasma protein molecule which mediates humoral immunity. Their production by B cells generally requires the collaboration of T helper (T_H) lymphocytes and macrophages. T lymphocytes, similar in size to small B cells, can not be distinguished morphologically from B cells.

T cells, however, differ from B cells in many surface molecules. The most prevalent method to recognize T cells is to detect CD3 molecules expressed on the surface. In terms of their functions, T cells are further classified into four different subsets (1) T_H which help differentiation/maturation of B cells to form and secrete specific antibodies, (2) cytotoxic or killer T cells (T_C) which kill virus-infected cells, tumor cells or grafted tissue, (3) suppressor T cells (T_S) which suppress humoral and cell-mediated immune responses by unknown mechanisms and (4) T cells that mediate delayed hypersensitivity (T_{DTH}). These functional subsets can be segregated to a certain degree, though not absolutely, by using monoclonal antibodies against surface marker molecules. Recently, based on the computor analysis of these monoclonal antibodies, systematic naming of the lymphocyte surface marker molecules (the CD system = cluster designation system) was established. T_H and T_{DTH} cells are $CD4^+$, whereas T_C cells T_S cells are $CD8^+$. The growth of T cells require at least two principal T cell growth factor, interleukines (IL) viz. IL-2 and IL-4. According to the results of clonal analysis, T_H cells are further classified into IL-2 producing T_H1 and IL-4 producing T_H2 subsets. T_H1 cells produce IL-2, IFN-γ, lymphotoxin, GM-CSF, and IL-3. On the other hand, T_H2 cells produce IL-4, GM-CSF, IL-3, IL-5, IL-6 and IL-10 (see further).

To express the functions, T_H and T_C cells should recognize antigen. The T cells recognize the determinant groups in association with major histocompatibility (MHC) antigen that are located on the cell surface of antigen presenting cells (APC). Such a way of antigen recognition is called MHC restriction. In general, T_H and T_{DTH} cells recognize antigen associated with class II MHC on APC, while T_C cells cells recognize antigen associated with class I MHC.

HUMORAL FACTORS INVOLVED IN HOST'S DEFENCE SYSTEM

Responding to the invaded parasites, host animals elicit a series of emergency reactions, acute nonspecific inflammatory responses, specific immune responses, and chronic inflammatory responses. The magnitude and the nature of these responses depends on the nature of invaders, the route of invasion and the duration of survival of the invaders. Regardless of the type of the response, all require rapid growth/differentiation of inflammatory cells and/or immunocompetent cells and are largely under the regulation of humoral factors which are now known as "cytokines". Among cytokines, the molecules involved in intercellular signalling and of which amino acid and/or DNA sequence have been determined, are under the agreement of an international workshop, termed "interleukines (ILs)". Up to now, IL-1 to IL-10 have been identified.* Cytokine pathways in the immune system are shown in Fig. 3.

IL-3, which is known as multi-colony stimulating factor to signify growth differentiation of an array of hemopoietic progenitor cells, has mast cell-growth factor activity and is known to play an important role in helminth rejection from the intestine. IL-5, which is originally identified as a B-cell growth factor, is known to have eosinophil differentiation factor and eosinophil chemotactic factor activities, so that it seems to be important in the mechanisms of eosinophilia in filarial and various other helminth infections. Possible role of IL-3 and IL-5 in helminth infections will be discussed later on in this review.

---

* IL-12, a recently described cytokine (see Scott, P.: Science **260**: 496–497, 1993) which stimulates IFN-γ production, induces the differentiation of T_H1 cells from an uncommitted T cell and, consequently, initiates cell-mediated immunity.

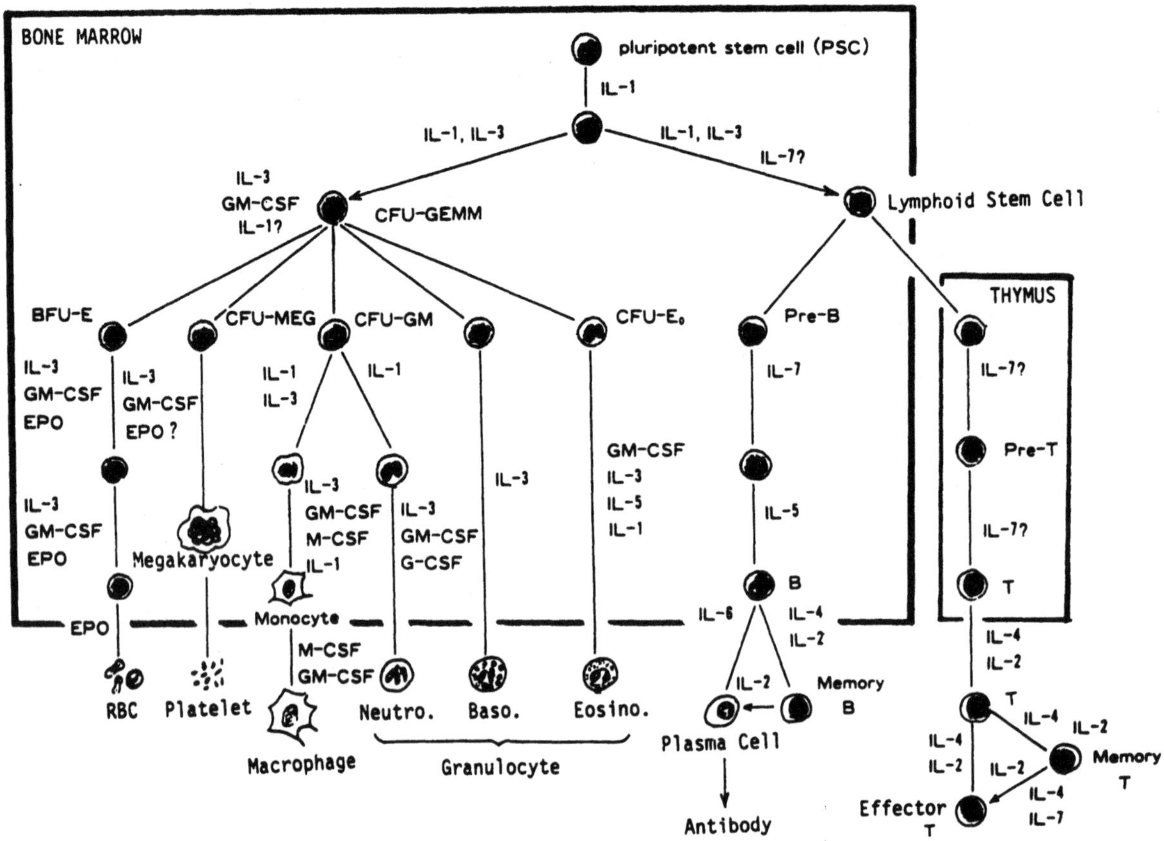

Fig. 3. Diagramatic representation of cytokine pathways in the immune system.

Antibody

Antibodies are glycoproteins produced by mature B lymphocytes—or plasma cells—and constitute the group of plasma protein termed "immunoglobulin (Ig)". Immunoglobulin having specific binding activity to a known antigenic determinant is called "antibody". Production of immonoglobulin by lymphocytes is stimulated by antigen with the help of T_H cells and macrophages. Each B cell is genetically predetermined to react with only one antigenic determinant (epitope) and, after being stimulated with antigen, clonal expansion of B cells takes place to produce immunoglobulin having specificity to react with only one antigenic determinant. There are five isotypes of immunoglobulin, viz. IgG, IgM, IgA, IgD and IgE, in order to their relative amount in the plasma. Basically, these immunoglobulins have two heavy and two light chains. IgM is a pentamer and the secretory form of IgA is a dimer, which are joined to each other through J-chains. Probably because of the nature of antigenicity of helminth parasites, in addition to IgM, IgG and IgA antibodies, specific and nonspecific IgE antibody production is commonly observed in helminthiasis. These immunoglobulins *per se* do not kill the invaders, but bind to their surface, facilitate the adherence of inflammatory cells and trigger the release of potentially helminthotoxic substances from the adherent cells.

Complement

The special set of plasma proteins termed "complement" is not a single substance but is a complex series of some 20 proteins which form a cascade of triggering hydrolytic enzymes. When the complement system is triggered by antigen-antibody complex (classical pathway) or by microbial polysaccharide or the cuticular surface of helminths (alternate pathway), the product of one reaction becomes the enzymatic catalyst for the next. Complement has three different biological roles. First of all, complement coats microbes so that the adherence of phagocytic or inflammatory cells having complement receptors is facilitated. Such an effect of complement is called "opsonization" and is a characteristic of C3b. Secondly, among complement cleavage products, C3a and C5a have potent biological activities such as causing the respiratory burst of neutrophils or triggering chemical mediator release from neutrophils. C5a is known to be a potent chemoattractant for neutrophils and eosinophils. C5a is also known as a potent vasodilator that increases vascular permeability. Because of such a wide range of biological activities, C3a and C5a are called anaphylatoxins (toxic substances that mimic the symptoms of anaphylaxis). The third function of complement is to cause cytolysis. Attachment of C5b-C9 complex (membrane attack complex) on microbial cell surface causes membrane damage and resultant cytolysis. This third funtion of complement is applicable to bacteria to protozoan parasites but is not applicable to helminths because the surface of helminths is usually covered by a tough layer such as the "cuticle" in nematodes.

Antibody-dependent or Complement-dependent Cell-mediated Cytotoxicity (ADCC or CDCC)
Since helminth parasites are too large to be affected by phagocytosis, and T cells are most unlikely to have direct effects on helminths, host animals must destroy and kill helminths in alternate ways. As mentioned above, once the surface of the parasite is coated by antibody, attachment of inflammatory cells having Fc receptors is facilitated. Similarly, when the parasite surface is coated with the complement component C3b directly or indirectly by bridging with antibody, attachment of inflammatory cells having C3b receptors is facilitated. After attachment, the inflammatory cells release various potentially helminthotoxic substances such as superoxide anion, proteolytic enzymes and neurotoxins. This type of damage and killing of parasites *in vitro* is called ADCC or CDCC depending on the involvement of either antibody or the complement. Both ADCC and CDCC seem to occur *in vivo* in the host animals anound invading helminths, though their relevance as immune-effector mechanism is difficult to establish unequivocally (see Chapter 15).

Granulomatous Response
Although some parasites or their eggs can be destroyed and digested within a short period by inflammatory cells, the majority of helminths and their eggs are, because of their physico-chemical nature, resistant to extracellular digestion. In such cases, the inflammatory reaction of the host against parasites becomes chronic leading to their encirclement by fibroblasts and their products including collagen. This type of reaction of the host is called a granulomatous response or encapsulation. Although the granulomatous response is fundamentally a result of host's protective reaction, this reaction is often harmful to the host because the granulomas cause mechanical damage to visceral organs of the host. Major pathological changes in schistosomiasis are caused by periovular granulomas. If the parasites entrapped in the granulomas produce proteolytic enzymes, the granulomas are liquefied from the inside so that they become cysts. Typical example of such cysts are seen in pulmonary paragonimiasis.

CELLULAR RESPONSES UNIQUE TO HELMINTH INFECTION
Helminth parasites are multicellular organisms and thus they are not ingested by phagocytic cells such as macrophages or neutrophils. Therefore, extracellular digestion and/or granulomas are the available ways for the host to protect itself against helminth invaders. Although neutrophils and macrophages contain some chemical mediators and hydrolytic enzymes, other specialized cells are involved in extracellular digestion. Eosinophils and mast cells are considered as "appropriate" cells because they contain large

amounts of various chemical mediators, hydrolytic enzymes, and cytotoxic substances in their granules and also in their cytoplasm. They release these substances by degranulation upon stimulation. This may be the major reason why eosinophils and mast cells are associated with helminth infections, if not playing more direct roles in *in vivo* protection.

EOSINOPHILS IN PARASITIC INFECTION

Peripheral blood eosinophilia and tissue eosinophil infiltration are the most characteristic responses of the host infected with helminth parasites. Eosinophils are the minor components of white blood cells in normal animals. In response to the increased demand, their numbers increase rapidly following infection with helminths, particularly with tissue-invading ones. Eosinophils are, like other blood cells, derived from pluripotent hemopoietic stem cells. Therefore, blood eosinophilia and tissue eosinophil infiltration are the results of sequential acceleration of (1) eosinophil production in hemopoietic tissues such as bone marrow, liver, and spleen, (2) mobilization of eosinophils from the site of production to the circulation, (3) emigration of eosinophils from the circulation to the extravascular inflamed site. Schematic diagram of these events and the factors involved in each step are summarized in Fig. 4.

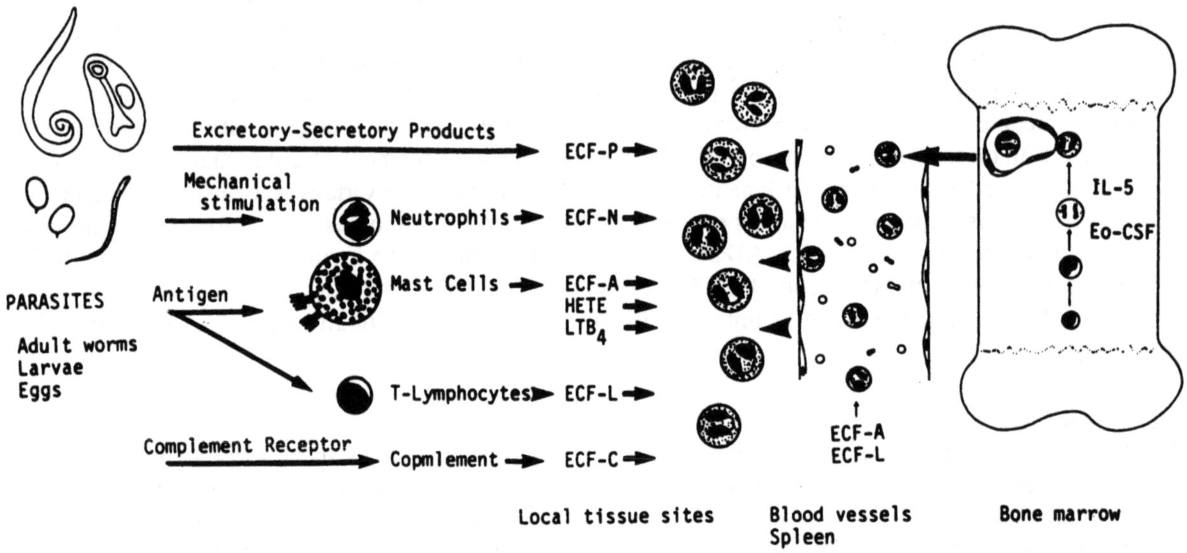

Fig. 4. Schematic diagram of the mechanisms of eosinophil mobilization (Reproduced with permission from J. Clin. Exp. Med. (IGAKU NO AYUMI) Vol. 156: No. 3, 1991).

Eosinophilopoiesis

Eosinophils are usually produced in bone marrow. In small animals like mice, extramedullary eosinophilopoiesis is observed in the liver after infection with *Schistosoma mansoni* [15, 19, 30] or with *S. japonicum* [44]. Such extramedullary eosinophilopoiesis is associated with the accumulation of pluripotent hemopoietic stem cells [44]. Even more prominent extramedullary eosinophilopoiesis in the liver and the accumulation of hemopoietic stem cells were also observed in *Toxocara canis*-infected mice [45]. Mice having a hemopoietic stem cell-deficiency ($WBB6F_1$-W/W^V) are poor responders to eosinophilopoietic [55] or neutrophilopoietic stimuli [14]. The magnitude of acute granulopoiesis is, thus determined primarily by the number of pluripotent hemopoietic stem cells.

The differentiation of eosinophils is under the influence of humoral factors such as "eosinophilopoietin"

[43], eosinophil colony-stimulating factor (Eo-CSF) [47, 56], or eosinophil differentiating factor (EDF) [71, 72]. Eosinophilopoietin, a low molecular weight peptide, was first detected in the serum from eosinophil-depleted animals, and then in the serum from *Trichinella spiralis*-infected rats. Ruscetti et al [70] and Bartelmez et al [8] reported an eosinophil growth stimulating factor (Eo-GSF) in the media conditioned by *T. spiralis*-infected murine spleen cells stimulated with *T. spiralis* antigen. Eo-GSF is similar to eosinophilopoietin in its physico-chemical and functional nature [8]. Eo-CSF, which can support clonal growth/differentiation of eosinophils in the semisolid culture system, was found in media conditioned by spleen cells stimulated with pokeweed mitogen [39, 46] and its physico-chemical nature was characterized [17, 56]. EDF also has B-cell growth factor activity [72] and is now known to be identical to IL-5 [41]. The *in vitro* generation of eosinophils from primitive hemopoietic stem cells seems to require the concatenate action of IL-1, IL-3 and IL-5 [78].

Eosinophil Mobilization

After being produced in the hemopoietic tissues, eosinophils are released into the circulation and emigrate into tissue compartment or inflamed sites. When the mechanisms of eosinophil mobilization are considered, two different steps should be distinguished:

(1) From the site of production to the circulation

The factors controlling this step of eosinophil movement are poorly understood [9]. Although an eosinophil-releasing factor has been reported in a series of passive transfer experiments involving plasma from rats 12–24 h after a single injection of *T. spiralis* larvae [73], whether this eosinophil release is from bone marrow or from the marginating pool [13] of the vascular compartment was not studied critically. We have found that both parasite-derived high m.w. ECF [58] and host mast cell-derived low m.w. ECF-A [59] were detected in the serum during the acute stage of murine schistosomiasis. Since both of these ECFs were chemotactic to eosinophils obtained from bone marrow, they could be possible candidates for eosinophil-releasing factors from the site of production to the circulation.

(2) From circulation to tissue

Various parasites contain and/or release ECF [36], some of which have been isolated and characterized physico-chemically. In addition, various host-derived ECFs such as mast cell-derived ECF-A, complement-derived ECF, and lymphocyte-derived ECF are, in one way or another, involved in the mediation of eosinophil emigration from circulation to tissue [81]. In parasitic infections, the generation, production and/or release of these host-derived ECFs are obviously dependent on the antigenic stimuli of the parasites. We have reported that splenic T cells [62] and granuloma T cells [60] could produce and release ECF lymphokine upon stimulation with a specific antigen. All these parasite-derived and host-derived ECFs are known to act as chemoattractants that cause directional migration of eosinophils from the vascular compartment to tissue compartment.

Since some of the parasite-derived and host-derived ECFs can be detected in the circulation, they might act as eosinophil-releasing factor as well. To study this point, the use of different maturation stages of eosinophils seems to be helpful. Recently, we have found that, using eosinophils obtained from the bone marrow and peritoneal cavity, ECF-lymphokine directed against bone marrow eosinophils and against peritoneal eosinophils were apparently different molecules [62, 63, 64].

Immunological Regulation of Eosinophilia

Since an array of cytokines and ECFs involved in the production and mobilization of eosinophils are produced by the host's immune system, there is little doubt about the regulatory role of the immune system in the mechanism of eosinophilia. However, the strict T cell dependency is still controversial in helminth infections. T cell-dependent eosinophilia was reported in *T. spiralis* [77] and *S. mansoni* [67] infections. Vadas [75] reported that eosinophilia induced by the non-parasite antigen, key-hole limpet hemocyanin (KLH), was also dependent on T cells. On the other hand, Sugane & Oshima [74] found that, after *T. canis* infection, the magnitude of the major peak of peripheral blood eosinophilia was comparable between nu/+ and nu/nu mice. Thymus-independent eosinophilia induced by a non-parasite antigen (KLH) was reported recently by using congenitally athymic rnu/rnu rats [12].

Although the role of macrophages in mediation of eosinophilia is not clearly understood, Hirashima et al [35] reported a regulatory role of macrophage products on ECF lymphokine production. In our recent study, when isolated intact granulomas of *S. japonicum*-infected mice were cultured under the presence of granulocyte-macrophage colony-stimulating factor (GM-CSF), which was partially purified from the media conditioned by spleen cells of *S. japonicum*-infected mice, significant augmentation of ECF lymphokine production was observed, and this enhanced ECF lymphokine production was due to the activation of macrophages by GM-CSF [61].

Functions of Eosinophils in Helminth Infection

Since eosinophils have Fc-gamma, Fc-epsilon, C3b and C5a receptors on their surface, when the invaded helminths are coated with antibody and/or complement, eosinophils attach to the surface of the parasites and release various helminthotoxic substances such as superoxide anion, proteolytic enzymes, neurotoxins etc. Such ADCC-type helminthotoxicity was first reported by using schistosomula of *S. mansoni* as the target [7, 18]. Subsequently, eosinophil-mediated ADCC was reported by using *S. mansoni* eggs [37], newborn larvae of *T. spiralis* [40], or microfilariae of *Onchocerca volvuls* [28]. Capron et al [20] reported functional heterogeneity of eosinophils in that only 'hypodense' eosinophils could mediate high levels of IgE-dependent schistosomula killing. Concerning functional heterogeneity, we found that murine eosinophils in bone marrow were strongly attracted by mast cell-derived ECF-A, whereas those in the peritoneal cavity did not react to ECF-A [54]. When eosinophils in bone marrow were exposed to mast cell-derived ECF-A, their chemotactic reactivity to parasite-derived ECF was enhanced [54]. These results indicate that the functions of eosinophils are potentiated by various chemoattractants and/or cytokines.

MAST CELLS IN PARASITIC INFECTION

Mast cells can be classified into connective tissue type (CTMC) and mucosal type (MMC). CTMC is known as the source of chemical mediators released during an immediate type hypersensitivity response. Since eosinophilia and elevation of parasite-specific and non-specific IgE production are the characteristic responses of the host during helminth infections, CTMC is considered to be involved in the host's defence and/or pathogenesis. The presence and the nature of MMC has been studied mainly by using *Nippostrongylus brasiliensis*-infected rats, and now MMC is known to be closely related to intestinal helminthiasis and is considered as one of the important candidate cells involved in immune-mediated worm expulsion from the intestine (see Chapters 1 and 15).

Although ample evidence is available that local mastocytosis is associated with parasitic infections, much of it is derived from observations on MMC responses in the small intestine during intestinal helminthiasis in experimental animals. For example, intestinal mastocytosis is caused not only by infection with intestinal helminths such as *N. brasiliensis* [48], *T. spiralis* [69], or *Strongyloides ratti* [21] but also with intestinal cestode *Hymenolepis nana* [16]. In human cases, intestinal mastocytosis is reported in *T. spiralis* infection [31]. In rats and human, MMC are observed in the small intestine even without parasitic [10, 27] infections, though their distribution is limited to the intestinal lamina propria. After parasitic infection, MMC increase in number not only in the lamina propria but also in the intestinal epithelial cell layer. MMC are rare in the small intestine of mice and, after *S. ratti* infection in mice, the majority of them are seen in the epithelial cell layer [2].

Concerning CTMC, mast cells are seen in periovular granulomas in the liver of mice chronically infected with *S. mansoni* [80]. Both CTMC and MMC are seen in the granulomatous lesions formed in the liver or pancreas of rats by penetration of cysticercoids of *Mesocestoides corti* [22]. The contribution of CTMC to granulomatous inflammation remains unknown.

Mast Cell Differentiation

The growth/differentiation of MMC is obviously dependent on T lymphocytes. The number of CTMC in athymic nude mice is comparable to that in euthymic control mice, whereas MMC are absent in nude mice even after parasitic infection [69]. When mesenteric lymph node cells obtained from parasitised rats or mice are stimulated *in vitro* with specific parasite antigen, the conditioned media contains mast cell-growth factor (= IL-3) activity [3, 33], and this activity is produced by helper T cells [29]. Thus, as the mechanism of intestinal mastocytosis, IL-3, which is produced by heper T cells upon stimulation with parasite antigen, promotes growth/differentiation of locally homed MMC precursors. Nude mice have normal numbers of MMC precursor but lack IL-3 producing cells, so that they are unable to elicit MMC responses even following helminth infection. However, MMC responses can be induced in nude mice by repeated injection of IL-3 without parasite infection, indicating that IL-3 alone is sufficient to induce an MMC response [5]. Although the majority of intestinal helminth infections cause intestinal mastocytosis in rats and mice, infection with adult worm of *Nematospiroides dubius* causes depression of the number of MMC in the intestine. This depression is assumed to be caused by depression of mast cell-growth factor production [26].

The growth/differentiation of CTMC has been thought to be independent of the presence of thymus, because nude mice have normal numbers of CTMC. However, *in vitro* growth/differentiation requires the presence of IL-4, a lymphokine mainly produced by T cells [34]. Furthermore, IL-3 dependent bone marrow-derived cultured mast cells can differentiate into CTMC upon contact with fibroblasts [42]. Therefore, under certain circumstances, growth/differentiation of CTMC seems to be regulated by T cells. In fact, the number of CTMC in the periovular granulomas of *S. mansoni*-infected mice is regulated by T cells [80]. From these facts, it can be postulated that, CTMC that are present in the connective tissue, and those developing in an inflammatory site, are different in terms of T cell or growth factor dependency.

Eosinophil and Mast Cell-derived Mediators

Antigen-specific and nonspecific IgE responses are induced by many helminth infections. In such infections, mast cells of the parasitised host have been sensitised by IgE antibody [38] and, upon stimulation with parasite antigen, they release an array of chemical mediators. Some of the mast cell-derived mediators, such as ECF-A, HETE, LTB_4 are known to have eosinophilotactic activity and seem to paly an important role in the mechanism of eosinophilia in parasitic infections. In *S. japonicum* infection, at around the time of periovular eosinophilic granuloma formation, both ECF-A and IgE antibody specific to soluble egg antigen become detectable in the serum [59]. In *S. japonicum*-infected mice, eosinophilic granulomas are smaller in size in mast cell deficient W/W^V [66] or IgE-deficient SJA/9 mice [65] than in their littermate controls. ECF-A released from mast cells acts on eosinophils not merely as a chemotactic but also as the functional potentiator. Once eosinophils are exposed to ECF-A *in vitro*, they lose chemotactic reactivity to homologous attractant [79], but their reactivity to parasite-derived ECF is enhanced [54]. In such *in vitro* phenomena are applied to *in vivo* situation, eosinophils produced in the hemopoietic organ mobilize into systemic circulation by ECF-A, ECF-lymphokine or other eosinophilotactic factors, and there they are exposed to ECF-A. After exposure to ECF-A in the circulation, chemotactic reactivity of eosinophils to parasite-derived ECF is potentiated so that they reach to the site where parasites or their eggs exist. The eosinophils thus mobilized to the inflamed site act as the effector cells to kill or destroy parasites or eggs in the presence of complement and/or antibody.

Role of Mucosal Mast Cells in Worm Expulsion

MMC have been thought to play important roles in warm expulsion [11, 82] because the peak of intestinal mastocytosis coincides well with the time of immune mediated worm expulsion in *T. spiralis* [6] or *S. ratti* [50] infection. In *N. brasiliensis* infection, MMC response is observed at around the terminal stage of worm expulsion [53] so that the importance of MMC in worm expulsion was discounted. However, the appearance of rat mast cell protease-II in the serum preceded to the appearance of MMC and coincided with the time of worm expulsion [82], supporting the importance of MMC-derived mediators in the expulsion mechanism.

To study the role of mast cells in the defence mechanism of the host, mast cell-deficient W/W^V mice are a fascinating model. When W/W^V mice were infected with *N. brasiliensis*, the MMC response was not observed and worm expulsion was delayed. Although defective MMC response of W/W^V mice was restored by reconstitution with normal bone marrow cells or spleen cells, delayed worm expulsion was not restored [25]. Defective MMC response and delayed worm expulsion were also observed in W/W^v mice after infection with *T. spiralis* [32, 57] or with *S. ratti* [52]. In these cases, both delayed worm expulsion as well as defective MMC response was restored by bone marrow reconstitution. These results suggest that MMC are important in the expulsion mechanism at least against *T. spiralis* and *S. ratti*.

Since W/W^V mice are, because of their hemopoietic stem cell deficiency, not merely deficient in mast cells but also have a wide range of hematologic disorders, a bone marrow reconstitution experiment does not rule out the contributions of cells other than mast cells. To examine directly the role of mast cells in worm expulsion, reconstitution experiment using pure mast cells or their precursors needs to be performed. When W/W^V mice were reconstituted with bone marrow-derived cultured mast cells (> 99% pure) and infected with *S. ratti*, MMC increased in number without causing accelerated worm expulsion [1]. The ineffectiveness of mast cell transfer to cause worm expulsion was subsequently explained by the fact that adoptively transfered mast cells located mainly in the intestinal lamina propria, whereas MMC induced in mast cell-sufficient mice by *S. ratti* infection were mainly seen in the epithelial layer [2]. These results indicate that, although cultured mast cells have many functional and morphological similarities to MMC, they are unable to migrate into the mucosal epithelium.

As an alternative approach to elucidate the role of mast cells in worm expulsion, *S. ratti*-infected nude mice were treated with repeated injection of IL-3. Nude mice were unable to expel *S. ratti* and unable to mount an MMC response. Defective MMC response in nude mice is known not due to the defects in MMC precursors but due to a defect in IL-3 production. If IL-3-dependent MMC response is critical for worm expulsion, IL-3 administration to parasitised nude mice should cause worm expulsion. In fact, when *S. ratti* -infected nude mice were given a total of 6×10^5 U IL-3, adult worms of *S. ratti* in the intestine were almost completely expelled in association with intestinal mastocytosis [4]. To see whether IL-3-dependent MMC response is effective only against certain parasite species, nude mice were concurrently infected with two species of intestinal helminths, *N. brasiliensis* and *S. ratti*, and then treated with IL-3. The results showed that only *S. ratti* was expelled from the intestine, whereas *N. brasiliensis* remained unaffected (Abe et al: Immunology **76**: 10–14, 1992). Thus, mast cells in the intestinal mucosa although play a key role in worm expulsion, their effect is highly selective and only certain species of parasites could be expelled form the gut lumen by this effector mechanism (Fig. 5). Other expulsive mechanisms such as mucus trapping needs to be investigated in detail.

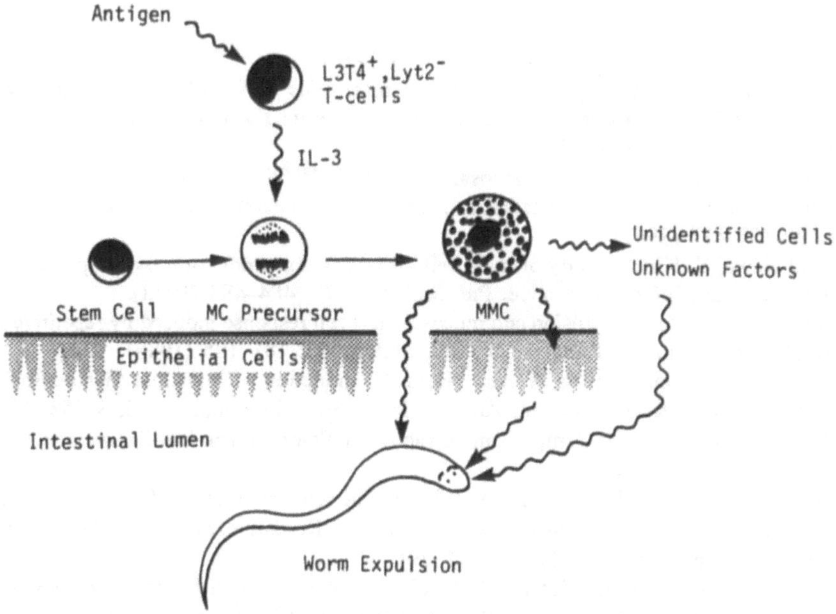

Fig. 5. Schematic diagram of the effect of IL-3 mucosal mast cell system on immune mediated worm expulsion (Reproduced with permission from "Mast Cells—Physiology and Pathology" (M. Kurosawa ed), Medical Review Co. Ltd., Tokyo, 1990).

CONCLUSION

In this review, the first section portraited a rough sketch of the host's defence system. There is little doubt about the involvement of all these cellular mechanisms against invasion of helminths. Specific antibodies against somatic components and/or excretory/sectetory products of parasites are commonly detected in the serum of the hosts infected with helminths. Sensitised T cells are also commonly detectable by blastogenic responses to helminth antigens in infected animals. In fact, detection of specific antibody or sensitised lymphocytes are often useful tool for the diagnosis of helminthiasis. However, due to the extraordinary large size of the helminths, both antibody and lymphocytes are, even with the cooperation of macrophages and neutrophils, unable to act as the direct effector to invaded multicellular organisms. Unique cellular responses, usually called "non-specific" inflammatory responses, observed in helminth infections are, therefore, generated to protect and maintain homeostatic integrity of the host animals. Since eosinophils and mast cells are, among these "nonspecific" inflammatory cells, thought to play important roles in the host defence system against helminths, these two cells were highlighted in the second section of this review. Roughly, eosinophils are considered as the effector/regulator cells to fight against tissue-migrating helminths and to encapsulate them, whereas mucosal mast cells are the effector/regulator cells to expel lumen dwelling parasites from the intestinal tract. To strengthen this idea, further systematic study is required for the identification and characterization of parasite antigens to stimulate eosinophilopoiesis or mastocytopoiesis and also for the understanding of the host's switching mechanisms to induce such unique cellular responses.

REFERENCES

1. Abe, T. and Nawa, Y. Reconstitution of mucosal mast cells in W/W^V mice by adoptive transfer of bone marrow-derived cultured mast cells and its effects on the protective capacity to *Strongyloides ratti* infection. Parasite Immunol. **9**: 31–38 (1987).
2. Abe, T. and Nawa, Y. Localization of mucosal mast cells in W/W^V mice after reconstitution with bone marrow cells of cultured mast cells, and its relation to the protective capacity to *Strongyloides ratti* infection. Parasite Immonol. **9**: 477–485 (1987).
3. Abe, T. and Nawa, Y. Kinetic study of mast cell growth factor production by lymphocytes during the course of *Strongyloides ratti* infection in mice. Parasitol. Res. **74**: 484–488 (1988).
4. Abe, T. and Nawa, Y. Worm expulsion and mucosal mast cell response induced by repetitive IL-3 administration in *Stronglyloides ratti*-infected nude mice. Immunology **63**: 181–185 (1988).
5. Abe, T., Ochiai, H., Minamishima, Y. and Nawa, Y. Induction of intestinal mastocytosis in nude mice by repeated injection of interleukin-3. Int. Archs Allergy Appl. Immunol. **86**: 356–358 (1988).
6. Alizadeh, H. and Wakelin, D. Comparison of rapid expulsion of *Trichinella spiralis* in mice and rats. Int. J. Parasitol. **12**: 65–73 (1982).
7. Anwar, A.R.E., Smithers, S.R. and Kay, A.B. Killing of schistosomula of *Schistosoma mansoni* coated with antibody and/or complement by human leukocytes *in vitro*: Requirement for complement in preferential killing by eosinophils. J. Immunol. **122**: 628–637 (1979).
8. Bartelmez, S.H., Dodge, W.H., Mahmoud, A.A.F. and Bass, D.A. Stimulation of eosinophil production *in vitro* by eosinophilopoietin and spleen cell derived eosinophil growth-stimulating factor. Blood **56**: 706–711 (1980).
9. Beeson, P.B. and Bass, D.A. Life cycle. In: *The eosinophil. Major problems in internal medicine* (L.H. Smith Jr. ed), **15**: 3–9. Saunders: Philadelphia (1989).
10. Befus, A.D., Pearce, F.L., Gauldie, J., Horsewood, P. and Bienenstock, J. Mucosal mast cells. I. Isolation and functional characteristics of rat intestinal mast cells. J. Immunol. **128**: 2475–2480 (1982).
11. Bienenstock, J., Befus, A.D., Pearce, F., Denburg, J. and Goodacre, R. Mast cell heterogeneity: derivation and function, with emphasis in the intestine. J. Allergy Clin. Immunol. **70**: 407–412 (1982).
12. Blomjous, F.J.E.M., Elegersma, A., Kruzinga, W. and Ruitenburg, E.J. Thymus independence of eosinophilia induced by a non-parasite antigen. Int. Archs Allergy Appl. Immunol. **79**: 376–379 (1986).
13. Boggs, D.R. and Winkelstein, A. *White Cell Manual.* 3rd edn. (F.A. Davis ed). Saunders: Philadelphia (1978).
14. Boggs, S.S., Wilson, S.M. and Smith, W.W. Effects of endotoxin on hematopoiesis in irradiated and non-irradiated W/W^v mice. Radiat. Res. **56**: 481–493 (1973).
15. Borojevic, R., Srocker, S. and Grimaud, J.A. Hepatic eosinophil granulocytopoiesis in murine experimental schistosomiasis mansoni. British J. Exp. Pathol. **62**: 480–489 (1981).
16. Bortoletti, G., Gabriele, F. and Palmas, C. Kinetics of mast cells, eosinophils and phospolipase activity in the spontaneous-cure response of two strains of mice (rapid and slow responder) to the cestode *Hymenolepis nana*. Parasitol. Res. **75**: 465–469 (1989).
17. Burgess, A. and Nicola, N. (eds). *Growth factors and stem cells.* pp. 43–91. Academic Press: Sydney (1983).
18. Butterworth, A.E., Sturrock, R.F. and Houba, V. Antibody-dependent cell mediated damage to schistosomula *in vitro*. Nature (Lond.) **252**: 503–505 (1974).
19. Byram, J.E., Imohiosen, E.A.E. and Von Lichtenberg, F. Tissue eosinophil proliferation and maturation in schistosome-infected mice and hamsters. Am. J. Trop. Med. Hyg. **27**: 267–270 (1978).
20. Capron, M., Capron, A., Dessaint, J.P., Torpier, G., Johansson, S.G.O. and Prin, L. Fc receptors for IgE on human and rat eosinophils. J. Immunol. **126**: 2087–2092 (1981).
21. Carroll, S.M., Mayrhofer, G., Dawkins, H.J.S. and Grove, D.I. Kinetics of intestinal lamina propria mast cells, globule leucocytes, intraepithelial lymphocytes, goblet cells and eosinophils in murine strongyloidiasis. Int. Archs Allergy Appl. Immunol. **74**: 311–317 (1984).
22. Chernin, J., Miller, H.R.P., Newlands, G.F.J. and Mclaren, D.J. Proteinase phenotypes and fixation properties of rat mast cells in parasitic lesions caused by *Mesocestoides corti*: selective and site-specific recruitment of mast cell subsets. Parasite Immunol. **10**: 422–433 (1988).

23. Cohen, S. and Sadun, E.H. (eds). *Immunology of parasitic infections.* Blackwell Scientific Publications: Oxford (1976).
24. Cox, F.E.G. (ed). *Modern Parasitology.* Blackwell Scientific Publications: Oxford (1982).
25. Crowle, P.K. Mucosal mast cell reconstitution and *Nippostrongyulus brasiliensis* rejection by W/W^v mice. J. Parasitol. **69**: 66–69 (1983).
26. Dehlawi, M.S. and Wakelin, D. Suppression of mucosal mastocytosis by *Nematospiroides dubius* results from an adult worm-mediated effect upon host lyumphocytes. Parasite Immunol. **10**: 85–95 (1988).
27. Fox. C.C., Dvorak, A.N., Peters S.P., Kagey-Sobotka, A. and Lichtenstein, L.M. Isolation and characterization of human intestinal mucosal mast cells. J. Immunol. **135**: 483–491 (1985).
28. Greene, B.M., Taylor, H.R. and Aikawa, M. Cellular killing of microfilariae of *Onchocerca volvulus*: Eosinophil and neutrophil-medaited immune serum-dependent destruction. J. Immunol. **127**: 1611–1618 (1981).
29. Grencis, R.K., Riedlinger, J. and Wakelin, D. Lymphokine production by T cells generated during infection with *Trichinella spiralis*. Int. Archs Allergy Appl. Immunol. **83**: 92–95 (1987).
30. Grimaud, J.A. and Borojevic, R. Mesenchyme et parenchyma hepatique dans la bilharziose experimentale a *Schistosoma mansoni*: Metaplasie Myeloide. C.R. Acad. Sci. Paris **274**: 897–899 (1972).
31. Gutowska, L., Ruitenberg, E.J., Elgersma, A., and Kocieka, W. Increase of mucosal mast cells in the jejunum of patients infected with *Trichinella spirallis*. Int. Archs Allergy Appl. Immunol. **71**: 304–308 (1983).
32. Ha, T.Y., Reed N.D. and Crowle, P.K. Delayed expulsion of adult *Trichinella spiralis* by mast cell-deficient W/W^V mice. Infect. Immun. **41**: 445–447 (1983).
33. Haig, D.M. Mckee, T.A., Jarrett, E.E., Woodbury, R. and Miller, H.R.P. Generation of mucosal mast cells is stimulated *in vitro* by factors derived from T cells of helminth-infected rats. Nature (lond.) **300**: 188–190 (1982).
34. Hamaguchi, Y. Kanakura, Y., Fujita, J., Takeda, S. Nakano, T., Tarui, S., Honjo, T. and Kitamura, Y. Interleukin 4 as an essential factor for *in vitro* clonal growth of murine connective tissue-type mast cells. J. Exp. Med. **165**: 268–273 (1987).
35. Hirashima, M. Tashiro, K., Sakata, K., Yoshimura, T. and Hayashi, H. Difference in requirement of macrophage products for concanavalin A-induced production of eosinophil chemotactic factor and macrophage chemotactic factor from guinea pig lymphocytes *in vitro*. J. Immunol. **133**: 2012–2017 (1984).
36. Horii, Y., Owhashi, M., Fujita, K., Nakanishi, H. and Ishii, A. A comparative study on eosinophil and neutrophil chemotactic activites of various helminth parasites. Parasitol. Res. **75**: 76–78 (1988).
37. James, S.L. and Colley, D.G. Eosinophil-mediated destruction of *Schistosoma mansoni* eggs *in vitro*. II. The role of cytophilic antibody. Cell. Immunol. **38**: 35–47 (1978).
38. Jarrett, E.E.E., Orr, T.S.C. and Riley, P. Inhibition of allergic reactions due to competition for mast cell sensitization sites by two reagins. Clin. Exp. Immunol. **9**: 585–594 (1971).
39. Johnson, G.R. and Metcalf, D. Detection of a new type of mouse eosinophil colony by Luxol-Fast-Blue staining. Exp. Hematol. **8**: 549–561 (1980).
40. Kazura, J.W. and Aikawa, M. Host defense mechanisms against *Trichinella spiralis* infection in the mouse: Eosinophil-mediated destruction of newborn larvae *in vitro*. J. Immunol. **124**: 355–361 (1980).
41. Kinashi, T., Harada, N., Severson, E., Tanabe. T., Sideras, M., Konishi, M., Azuma, C., Tominaga, A., Bergstedt-Lindqvist, S., Takahashi, M., Matsuda, F., Yaoita, Y., Takatsu, K.,. and Honjo, T. Cloning of complementary DNA encoding T-cell replacing factor and identity with B-cell growth factor II. Nature (Lond.) **324**: 70–73 (1986).
42. Levi-Schaffer, F., Austen, K.F., Gravallese, P.M. and Stevens, R.L. Co-culture of interleukin 3-dependent mouse mast cells with fibroblasts results in a phenotypic changes of the mast cells. Proc. Nat. Acad. Sci., U.S.A. **83**: 6485–6488 (1986).
43. Mahmoud, A.A.F., Stone, M.K. and Kellermeyer, R.W. Eosinophilopoietin. A circulating low molecular weight peptide-like substance which stimulates the production of eosinophils in mice. J. Clin. Invest. **60**: 675–682 (1977).
44. Maruyama, H., Higa, A., Asami, M., Owhashi, M. and Nawa, Y. Extrameduallary eosinophilopoiesis in the liver of *Schistosoma japonicum*-infected mice, with special reference to hemopoietc stem cells. Parasitol. Res. **76**: 461–465 (1990).

45. Maruyama, H., Higa, A., Asami, M., Owhashi, M. and Nawa, Y. Hepatic eosinophilopoiesis from multipotent hemopoietic stem cells in *Toxocara canis*-infected mice. Exp. Hematol. **19**: 77–80 (1990).
46. Metcalf, D. and Johnson, G.R. Production by spleen and lymph node cells of conditioned medium with erythroid and other hemopoietic colony-stimulating activity. J. Cell Physiol. **96**: 31–42 (1978).
47. Metcalf, D., Parker, J., Chester, H.M. and Kincade, P.W. Formation of eosinophil-like granulocytic colonies by mouse bone marrow cells *in vitro*. J. Cell Physiol. **84**: 275–289 (1974).
48. Miller, H.R.P. and Jarrett, W.F.E. Immune reactions in mucus membranes I. Intestinal mast cell response during helminth expulsion in the rats. Immunology **20**: 277–288 (1971).
49. Miller, H.R.P. Gastrointestinal mucus, a medium for survival and for elimination of parasitic nematodes and protozoa. Parasitology **94**: S77–S100 (1987).
50. Mimori, T., Nawa, Y., Korenaga, M. and Tada. I. *Strongyloides ratti*: mast cell and goblet cell response in the small intestine of infected rats. Exp. Parasitol. **54**: 366–370 (1982).
51. Mims, C.A. (ed). *The pathogenesis of infectious disease*. Academic press: New York (1976).
52. Nawa, Y., Kiyota, M., Korenaga, M. and Kotani, M. Defective protective capacity of W/W^V mice against *Strongyloides ratti* infection and its reconstitution with bone marrow cells. Parasite Immunol. **7**: 429–438 (1985).
53. Nawa, Y. and Korenaga, M. Mast and goblet cell responses in the small intestine of rats concurrently infected with *Nippostrongylus brasiliensis* and *Strongyloides ratti*. J. Parasitol. **69**: 1168–1170 (1983).
54. Nawa, Y., Owhashi, M., Imai, J. and Abe, T. Chemotactic reactivity of eosinophils obtained from bone marrow and peritoneal cavity of cyclophosphamide-treated, *Toxocara canis*-infected mice. Int. Archs Allergy Appl. Immunol. **80**: 412–416 (1986).
55. Nawa, Y., Owhashi, M., Imai, J. and Abe, T. Eosinophil response in mast cell-deficient W/W^V mice. Int. Archs Allergy Appl. Immunol. **83**: 6–11 (1987).
56. Nocola, N.A., Metcalf, D., Johnson, G.R. and Burgess, A.W. Separation of functionally distinct human granulocyte-macrophage colony stimulating factors. Blood **54**: 614–627 (1979).
57. Oku, Y., Itayama, H. and Kamiya, M. Expulsion of *Trichinella spiralis* from the intestine or W/W^V mice reconstituted with haematopoietic and lymphopoietic cells and origin of mucosal mast cells. Immunology **53**: 337–344 (1984).
58. Owhashi, M., Horii, Y., Ishii, A. and Nawa, Y. Detection of high molecular weight eosinophil chemotactic factor in murine schistosomiasis sera. Am. J. Trop. Med. Hyg. **35**: 1192–1197 (1986).
59. Owhashi, M., Horii, Y., Ishii, A. and Nawa, Y. Low molecular weight eosinophil chemotactic factor (ECF) in the serum of murine Schistosomiasis japonica. Int. Archs Allergy Appl. Immunol. **79**: 178–181 (1986).
60. Owhashi, M., Maruyama, H. and Nawa, Y. Eosinophil chemotactic lymphokine produced by egg-associated granulomas in murine schistosomiasis japonicum. Infect. Immun. **54**: 723–727 (1986).
61. Owhashi, M., Maruyama, H. and Nawa, Y. Granulocyte-macrophage colony-stimulating factor enhances the production of eosinophil chemotactic lymphokine by egg-associated granulomas of *Schistosoma japonicum*-infected mice. Infect. Immun. **55**: 2042–2048 (1987).
62. Owhashi, M. and Nawa, Y. Eosinophil chemotactic lymphokine produced by spleen cells of *Schistosoma japonicum*-infected mice. Int. Archs Allergy Appl. Immunol. **82**: 20–25 (1987).
63. Owhashi, M. and Nawa, Y. Eosinophil chemotacitc lymphokine produced by spleen cells of *Schistosoma japonicum*-infected mice. II. Physico-chemical heterogeneity of eosinophil chemotactic lymphokines selective to bone marrow- or peritoneal exudate-eosinophils. Int. Archs Allergy Appl. Immunol. **83**: 290–295 (1987).
64. Owhashi, M. and Nawa, Y. Eosinophil chemotactic lymphokine produced by spleen cells of *Schistosoma japonicum*-infected mice. III. Isolation and characterization of two distinctive eosinophil chemotactic lymphokines directed against difference maturation stages of eosinophils. Int. Archs Allergy Appl. Immunol. **84**: 185–189 (1987).
65. Owhashi, M., Nawa, Y. and Watanabe, N. Granulomatous response in selective IgE-deficient SJA/9 mice infected with *Schistosoma japonicum*. Int. Archs Allergy Appl. Immunol. **90**: 310–312 (1989).
66. Owhashi, M. Horii, Y., Ikeda, T., Maruyama, H., Abe, T. and Nawa, Y. Importance of mast cell-derived ECF-A on granuloma formation in murine schistosomiasis japonica: evaluation using mast cell-deficient W/W^V mice. Int. Archs Allergy Appl. Immunol. **42**: 64–68 (1990).

67. Phillips, S.M., Diconza, J.J., gold, J.A. and Reid, W.A. Schistosomiasis in congenitally athymic (nude) mouse. I. Thymic dependency of eosinophilia, granuloma formation and host morbidity. J. Immunol. **118**: 584–599 (1977).
68. Roitt, I., Brostoff, J. and Male, D. (eds). *Immunology*. 2nd edn. Churchill Livingstone, Gower Medical Publishing: London, New York (1989).
69. Ruitenberg. E.J. and Elgersma, A. Absence of intestinal mast cell response in congenitally athymic mice during *Trichinella spiralis* infection. Nature (Lond.) **264**: 258–260 (1976).
70. Ruscetti, F.W., Cypes, R.H. and Chervenick, P.A. Specific release of neutrophilc- and eosinophilc-stimulating factors from sensitized lymphocytes. Blood **47**: 75–765 (1976).
71. Sanderson, C.J., Warren, D.J. and Strath, M. Identification of a lymphokine that stimulates eosinophil differentiation *in vitro*. Its relationship to interleukin 3, and functional properties of eosinophils produced in cultures. J. Exp. Med. **162**: 60–74 (1985).
72. Sanderson, C.J., O'Grra, A., Warren, D.J. and Klaus, G.G.B. Eosinophil differentiation factor also has B-cell growth factor activity: Proposed name interleukin 4. Proc. Nat. Acad. Sci., U.S.A. **83**: 437–440 (1986).
73. Spry, C.J.F. Mechanism of Eosinophilia. VI. Eosinophil mobilization. Cell Tissue Kinet. **4**: 365–374 (1971).
74. Sugane, K. and Oshima, T. Eosinophilia, granuloma formation and migratory behaviour of larvae in the congenitally athymic mouse infected with *Toxocara canis*. Parasite Immunol. **4**: 307–318 (1982).
75. Vadas M.A. Cyclophoshamide pretreatment induces eosinophilia to non-parasite antigens. J. Immunol. **127**: 2083–2088 (1981).
76. Wakelin, D. *Immunity to Parasites: How animals control parasite infections*. Edward Arnold: London (1984).
77. Walls, R.S., Carter, R.L., Leuchers, E. and Davies, A.J.S. The immunopathology of trichiniasis in T-cell deficient mice. Clin. Exp. Immunol. **13**: 231–242 (1973).
78. Warren, D.J. and Moore, M.A.S. Synergism among interleukin 1, interleukin 3, and interleukin 5 in the production of eosinophils from primitive hemopoietic stem cells. J. Immunol. **140**: 94–99 (1988).
79. Wasserman S.I., Whitmer, D., Goetzl, E.J. and Austen, K.F. Chemotactic deactivation of human eosinophils by the eosinophil chemotactic factor of anaphylaxis. Proc. Soc. Exp. Biol. Med. **148**: 301–306 (1975).
80. Weinstock, J.V. and Boros, D.L. Modulation of granulomatous hypersensitivity. VI. T lymphocyte subsets influence mast cell density in liver granulomas of *Schistosoma mansoni*-infected mice. J. Immunol. **131**: 959–961 (1983).
81. Weller, P.F. and Goetzel, E.J. The regulatory and effector roles of eosinophils. Adv. Immunol. **27**: 339–371 (1980).
82. Woodbury, R.G., Miller, H.R.P. Huntley, J.F., Newlands, G.F.J., Palliser, A.C. and Wakelin, D. Mucosal mast cells are functionally active during spontaneous expulsion of intestinal nematode infection in rat. Nature (Lond.) **312**: 450–452 (1984).

10. Antigens of Parasitic Helminths in Protection and Pathology

L.J.S. Harrison and R.M.E. Parkhouse*
Centre for Tropical Veterinary Medicine, University of Edinburgh
Easter Bush, Roslin, Midlothian, Scotland

*Institute for Animal Health, Pirbright Laboratories
Pirbright, Working, Surrey, UK

INTRODUCTION

The low mortality-high morbidity typical of helminth infections, presumably reduces interest and thus well focussed investigation into their control and treatment, despite their worldwide prevalence [22]. In addition to causing economic losses to livestock industries, these infections typically provoke severe disablement in the human population through pathological lesions and malnutrition. Elephantiasis and blindness in filarial infections constitute the most profoundly disturbing examples of this.

AIDS, currently spreading worldwide, but particularly rapidly in less developed countries, where helminth and many other infections abound, threatens to profoundly change the mortality/morbidity balance in helminth infections. In the face of the compromised immune system associated with AIDS, it is not unreasonable to predict an exacerbation of the consequences of these chronic helminth infections.

Recent technical innovations including immunochemical analysis and monoclonal antibodies, molecular biology with recombinant antigens, cellular immunology with T cell clones and anti-idiotypic reagents are ideally suited to the immediate objectives of obtaining a better understanding of the host-parasite interphase in helminth infections, leading to vaccine production and alleviation of immunopathological consequences. Progress in these areas is very rapid so this chapter attempts to present a general overview of the field, concentrating on protection and pathology while emphasizing current principles and drawing upon systems with defined molecular correlates as the major examples. Particular emphasis will be paid to the nematode *Trichinella spiralis*, as it serves as a focus and stimulus for the research objectives outlined in this article.

Relevant reviews, focussing of defined helminth antigens have been published for cestodes [34]. nematodes [2], schistosomes [87] and trematodes [39]. Since then an excellent general review on helminth antigens including those of gastro-intestinal nematodes has been presented [61]. Molecular aspects of recent studies on schistosomes [17, 62, 85], filarial nematodes [21, 64], nematodes [63, 80], and other intestinal helminths [77, 78] have also been well covered in the last few years, along with general immunity to parasites [48, 96].

GENERAL CONSIDERATIONS

Characterization of the immune responses of hosts to their parasites is basic to the understanding of all aspects of parasitic diseases. In essence, this involves determining which parasite components are immunogenic and defining the range of immune responses and effector mechanisms called into play. The clinical status of the host, such as resistant versus susceptible, or degree of pathological damage, will depend on the interactions within this presumably complex network of immunological phenomena. An additional problem which needs to be considered is the possible existance of antigenically dissimilar parasite races, or strains (see Chapters 5 and 15).

For convenience, the variety of different antigens presented to the host in helminth infections may be classified into four major levels: parasite stage, antigenic compartment within a stage (e.g. surface, secretions or somatic antigens), antigenic components within a compartment and antigenic epitopes of a single defined antigen. A further level of complexity in the serological response to a parasite is the variable immunoglobulin isotype profile elicited by each individual antigen or antigenic epitope.

The basic point to be emphasized is that each individual antigenic epitope of all the antigens, presented by a parasite to its host, elicits its own independent response, both in terms of quantity, Ig isotype profile and antibody affinity [3, 4, 5, 8, 29, 40, 42]. The net result is an enormously complex array of antibody specificities, presenting us with the problem of elucidating which antibody specificities are relevant to protection and pathology.

The fact that the quantity and quality (Ig isotype) of antibodies synthesized to parasite antigens may vary, sometimes due to genetic differences between outbred hosts and often with time-dependent characteristics, has two important consequences for the construction of vaccines. First, the probability of a given antigenic determinant being universally recognised by all animals from infected individuals at all times during and after infection must be low, and thus cloned antigen mixtures will probably be required for vaccines. Second, a thorough knowledge of the kinetics and classes of antibodies synthesized within a population will allow the construction of suitable vaccines while avoiding pathological consequences in a particular host. Thus an immunochemical dissection of humoral anti-parasite responses may help in the understanding of protective versus pathological responses.

An important consideration is that many antigens, in particular, surface and secreted components, may be stage (or age) specific. These provide a basis for stage specific diagnosis, protection and pathological reactions. A systematic study of parasite antigens is thus a prerequisite for a programme of investigation aimed at understanding the immunological dialogue between parasite and host.

The principal subjects of vaccines and immunomodulation to diminish pathological damage, are integrally related to the three defined parasite antigen compartments: the surface, the secretions and the somatic antigens. In larval cestodes, the cyst fluid can be considered as a fourth compartment. Surface antigens form an obvious focus for vaccines. The approach of first defining, then cloning and testing of these surface antigens is essentially routine. Similarly, stage and species-specific antigens are found in the excretory-secretory (E/S) compartment of many helminths. These may well also contain possible vaccine targets [54], while also providing stimuli for pathological responses. They also form a logical focus as targets for parasite detection.

It is important to stress that only the surface and secretions provide an interphase between the viable parasite and its host. It follows that all means of parasite evasion, and in turn host protective mechanisms, must act principally through the molecules of these compartments. The remaining somatic antigens are only recognised by the host when the parasite's surface has been breached by a potentially protective immune recognition process. Once this has occurred, however, parasite rejection may be favoured by immune responses to internal antigens, for example the enzyme glutathione transferase, which may play a role in schistosome membrane repair after granulocyte-mediated attack of the worm's surface [89]. On the death of the parasite, by whatever means, the host is immediately presented with a multiplicity of somatic antigens, and so there is always the danger that some of these will provoke pathological reactions.

The paucity of parasite material frequently dictates the use of the special approaches of molecular biology to supply materials on the scale necessary for vaccines. For this reason, the emphasis must be on cloned protein antigens, rather than carbohydrate or lipid antigens. Many of the relevant parasite antigens will be proteins and thus their commercial production via recombinant nucleic acid technology is feasible. For non-protein (i.e. non-clonable antigens) however, the preparation of anti-idiotype reagents to complementary monoclonal antibodies provides a possible mechanism for "cloning the unclonable". Although as yet entirely unproven in the parasite world, such anti-idiotypic reagents could make possible vaccines, if used as subsitutes for antigen.

An important point is that most helminths share common anabolic and catabolic activities, and the necessary biological components for these processes are likely to be conserved in evolution and thus antigenically, cross-reacting. Similarly, many structural proteins, for example cytoskeletal and mitotic spindle elements, as well as many carbohydrate prosthetic groups, are likely to be molecules which are conserved in nature and thus cross-reacting. For protective antigens, some degree of cross-reaction with a variety of different parasites might be acceptable provided of course they did not result in pathological sequelae say due to cross reactions with the host invoking an autoimmune response.

The most well known immunological characteristics of helminth infections, stimulation of IgE synthesis [41] and eosinophilia [14] still await elucidation at the molecular level and explanation of their biological significance. Certainly lymphokines will be found to play a crucial part in these phenomena [23], and the demonstrated binding of IgE to specific receptors on eosinophils, macrophages and platelets [18, 19], [92], must be highly relevant. The original idea that nematode allergens (or other allergens) might elicit IgE synthesis through a common structural characteristic is probably naive, but it did stimulate a great deal of work into the isolation and characterisation of these allergens. As a result many of them are still the most highly defined molecules of many nematodes [2]. In addition, nematodes such as *Nippostrongylus brasiliesnsis* and *Ascaris* spp. are as much associated with research into the control of IgE productions and allergy as in the control of parasitosis [2].

One recent development in this area, predictable from a consideration of basic immunology, is the demonstration of a major histocompatibility-linked control of the serological response to a defined helminth antigen [49]. In these experiments, an H-2-linked control of the antibody response in mice to secreted antigens of *Ascaris suum* was reported. If such a linkage also occurs in man, it could have consequences for the interpretation of serodiagnosis and the understanding of variable pathology, development of resistance and predisposition to infection.

An equally important but much more neglected area is the repertoire and subpopulation balance of T cells elicited in response to parasitic infections.

Trichinella spiralis: A model parasite

Over the last few years, a considerable amount of work has been performed with *Trichinella spiralis* on the biochemical characterization of molecules derived from different compartments of the parasite. A very recently published book covers antigens and other aspects of *Trichinella* spp. in great detail [94]. Interest in *T. spiralis* is two-fold. Not only is it a potential threat to human health in some parts of the world, but it also provides an admirable model for nematodiasis in general. It is convenient because it completes its life cycle in laboratory rodents in a relatively short period of time. Moreover, during its life history the worm passes through three entirely different locations in the host. The adult exists as an intestinal parasite, the newborn larvae first penetrate the intestinal wall and then become systemic, circulating in the blood and lymph, and the infective larvae reside as tissue parasites within muscle tissue. One parasite in a single inbred host thus provides potential models for the study of host-parasite interactions in these very different host environments. The ease of obtaining parasite material and wide host range facilitates a thorough investigation of serological and cellular host responses in inbred laboratory rodents. Armed with this information, a rational design for the development of potential vaccines may be possible, not only for this parasite but also for other nematodes affecting humans.

A fundamentally important conclusion is that *T. spiralis*, and nematodes in general, have both stage-specific and inter-stage specific surface and secreted proteins [60, 73, 75, 76, 77]. Some, but not all somatic antigens are also stage-specific [73]. This simple approach of biochemical characterization forms a general basis, or antigen catalogue, for the design of vaccines, and for the understanding of immunopathological reactions.

Stage-specific labelled surface and secreted proteins of *T. spiralis* have been used as probes in Ig isotype-specific immune coprecipitation and surface binding studies, in order to determine the kinetics of murine Ig class responses to individual antigenic component of *T. spiralis* [3, 4, 7], and also *Dipetalonema viteae* [8]. The overall picture that emerged from these studies is one of independent variation of the immune responses of each individual Ig class to each individual antigenic epitope of each and every parasite compartment and stage. At a practical level these studies indicate the complexity of the immune response and that candidate vaccine proteins will have to be chosen with care. The enormous variation in the kinetics of responses to individual antigens of these nematodes makes it unlikely that only one antigen will serve as a suitable vaccine given the obvious variation (genetic, parasite stage or infection history). This may be of particular relevance in nematodes, with their marked stage differences.

The most interesting correlation arose from the systematic study of two strains of mice which differ in the kinetics of adult worm (*T. spiralis*) rejection from the intestine. The NIH strain of mice are resistant, since they expel all adults from the intestine by day 15 after a primary infection, and the C3H strains are relatively susceptible, permitting adult survival for more than 30 days [46]. Against this background it may be particularly relevant that the resistant NIH mice produced a peak of IgA antibodies to adult worm surface antigens on days 10–20 post infection which is absent in susceptible C3H mice. The most obvious conclusion, albeit not formally proved, is that rejection of the *T. spiralis* adults from the intestine is mediated by an IgA anti-adult surface antigen response. A further important consideration raised by this series of experiments is that antibody responses are dynamic events, not only in their specific profiles, but also in the particular isotypes synthesized at a given time to a given antigenic epitope. For example, the above mentioned IgA anti-adult surface antigens synthesized by resistant NIH mice are present at high levels only between days 10-20 post infection. Therefore, the ability to passively transfer infection may be temporally restricted.

The importance of nematode surface antigens as possible vaccines is also seen with the other two recoverable stages of *T. spiralis*, the infective larvae and the adult. Thus, purified surface antigens of the infective larvae have been shown to be protective in mice [28, 71, 86, 87]. Surface antigens of newborn larvae may also play a role in protection, since a monoclonal antibody to a newborn larvae surface protein transferred passive immunity to infection (with newborn larvae) [72]. The same monoclonal antibody was also able to function in an eosinophil-mediated, antibody-dependent killing reaction *in vitro*, providing for the first time strong circumstantial evidence for the relevance of eosinophil killing mechanisms in control of nematode infections (see Chapter 9).

From the viewpoint of pathology, *Trichinella* spp. also provide an excellent model. *Trichinella pseudospiralis* is an isolate (or species) of *Trichinella* which undergoes a markedly different pathobiological interaction with its host than *T. spiralis*. Adult *T. pseudospiralis* worms survive longer in the gastrointestinal tract, but more striking is the failure of *T. pseudospiralis* infective larvae to encyst within muscle tissue (as does *T. spiralis*), although the rest of its intracellular development is very similar to *T. spiralis*. The host's reaction to surface components could be critical in determining the course and pathology of infection. Therefore, a comparison of the surface labelled proteins of each stage of *T. pseudospiralis* and

T. spiralis was made [6]. Although both worms exhibited very similar profiles of stage-specific labelled surface proteins, small differences between individual components of infective larvae and adults of the two species were noted. Whether these actually account for the markedly different pathologies is as yet unknown, but certainly these two *Trichinella* spp. provide a wonderful example of differential pathology whose immunology requires explanation at the molecular and cellular immunological levels.

Finally, perhaps the two most urgent questions relating to the parasite-host interphase are: what are the functions of parasite surface proteins, and how is there stage-specific expression controlled at the genomic level? The first question is as yet answered in the *T. spiralis* system, as it requires novel experimental approaches, but the second question has recently been addressed. In *T. spiralis* control of stage-specific surface antigen expression is not necessarily exercised at the level of selective gene activation. Here, regulation of expression of stage-specific surface proteins can occur at the level of transport to the surface or at the level of mRNA transcription and/or processing. The gene coding (see Chapters 7 and 12) for the principle stage-specific surface protein of the infective larvae [76] is transcribed in both the other two easily recoverable stages, the adult and the newborn larvae. Thus, the appropriate gene product is also expressed in the intestine and stichosome of the adult, but never transported to the surface [65]. A different mechanism ensures negative stage-specific regulation in the newborn larvae. Here, transcription of the same gene occurs, but yields two "aberrant" mRNA species (mol. wts. 1.4 and 2.6 Kb) instead of the 1.8 Kb component transcribed in adults and infective larvae [20]. Possible mechanisms for expression of these two aberrant mRNA transcripts in newborn larvae may, perhaps, be sought in either differential RNA splicing or alternative transcription initiation and/or termination sites. Should either of these two mechanisms (i.e. control at the level of transport to the cuticular surface or species of mRNA transcribed) operate in filarial nematodes, then it should be possible to clone and express the surface antigens of the very scarce infective larvae, but starting perhaps with the more easily recoverable life cycle stages, such as the adult.

PROTECTION

Perhaps the most fascinating aspect of parasites is their continued survival in the face of all the defensive mechanisms at the disposal of the vertebrate host. Simultaneously, many parasites exhibit an exquisitely tuned specificity in terms of host range, not only in their primary hosts, but also in their intermediate carriers, or vectors. This quality of resistance in appropriate host systems versus susceptibility in others seeks its primary explanation in the interphase between parasite and host. At the same time other parasites such as *Fasciola hepatica* or *Echinococcus* spp. are much less host restricted. The strategies by which an unrestricted host range is achieved is also highly interesting (see Chapter 1).

If one considers the biology of parasites and the methods they employ to gain entry into their hosts, it is clear that candidate vaccines may be directed against various parasite components in order to interfere with invasion via the skin or intestinal wall, in the first instance, or establishment in host tissues, in the second. Not surprisingly therefore, the study of parasite antigens has focused on the surface and secretions as they are clearly the critical compartments to search for the target antigens of effective vaccines. The excretory/secretory products of helminths have been extensively reviewed [54]. Alternative "functional" antigens, such as proteases involved in tissue invasion [95], and enzymes involved in repair of granulocyte-mediated surface damage [89] have also been suggested as possible foci for a vaccine approach. Probably the final effective vaccines will be multivalent, including selected surface and secreted products and perhaps, enzymes such as glutathione transferase and proteases. To optimise the chances of success, such vaccines would have to be closely modelled on each individual parasite life style.

Surface and secreted antigens are easy to characterise, simply by first using radio-labelling techniques selected for the surface and biosynthetic labelling for secreted antigens. Antigens are then defined as any radio-labelled component recognised in the sera of infected individuals. In the absence of an animal model, the first approach is to dissect serological immune responses to the antigens in clinically well defined groups, for example resistant and susceptible.

Resistance to parasites is not an inevitable result of a humoral immune response to surface, or any other antigens. The immune response must call into play appropriate destructive effector mechanisms. This does not always occur, and ineffective 'blocking' antibody responses, for example, can certainly arise during the course of parasite infections [29, 32]. The factors responsible for controlling the balance between destructive and 'blocking' responses still remain largely mysterious, and their exploration is an urgent research priority. Investigation here is clearly in the realm of humoral and cellular immunology, and its importance becomes more pressing as the number of defined parasite surface antigens expressed through recombinant nucleic acid techniques increases.

A major consideration is that a conventional infection may yield a totally different mosaic of immune responses to a given surface antigen than the injection of the same antigen in its purified or 'expressed' form. Thus, protective antigens, once cloned via recombinant nucleic acid technology may not protect as efficiently as intact parasites via a natural infection. For all of the above mentioned reasons a thorough understanding of the immunobiology of protection will be an essential basis for vaccine design. The major priorities include careful dissection of the different functional Ig classes stimulated by a given potentially protective antigen and systematic studies on the T cell responses and their lymphokine profiles.

The use of animal models in the establishment of a vaccine may allow rapid development since experimental immunisation, both active and passive can be done. However, the use of such models is not always possible and filariasis, with an estimated human population of 130 million at risk is one prime example. In the absence of suitable animal models, the only approach is to study immune responses in clinically well-defined groups, for example, resistant and susceptible, within endemic areas. Here again, the availability of defined antigens and clinically defined blood samples as probes for systematic serological and cellular investigations, provides the major approach in the search for clues to the mechanisms of resistance.

Of all the helminth infections, development of a vaccine against cestodes is the most feasible as there is such strong evidence that larval taeniid infections can be eliminated by immune mechanisms [33, 83]. Resistance to secondary infection can be produced either by primary infection with the parasite or by immunization using parasite extracts [24, 25, 26, 27, 55, 56, 83]. The oncosphere, or material secreted or shed by the oncosphere, is thought to be one potent source of immunogens. However, until recently, little work has been done on the characterization at the molecular level of the antigens involved [34].

A considerable amount of evidence points to the importance of humoral immunity in resistance to the early stages of taeniid parasites. Protection by passive transfer of immune sera in systems such as *Taenia taeniaeformis* indicated that antibodies play a definitive role in immunity and that it is the invasive larvae or oncosphere which are vulnerable to such attack [67, 68, 91]. Protection from *T. saginata* infection can similarly be transferred to neonatal cattle [57].

More recently, a Western blotting study on *T. saginata* oncospheres in conjunction with sera from resistant and susceptible cattle identified certain oncospheral components which were recognised by antibodies from resistant animals. BALB/c mice responded to the same antigens so the production of monoclonal antibodies directed against these determinants is feasible [37].

An analysis of the surface, excretory/secretory and somatic compartments of *T. saginata* oncospheres by a combination of direct and biosynthetic radio-labelling and SDS-PAGE revealed a restricted number of components in all three compartments. Mouse derived monoclonal antibodies directed against the oncospheral stage demonstrated both stage species and common determinants on the surface of oncospheres and cysticerci. One of these monoclonal antibodies, when injected into cattle conferred protection against oral challenge with *T. saginata* eggs [35]. Studies indicated that the antigenic component identified by this monoclonal, once purified from an extract of whole oncospheres, may be involved in stimulating immunity in cattle to oral challenge with eggs [36]. More recently, an elegant series of studies resulted in the vaccination of sheep against the closely related parasite *Taenia ovis* using a defined recombinant antigen [43, 82].

Although the invasive oncosphere is considered the prime 'target' of the protective immune response, components of the metacestode stage may also constitute possible 'targets' [33, 84]. Extracts of cysticerci have been used to vaccinate pigs against *T. solium* infection [70], as have extracts of *T. saginata* proglottids to immunize cattle against *T. saginata* infection [26]. There would be clear advantages in having a vaccine which was directed against both the early invasive and later developmental stages of the parasite. If any invasive parasites survived they could then be killed at a later stage in development.

As shown by the monoclonal antibodies raised against *T. saginata* oncospheres [35], some components may be shared between oncospheres and metacestodes. This also appears to be the case with *Taenia taeniaeformis*. However, other metacestode components may be unique [29]. Recent studies into the surface, excretory/secretory, somatic and cyst fluid compartments of *T. saginata* so far indicate that an antibody response to certain cysticercal components may indeed be associated with immunity [44, 45, 46, 74].

Studies on *T. solium* have suggested that resistance against mature cysticerci may have a cellular component [69]. Evidence from other systems such as *T. taeniaeformis* in mice also points to a cellular involvement in the killing of larval cestodes [52, 53]. It is worth noting that most work up till now has concentrated either on invasive oncospheres or on the fully developed metacestodes. Intermediate developmental forms also require study.

In summary, the biological evidence of effective humoral immunity versus taeniid parasites suggests that the only impediment to the development of effective vaccines, perhaps even a combined *T. solium-T. saginata-T. ovis-T. hydatigena* vaccine, is failure to clone the protective antigen(s). A start has now been made in this direction [43, 82] and so one may confidently predict the first trials in the near future.

Vaccines against other helminths remain a long term goal either for reasons of low efficiency of protection (schistosomes), poor evidence of long lasting and sterilizing immunity (may intestinal nematodes), or absence of animal models for the necessary preliminary evaluation (filarias). In addition, the combined, but unrelated realities that these diseases affect the less developed countries and, are largely susceptible to the present armoury of modern drugs, provide a depressing argument against the extensive commerical development of anti-helminth vaccines in the near future. This subject of possible vaccination against parasites and also probable future strategies were considered in some detail [63, 65], and these publications are a useful source of general information.

PATHOLOGY

The majority of helminth infections are rarely fatal but characteristically debilitating due to pathological changes provoked by the parasite. It is perhaps wiser to consider the pathology of helminth infections in a broad framework, including physical damage, trauma and disruption caused by the parasite *in situ*, but undoubtedly many debilitating pathological syndromes are a result of immune responses to the excretory/secretory and/or surface compartments of the parasite. Internal or somatic antigens, of course, can only interact with the immune system following the breach of the parasite's surface.

Those parasite antigens which provoke the interacting immune responses responsible for the immunopathology, require definition. In addition, studies of the immune state, both serological and cellular in infected subjects, might provide clues as to the basic underlying mechanisms responsible for the spectrum of the pathology (see Appendix, Chapter 3). An additional contributory factor to the pathological inflammatory responses which occur during parasitic infections is the interaction between the parasite and the host complement system [51].

Of the major groups of helminths, the nematodes and trematodes are probably responsible for the most severe immunopathology. One prime example is *Onchocerca volvulus*, where dermatitis and, less frequently, blindness can result from immunopathological reactions of humans to as yet undefined microfilarial antigens. The dermatitis may range from mild to severe. Hyperreactive, or severe, onchodermatitis may be localised, for example to one leg, when it is often referred to as Sowda. In Sowda, there is a well characterized, localized dermal and lymphatic pathology associated with low skin microfilarial density and humoral and cellular hyperimmune responsiveness [10, 13, 15, 30, 58]. The reverse is true of generalized onchocerciasis where microfilariae survive in greater numbers, possibly due to a lowered immune responsiveness [31, 59].

In a recent study [16], individual human Ig class response to *Onchocerca volvulus* antigens were examined in patients with generalized onchocerciasis or with the localised, chronic hyperreactive Sowda type onchodermatitis. With the marked exception of the IgG_3 responses, the patterns of recognition of human Ig classes were similar in Sowda and generalised onchocerciasis. The results with IgG_3, however, was striking. Here, two major antigens (9 kDa and 72 kDa) were recognized by IgG_3 antibodies in sera from Sowda patients but not sera from generalized onchocerciasis cases. Furthermore, these two antigens were not recognised by another Ig class, either in generalized or Sowda onchocerciasis, nor were they detected by antibodies of any class present in a collection of sera representative of other nematode infections. This difference in the IgG_3 response was so pronounced that sera from Sowda patients could be distinguished from generalized onchocerciasis cases by an IgG_3-specific ELISA assay even using a crude PBS parasite extract as the "antigen". Therefore, it has been established that one particular pathological condition, localised onchodermatitis or Sowda, is associated with an immune response restricted to antibodies of one immunoglobulin class which are directed against two defined parasite antigens. Preliminary findings indicate that genetic differences between infected humans are associated with the development of either localised or generalized onchocerciasis [12]. Thus, a possible association between pathology and host genetic differences may be further elucidated by the study of the host IgG_3 response to the 9 kDa and 72 kDa *O. volvulus* antigens.

The characteristic properties of nematode infections of promoting the elevated IgE responses in their host responsible for allergy has provoked a great deal of research into their molecular nature [2, 80]. The 14 kDa allergen of *Ascaris suum* is one example [9]. As stated earlier, the pathology of parasite infections should not necessarily be viewed in isolation simply as the result of immunological responses, other factors, may come into play [48 93, 98]. The final word on nematodes, however, is that although the filarial parasites account for perhaps the most spectacular and profoundly shocking example of tropical immunopathological lesions, we still have yet to unequivocally identify a single causal antigen or immune response.

Of the trematodes, the schistosomes probably cause the greatest degree of immunopathology. The granulomatous responses induced by egg antigens are undoubtedly the major factor in this pathology [90, 97, 98]. Despite the fact that some of these antigens have been characterized at the molecular level, a definite causal link between any given parasite component and the pathology has not been established. Possible mechanisms responsible for the granuloma formation have been reviewed recently [11, 54].

The pathology associated with cestodes is often that of space occupying lesions such as in *T. solium* neurocysticercosis [81] or hydatid disease [50, 84]. Patho-morphological changes also occur in the intestine caused by adult tapeworm [77, 78]. The molecular nature of parasite antigens which may contribute to this pathology is not well studied, however, as with the other helminths, certain antigens, which are wide ranging in their molecular nature, provoke immediate-type hypersensitivity [99]. A potentially dangerous form of this is the risk to *Echinococcus granulosus* hydatiodosis patients who react to the well characterised hydatid lipoprotein allergen (antigen B). If a hydatid cyst ruptures, either due to trauma or surgical intervention, the patient runs the risk of developing anaphylactic shock as a consequence of exposure to cyst fluid antigens. Patients can be checked for sensitivity preoperatively. For this and other reasons great care is taken during surgery not to allow cyst fluid into contact with the patient [1].

CONCLUSIONS AND FUTURE PROSPECTS

A thorough study of parasitic helminth antigens is a prerequisite for vaccine development and perhaps immune modulation to diminish pathological sequelae. Studies should be directed at the identification of those stage or age specific surface, secreted and somatic antigens which are involved in the host/parasite interactions responsible for immunity and/or pathology. Surface and secreted antigens are the most likely targets for protective immune responses and thus form a logical focus for vaccine design. The cestodes which present such strong evidence for immunity following natural infection, are likely to yield effective vaccines by modern procedures if the protective antigens are proteins. Certain antigens must, however, stimulate the humoral and/or cellular responses which are responsible for the undesirable immunopathologic consequences of may helminthic diseases. The nematodes furnish some extreme examples of such pathology. The ultimate objective in identifying these particular antigens is to utilise them in the appropriate down regulation of the immune response responsible for such pathology. As an illustration we have presented an interesting correlation between one particular clinical condition of onchocerciasis (Sowda) and the serological response defined both in terms of the parasite antigens and a immunoglobulin class restricted antibody response.

The complexity of these parasite systems and the host response to the parasite should not be underestimated. Consideration should also be given to the many extrinsic factors affecting the possible range of immune responses to a parasitic infection in a host population. Modern analytical techniques allow detailed analysis in terms of the humoral antibody responses and afford the possibility of the future development of control and disease management procedures taylored to each individual host/parasite system. However, novel systems are required to complete the analysis of the cellular components of the immune response to parasite antigens and functional studies are needed to determine the role that these parasite antigens play in the complex interaction between parasite and host. Finally detailed epidemiological data are required in order to assess the likely success of any particular vaccine as part of a general control programme.

Although certain groups such as the cestodes tend to lag behind the more fashionable nematodes and trematodes, many parasitic helminth antigens are defined both in terms of molecular weight and in their chemical nature. Serological responses to these are beginning to be dissected in terms of antigenic epitopes and the profile of Ig classes elicited, but T cell recognition patterns urgently require investigation. What appears to be largely lacking is a clear understanding of what function these antigens play in the host/parasite dialogue. Such an understanding would surely be of value in the development of novel control and/or disease management procedures, whether this be chemotherapeutic, immunomodulatory or immunoprophylactic in nature.

ACKNOWLEDGEMENTS

RMEP's work on parasitic helminths was funded by grants from the McConnel Clark Foundation and the European Communities R and D Programme "Science and Technology for Development", Subprogrammes "Medicine, Health and Nutrition in the Tropics", while JSH's work was funded through grants from the Overseas Development Administration of the Foreign and Commonwealth Office and by the Wellcome

Trust. We would like to thank these funding bodies for their support. The contributions of Drs NM Almond, Z Cabrera, W Harnett, G Ortega-Pierres and MMH Sewell are gratefully acknowledged.

REFERENCES

1. Aarons, B.J. Modern surgical treatment of human hydatidosis. Aust. Vet. J. **55**: 164–148 (1979).
2. Almond, N.M. and Parkhouse, R.M.E. Nematode antigens. Curr. Topics Microbiol. Immunol. **120**: 173–203 (1985).
3. Almond, N.M. and Parkhouse, R.M.E. Immunoglobulin class specific responses to biochemically-defined antigens of *Trichinella spiralis*. Parasite Immunol. **8**: 391–406 (1986).
4. Almond, N.M. and Parkhouse, R.M.E. The Ig class distribution of anti-phosphorylcholine responses in mice infected with parasitic nematodes. Immunology **59**: 633–635 (1986).
5. Almond, N.M., and Parkhouse, R.M.E. The importance of antibody class in helminth infections. In: *Progress in Vaccinology* (G.P. Talwar ed), pp. 261–276. Springer-Verlag: Berlin (1989).
6. Almond, N.M., McLaren, D.J. and Parkhouse, R.M.E. Comparison of the surface and secretions of *T. pseudospiralis* and *T. spiralis*. Parasitology **93**: 163–176 (1986).
7. Almond, N.M. Parkhouse, R.M.E., Chapa Ruiz, M.R. and Garcia Ortigosa, E. The response of humans to surface and secreted antigens of *Trichinella spiralis*. Trop. Med. Parasitol. **37**: 381–384 (1986).
8. Almond, N.M., Worms, M.J., Harnett, W. and Parkhouse, R.M.E. Variations in specific humoral immune responses of different mouse strains to microfilariae of *Dipetalonema viteae*. Parasitology **95**: 559–568 (1987).
9. Ambler, J., Croft, A.R., Doe, J.E., Gemmel, D.K., Miller, J.N. and Orr, T.S.C. Biological techniques for studying the allergenic components of nematodes. II. The characterisation of the allergen released by *Ascaris suum* maintained in mice. J. Immunol. Meth. **2**: 315–323 (1973).
10. Bartlett, A., Turk, J., Ngu, N.L., MacKenzie, C.D., Fuglsang, H. and Anderson, J. Variation in delayed hypersensitivity in onchocerciasis. Trans. R. Soc. Trop. Med. Hyg. **72**: 372–377 (1978).
11. Boros, D.L.: Immunopathology of *Schistosoma mansoni* infection. Clin. Microbiol Rev. **2**: 250–269 (1989).
12. Brattig, N.W., Tischendorf, F.W., Reifegerste, R., Albeiz, A.J. and Berger, J. Differences in the distribution of HLA antigens in localised and generalised form onchocerciasis. Tropenmed. Parasitol. **37**: 271–275 (1986).
13. Brattig, N.W., Tischendorf. F.W., Reifegerste, R., Albeiz, A.J. and Berger, J. Distribution pattern of peripheral lymphocyte subsets in localised and generalised onchocerciasis. Clin. Immunol. Immunopath. **44**: 149–159 (1987).
14. Buijs, J. and Ruitenberg, E.J. Immunopathology of intestinal nematode infections. *Bailliere's Clinical Tropical Medicine and Communicable Diseases* **2**: 535–551 (1987).
15. Buttner, D.W., Von Laer, G., Mannweiller, E. and Buttner, M. Clinical, parasitological and serological studies on onchocerciasis in the Yemen Arab Republic. Tropenmed. Parasitol. **33**: 201–212 (1982).
16. Cabrera, Z., Buttner, D.W. and Parkhouse, R.M.E. Unique recognition of a low molecular weight *Onchocerca volvulus* antigen by IgG3 antibodies in chronic hyperreactive onchodermatitis (Sowda). Clin. Exp. Immunol. **74**: 223–229 (1988).
17. Capron, M. and Capron, A. Rats, mice and men—models for immune effector mechanisms against schistosomiasis. Parasitol. Today **2**: 69–75 (1986).
18. Capron, A., Dessaint, J.P., Capron, M., Joseph, M., Ameisen, J.C., and Tonnel, A.B. From parasites to allergy: a second receptor for IgE. Immunol. Today **7**: 15–18 (1986).
19. Capron, A. Dessaint, J.P., Haque, A. and Capron, M. Antibody-dependent cell-mediated cytotoxicity against parasites. Prog. Allergy **31**: 234–267 (1982).
20. Chambers, A.E., Simpson, A.J.G. and Parkhouse, R.M.E. Unpublished observations (1988).
21. CIBA Foundation Symposium Filariasis. 127. John Wiley & Sons: London (1987).
22. Crompton, D.W.T. Human helminthic populations. *Bailliere's Clinical Tropical Medicine and Communicable Diseases* **2**: 489–510 (1987).
23. Finkelman, F.D., Katona, I.M., Urban, J.F., Snapper, C.M., O'Hara, J. and Paul, W.E. Suppression of *in vivo* polyclonal IgE responses by monoclonal antibody to the lymphokine B-cell stimulatory factor. 1. Proc. Nat. Acad. Sci., U.S.A. **83**: 9675–9678 (1986).

24. Gallie, G.J. and Sewell, M.M.H. The survival of *Cysticercus bovis* in resistant calves. Vet. Rec. **91**: 481–482 (1972).
25. Gallie, G.J. and Sewell, M.M.H. The serological response of three month old calves to infection with *Taenia saginata (Cysticercus bovis)* and their resistance to reinfection. Trop. Anim. Hlth. Prod. **6**: 163–171 (1974).
26. Gallie, G.J. and Sewell, M.M.H. Experimental immunisation of six month old calves against infection with the cysticercus stage of *Taenia saginata*. Trop. Anim. Hlth. Prod. **8**: 233–242 (1976).
27. Gallie, G.J. and Sewell, M.M.H. Inoculation of calves and adult cattle with oncospheres of *Taenia saginata* and their resistance to reinfection. Trop. Anim. Hlth. Prod. **13**: 147–154 (1981).
28. Gamble. H.R. *Trichinella spiralis*: immunization of mice using monoclonal antibody affinity isolated antigens. Exp. Parasitol. **59**: 398–404 (1985).
29. Gibbons, J.C., Harrison, L.J.S. and Parkhouse, R.M.E. Immunoglobulin class responses to *Taenia taeniaeformis* in susceptible and resistant mice. Parasite Immunol. **8**: 491–502 (1986).
30. Gibson. D.W., Connor, D.H., Brown, H.L., Fuglsang, H., Anderson, J., Duke, B.O.L. and Buck, A.A. Onchocercal dermatitis: ultrastructural studies of microfilarial and host tissue, before and after treatment with diethylcarbamazine (Hetrazan). Am. J. Trop. Med. Hyg. **25**: 74–87 (1976).
31. Greene, B.M., Fanning, M.M. and Ellner, J.J. Non-specific suppression of antigen-induced lymphocyte blastogenesis in *Onchocerca volvulus* infection in man. Clin. Exp. Immunol. **52**: 259–265 (1983).
32. Grzych, J.M., Capron. M., Dissous, C. and Capron, A. Blocking activity of rat monoclonal antibodies in experimental schistosomiasis. J. Immunol. **133**: 998–1004 (1984).
33. Harrison, L.J.S. and Joshua, G.W.P. Immunoprophylaxis of *Taenia saginata* cysticercosis. In: *Helminth Zoonoses* (S. Geerts, V. Kumar and J. Brandt eds), pp. 81–84. Martinus Nijhoff: Dordrecht, The Netherlands (1986).
34. Harrison, L.J.S. and Parkhouse, R.M.E. Antigens of Taeniid cestodes in protection, diagnosis and escape. Curr. Topics Microbiol. Immunol. **120**: 159–172 (1985).
35. Harrison, L.J.S and Parkhouse, R.M.E. Passive protection against *Taenia saginata* infection in cattle by a mouse monoclonal antibody reactive with the surface of the invasive oncosphere. Parasite Immunol. **8**: 319–332 (1986).
36. Harrison, L.J.S and Parkhouse, R.M.E. Identification of protective antigens in *Taenia saginata* cysticercosis. *Immunobiology and Molecular Biology of Cestode Infections*. Cooper's Animal Health Symposium, Melbourne, **A2** (1986).
37. Harrison, L.J.S., Parkhouse, R.M.E. and Sewell, M.M.H. Variation in 'target' antigen between appropriate and inappropriate hosts of *T. saginata* metacestodes. Parasitology **88**: 659–663 (1984).
38. Holmes, P.H Pathophysiology of nematode infections. In: *Parasitology-Quo Vadit*? Proc. Sixth Int. Cong. Parasitol. (M.J. Howell ed), pp. 443–451. Australian Academy of Science, Canberra, Australia (1986).
39. Hughes, D.L. Trematodes, excluding schistosomes with special emphasis on *Fasciola*. Curr. Topics Microbiol. Immunol. **120**: 241–260 (1985).
40. Hussain, R., Groge, M. and Ottesen, E. IgG antibody subclasses in human filariasis. Differential subsclass recognition of parasite antigens correlates with different clinical manifestations of infection. J. Immunol. **139**: 2794–2798 (1987).
41. Jarrett, E.E. and Miller, H.R. Production and activities of IgE in helminth infection. Prog. Allergy **31**: 178–233 (1982).
42. Jassim, A., Hassam, K. and Catty, D. Antibody isotypes in human schistosomiasis mansoni. Parasite Immunol. **9**: 627–650 (1987).
43. Johnson, K.S., Harrison, G.B.L, Lightowlers, M.W., O'Hoy, K.L., Cougle, W.G., Dempster, R.P., Lawrence, S.B., Vinton, J.G., Heath, D.D. and Rickard, M.D. Vaccination against ovine cysticercosis using a defined recombinant antigen. Nature (Lond.) **338**: 585–587 (1989).
44. Joshua, G.W.P., Harrison, L.J.S. and Sewell, M.M.H. Excreted/secreted proteins of *Taenia saginata* metacestodes. Parasitology **97**: 477–87 (1988).
45. Joshua, G.W.P., Harrison, L.J.S. and Sewell, M.M.H. Developmental changes in proteins and glycoproteins revealed by direct radio-iodination of viable *Taenia saginata* larvae. Parasitology **99**: 265–274 (1989).
46. Joshua, G.W.P., Harrison. L.J.S. and Sewell, M.M.H. Protein antigens in the cyst fluid of *Taenia saginata* cysticerci. Parasitology **100**: 463–467 (1990).

47. Jungery, M. and Ogilvie, B.M. Antibody response to stage specific *Trichinella spiralis* surface antigens in strong and weak responder mouse strains. J. Immunol. **129**: 839–843 (1982).
48. Kassai, T. Effector mechanisms of the protective immunity induced by intestinal helminths—Review article. Parasitol. Hungary **22**: 5–20 (1989).
49. Kennedy, M.W., Gordon, A.M.S., Tomlinson. L.A. and Qureshi, F. Genetic (major histocompatability complex?) control of the antibody repertoire to the secreted antigens of *Ascaris*. Parasite Immunol. **9**: 269–273 (1986).
50. Lawson, J.R. and Gemmel, M.A. Hydatidosis and cysticercosis. Adv. Parasitol. **22**: 261–296 (1984).
51. Leid, R.W. Parasites and complement. Adv. Parasitol. **27**: 131–168 (1988).
52. Letonja, T. and Hammerberg, C. *Taenia taeniaeformis*: early inflammatory response around developing metacestodes in the liver of resistant and susceptible mice. I. Identification of leucocyte response with monoclonal antibodies. J. Parasitol. **73**: 962–970 (1987).
53. Letonja, T. and Hammerberg, C. *Taenia taeniaeformis*: early inflammatory responses around developing metacestodes in the liver of resistant and susceptible mice. II. Histochemistry and cytochemistry. J. Parasitol. **73**: 971–979 (1987).
54. Lightowlers, M.W. and Rickard, M.D. Excretory/secretory products of helminth parasites: Effects on host immune responses. Parasitology **96**: S123–S166 (1988).
55. Lloyd, S. Homologous and heterologous immunisation against the metacestodes of *Taenia saginata* and *Taenia taeniaeformis* in cattle and mice. Zentbl. Parasitenkunde **60**: 87–96 (1979).
56. Llyod, S. Passive and active immunisation against cysticercosis. In: *Agriculture. Some important parasitic infections in bovines considered from economic and social (zoonosis) points of view* (J. Euzeby and J. Grevrey eds), pp. 187–198. EEC Publications: Brussels (1984).
57. Lloyd, S. and Soulsby, E.J.L. Passive transfer of immunity to neonatal calves against metacestodes of *Taenia saginata*. Vet. Parasitol. **2**: 355–362 (1976).
58. Lucius, R., Buttner, D.W., Kirsten, Ch. and Diesfeld. H.J. A study on antigen recognition by onchocerciasis patients with different clinical forms of disease. Parasitology **92**: 569–580 (1986).
59. Mackenzie, C.D. and Williams, J.F. Variation in clinical presentation in onchocerciasis and their relationship to host-parasite interaction. Sudan Med. J. (Suppl.) **21**: 41–48 (1985).
60. MacKenzie, C.D., Preston, P.M. and Ogilvie, B.M. Immunological properties of the surface of parasitic nematodes. Nature (Lond.) **276**: 826–828 (1978).
61. MacInnis, A.J. *Molecular paradigms for eradicating helminthic parasites*. U.C.L.A. Symposia on Molecular and Cellular Biology, New Series **60**. A.R. Liss Inc.: New York (1987).
62. Mahmoud, A.A.F. Schistosomiasis. *Bailliere's Clinical Tropical Medicine and Communicable Diseases*. **2** No. 2. Bailliere & Tindall: London (1987).
63. Maizels, R.M. and McKerrow, J.H. Future molecular strategies for the study of nematode trematode and cestode parasites. In: *Molecular paradigms for eradication of helminth parasites* (A.J. MacInnis ed), pp. 561–566. A.R. Liss Inc.: New York (1987).
64. Maizels, R.M. and Selkirk, M.E. Antigens of filarial parasites. ISI Atlas of Science, pp. 1–5 (1988).
65. McLaren, D.J. and Chappell, L.H. Vaccines and vaccination strategies. *Symposium of the British Society of Parasitology*. Parasitology **98**: S1–S100 (1989).
66. McLaren, D.J., Ortega-Pierres, M.G. and Parkhouse, R.M.E. *Trichinella spiralis:* immunocytochemical localisation of surface and intracellular antigens using monoclonal antibody probes. Parasitology **94**: 101–114 (1987).
67. Mitchell, G.F., Goding, J.W. and Rickard M.D. Studies on immune responses to larval cestodes in mice. Increases susceptibility of certain mouse strains and hypothymic mice to *Taenia taeniaeformis* and analysis of passive transfer of resistance with serum. Aust. J. Exp. Biol. Med. Sci., **55**: 165–186 (1977).
68. Mitchell, G.F., Anders, R.F., Brown, G.V. Handyman, E., Roberts-Thomas, I.C., Chapman, K.P. Forsyth, Kahl, L.P. and Cruise, K.M. Analysis of infection characteristics and antiparasite immune responses in resistant compared with susceptible hosts. Immunol. Rev. **61**: 137–187 (1982).
69. Molinari, J.L., Meza. R. and Tato, P. *Taenia solium*: cellular reactions in the larva *(Cysticercus cellulosae)* in naturally parasitized, immunized hogs. Exp. Parasitol. **56**: 327–338 (1983).
70. Molinari, J.L., Meza, R., Suarez, B., Palacias, S., Tato, P. and Retano, A. *Taenia solium*: immunity in hogs to the cysticercus. Exp. Parasitol. **55**: 340–357 (1983).

71. Ortega-Pierres, M.G., Almond, N.W. and Parkhouse, R.M.E. Applications of biochemically defined antigens of *Trichinella spiralis* in host immunity, protection and diagnosis. Wiad. Parazyt. **33**: 243–452 (1987).
72. Ortega-Pierres, M.G., MacKenzie, C.D. and Parkhouse, R.M.E. Protection against *Trichinella spiralis* induced by a monoclonal antibody that promotes killing of newborn larvae by granulocytes. Parasite immunol. **6**: 275–284 (1984).
73. Parkhouse. R.M.E. and Clark, N.W.T. Stage-specific secreted and somatic antigens of *Trichinella spiralis*. Mol. Biochem. Parasitol. **9**: 319–327 (1983).
74. Parkhouse, R.M.E. and Harrison, L.J.S. Cyst fluid and surface associated glycoprotein antigens of *Taenia* sp. metacestodes. Parasite Immunol. **9**: 263–268 (1987).
75. Parkhouse, R.M.E., and Ortega-Pierres, M.G. Stage-specific antigens of *Trichinella spiralis*. Parasitology **88**: 623–630 (1984).
76. Parkhouse, R.M.E., Phillip, M. and Ogilvie, B.M. Characterization of surface antigens of *Trichinella spiralis* infective larvae. Parasite Immunol. **3**: 339–352 (1981).
77. Pawlowski, Z.S. Intestinal helminthiasis and human health: recent advances and future needs. In: *Parasitology-Quo Vadit*? Proc. Sixth Int. Congr. Parasitol. (M.J. Howell ed), pp. 159–167. Australian Academy of Science, Canberra, Australia (1986).
78. Pawlowski, Z.S. Intestinal helminthic infections. *Bailliere's Clinical Tropical Medicine and Communicable Diseases*. **3**, No. 2. Bailliere & Tindall: London (1987).
79. Phillip, M., Parkhouse, R.M.E. and Ogilvie, B.M. Changing proteins on the surface of a parasitic namatode. Nature (Lond.). **287**: 538–540 (1980).
80. Pritchard, D.I. The molecular biology of gastrointestinal nematodes. *Bailliere's Clinical Tropical Medicine and Communicable Diseases* **2**: 511–534 (1987).
81. Rabiella-Cervantes, M.T., Rivas-Hernandez, A., Rodrigues-Ibarra, J., Castillo-Medina, S. and Cancino, F. de M. Anatomopathological aspects of human brain cysticercosis. In: *Cysticercosis: present state of knowledge and perspectives* (A. Flisser et al eds), pp. 179–200. Academic Press: London (1982).
82. Rickard, M.D. A success in veterinary parasitology: cestode vaccines. In: *New Strategies in parasitology*. (K.P.W.J. McAdam ed), pp. 3–16. Churchill Livingstone: Edinburgh, Scotland (1989).
83. Rickard, M.D. and Brumley, J.L. Immunisation of calves against *Taenia saginata* infection using antigens collected by *in vitro* incubation of *Taenia saginata* oncospheres or ultrasonic disintegration of *T. saginata* and *T. hydatigena* oncospheres. Res. Vet. Sci., **30**: 99–103 (1981).
84. Rickard, M.D. and Williams, J.F. Hydatidosis/cysticercosis: immune mechanisms and immunization against infection. Adv. Parasitol. **21**: 230–296 (1983).
85. Rollinson, D. and Simpson, A.J.G. The biology of schistosomes—from gene to latrine. Academic Press: London, England (1987).
86. Silbertein, D.S. and Despommier, D.D. Antigens from *Trichinella spiralis* that induce a protective response in the mouse. J. Immunol. **132**: 898–904 (1984).
87. Silberstein, D.S. and Despommier, D.D. Effects on *Trichinella spiralis* of host responses to purified antigens. Science **227**: 948–950 (1985).
88. Simpson, A.J.G. and Smithers, S.R. Schistosomes: surface, egg and circulating antigens. Curr. Topics microbiol. Immunol. **120**: 205–239 (1985).
89. Smith, D.B., Daverm, K.M. Board, P.G., Tui Wu, Garcia, E.G. and Mitchell, G.F. Mr 26,000 antigen of *Schistosoma japonicum* recognised by resistant WEHI 129/J mice is a parasite glutathione-S-transferase. Proc. Nat. Acad. Sci. U.S.A. **83**: 8703–8707 (1986).
90. Smithers, S.R. and Doenhoff, M.J. Schistosomiasis. In: *Immunology of Parasitic Infections*. 2nd edn. (S. Cohan and K.S. Warren eds), pp. 527–607. Blackwell Scientific Publications: Oxford, England (1982).
91. Soulsby, E.J.L. and Lloyd, S. Passive immunization in cysticercosis: characterization of the antibodies concerned. In: *Cysticercosis: present state of knowledge and perspectives*. (A. Flisser et al eds), pp. 539–548. Academic Press: London (1982).
92. Spiegelberg. H.L. Structure and function of Fc receptors for IgE on lymphocytes, monocytes and macrophages. Adv. Immunol. **35**: 61–88 (1984).
93. Symons, L.E.A. Anorexia: occurrence, pathophysiology, and possible causes in parasitic infections. Adv. parasitol. **24**: 103–133 (1985).

94. Tanner, C.E., Martinez-Fernandez, A.R. and Bolas-Fernandez F. Trichinellosis. Proc. Seventh Int. Conf. Trichinellosis. Concejo Superior de Investigaciones Scientificas, Madrid (1988).
95. Toy, L., Pettit, M., Wang, Y.F., Hedstrom, R. and McKerrow, J.H. The immune response to stage-specific proteolytic enzymes of *Schistosoma mansoni*. In: *Molecular Paradigms for Eradicating Helminthic parasites.* UCLA Symposia on Molecular and Cellular Biology, New Series **60**, pp. 85–103. Alan R. Liss Inc.: New York (1987).
96. Wakelin, D. *Immunity to parasites: How animals control parasite infections*. Edward Arnold: London (1984).
97. Warren, K.S. The pathology, pathobiology and pathogenesis of schistosomiasis. Nature (Lond.) **273:** 608–612 (1978).
98. Warren, K.S. The secret of the immunopathogenesis of schistosomiasis: *in vivo* models. Immunol. Rev. **61**: 189–213 (1982).
99. Williams, J.F. and Sandeman, M.R. Antigens of taeniid cestodes. In: *Cysticercosis: present state of knowledge and perspectives* (A. Flisser et al eds), pp. 525–537. Academic Press: London (1982).

11. Application of Radioisotopes in Pathophysiology and Diagnosis of Helminthic Infections

J.M. Maclean
Department of Veterinary Physiology, University of Glasgow Veterinary School
Bearsden Road, Glasgow, UK

INTRODUCTION

Radioisotopic methods have been applied extensively to study host-parasite interactions and immunity. Much of the accumulated knowledge on the aetiology of the physiological and biochemical changes occurring in the host as a result of helminth infections, particularly in relation to anaemia and hypoproteinaemia, has come from the application of radioisotopic techniques. Conventional laboratory measurements such as packed cell volume (PCV), blood haemoglobin and plasma protein concentration provide useful information on changes in body fluids, yet can obscure a situation where minor changes in concentration may be accompanied by significant alterations in kinetics or pool size. Radioisotopic tracer methods on the other hand, permit changes in body fluid pools to be measured and can show if these result from alterations in the rate of synthesis or catabolism or both.

The energy emissions of a particular decaying radionuclide have characteristic values. This allows the identification of more than one radioisotope in a single sample, e.g. red cell and plasma protein turnover rates may be simultaneously measured *in vivo* since ^{51}Cr, ^{59}Fe and ^{125}I may be counted in a single sample. This phenomenon has proved very useful in the study of the pathophysiology of several helminthic infections.

Radioimmunoassays (RIA) have been devised to detect very low concentrations of substances, usually proteins in body fluids, which are difficult or impossible to measure by conventional methods. RIA techniques use a radiolabelled form of the protein of interest. The labelled compound competes with a binding protein specific for the substance of interest, thus giving an indirect measure of the amount of labelled material present in the sample. In parasitology, a RIA is usually used as a sensitive technique for detecting anti-parasite antibody rather than parasite antigen.

Radiolabelled parasites have been used in studies of helminth dynamics in investigations of resistance to anthelmintics and for following the migration of parasites in the host. Methods of trace labelling helminths have been described by several workers. These techniques, however, have not had a wide application.

RADIOTRACERS IN HELMINTHIASIS

Red Cell Kinetics and Iron Metabolism

In studies of the aetiology of anaemias due to helminthic infections, two radioisotopes, ^{51}Cr and ^{59}Fe, have been used to great advantage. Erythrocytes trace-labelled with ^{51}Cr have been used to study changes in circulating red cell volumes and to estimate erythrocyte loss or breakdown. In 1950 Gray & Sterling described the labelling of red cells with ^{51}Cr [26]. After injection of ^{51}Cr-labelled erythrocytes blood samples are taken daily for about 14–21 days. From the haematocrit and the radioactivity of the blood sample taken 5–15 minutes after injection the circulating red cell volume is calculated. The 'apparent' half-life ($T_{1/2}$) of the red cells is found from a semi-log plot of the ^{51}Cr radioactivity of the erythrocytes against time. This value can not be regarded as quantitative since there is considerable elution of the chromium label from the cells in the first few days post-injection particularly in ruminants. Nevertheless, the $T_{1/2}$ of the ^{51}Cr-erythrocytes provides a useful index for comparing red cell survival in normal and disease states. By dividing the ^{51}Cr radioactivity of the total daily urine or faecal output by the ^{51}Cr activity per ml of red cells at the beginning of the collection period, daily urinary or faecal red cell clearances may be calculated. The appearance of significant quantities of the ^{51}Cr label in faeces indicates red cell loss into the gastrointestinal tract and ^{51}Cr activity in the urine indicates intravascular breakdown of erythrocytes [28]. It is, however not possible to monitor changes in erythropoiesis using ^{51}Cr-labelled red cells.

If the plasma iron concentration and the rate of disappearance of intravenously injected ^{59}Fe-ferric citrate are known the plasma iron turnover rate and hence the rate of haemoglobin synthesis may be calculated. The rate of red cell iron utilisation and an estimate of haemoglobin iron reabsorption may also be calculated [15].

The above techniques have been applied to a number of host/parasite systems.

Trematoda

Schistosoma mattheei

The fate of ^{51}Cr-labelled erythrocytes in *S. mattheei* infected sheep was monitored between weeks 4 and 16 of infection [48]. In spite of evidence of a mild anaemia between weeks 4 and 6, erythrocyte survival and the low level of intestinal red cell loss were similar in infected and control animals. More severe haematological disturbances were observed after week 6 of infection when the radioisotope study showed an excellent correlation between the severity of anaemia, the reduction of red cell survival and the magnitude of gastrointestinal red cell leak. Two animals which died of acute gastrointestinal haemorrhage were shown to be suffering daily losses of 50-100 ml red cells in the terminal stages. The absence of any significant amount of ^{51}Cr in the urine showed that the anaemia of ovine schistosomiasis was not haemolytic in origin.

The same authors [18] monitored the erythropoietic response of sheep following exposure to 10,000 *S. mattheei* cercariae over a 16 week period. Anaemia was shown to be due mainly to gastrointestinal haemorrhage. There was also evidence of reduced red cell production in the infected animals. It was also demonstrated that the low serum iron levels recorded between weeks 4 and 8 of infection were due to increased plasma volumes, i.e. there was no iron deficiency. The ability of sheep to reutilise iron lost into the gut was estimated from the faecal excretion of an oral dose of lysed red cells labelled with ^{51}Cr and ^{59}Fe. Both normal and infected animals were unable to reutilise significant amounts of such iron, suggesting that in severe or chronic infections iron levels could be a limiting factor in erythropoiesis.

S. mansoni

Holmes et al [32] found that reduced red cell survival in baboons infected with *S. mansoni* correlated with significantly increased faecal blood loss and haemodilution in the infected animals. There was no evidence of depressed erythropoiesis or haemolytic factors contributing to the anaemia.

S. incognitum
Erythrocytes labelled with ^{51}Cr were used to study red cell loss in *S. incognitum* infected rabbits [53]. Mean daily faecal blood clearance of 2 ml in the infected rabbits compared with 0.65 ml in the controls was observed. Reduced packed cell volumes and red cell half-lives were also observed in the infected animals.

Fasciola hepatica
The application of studies with ^{51}Cr-erythrocytes and ^{59}Fe-ferric citrate played the major part in resolving the controversy surrounding the aetiology of the anaemia in fascioliasis. It is now accepted that the liver fluke is haematophagic.

Holmes et al [27] found a reduced ^{51}Cr-red cell half-life and the appearance of significant amounts of ^{51}Cr in the faeces of infected sheep. There was no evidence of increased ^{51}Cr in the urine of the infected sheep indicating the absence of haemolytic anaemia. Further evidence of the haemotophagic nature of liver flukes came in studies on ferrokinetics in infected rabbits [15]. Increased rates of ^{59}Fe disappearance from plasma and increased incorporation of ^{59}Fe into the erythrocytes of the infected animals compared with the controls were noted. There was also an increase in the rate of disappearance of labelled red cells from the circulation and increased faecal clearance of both ^{51}Cr and ^{59}Fe. Neither infected nor control animals were able to re-utilise iron from red cells lost into the gastrointestinal tract. The same authors [16] reported increased gastrointestinal red cell loss in *F. hepatica* infected rabbits using ^{51}Cr-red cells.

Using erythrocytes labelled with ^{51}Cr and ^{59}Fe, it was demonstrated that a considerable loss of iron occurred into the gut of *F. hepatica* infected sheep, between 15.2 mg/day compared with 0.7 mg/day in the controls, and iron reabsorption occurred only in the most severe infections [28]. It was also shown that the underlying cause of the anaemia in all animals was a combination of haemodilution and haemmorhage both appearing earlier in poorer-fed animals.

Nematoda

Haemonchus contortus
Dargie and Allonby [9] infected sheep with two doses of 1,000 *H. contortus* larvae given 8 weeks apart. In single infections gastrointestinal red cell loss measured with ^{51}Cr-labelled erythrocytes commenced between the seventh and tenth day of infection and increased over the following two weeks, one animal losing an average of 12 ml red cells per day. Between weeks 3 and 8 of infection, it was observed that although there was a high daily loss of red cells into the gut packed cell volumes recovered slightly. Rate of red cell loss and rate of production (measured with ^{59}Fe-ferric citrate) were about equal. Plasma iron clearance and turnover was reduced in sheep which self-cured after challenge. In animals which became reinfected, there was also a fall in iron turnover which was shown to be due to low serum iron levels and high plasma ^{59}Fe clearances; i.e. exhaustion of iron stores. These results showed that anaemia is largely determined by the host's erythropoietic capability and that packed cell measurement alone is not a good index of blood loss.

In chronic haemonchosis of Marino and Blackface lambs, it was demonstrated that the resulting anaemia was associated with a modest but significant increase in abomasal blood loss and elevated erythropoiesis in both breeds relative to the controls [2]. The infected Blackface lambs lost an average of 3 ml red cells per day into the gut, the infected Merinos 5 ml compared with a loss of 0.3 ml in the controls. Increased iron turnover and red cell loss was also observed in the infected animals.

Trichuris suis

Using ^{59}Fe-labelled erythrocytes Beer et al [7] observed blood losses of 7 ml to 35 ml per day in pigs infected with between 15,000 and 100,000 *T. suis* ova, although anaemia was not detected in any of the animals. The blood losses were approximately proportional to the infective dose.

Hypoproteinaemia and Protein Turnover

Since hypoalbuminaemia is a common feature of most helminth infections many studies have centred round albumin metabolism using plasma albumin labelled with ^{125}I or ^{131}I [37]. Following intravenous injection of iodine-labelled albumin blood samples are taken for 14–21 days and the daily output of faeces and urine is measured. The radioactivity of aliquots of plasma, urine and faeces and of a standard prepared from the injection material is determined.

From the radioactivity of the first plasma sample and the application of the 'dilution principle' the plasma volume is determined. The intravascular albumin pool is obtained by multiplying the plasma volume by the plasma albumin concentration. The total body albumin is calculated either indirectly by extrapolation of the plasma radioactivity curve [50] or by a direct method [9] which requires collection of urine and faeces. The difference between total and intravascular albumin equals the extravascular albumin pool. The fraction of the intravascular albumin broken down each day (fractional catabolic rate) may be found indirectly by analysis of the plasma radioactivity [40]. The most significant of the above measurements is the fractional catabolic rate of albumin. In many helminth infections it has been shown that apparently normal albumin concentrations and intravascular pool sizes can conceal a highly abnormal state when albumin hypercatabolism is compensated for by increased synthesis. The concept of 'albumin turnover', therefore, became an important parameter in the study of the pathophysiology of helminthiases (see Chapter 8; Appendix, Chapter 3).

In many cases, hypercatabolism of albumin is the result of plasma protein loss into the gastrointestinal tract. Iodine-labelled proteins may not be used to quantify this loss since iodine is substantially reabsorbed following degradation of the labelled protein in the gut. Such protein loss may only be measured by injecting a labelled macromolecule which is not subject to enzymatic degration in the intestine. Substances which have been used include polyvinylpyrrolidone (PVP) labelled with ^{131}I [25] and ^{51}Cr-labelled plasma proteins [26]. Faecal 'clearances' which represent the volume of plasma which would have to leak into the gut to account for the daily faecal radioactivity is calculated by dividing the daily faecal radioactivity by the radioactivity of 1.0 ml of plasma at the beginning of the collection period.

Trematoda

Schistosoma mattheei

Severe hypoalbuminaemia was observed in sheep between the 4th and 14th week after infection with *S. mattheei* [17]. Studies with ^{125}I-albumin showed that this was associated with a fall in total exchangeable and extravascular albumin pools. Reductions in the intravascular albumin pools were observed only in severely affected cases. High catabolic rates of albumin and reduced ^{125}I-albumin $T_{1/2}$, showed that these changes were the result of hypercatabolism of albumin.

S. bovis

A study of albumin turnover in Ethiopian Highland sheep infected with *S. bovis* using ^{125}I-albumin found hypoalbuminaemia, significantly increased albumin catabolism, increased plasma volume and decreased extravascular albumin in the infected animals [31].

Fasciola hepatica
As might be expected elevated gastrointestinal plasma leak was demonstrated in *F. hepatica* infected rabbits [14]. Faecal clearance of ^{131}I-PVP in infected animals ranged from 20–40 ml/day compared with 2–5 ml/day in the control rabbits. The fractional catabolic rate of albumin in the infected rabbits was also significantly greater than in the controls. Reduced ^{131}I-albumin $T_{1/2}$, increased fractional catabolic rate of albumin and an increased ^{131}I-PVP faecal clearance were demonstrated in fluke-infected sheep by Holmes et al [27]. Other workers using ^{131}I-albumin and ^{125}I-IgG also demonstrated that the turnover rate of albumin and IgG was greatly accelerated in fluke infected sheep [44, 45, 46]. In further studies using radioisotopic methods, hypoalbuminaemia, increased albumin turnover and gastrointestinal plasma loss were found to be a feature of chronic ovine fascioliasis [20, 30].

Nematoda

Trichostrongylus colubriformis
Elevated losses of plasma protein into the gut of lambs infected with *T. colubriformis* were measured by Barker [11] using ^{51}Cr labelled proteins. In a series of studies on *T. colubriformis* infected sheep and guinea-pigs, using L-^{14}C leucine and L-^{14}C tyrosine, Symons and co-workers showed that skeletal muscle protein was reduced in infected animals whilst synthesis of protein in the small and large intestines and in the liver was greatly increased (see review by Symons [51]).

Cooperia oncophora
Some species of *Cooperia* cause severe enteritis in calves in tropical and sub-tropical regions. Increased gastrointestinal proteins loss at 5–7 weeks post-infection and more significantly between 10 and 12 weeks after infection, were observed in *C. oncophora* infected calves using ^{51}Cr-labelled proteins [3].

Hyostrongylus rubidus
Studies using ^{131}I-albumin and ^{131}I-PVP have shown that the observed hypoproteinaemia in *H. rubidus* infected pigs was the result of gastrointestinal plasma protein leakage. Between 14 and 23 days post-infection 17.6% of the injected ^{131}I-PVP was excreted in the infected pigs compared with 1.67% in control animals [54].

Haemonchus contortus
Dargie [13] found a degree of hypoalbuminaemia in chronic haemonchosis in sheep, 18 g/l in the infected animals compared with 26.3 g/l in the controls. This did not indicate any significant disturbance in albumin metabolism. However the rate of disappearance of albumin and the catabolic rate of albumin in the infected animals was more than twice that of the controls. The loss of albumin was shown to correspond to increased macromolecular leak into the gut. Mean daily ^{131}I-PVP clearances of 273 ml and 38 ml were found in the infected and control sheep, respectively. A study of albumin metabolism in chronic low level of *H. contortus* infections in Merino and Scottish Blackface sheep by Abbott et al [1] showed that in the former animals, there was a slight increase in plasma volume and intravascular albumin. Increased albumin breakdown and increased albumin catabolic rates were also observed in the Merinos. The only significant disturbance in albumin metabolism in the Blackface group was a slightly increased albumin catabolic rate. These workers [2] also compared albumin metabolism in lambs maintained on a high or low plane of nutrition and found reduced ^{125}I-albumin $T_{1/2}$ values particularly in the lambs on the low protein diet. Fractional catabolic rates were increased in both infected groups, more significantly in the lambs on the high nutritional plane.

Ostertagia ostertagi

A comparison of the half-life and fractional catabolic rate of ^{131}I-albumin in ostertagiasis of cattle showed increased albumin catabolism in the infected animals to be a significant factor in the hypoalbuminaemia [43]. Later using ^{125}I-labelled albumin and $^{51}CrCl_3$, it was demonstrated that *O. ostertagi* infection in calves resulted in an increase in albumin disappearance, catabolic rates and plasma faecal clearance [52].

O. circumcincta

Studies with ^{131}I-albumin and ^{51}Cr-labelled proteins in sheep infected with 300,000 or 900,000 larvae showed that the hypoalbuminaemia and increased turnover of albumin was related to the level of infection [29]. Faecal plasma clearance rates of between 55 and 130 ml per day were observed in the infected animals and were related to parasite burden compared with a mean of 29 ml in the control sheep.

Oesophagostomum radiatum

It was shown that protein-losing enteropathy was a significant factor in the hypoproteinaemia of bovine oesophagostomosis. Bremner [8] using ^{51}Cr-labelled plasma proteins found a mean daily plasma clearance of 927 ml into the gut of calves three weeks post-infection compared with a comparable value of 144 ml in the controls. The ^{51}Cr-plasma $T_{1/2}$ in the infected animals was also significantly shorter than in the controls.

Strongyloides ransomi

Using albumin trace-labelled with Niobium-95, it was shown that there was a greatly elevated plasma leak into the gut of piglets infected with *S. ransomi*, only 4 ml in the controls, compared with 112 ml in infected animals [23].

RADIOIMMUNOASSAY

There have been few applications of RIA in helminthology. The general aspects and application of RIA in parasitic diseases have been reviewed [22].

Trematoda

Schistosoma mansoni

Williams et al [55] described a ^{125}I-antigen labelled assay for *S. mansoni* antibodies. This test was not suitable for diagnostic purposes.

Cestoda

A diagnostic RIA for hydatid disease in man, similar to *Trichinella* RIA was developed by Matossian [39].

Nematoda

Dictyocaulus filaria

A RIA for ovine dictycauliasis was developed by Movsesijan and Lalic [41]. Immunoglobulin from *D. filaria* infected lambs and anti-sheep gamma globulins were labelled with ^{125}Iodine and juvenile stages of the parasite served as antigen. Levels of specific antibody were estimated quantitatively.

Trichinella spiralis
A ^{125}I-antigen labelled RIA for *T. spiralis* in humans and a number of animal species has been described [56] and a RIA for the diagnosis of Trichinellosis in man using ^{125}I-labelled sheep antibody specific to human immunoglobulin has also been described (see Chapter 4).

Haemonchus contortus
A RIA for detecting IgA and IgG anti-*H. contortus* larval antibodies using ^{125}I-labelled anti-serum was described by Smith [49]. Movsesijan et al [42] developed an assay for measuring circulating antibody in *H. contortus* infected sheep. Specific antibodies derived from a complex formed from solid phase whole parasite antigen incubated with IgG from an infected sheep, were labelled with ^{125}I and used to measure the amount of circulating antibody in infected animals by competitive reaction with antigen. The same method is used to measure the amount of soluble antigen.

RADIOLABELLING OF PARASITES

Trematoda

Fasciola hepatica
Nansen et al [47] labelled *F. hepatica* miricidia by intraperitoneal injection of ^{75}Se-methionine. The radiolabelling was shown to have no adverse effects on fluke development. Miricidia labelled with ^{75}Se were used to study the influence of temperature on host finding capacity of *F. hepatica* [10].

Schistosoma spp.
S. japonicum were labelled with ^{59}Fe by adding ferric chloride to cercariae [21]. These organisms were detected in mice 30–40 days later by autoradiography. Snails infected with *S. mansoni* were labelled with ^{14}C by placing the snails in water containing ^{14}C-glucose [36]. The labelled schistosomes were similarly detected in mice 36 days later by autoradiography. *S. mansoni* infected snails were incubated with $H_2{}^{75}SeO_3$ or ^{75}Se-methionine and the radiolabelled cercariae shed were used to study cercarial predators [33]. The *in vitro* uptake of ^{3}H-labelled amino acids, by *S. mansoni* was used to study amino acid transport in schistosomes [4]. *S. mansoni* infected snails were exposed to ^{3}H labelled thymidine, amino acids or formate. The labelled schistosomes could be detected by autoradiography 15–19 days after infection in mice. Christensen et al [12] described an *in vivo* method for labelling cercariae of *S. mansoni* and *S. intercalatum* with ^{75}Se-methionine. The labelled cercariae were used to study the host finding capacity of the cercariae.

Cercariae of *S.bovis*, *S. intercalatum* and *S. haematobium* were labelled with ^{75}Se by exposure of their respective snails to ^{75}Se-methionine. Maximum cercariae-bound radioactivity was observed 4 to 7 days later and detectable levels of ^{75}Se were present on day 32 after labelling [11].

Nematoda

Haemonchus contortus
H. contortus larvae may be labelled with ^{75}Se by adding ^{75}Se-methionine to a mixture of eggs and faecal sediment [24]. Seventeen days after infecting sheep with labelled larvae worms recovered showed no damage in terms of either survival or sex ratio. The establishment and regulation of ^{75}Se-*H. contortus* larvae in sheep was monitored by Barger et al [5].

Dirofilaria immitis

After testing more than 20 radionuclides Lengemann et al [34] found that ^{203}Hg showed an *in vitro* retention of about 90%. Labelling was achieved by adding mercuric chloride to a suspension of microfilariae in culture medium. Incorporation of the mercury was low and the toxicity of mercury to both the parasite and the host limited the usefulness of this method. The fate of microfilariae labelled with ^{203}Hg in mice and dogs was investigated [35] and the lungs were shown to be an important location for the microfilariae soon after reinjection into the host.

Strongyloides ratti

Infective *S. ratti* larvae were trace-labelled with ^{32}P or ^{75}Se by adding the radioisotopes to the faecal culture. The use of these trace-labelled larvae as a tool in the study of migratory behaviour has been discussed [57].

Nippostrongylus brasiliensis

A method for measuring the *in vivo* uptake of ^{32}P, ^{14}C or ^{75}Se by *N. brasiliensis* worms following the intravenous injection of ^{32}P-sodium phosphate, ^{14}C-glucose or ^{75}Se-selenomethionine into the host rat has been described [38]. The technique allows changes in worm metabolic activity to be accurately quantified at any time during an infection and, although this host/parasite system is of no economic importance, the method could be adapted to monitor changes in parasite function in other systems, e.g., resistance to an anthelmintic following its administration.

CONCLUSIONS

Radioisotopic studies have been extensively applied to the study of the pathophysiology and immunology of host-parasite interactions and have led to a better understanding of the aetiology of several helminth infections. The techniques have allowed changes in plasma protein metabolism and red cell kinetics to be monitored in infections such as fascioliasis and schistosomiasis and to provide information which would not otherwise have been obtained by non-radioisotopic studies.

Much useful information on the pathophysiology of nematodes in sheep and cattle, principally infections with *Trichostrongylus* spp., *Haemonchus* spp. and *Ostertagia* spp., has been obtained from radioisotopic studies.

Other techniques, such as radioimmunoassays and radiolabelling of parasites have not had a wide application in helminthological research, but where they have been used have proved to be useful research tools.

REFERENCES

1. Abbott, E.M., Parkins, J.J. and Holmes, P.H. Studies on the pathophysiology of chronic ovine haemonchosis in Merino and Scottish Blackface lambs. Parasitology **89**: 585–596 (1984).
2. Abbott, E.M., Parkins, J.J. and Holmes, P.H. Influence of dietary protein on the pathophysiology of haemonchosis in lambs given continuous infections. Res. Vet. Sci. **45**: 41–49 (1988).
3. Armour, J., Bairden, K., Holmes, P.H., Parkins, J.J., Ploeger, H. and Salman, S.K. Pathophysiological and Parasitological studies on *Cooperia oncophora* infections in calves. Res. Vet. Sci. **42**: 373–381 (1987).
4. Asch. H.L. and Reid, W.A. Membrane transport in *Schistosoma mansoni*: transport of amino acids by adult males. Exp. Parasitol. **38**: 123–135 (1980).
5. Barger, I.A., Le Jambre, L.F., George, J.R. and Davies, H.I.: Regulations of *Haemonchus contortus* populations in sheep exposed to continuous infection. Int. J. Parasitol. **15**: 529–533 (1985).

6. Barker, I.K. A study of the pathogenesis of *Trichostrongylus colubriformis* infection in lambs with observations on the contribution of gastrointestinal plasma loss. Int. J. Parasitol. 3: 743–757 (1973).
7. Beer, R.J. and Sansom, B.F. Erythrocyte losses from pigs with experimental *Trichuris suis* infections measured with a whole-body counter. J. Comp. Path. 84: 331–346 (1974).
8. Bremner, K.C. Pathogenic factors in experimental bovine oesophagostomosis. III. Demonstration of protein losing enteropathy with ^{51}Cr-albumin. Exp. Parasitol. 24: 364–374 (1969).
9. Campbell, R.M., Cuthbertson, D.P., Matthews, C.M.E. and MacFarlane, A.S. The behaviour of ^{14}C and ^{131}I-labelled plasma proteins in the rat. Int. J. Appl. Radiat. Isot 1: 66–84 (1956).
10. Christensen, N.O. and Nansen, P. The influence of temperature on the infectivity of *Fasciola hepatica* miracidia to *Lymnaea truncatula*. J. Parasitol. 62: 698–701 (1976).
11. Christensen, N.O. *Schistosoma bovis, S. intercalatum* and *S. haematobium*: methods for *in vivo* radiolabelling of cercariae with radioselenium. J. Helminthol. 55: 9–12 (1981).
12. Christensen, N.O. and Frandsen, F. Host-finding capacity of schistosome cercariae: comparative efficiency of methods of mouse infection and a radioisotope assay system. J. Helminthol. 51: 105–113 (1977).
13. Dargie, J.D. Ovine haemonchosis: Pathogenesis. In: *Helminth Diseases of Cattle, Sheep and Horses in Europe* (G.M. Urquhart and J. Armour, eds), pp. 63–71. MacLehose & Co. Ltd.: Glasgow, Scotland (1974).
14. Dargie, J.D., Holmes, P.H., Maclean, J.M. and Mulligan, W. Pathophysiology of fascioliasis in the rabbit. Studies on albumin turnover. J. Comp. Path. 78: 101–105 (1968).
15. Dargie, J.D. and Mulligan, W. Ferrokinetic studies on normal and fluke-infected rabbits. J. Comp. Path. 80: 37–45 (1970).
16. Dargie, J.D. and Mulligan, W. The onset and development of anaemia and hypoalbuminaemia in rabbits infected with *Fasciola hepatica*. J. Comp. Path. 81: 187–202 (1971).
17. Dargie, J.D., Maclean, J.M. and Preston, J.M. Pathophysiology of ovine schistosomiasis. III. Study of plasma protein metabolism in experimental *Schistosoma mattheei* infection. J. Comp. Path. 83: 543–557 (1973).
18. Dargie, J.D. and Preston, J.M. Pathophysiology of ovine schistosomiasis. VI. Onset and development of anaemia in sheep experimentally infected with *Schistosoma mattheei*. Ferrokinetic studies. J. Comp. Path. 84: 83–91 (1974).
19. Dargie, J.D. and Allonby, E.W. Pathophysiology of single and challenge infections of *Haemonchus contortus* in Merino sheep: studies on red cell kinetics and the 'self-cure' phenomenon. Int. J. Parasitol. 5: 147–157 (1975).
20. Dargie, J.D. and Berry, C.I. The hypoalbuminaemia of ovine fascioliasis: the influence of protein intake on the albumin metabolism of infected and of pair-fed control sheep. Int. J. Parasitol. 9: 17–25 (1979).
21. De la Cruz, B., Pesigan, T.P., Banzon, T.C., Sevilla, J., Donata, C., Cruz, A. and Castro, B. A report on the labelling of *Schistosoma japonicum* cercariae with radioiron. Am. J. Trop. Med. Hyg. 13: 545–547 (1964).
22. Dessaint, J.P. and Capron, A. Nuclear techniques in seroepidemiology and diagnosis of parasitic diseases. In: *Nuclear Techniques in the Study of Parasitic Infections*. pp. 317–330. IAEA, Vienna (1982).
23. Enigk, K. and Dey-Hazra, A. Intestinal plasma and blood loss in piglets infected with *Strongyloides ransomi*. Vet. Parasitol. 1: 69–75 (1975).
24. Georgi, J.R. and Le Jambre, L.F. Radiolabelling of infective larvae of *Haemonchus contortus* (Nematode Trichostrongyloidea) with ^{75}Se-methionine and their performance as tracers in sheep. J. Parasitol. 69: 875–882 (1983).
25. Gordon, R.S. Jr Exudative enteropathy. Abnormal permeability of the gastrointestinal tract demonstrable with labelled polyvinylpyrrolidone. Lancet 1: 325–326 (1959).
26. Gray, S.J. and Sterling, K. The tagging of red cells and plasma proteins with radioactive chromium. J. Clin. Invest. 29: 1604–1619 (1950).
27. Holmes, P.H., Dargie, J.D., Maclean, J.M. and Mulligan, W. The anaemia in fascioliasis: studies with ^{51}Cr labelled red cells. J. Comp. Path. 78: 415–420 (1968).
28. Holmes, P.H. and Maclean, J.M. Intestinal loss of iron and its possible reabsorption in chronic ovine fascioliasis. Res. Vet. Sci. 10: 488–489 (1969).
29. Holmes, P.H. and Maclean, J.M. The pathophysiology of ovine ostertagiasis: a study of the changes in plasma protein metabolism following single infection. Res. Vet. Sci. 12: 265–271 (1971).

30. Holmes, P.H., Maclean, J.M. and Mulligan, W. A study of the onset and development of anaemia and hypoproteinaemia in chronic ovine fascioliasis. In: *Pathology of Parasitic Diseases* (S.M. Gaafar ed.), pp. 69–81. Purdue University Press, Indiana (1971).
31. Holmes, P.H., Mamo, E. and Lema, A. Preliminary studies on albumin metabolism: Ethiopian sheep infected with *Schistosoma bovis*. In: *Nuclear Techniques in Helminthology Research.* pp. 115–120. IAEA, Vienna (1975).
32. Holmes, P.H., James, E.R., Maclean, J.M., Nelson, G.S. and Taylor, M.G. Erythrokinetic studies on baboons (*Papio anibus*) with acute experimental schistosomiasis mansoni. J. Helminthol. **51**: 95–104 (1977).
33. Knight, W.B., Laird, F., Ritchie, L.S., Pellegrino, J. and Chiroboga, J. Labelling of *Biomphalaria glabrata* and cercariae of *Schistosoma mansoni* with radioselenium. Exp. Parasitol. **22**: 309–315 (1968).
34. Lengemann, F.W., Grieve, R.B., Chmielewicz, M., Fallfelz, F.A. and Georgi, J.R. External monitoring of ^{203}Hg-labelled *Dirofilaria immitis* microfilaria in mice and dogs. Parasitology **92**: 463–469 (1986).
35. Lengemann. F.W., Grieve, R.B., Chmielewicz, M. and Georgi, J.R. ^{203}Hg and other gamma-emitting radio-isotopes as labels for *Dirofilaria immitis microfilaria.* Parasitology **92**: 451–461 (1986).
36. Lewert, R.M. and Para, B.J. The physical incorporation of carbon in *Schistosoma mansoni* cercariae. J. Infect. Dis. **116**: 172–182 (1966).
37. McFarlane, A.S. Efficient trace labelling of proteins with iodine. Nature (Lond.) **182**: 53 (1958).
38. Maclean, J.M. Radioisotope techniques in the study of *in vivo* parasite metabolism. In: *Nuclear Techniques in the Study of Parasitic Infections.* pp. 483–495. IAEA, Vienna (1982).
39. Matossian, R.M. A simplified radioimmunoassay technique for hydatid disease and human trichinosis. J. Helminthol. **55**: 49–57 (1981).
40. Matthews, C.M. The theory of tracer experiments with ^{131}I-labelled plasma proteins. Phys. Med. Biol. **2**: 36–53 (1957).
41. Movsesijan, M. and Lalic, R. Radioactive antibody studies on *Dictyocaulus filaria* infection. Res. Vet. Sci. **12**: 282–284 (1973).
42. Movsesijan, M., Janovic, B., Borojevic, D. and Petrovic, M. The preparation and use of radiolabelled specific helminth antibodies. Int. J. Appl. Rad. Istol **34**: 805–807 (1973).
43. Mulligan, W., Maclean, J.M., Halliday, G.J. and Dalton, R.G. Plasma protein metabolism in Ostertagiasis of cattle. In: The *Reaction of the Host to Parasitism.* pp. 43–49. Proc. Third Int. Conf. Wld Assoc. Adv. Vet. Parasitol., Lyon (1967).
44. Nansen, P. Hypoproteinaemias associated with helminthic infections, with special reference to gastrointestinal parasites. In: *Isotopes and Radiation in Parasitology.* Vol. I, pp. 125–132. IAEA, Vienna (1968).
45. Nansen, P. Immunoglobulin synthesis in chronic bovine and ovine fascioliasis. In: *Isotopes and Radiation in Parasitology.* Vol. II, pp. 71–76. IAEA, Vienna (1970).
46. Nansen, P., Eriksen, L., Simeson, M.G. and Neilsen, K. Chronic fascioliasis in sheep. II. Metabolism of ^{125}I-labelled immunoglobulin. Nord Veterinaemed **20**: 651–656 (1968).
47. Nansen, P., Christensen, N.O. and Frandsen, F. A technique for *in vivo* labelling of *Fasciola hepatica* miricidia with radioselenium. Z. Parasitenkunde **49**: 73–80 (1976).
48. Preston, J.M. and Dargie, J.D. Pathophysiology of ovine schistosomiasis. V. Onset and development of anaemia in sheep experimentally infected with *Schistosoma mattheei*—Studies with ^{151}Cr-labelled erythrocytes. J. Comp. Path. **84**: 73–81 (1974).
49. Smith, W.D. Anti-larval antibodies in the serum and abomasal mucus of sheep hyperinfected with *Haemonchus contortus.* Res. Vet. Sci. **22**: 334–338 (1977).
50. Sterling, K. The turnover rate of serum albumin in man as measured by ^{131}I-tagged albumin. J. Clin. Invest. **30**: 1228–1237 (1951).
51. Symons, L.E.A. Anorexia: occurrence, pathophysiology and possible causes in parasitic infections. Adv. Parasitol. **24**: 103–133 (1985).
52. Taylor, L.M., Parkins, J.J., Armour, J., Holmes, P.H., Bairden, K., Ilbarra-Silva, A.M. and Salman, K.S. Pathophysiological and parasitological studies on *Ostertagia ostertagi* infections in calves. Res. Vet. Sci. **46**: 218–225 (1989).
53. Tewari, H.C., Singh, K.S. and Sharma, B.K. Blood loss estimation in *Schistosoma incognitum* by use of ^{51}Cr-labelled red cells. J. Nucl. Agric. Biol. **5**: 71–73 (1976).

54. Titchener, R.N., Herbert, I.V. and Probert, A.J. Plasma protein loss in growing pigs during the prepatent and early patent periods of infection with high doses of *Hyostrongylus rubidus* larvae. J. Comp. Path. **84**: 399–407 (1974).
55. Williams, J.S., Sadun, E.H. and Gore, R.W. A radioactive antigen microprecipitin (RAMP) assay for schistosomiasis. J. Parasitol. **57**: 220–232 (1971).
56. Williams, J.S., Gore, R.W., and Sadun, E.H. *Trichinella sprialis*: antigen-antibody interaction assayed by radioactive iodinated antigen. Exp. Parasitol. **31**: 299–306 (1972).
57. Wilson, P.A.G. Tracking radioactive larvae of *Strongyloides ratti* in the host. Parasitology **79**: 29–38 (1979).

12. Immunodiagnosis of Helminthic Infections*

S.P. Ahuja and M. Tsuji[1]
Department of Biochemistry, College of Basic Science and Humanities
Punjab Agricultural University, Ludhiana, India
[1]Department of Parasitology, Kyorin University, School of Medicine
Mitaka, Tokyo, Japan

INTRODUCTION

The traditional concept of clinico-helminthological diagnosis involves direct visualisation of eggs or recovery of the parasites in different stages of development. Such examinations are difficult in case of tissue parasites and in case of mild, latent/early helminthiases. The development of chromatographic techniques for characterization of antigens has been paralleled by the development of immunoassay, hybridoma technology and recombinant-DNA technology. Now, the additional criterion involves the detection of stage specific nucleic acid sequences and amplification of DNA by polymerase chain reaction (PCR). Such techniques can enlarge the choice of methods for diagnostics as well as to study the epidemiology, and pathogenesis of helminthiases.

The sensitive and specific immunoassay involving the use of polyclonal/monoclonal antibodies (MAb) and of antigens purified by chromatography or produced by recombinant-DNA technology have become acceptable in the diagnosis of infectious diseases (including the helminthic diseases), assessment of intensity of helminthiasis and effectiveness of anthelmintic and surgical measures of treatment. The usefulness of immunoassays viz. immunodiffusion (ID), immunoelectrophoresis (IE), complement fixation test (CFT), indirect haemagglutination test (IHA), latex agglutination test (LAT), western blotting, enzyme linked immunosorbent assay (ELISA), cathodic antigen assay and anodic antigen assay in the diagnosis of trematodiases, cestodiases and nematodiases, have been investigated and some of these have been reviewed ([38]; see Chapter 4). The diagnostic reagents for such assay either do not exist or are inadequate for accurate identification of the causative agent.

It is well known that positive serological tests are closely related with the survival of the worms in the patients. In post-treatment follow up studies, the antibody titres in the cured patients tend to decrease and can be used to monitor the efficacy of the therapy.

PRODUCTION OF REAGENTS AND DESCRIPTION OF IMMUNOASSAY

Immunoassays require the purified antigens or highly specific antibodies as such or as conjugates with enzyme so as to detect the circulating antigen/antibodies produced after the infection, and ELISA has emerged as the sensitive and specific immunoassay.

Antibodies (Fig. 1) are immunoglobulins in nature, having specific, complementary antigen binding sites in their Fab region. This specificity of interaction between the antigens and the antibodies is the

---

*Parasitological data of the manuscript were compiled by Professor M. Tsuji before his retirement from Hiroshima University—*Editors*

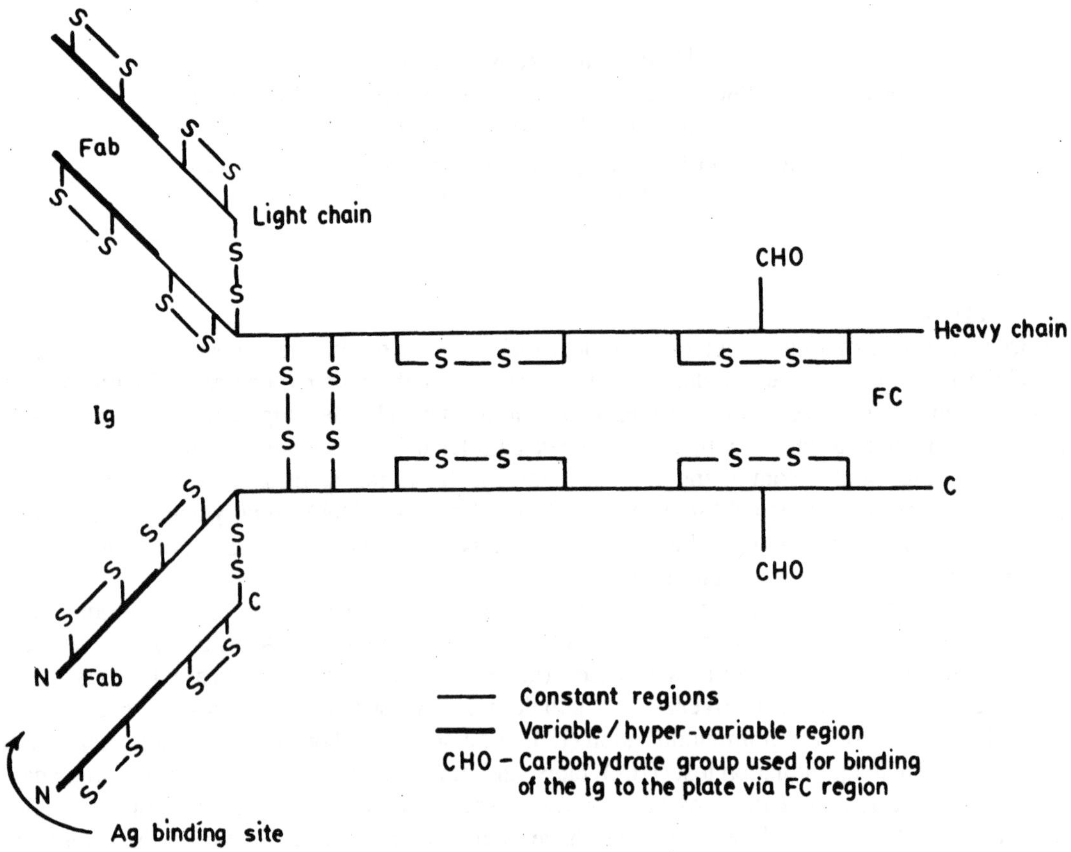

Fig. 1. Structural features of an antibody (Ig) molecule.

basis of immunoassay. The availability of highly purified antigen and/or the antibodies with high specificity, selectivity and affinity determines the development of a useful immunoassay.

Recombinant DNA Technology

The purified antigens can be produced by chromatographic methods. One of the emerging techniques for this purpose is the immunoaffinity chromatography, in which specific antibodies immobilized on a matrix can be used to purify required antigens from helminths. Such antigens in adequate amounts can also be generated by recombinant-DNA technology (Fig. 2). The DNA coding for the antigen of interest is obtained from genomic cDNA library, inserted into a vector, got expressed in *Escherichia coli* and the antigen synthesized by the bacteria is purified for use in immunoassay.

Monoclonal Antibodies

The antibodies are either produced by hybridoma or by immunising rabbits/sheep with crude or partially/ highly purified antigen. The latter method suffers from the disadvantages that the antibodies formed are polyclonal and their spectrum/titre varies from animal to animal. Thus, such antibodies lack constant specificities/affinities. These shortcomings can be overcome by raising MAb [28, 29]. The typical procedure to generate MAb (Fig. 3) involves fusion of genes of a cell (obtained from spleen of sensitized animal), capable of producing specific antibodies with those of a myeloma cells with ability to multiply continously.

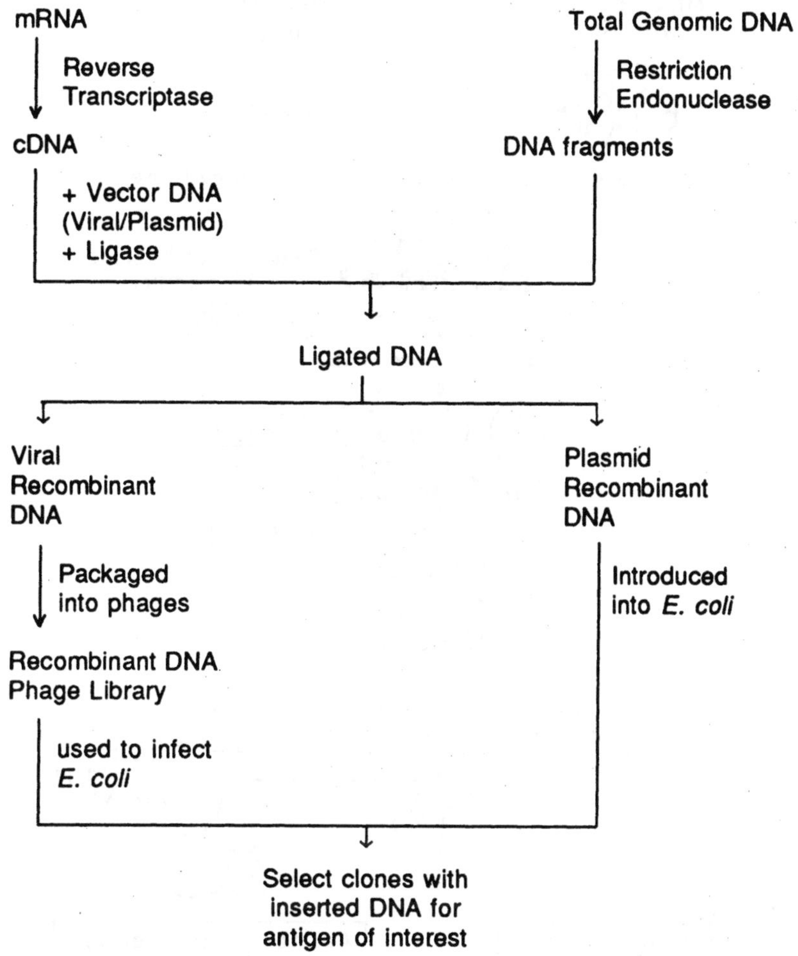

Fig. 2. Construction of genetic and cDNA libraries for production of helminth antigen of interest.

These myeloma cell lines are deficient in the enzyme hypoxanthine phosphoribosyl transferase (HGPRT) and do not grow in medium containing Hypoxanthine, Aminopterin, Thymidine (HAT) because aminopterin inhibits DNA biosynthesis. Only the hybrid cells can grow in the HAT medium as the splenocyte provides HGPRT gene which on expression overcomes the inhibition by aminopterin. The myeloma cells provide genes for continuous multiplication in tissue culture. The hybrid cells producing antibodies of the desired specificity are selected, propagated and preserved. The propagation can be done either by culturing in the growth medium or in the peritoneal cavity of BALB/c mice. This produces MAb with constant affinity and monospecificity.

Enzyme Linked Immunosorbant Assay

The ELISA involves covalent linking of the enzyme to antibodies with N, N′, O-phenylenedimaleimide or with succinimidyl-3-(2-pyridyldithio)-propionate (Fig. 4).

Reaction of the enzyme-antibody/anti-antibody conjugate with the immobilized antigen (Figs. 5-7) is followed by substrate hydrolysis and the estimation of the product formed is done visually or by spectrophotometry (Table 1). If the antigen is immobilized in the microtiter plate through the antibodies

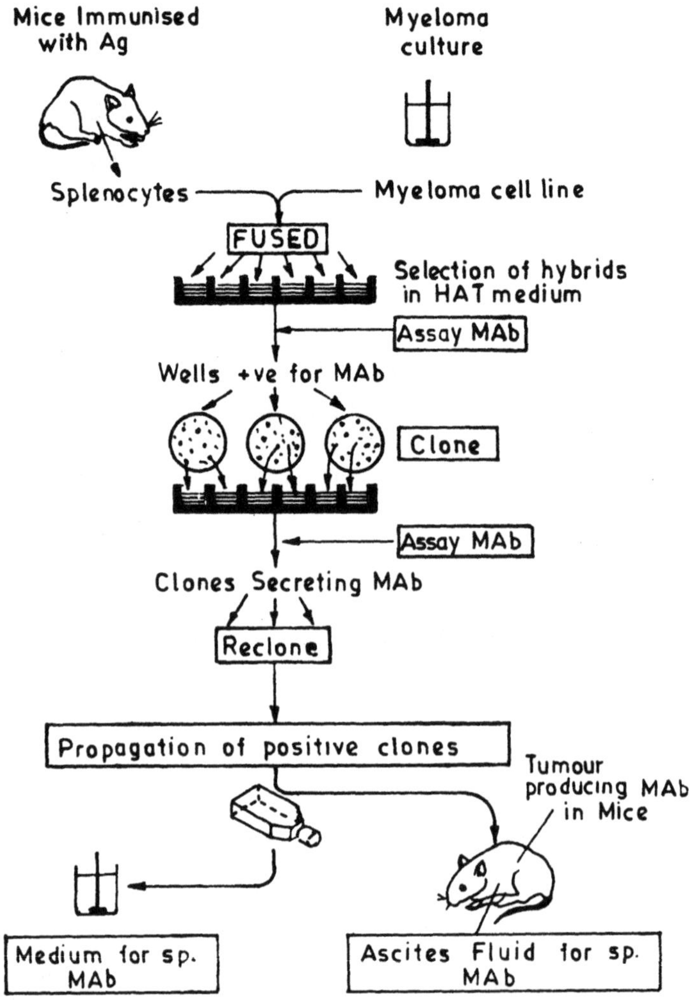

Fig. 3. Basic protocol for the production of monoclonal antibodies.

(Figs. 5 and 7) then the latter should be fixed specifically through the Fc region of the antibody (Fig. 5). The enzymes commonly used for conjugation are horse radish peroxidase; *E. coli* alkaline phosphatase and *Staphylococcus aureus* penicillinase (6).

The sandwich ELISA has been readily accepted for epidemiological assessment of the major helminthic diseases. The specificity of the test depends on the antigen employed. The high background values on the negative sera and antigen cross reactivity between parasites can be reduced [61] by using purified antigens, by post coating of antigen-coated microtiter plates with bovine serum albumin and by incorporating nonspecific immunoglobulin in the solution used for diluting the conjugate. In addition to immunoassay, DNA probes to identify organism specific DNA sequences are available (Fig. 8). The sequence of steps in the ELISA and immunoblotting are given in Table 1. The computerized Axial Tomography (CAT) scanning and Magnetic Resonance Imaging (MRI) are also being used in conjunction with ELISA for the diagnosis of larva migrans and in assessing the size of the lesions before and after surgery (see further).

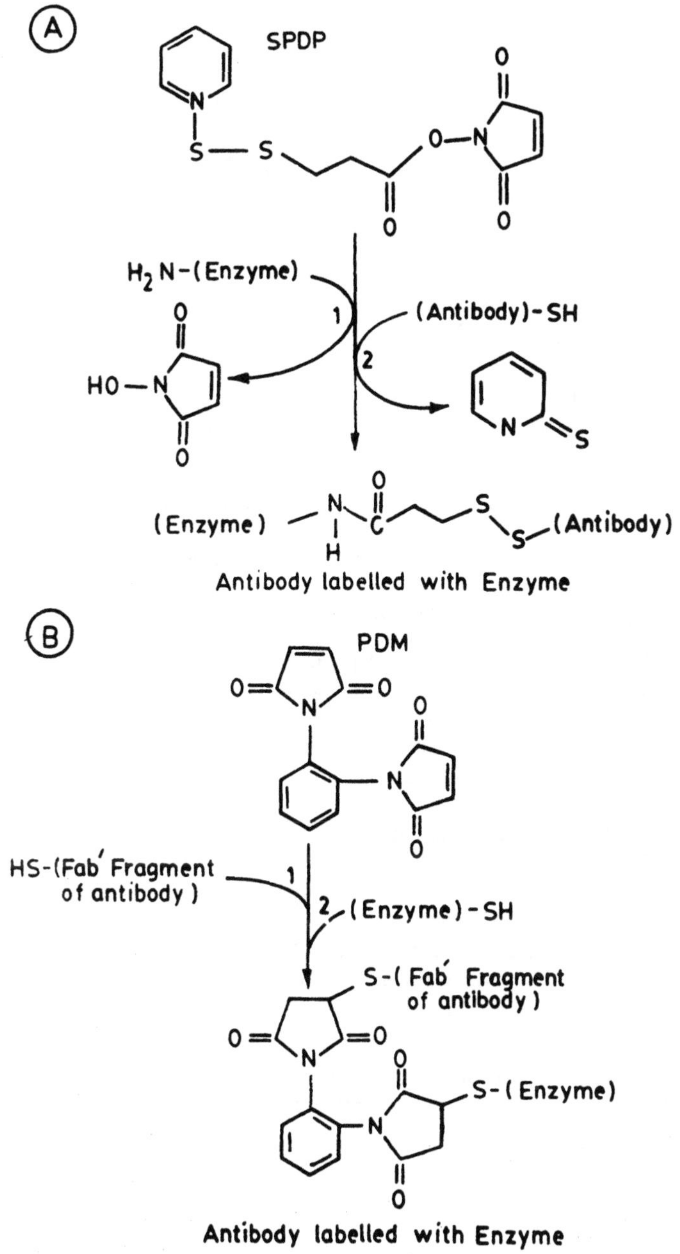

Fig. 4. Conjugation of enzyme(s) to antibodies (Ig) through N-succinimidyl-3-(2-pyridyldithio)-propionate (SPDP) and N, N'-O-phenylenedimaleimide (PDM).

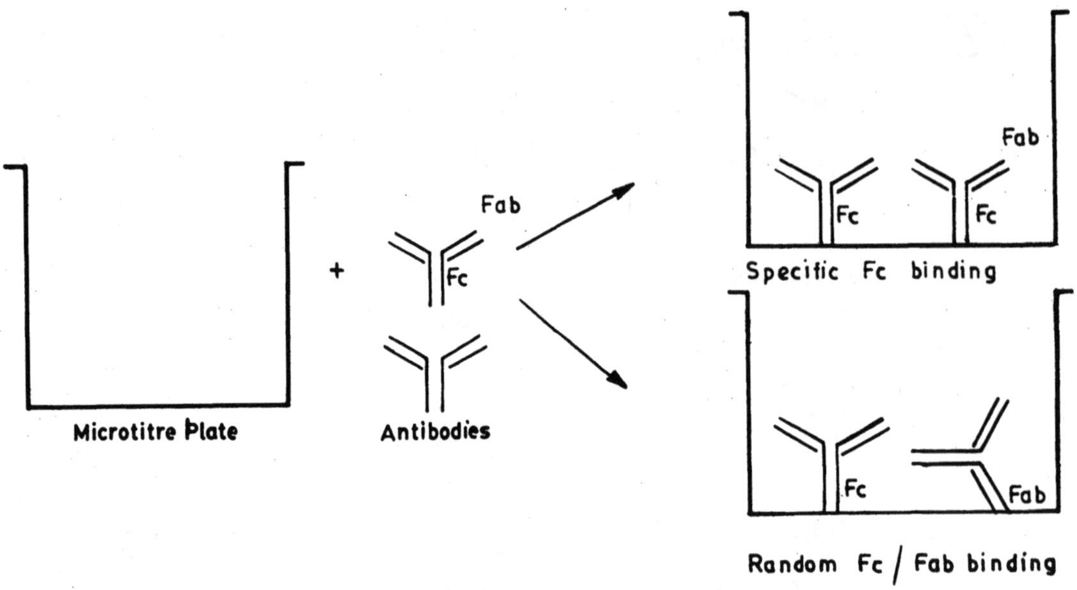

Fig. 5. Modes of binding of antibodies to the walls of the microtitre plate wells.

Table 1. Procedure for ELISA and Immunoblotting

- Coating of microtiter plate with antigen.
- Non-specific binding site blocked with bovine serum albumin.
- Incubation and washing off unbound antigen.
- Reaction with primary antibody.
- Incubation and washing off unbound antibody.
- Reaction with enzyme labelled anti-antibody.
- Incubation and washing off unbound anti-antibody.
- Reaction of fixed enzyme with substrate.
- Evaluation of product formed (visual/photometric).
- In immunoblotting following two steps are performed instead of the first step of coating the microtiter plates with antigen:

* Electrophoretic separation of antigen
* Transfer of antigens to nitrocellulose membrane.

PRACTICAL VALUE OF IMMUNODIAGNOSIS

Trematodiases

Paragonimiasis

Trematode eggs can be readily demonstrated by faecal examination, especially in cases of gastrointestinal infection viz., with *Metagonimus* spp. and *Clonorchis* spp. However, eggs of tissue parasites such as *Paragonimus* spp. and *Schistosoma* spp. are difficult to detect in stools. Immunodiagnostic methods are more sensitive and specific for these infections (Table 2). One of the immunoassay viz. CFT, is 3 times more sensitive than faecal examination (Table 3) for the diagnosis of paragonimiasis. The percentage of CFT positive paragonimiasis cases confirmed by faecal examination increased with increasing antibody titre against *P. westermanii* antigens (Table 3) and only 0.5% of CFT negative cases showed *P. wetermanii*

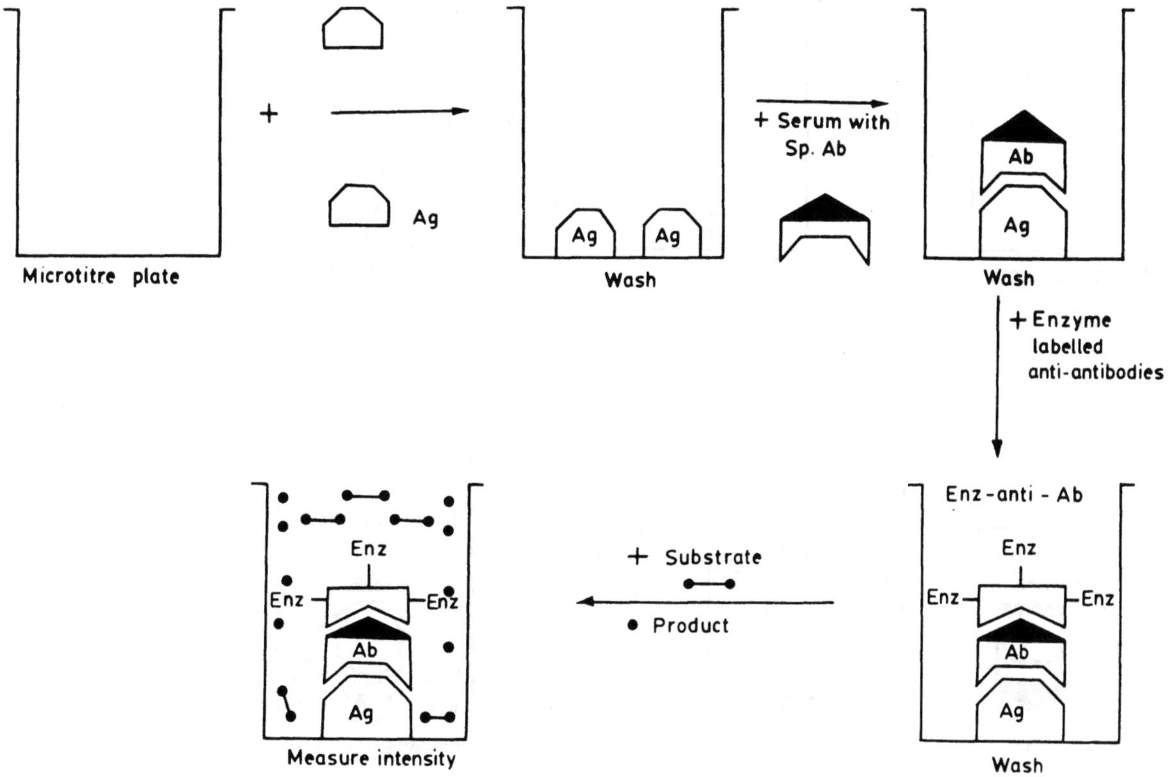

Fig. 6. ELISA for detection/assay of antibodies from serum by immobilizing antigen and followed by reaction with enzyme labelled antibodies/anti-antibodies (Enz-anti-Ab).

eggs on faecal examination. Among the trematodes antigen antibody cross-reactions were to the extent of 13–26% (except 0–4 % for ascariasis) by CFT (Table 4). The number of precipitation bands observed by ID and IE of sera from patients infected with *P. westermanii* was 1–9 and 2–12, respectively. One specific band was persistent in samples which were positive by CFT and ELISA. Bands specific for several helminths have been demonstrated by IE [57]. LAT [27] is also sensitive and could be used effectively in field surveys. The excretory-secretory (E-S) antigens and adult worm antigens of *P. miyazakii* do not cross-react by ELISA with antigens of *Fasciola hepatica* or *Schistosoma japonicum* [54]. Such E-S antigens are sensitive to cysteine proteases and have molecular weights of 27 KDa, 33–37 KDa, and 120 KDa. The 27 KDa and 33–37 KDa antigens were genus specific and the 120 KDa antigen was species specific. The CFT antibody titres in paragonimiasis patients are decreased immediately after treatment with bithionol and become negative within a period of 1–24 months (Table 5), as evaluated by immunofluorescence test (IFA), CFT, ELISA, ID, IE and IHA [59].

Schistosomiasis

Faecal detection of *S. japonicum* eggs corresponds with ID, IE, ELISA, IFA and CFT positive sera from patients infected with *S. japonicum*. The levels of gut associated 70 KDa anodic-antigens are strongly correlated with urinary excretion of antigens and faecal excretion of eggs of *S. Mansoni* and *S. haematobium* [16]. The level of antigens drops rapidly after chemotherapy. Time resolved fluoroimmunoassay (TRFIA)—a new immunoassay for urinary schistosomiasis—was more specific then ELISA and gave negligible false positive tests [1].

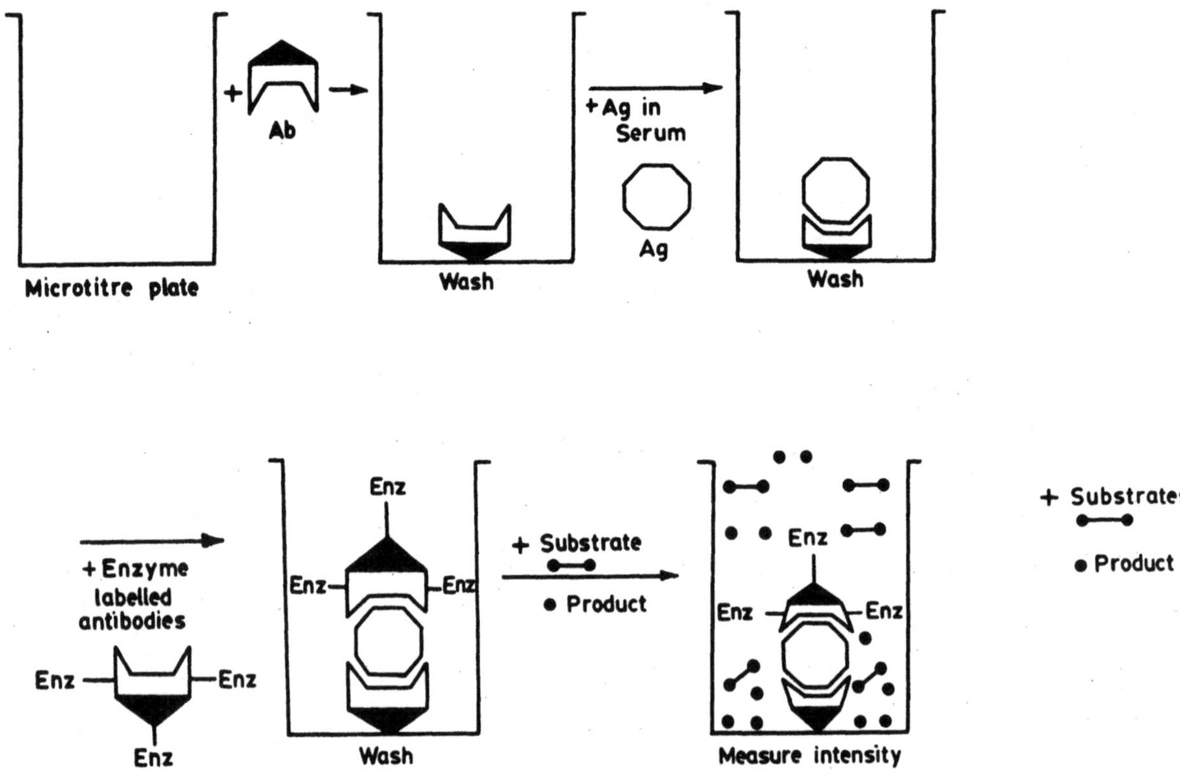

Fig. 7. ELISA for detection/assay of antigen in serum.

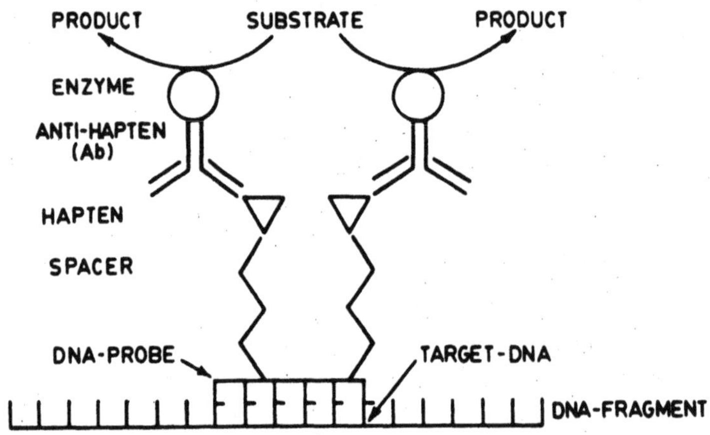

Fig. 8. Detection of specific DNA-sequence from genomic DNA of helminth by using ELISA.

A schistosomal protein produced by recombinant-DNA technology has been found useful for immunodiagnosis of schistosomiasis [11]. The antibodies against 32–35 KDa E-S antigens of *S. mansoni* act as marker of recent infection in non-exposed visitors to endemic areas [17] and serodiagnostic methods have been used for detection of early schistosomal infection in Egyptians [44]. Similarly, a sandwich ELISA with cathodic antigens and MAb has been used to differentiate [18] current and past infection with *S. japonicum* in humans.

Table 2. Utility of various methods for serodiagnosis of helminthiases

	CFT	IHA	IE	IFT	ELISA	Western blotting
Trematodiases						
Schistosomiasis	90	90	95	92	95	98
Fascioliasis	98	95	98	98	98	
Paragonimiasis	95	90	98	95	95	
Clonorchiasis	45	30	50	50	50	
Metagonimiasis	5	5	5	15	10	
Cestodiases						
Taeniasis	20	15	20	25	20	
Cysticercosis	85	80	95	95	95	
Echinococcosis	90	85	95	98	98	
Diphyllobothriasis	20	15	20	25	25	
Sparganosis	90	85	95	98	95	
Nematodiases						
Ascariasis	0	0	20	20	20	
Ancylostomiasis	0	0	5	15	15	
Anisakiasis	85	85	95	95	95	
Gnathostomiasis	80	80	85	90	95	
Toxocariasis	90	85	90	85	90	
Filariasis bancrofti	75	85	85	85	85	
Onchocercosis	85	85	90	95	95	

Values represent percent positive cases detected by immunological methods.

Table 3. Comparison of complement fixation test (CFT) and stool examination (SE) for paragonimiasis

	CFT positive		SE positive from CFT positive	
	No.	% of total	No.	%
Positive*	1485	25.3	539	36.6
Antibody titre				
10–20	473	8.1	97	20.5
21–40	318	5.4	105	33.0
41–80	263	4.5	106	40.3
81–160	199	3.4	93	46.7
160	233	3.9	138	59.5

*0.5% of the CFT negative and 36.6% of the CFT positive cases were positive by SE; n = 5878.

Reports on serodiagnosis of schistosomiasis in animals are few. Only Cercarian-Hullen reaction has been used to detect *S. nasale* infection in cow and buffalo calves [42]. The reaction turned positive on 7th day post-infection in experimental schistosomiasis incognitum with rhesus monkeys. However, intensity of this reaction was not related to duration of the infection or number of worms recovered from the monekys.*

*See Das et al: J. Vet. Parasitol. 1: 63–65 (1987).

Table 4. Specificity of complement fixation test for diagnosis of paragonimiasis by using sera of confirmed patients and with antigen from *Paragonimus westermanii*

	n	Positive	
		No.	%
Reaction of *P. westermanii* antigen with sera from patients of			
Paragonimiasis	559	539	96.4
Clonorchiasis	32	6	18.8
Schistosomiasis	38	8	21.1
Fascioliasis	15	2	13.3
Ascariasis	10	0	0.0
Reaction of sera from paragonimiasis patients with antigen from			
Clonorchis sinensis	43	11	25.6
Schistosoma japonicum	37	8	21.6
Fasciola hepatica	35	9	25.7
Ascaris lumbricoides	26	1	3.8

Table 5. Comparison of serological tests in monitoring the therapeutic efficacy of bithionol for paragonimiasis

	First positive (DPI)	First negative (MPT)
Complement Fixation Test	10–25	1–12
Indirect Haemagglutination Test	10–20	1–18
ELISA	10–20	1–18
Immunoelectrophoresis	10–35	2–24
Indirect Immunofluorescence Test	10–30	1–24

DPI = Days post infection; MPT = Months post treatment.

Fascioliasis

Immunodiagnosis is the only specific antemortem method for the diagnosis of fascioliasis, as well as for chronic fascioliasis, where standard faecal examination could give false negative results due to low egg count or their absence in the stools [40]. Crossed immunoelectrophoresis (CIE) is more sensitive, rapid and without any false positive reaction. Barbet [8] reviewed the characteristics of E-S antigens and surface antigens of *F. hepatica* in different stages of development. By using MAb, 25–40 KDa and 50 KDa components were detected by this author. However, infected sheep serum revealed 23–26 KDa antigens. The low molecular weight antigens recognized by MAb and by sera of infected sheep may be E-S antigens and of diagnostic value.

CESTODIASES

Hydatidosis

The cestodiases can be established from the presence of eggs/proglottids in the faeces or anal swabs of the patients. The serological tests are useful for the diagnosis of hydatidosis and cysticercosis but not taeniasis (Table 2). The somatic tissues and E-S antigens of the parasites are potential antigens for serodiagnosis. The hydatid cyst fluid is superior source of antigens as compared to the antigens from protoscoleces and cyst membranes. The two antigens associated with hydatidosis are lipoproteins with molecular weight between 100–400 KDa and the antigen of the adult *Echinococcus* spp. have not been characterized [14]. For diagnosis of pulmonary hydatidosis in man, ELISA with human hydatid cyst fluid antigens and soluble antigens from scoleces is more sensitive than the IHA [62]. However, IE and CFT are more useful for post-therapeutic follow up work. Al-Yaman et al [3] reported that ELISA and IHA were more sensitive than CIE and ID and did not find any cross-reaction. It has generally been observed that the diagnosis is more accurate if both ELISA and IHA were used simultaneously and may be of use in seroepidemiological surveys of hydatidosis in man. False positive results have been observed where fascioliasis [9], schistosomiasis [22], filariasis [45], taeniasis [57] and cysticercocosis [49] are also common. Although immunoassay can detect low levels of anti-*Echinococcus* antibodies, yet non-specificity can be a problem [34]. One of the cross-reacting component may be P_1 blood group antigen [9]. Differentiation of specific infecting species can be achieved by using MAb or immuno-affinity purified antigen in the ELISA system. The false positive reaction can thus be avoided, while the false negative tests may be excluded by simultaneous use of 2 or more tests. Serological surveillance of patients after surgery but not after chemotherapy can be carried out by CFT.

While serodiagnostic tests have been employed successfully in the diagnosis of human hydatidosis, their use in detection of hydatid infection in animals has not been of much help. The proteins for hydatid cysts from buffaloes have been characterized [13] but the lack of detailed characterization of antigens [7], lower production of antibodies against *Echinococcus* spp. in animals and higher cross-reactivity with species of taeniid cestodes which have same definitive and intermediate hosts as the *E. granulosus*, are the causes of few diagnostic studies in animals. Therefore, the serological tests for hydatidosis in animals are difficult to interpret. Specific diagnosis by serological methods has not been possible in cattle [35], buffaloes [43], sheep [21, 33, 35], goats [35], pigs [35, 39] and camels [2, 15]. False positive reactions occur in these intermediate hosts [35, 43]. In sheep IHA and ELISA are not useful for diagnostic or epidemiological work in areas where more than one larval cestodes, are present [65, 66]. The detection of cathodic antigen in sheep by IE, is a better [21] tool for diagnosis of hydatidosis. ELISA with immuno-affinity purified antigen has been partly successful in reducing cross-reaction in sheep [33]. However, the combined use of IHA and ID for the diagnosis of hydatidosis in camels, using crude antigen preparation has been recommended [2] for the diagnosis of hydatidosis in camels. Such reports need further substantiation.

Although a reliable species specific immunodiagnostic test for *E. granulosus* infection in dogs would be of great value, little has been done to develop such tests for definitive hosts. Experimentally infected dogs showed antibodies to protoscolex antigens in their sera by ELISA [25]. Thus, possibility of serodiagnosis of *E. granulosus* in dogs is promising. The use of such tests for detection of *Echinococcus* spp. antigens in the faeces of dogs, prior to worm patency, may be more useful [7].

Cysticercosis

The natural and experimental infection of *Taenia solium* cysticercosis in pigs has been successfully diagnosed by CIE [31] using crude or partially purified antigens from scoleces of the larval stages. Subsequent comparison of ID, IE, CIE, IHA and ELISA [32] with partially purified antigens from cysticerci for the diagnosis of *T. solium* infection in pigs showed that CIE and ELISA may be used in field respectively for antemortem detection and for confirmation of *T. solium* cysticercosis. Although ELISA has been recommended for laboratory diagnosis [32], yet it may be used more efficiently in field as dot ELISA or other simple forms of ELISA.

CIE identifies and differentiates *Cysticercus cellulosae* from *C. tenuicollis* infection in pigs [39]. A highly specific band was observed within 90 minutes and without cross-reaction or false reaction.

NEMATODIASES

Common nematodes like *Ascaris* spp., *Ancylostoma* spp. develop into mature worms in the gastrointestinal tract and in the case of these nematode infections, the antibody titres are very low and immunodiagnostic tests are not useful (Table 2). However, such infections are easily diagnosed by faecal examination. On the other hand, visceral larva migrans are not detected by faecal examination, but the immunodiagnostic tests have been used by a number of workers (see [38]).

Wuchereriasis (Filariasis)

Due to quantitative variations in the number of parasites, light infections and periodicity of infection, it is difficult to perform epidemiological surveys for filariasis. In human population at Varanasi, India, ELISA and IHA have been used effectively for the diagnosis of filariasis when *Dirofilaria immitis* antigens or antibodies against them were used (50). In case of asymptomatic carriers of microfilaria, the detection was 86.6%, whereas in endemic normal cases the detection rate was upto 50% by IHA and ELISA. Relatively high titres by ELISA in some cases were attributed to repeated bites by infected mosquitoes. In neighbouring region of Lucknow, India, symptomatic cases of filariasis showed higher levels of IgG and titre of antibodies cross-reacting with adult *Litomosoides carinii* antigens by ELISA [48]. Moreover, ELISA was effectively used for the diagnosis of pulmonary dirofilariasis [37], because of its higher sensitivity. It showed cross-reaction to *Ascaris* spp. and *Toxacara* spp. when E-S antigen or extracted antigens from adult females were used, thus making it less specific [37]. The E-S antigens from microfilariae are more specific than the somatic antigens when used for immunodiagnosis of human filariasis [26, 41]. ELISA is specific and sensitive enough for immunodiagnosis of bancroftian filariasis [47] as are ID and IE for filariasis when *Setaria cervi* antigens are used [19]. There is still a need to identify antigens for early detection of *Wuchereria bancrofti* infection in endemic areas [12].

The serodiagnosis of filariasis in animals has received little attention. The IHA has been recommended [56] for immunodiagnosis of *Dictyocaulus filaria* infection in sheep.

Onchocercosis

Tada et al [52, 53] used *Onchocerca volvulus* antigens to test the blood samples taken on filter paper, for specific serodiagnosis of onchocerciasis by micro ELISA and could differentiate human onchocerciasis from loiasis (*Loa loa*) and dipetalonemiasis (*Dipetalonema perstans*). Micro ELISA was more sensitive than IFA and was not affected by the intestinal helminth infection. Roffi et al [46] reported cross-reaction between onchocerciasis and bancroftian filariasis but Tada et al [53] did not examine for this type of cross-reaction. Earlier, Tada et al [52] had reported that sensitivity/specificity both of IHA and ELISA was good for determination of intensity of microfilarodermia by using *O. volvulus* antigen and eluent of blood from filter paper. Nogami et al [36] identified four MAb (non-reactive to *A. suum*) which specifically reacted with *O. volvulus* and not with other 12 helminths. SDS-PAGE, PAGE and immunoblotting revealed that an epitope specific for *O. volvulus* existed on a 27 KDa polypeptide. It was concluded that non-reactivity to *A. suum* may be used for selection of MAb specific for *O. volvulus*.

Trichinellosis

The biochemical characteristics of surface and E-S antigens of different stages of *Trichinella spiralis* have been reviewed [8, 64]. In the newborn larvae, infective larvae and adult worms, antigens of 64, 58, 34, 30 KDa, 105, 90, 55, 47 KDa and 40, 33, 20 KDa respectively, have been shown by radiolabelling and on reaction with MAb preparations [8]. The 48–55 KDa antigens may be of diagnostic value as the hosts develop antibodies against these antigens. The efficacy of CFT, ID, IE, CIE, IHA and ELISA for the diagnosis of trichinosis has been compared [64]. CIE was more specific for such diagnosis, but ELISA gave cross-reactions with antigens of other nematodes. As the serological tests provide presumptive diagnosis, so the confirmation has to be done by demonstrating larvae in muscle biopsy [64] (see Chapter 4).

Toxocariasis

ELISA has been used to detect E-S antigens, for the diagnosis of experimental toxocariasis in rabbits [30] in which the antibody titre increased within two weeks after experimental infection. Whereas in experimentally infected rats the antibody titres were elevated from 9–18 weeks when antigen from embryonated eggs or larvae were used [23]. In humans, the antibody titres were higher in case of visceral larva migrans as compared to those in ocular larva migrans [10]. This may be due to isolated location, with poor blood supply in the ocular compartment.

The *Toxocara* antigens show high cross-reactivity with other nematodes making it difficult to differentiate *T. canis* infection from that of other nematodes like *Ascaris, Anisakis, Angiostrongylus, Gnathostoma, Capillaria* and *Spirometra* spp. (see Table 2). With ID and IE no specific precipitation bands were detected [60]. *T. canis* infected patients also show cross-reactivity with antigens from adult *Gnathostoma* spp. [4]. The cross-reactivity problem has been overcome, to some extent, by using immuno-affinity purified antigens in the immuno-assay. Species specific antigens from adult *T. canis, A. lumbricoides* and *D. immitis* have been purified [63] and successfully tested by polyacrylamide gel electrophoresis (PAGE), ID and IE. The crude antigen preparations do not show such specificity. However, with the availability of *T. canis*-specific reagents* accurate sero-diagnosis of this potentially common but clinically inapparent condition may be possible.

Ascariasis

By radioimmunoassay (RIA), Tanaka et al [55] detected high concentration of *A. suum* protein in patients suffering from inflammatory disease of the bowl. The concentration of this protein in sera of patients with other diseases of gastrointestinal tract was lower and similar to that in sera of control subjects. As already pointed out (Table 2), the immunodiagnostic methods are not of much use and parasitological examination may be preferred for diagnosis of ascariasis. However, the possibility of developing immunoassay on faecal extracts, for detecting nonpatent infections of helminths in the intestine needs to be investigated.

Ancyclostomiasis

The diagnosis of *Ancyclostoma duodenale* and *A. caninum* infections is presumptive but ELISA using highly specific MAb may be useful. ELISA with somatic antigens and E-S antigens is highly sensitive and specific for diagnosis of hookworm infection in man [20, 51].

The serological test IHA has been used [5] for diagnosis of experimental hookworm (*Gaigeria pachyscelis*) infection in lambs. Such tests have not been used extensively for the diagnosis of hookworm infection in man and animals [20, 51].

CONCLUSIONS AND FUTURE PROSPECTS

Despite the successful use of the immunoassays in the diagnosis of helminthic infections in man, there are many avenues for their improvement as well as for their use in diagnosis of helminthiases in animals. The use of immunoassay in diagnosis of helminthic infections in animals has not been successful due to high cross-reactivities and low production of antibodies. One of the reasons for cross-reactivities is the presence of multiple helminth infections. Immunoassay are relatively more specific and sensitive for the

*See Bowman et al: Exp. Parasitol. **64**: 548–465 (1987); Kennedy et al: Parasite Immunol. **9**: 407–420 (1987).

diagnosis of larva migrans than for the intestinal helminths. The possibility of developing immunoassay on faecal extracts, for detecting nonpatent infection of helminths in the intestine needs to be investigated. Amongst the immunoassays, ELISA has emerged as the most important tool in the diagnosis and epidemiological studies of helminthiases. The immunoassay in general are still inadequate to discriminate between various species of helminths and for cross-reactivities commonly observed between helminths. The use of antigens purified by immunoaffinity chromatography and of monoclonal antibodies in the immunoassay especially ELISA, demonstrates more specific diagnosis of helminthiases. Simpler improved forms of ELISA, which do not require sophisticated equipments or technical knowhow, need to be evolved for the diagnosis of helminthic infections as well as for epidemiological studies in man and animals. Immunoblotting has also shown better specificity and needs to be investigated broadly. Battery of monoclonal antibodies and helminth antigens are required to be generated by hybridoma technology and immunoaffinity chromatography/recombinant-DNA technology for use in immunodiagnosis of helminthic infections both in man and animal.

REFERENCES

1. Aceti, A., Pennica, A., Celestino, D., Paparo, S.B., Caferro, M., Sanguigni, S., Morangi, M. and Sebastiani, A. A new serological technique, time resolved fluoroimmuno assay (TRFIA) for the immunological diagnosis of urinary schistosomiasis. Trans. R. Soc. Trop. Med. Hyg. **82**: 445–447 (1988).
2. Ali, A., Muhammad, G., Hussain, I., Hayat, C.S. and Akhtar, M. Hydatidosis: II. Immunodiagnosis of hydatid disease in camels using indirect haemagglutination, latex agglutination, and double diffusion tests. Pakistan Vet. J. **10**: 175–179 (1990).
3. Al-Yaman, F.M., Abdel-Hafez, S.K., Assaf, L.M. and Malkawi, F.K. Evaluation of various serodiagnostic tests for human hydatidosis. Jpn. J. Parasitol. **37**: 133–138 (1988).
4. Anantaphruti, M.T. ELISA for diagnosis of gnathostomiasis using antigens from *Gnathostoma doloresi* and *G. spinigerum*. Southeast Asian J. Trop. Med. Pub. Hlth. **20**: 297–304 (1989).
5. Ansari, M.Z. and Singh, K.S. Indirect haemagglutination test in lambs vaccinated with *Gaigeria pachyscelis*. Indian J. Anim. Sci. **50**: 558–560 (1980).
6. Avrameas, S. Immunoenzymatic techniques for biomedical analysis. In: *Methods in Enzymology* (K. Mosbach ed), **44**: 709–716. Academic Press Inc.: New York (1979).
7. Babos, S. Untersuchungen über die serodiagnostic der Echinokokkose. Agnew. Parasitol. **3**: 2–4 (1962).
8. Barbet, A.F. Vaccines for parasitic infection. In: *Advances in Veterinary Science and Comparative Medicine* (J.L. Bittle and F.A. Murphy eds), pp. 345–375. Academic Press Inc.: San Diego, New York (1989).
9. Ben-Ismail, R., Carme, B., Niel, G. and Gentilini, M. Non-specific serological reactions with *Echinococcus granulosus* antigens: role of anti-P1 anti-bodies. Am. J. Trop. Med. Hyg. **29**: 239–245 (1980).
10. Brunello, F., Falagiani and Carrenna, G.M. Enzyme immunoassay (ELISA) for the detection of specific IgG antibodies to *Toxocara canis* ES antigens. Boll. Ist. Sieroter. Milan. **65**: 54–60 (1986).
11. Capron, A. Perspectives vaccinals centre la bilharziose (Prospect of vaccination against schistosomiasis). Bull. Acad. Nat. Med. **173**: 495–504 (1989).
12. Carvello, P.A., Rajasekariah, G.R., Deshpande, L., Parab, P.B., Rao. U.R. and Subrhamanyam, D. Evaluation of *in vitro* released *Wuchereria bancrofti* third stage larval antigens for detection of bancroftian filariasis. Trop. Med. Parasitol. **41**: 71–72 (1990).
13. Chowdhury, N., Kinger, S. and Ahuja, S.P. The chemical composition of secondary hydatid cysts of buffalo origin. Ann. Trop. Med. Parasitol. **80**: 469–471 (1986).
14. Chordi, A. and Kagan, I.G. Identification and characterization of antigenic components of sheep hydatid fluid by immunoelectrophoresis. J. Parasitol. **51**: 63–71 (1965).
15. Dada, B.J.O., Adegboye, S. and Mohammed, A.N. Experience in the northern Nigeria with countercurrent immunoelectrophoresis, double diffusion and indirect haemagglutination tests for diagnosis of hydatid cysts in camels. J. Helminthol. **55**: 197–202 (1981).

16. De Jonge, N. and Deelder, A.M. Immunodiagnosis of Schistosomiasis by ELISA for the detection of circulating anodic antigen using monoclonal antibodies. Twelveth Int. Congr. Trop. Med. Mal. Abstr. pp. 93 (1988).
17. Evengard, B., Hammarström, L., Smith, C.I., and Linter E. Early antibody responses in human schistosomiasis. Clin. Exp. Immunol. **80**: 69–76 (1990).
18. Fu, C. and Carter, C.E. Detection of a circulating antigen in human schistosomiasis japonica using a monoclonal antibody. Am. J. Trop. Med. Hyg. **42**: 347–351 (1990).
19. Gupta, S.K., Raychaudhury, C., Bandopadhyay, A.K. and Ghosh, D.K. An approach for immunodiagnosis of clinical case of filariasis. Life Science **46**: 1155–1165 (1990).
20. Ganguly, N.K., Mahajan, R.C., Sehgal, R., Shetty, P. and Dilawari, J.B. Role of specific IgG to E-S antigen in the diagnosis and prognosis of the hookworm infection, J. Clin. Microbiol. **26**: 739–742 (1986).
21. Hamel, K.L. and Ris, D.R. The use of a cathodic antigen in the immunoelectrophoretic serodiagnosis of *Echinococcus granulosus* in sheep. Vet. Immunol. Immunopath. **3**: 419–425 (1982).
22. Iacona, A., Pini, C. and Vicari, G. Enzyme-linked immunosorbent assay (ELISA) in the serodiagnosis of hydatid disease. Am. J. Trop. Med. Hyg. **29**: 95–102 (1980).
23. Inoue, H. and Tsuji, M. Studies on visceral larva migrans: Detection of anti-*Toxocara canis* IgG antibodies by ELISA in human and rat sera. Jpn. J. Parasitol. **38**: 68–76 (1989).
24. Ito, M. and Sato, S. An antigenic component for the serodiagnosis of *Paragonimus miyazakii*. Jpn. J. Parasitol. **37**: 347–352 (1988).
25. Jenkins, D.J. and Rickard, M.D. Specific antibody responses to *Taenia hydatigena, T. pisiformis*, and *Echinococcus granulosus* in dogs. Aust. Vet. J. **62**: 72–78 (1985).
26. Kaushal, N.A., Hussain, R. and Ottesen, E.A. Excretory-secretory and somatic antigens in diagnosis of human filariasis. Clin. Exp. Immunol. **56**: 567–576 (1984).
27. Kobayashi, M., Kojima, S., Yokogawa, M., Tsuji, M. and Tsubota, N. Application of the latex agglutination test for human paragonimiasis in South America. Trans. R. Soc. Trop. Med. Hyg. **82**: 300–302 (1988).
28. Köhler, G. and Milstein, C. Continuous cultures of fused cells secreting antibody of predefined specificity. Nature **256**: 295–297 (1975).
29. Köhler, G. and Milstein, C. Derivation of specific antibody producing tissue culture and tumour lines by cell fusion. Eur. J. Immunol. **6**: 511–519 (1976).
30. Kondo, K., Akao, N., Konishi, Y. and Yoshimura, H. Experimental studies on visceral larva migrans: 4. Examination of immunoglobulins in sera of infected rabbits with *Toxocara canis* by means of indirect flourorescent antibody (IHA) and enzyme-linked immunosorbent assay (ELISA). Jpn. J. Parasitol. **33**: 99–104 (1984) [In Japanese with English summary].
31. Kumar D. and Gaur, S.N.S. Counter-immunoelectrophoresis in the diagnosis of *Taenia solium* cysticercosis in pigs, using fractionated antigens. J. Vet. Parasitol. **1**: 43–46 (1987).
32. Kumar D. and Gaur, S.N.S. Comparative evaluation of various immunodiagnostic tests for the diagnosis of *Taenia solium* cysticercosis in pigs using fractionated antigens. J. Helminthol. **63**: 13–17 (1989).
33. Lightowlers, M.W., Rickard, M.D., Honey, R.D., Obendorf, D.L. and Mitchell, G.F. Serological diagnosis of *Echinococcus granulosus* infection in sheep using cyst fluid antigen processed by antibody affinity chromatography. Aust. Vet. J. **61**: 101–108 (1984).
34. Mahajan, R.C., Ganguly, N.K. and Chhabra, N.L. Evaluation of immunodiagnostic techniques for hydatid disease. Indian J. Med. Res. **64**: 405–409 (1976).
35. Martinez Gomez, F., Harnandez Rodriguez, S., Navarrote Lopez-Cozar, I. and Calero Carretero, R. Serological tests in relation to the viability, fertility and localization of hydatid cysts in cattle, sheep, goats and swine. Vet. Parasitol. **7**: 33–38 (1980).
36. Nogami, S., Hayashi, Y., Korenaga, M., Tada, I. and Tanaka, H. Monoclonal antibodies specific for *Onchocerca volvulus* as determined by immunofluorescence. Int. J. Parasitol. **18**: 503–507 (1988).
37. Ohnishi, Y. and Yoshimura, H. Crossreactivtiy in the sera of patients with human pulmonary dirofilariasis by means of a mixed passive haemagglutination (MPHA) test, in comparison with an enzyme-linked immunosorbent assay (ELISA). Jpn. J. Parasitol. **37**: 258–260 (1988).
38. Parija, S.C. (ed). *Review of Parasitic Zoonoses*. AITBS Publishers: New Delhi (1990).
39. Pathak, K.M.L., Gaur, S.N.S. and Garg, S.K. Counter-current immunoelectrophoresis, a new technique of the rapid serodiagnosis of porcine cysticercosis. J. Helminthol. **58**: 321–324 (1984).

40. Pathak, K.M.L., Kumar, D. and Gaur, S.N.S. Immunodiagnosis of fascioliasis by counter-immunoelectrophoresis in sheep. Indian J. Vet. Med. 6: 38–40 (1986).
41. Prasad, G.B.K.S., Kharat, I. and Harinath, R.C. Detection of anti-microfilarial E-S antigen antibody in immunecomplexes in Bancroftian filariasis by enzyme immunoassay. Trans. R. Soc. Trop. Med. Hyg. 77: 771–772 (1983).
42. Rajamohanan, K., Sundaram, R.K. and Peter, C.T. On "Cercarien- Hullen Reaktion" of Vogel and Mining 1949 in *Schistosoma nasale* infection. Kerala J. Vet. Sci. 3: 76–77 (1972).
43. Rao, B.V. and Mittal, K.R. Studies on hydatidosis in Indian buffaloes. 1. Some serological observations in natural infection of hydatidosis in buffaloes. Z. Tropenmed. Parasitol. 24: 476–480 (1973).
44. Rashed, S.M., Kamal, K. and Shaheen, H. Comparison between parasitological and serological techniques in assessment of schistosomiasis prevalence. J. Egypt. Soc. Parasitol. 20: 169–174 (1990).
45. Rickard M.D., Honey, R.D., Brumley, J.L. and Mitchel, G.B. Serological diagnosis and post-operative surveillance of human hydatid disease. II. The enzyme-linked immunosorbent assay (ELISA) using various antigens. Pathology 16: 211–216 (1984).
46. Roffi, J., Lafabrie, B., Chovet, M., Viale, H. and Digoutte, J.P. Application du test ELISA au depistage de lonchocercose. Med. Trop. 42: 297–306 (1982).
47. Romia, S.A., Abouzakham, A.A., Rozk, H. and Abou Shady, A.F. Immunodiagnosis of Bancroftian filariasis. J. Egypt. Soc. Parasitol. 20: 235–241 (1990).
48. Saxena, R.P., Saxena, R.K., Tandon, A., Srivastava, A.K., Bhatia, B., Srivastava, V.K. and Saxena, K.C. Immunological status in Bancroftian filariasis. Indian J. Parasitol. 10: 101–107 (1986).
49. Schantz, P.M., Shanks, D. and Wilson, M. Serologic cross-reactions with sera from patients with echinococcosis and cysticercosis. Am. J. Trop. Med. Hyg. 29: 609–612 (1980).
50. Sengupta, G., Mohapatra, T.M. and Sen. P.C. Detection of human filaria cases using enzyme-linked immunosorbent assay and indirect haemagglutination tests. Indian J. Parasitol, 9: 161–164 (1985).
51. Shetty, P., Dilawari, J.B., Virk, J.K., Sehgal, R., Ganguly, N.K. and Mahajan, R.C. ELISA using E-S and somatic antigen as a diagnostic test for human hookworm infection. Trans. R. Soc. Trop. Med. Hyg. 82: 736–738 (1988).
52. Tada, I., Korenaga, M., Mimori, T., Sakamoto, M., Yoshimura, T., Racinos, M.M., Flores, O.F., Lujan T.A., Ochoa, A.J.O., Castre, J.C. and Zea, F.G. A comparative study of several diagnostic measures applied in Guatemalan onchocerciasis. Jpn. J. Parasitol. 34: 261–271 (1985).
53. Tada, I., Korenaga, M., Shiwaku, K., Ogunba, E.O., Ufomadu, G.O. and Nwoke, B.E.B. Specific serodiagnosis with adult *Onchocerca volvulus* antigen in an enzyme-linked immunosorbent assay. Am. J. Trop. Med. Hyg. 36: 383–386 (1987).
54. Takemoto, Y. and Tsuji M. Characterization of antigens of *Paragonimus miyazakii* by ELISA. Southeast Asia J. Trop. Med. Pub. Hlth. 20: 87–93 (1989).
55. Tanaka, K., Kawamura, H., Tohgi, N., Tsuji, M., Miyachi, Y. and Miyoshi A. The measurement of *Ascaris suum* protein by radioimmunoassay in sera from patients with helminthiasis and with gastrointestinal diseases. Parasitology 86: 291–300 (1983).
56. Tewari, H.C., Mittal, R.R. and Gangadhar Rao, Y.V.B. A preliminary note on the passive haemagglutination test to detect circulating antibodies in sheep vaccinated with irradiated larvae of *Dictyocaulus filaria*. Indian vet. J. 49: 12–14 (1972).
57. Todorov, T., Dakov, I., Kosturkova, M., Tenev, S. and Dimitrov, A. Immunoreactivity in pulmonary echinococcosis. I. A Comparative study of immunodiagnostic tests. Bull. Wld Hlth. Org. 57: 735–740 (1979).
58. Tsuji, M. Comparative studies on the antigenic structure of several helminths by immunoelectrophoresis. Jpn. J. Parasitol. 24: 227–236 (1975) [In Japanese with English Summary].
59. Tsuji, M. Pre- and post-treatment serodiagnosis for paragonimiasis. Arzneim—Forsch/Drug Res. 34: 1204–1207 (1984).
60. Tsuji, M. and Yokogawa, M. Immunological diagnosis of helminthic infections. Proc. SEAMEO Trop. Med. Tech. Meeting (C. Harinasuta ed), pp. 180–196 (1974).
61. Voller, A. and Bidwell. D.E. The use of ELISA in viral and parasitic diseases. In: *Nonisotopic immunoassays* (G.P. Talwar ed), pp. 234–243. Vikas Publishing House: New Delhi (1983).

62. Wattal, C., Malla, N., Khan I.A. and Agarwal, S.C. Comparative evaluation of enzyme-linked immunosorbent assay for the diagnosis of pulmonary echinococcosis. J. Clin. Microbiol. **24**: 41–46 (1986).
63. Welch, J.S., Symons, M.H. and Dobson, C. Immunodiagnosis of Parasitic Zoonoses: Purification of *Toxocara canis* antigens by affinity chromatography. Int. J. Parasitol. **13**: 171–178 (1983).
64. Yamaguchi, T. Immuno-serological tests. *Trichinella and Triochinellosis in Japan.* pp. 280–403. Nanko-DO: Tokyo (1989) [In Japanese].
65. Yong, W.K., Heath, D.D. and Van Knapen, F. Comparison of cestode antigens in an ELISA for the diagnosis of *E. granulosus, T. ovis* and *T. hydatigena* infections in sheep. Res. Vet. Sci. **36**: 24–31 (1984).
66. Yong, W.K., Heath, D.D. and Parameter, S.N. *E. granulosus, T. ovis* and *T. hydatigena*: evaluation of cyst fluid as antigen for serodiagnosis of larval cestodes in sheep. N.Z. Vet. J. **26**: 231–234 (1978).

13. Application of Radiation in the Control of Helminthic Infections

J.M. Maclean
Department of Veterinary Physiology, University of Glasgow Veterinary School
Bearsden Road, Glasgow, UK

INTRODUCTION

Nuclear techniques have been of special value in several areas of helminthological research. Vaccines against a wide range of bacterial and viral diseases have been developed from whole killed organisms, less virulent or attenuated strains or products released from the organisms. Attempts to produce vaccines against helminths by similar means met with little success and from a variety of studies it appeared that live metabolising parasites were a prerequisite for protection.

Like all living organisms helminths are damaged by ionising radiation. If infective larvae are irradiated before infection reduced numbers of parasites develop to maturity, reproductive capacity is impaired and the pathogenicity is reduced. Radiation attenuated parasites can exert an immunogenic effect if a sufficient number develop and survive for an adequate period of time. The efficacy of any radiation attenuated vaccine will depend on choosing a radiation dose which will reduce the pathogenicity of the parasite without significantly reducing the immunogenicity and of selecting the appropriate vaccinating dose of irradiated larvae.

Despite pioneering work of Tyzzer and Honeij during first world war [87], when the immunogenicity of gamma-irradiated *Trichinella spiralis* was investigated, attempts to prepare radiation attenuated vaccines against helminths were not pursued seriously until four decades later when Jarrett and co-workers [45] at Glasgow University Veterinary School developed a vaccine against the important bovine lungworm *Dictyocaulus viviparus*. Irradiated vaccines have since been produced, with varying degrees of success, against a number of helminthic infections.

TREMATODA

Schistosomatidae

Schistosoma spp. are small flukes, usually found in the mesenteric veins, which infect mainly ruminants but also other animals in many parts of the world.

Taylor et al [80] studied immunity to *S. mattheei* which infects ruminants, and occasionally humans, in Africa. Sheep were immunised percutaneously four times at weekly intervals with either 10,000 cercariae irradiated at 6Kr or with 10,000 normal cercariae. Following challenge with 5,000 cercariae 9–11 weeks later, worm burdens were reduced by 74% and tissue egg counts by 80% in the sheep immunised with the irradiated cercariae compared with challenge controls. In a second experiment designed to evaluate double vaccination, 6 month-old wethers were immunised with 10,000 cercariae irradiated at 6Kr. Different groups were immunised once or twice at various times, and when challenged 16 weeks after the start of the experiment with 5,000 cercariae, no significant protection against challenge was observed in any of the groups. In the final part of the study sheep were vaccinated percutaneously with cercariae irradiated at 3Kr or 6Kr, and 6Kr skin-transformed schistosomula were given intramuscularly or intravenously. Syringe transformed schistosomula irradiated at 6Kr were also given intramuscularly. The degree of protection in terms of reduction in worm burden achieved by these different immunisation regimes was about 70%, thus indicating the possibility of developing a live radiation-attenuated vaccine against ovine schistosomiasis. The study also showed that irradiated skin and syringe transformed schistosomula gave similar degrees of protection as irradiated cercariae. It also appeared that adult schistosomes were not necessary for protection and other routes could be as effective as the percutaneous route.

In a later study designed to determine the duration of resistance conferred by this vaccine, sheep were vaccinated with 40,000 or 30,000 syringe-transformed *S. mattheei* schistosomula irradiated at 6Kr or 2.7Kr, respectively over a five month period [7]. Fifty-five weeks after percutaneous challenge with 3,400 normal *S. mattheei* cercariae, worm reductions of 60%–65% were found in the vaccinated sheep. Although the results of the above experiments were encouraging, further studies are required before a commercially produced vaccine could be prepared.

S. bovis infects ruminants in many parts of the world. Following a preliminary study [43] in which Sudanese desert sheep were shown to acquire partial protection against challenge 37 weeks after vaccination with 10,000 *S. bovis* cercariae irradiated at 4Kr, 15 month-old wethers were vaccinated with 6Kr attenuated *S. bovis* schistosomula [81]. One group was given 20,000 irradiated schistosomula over 210 days in 3 to 5 doses, the other group 10,000 irradiated schistosmula over 150 days in 4–5 doses. Following challenge the reduction in worm count in both groups was approximately 70%. An evaluation of this type of vaccine was carried out in Zebu cattle [12] when 9 month-old animals were immunised with 10,000 3Kr irradiated schistosomula, in one to three doses, intramuscularly or subcutaneously. Another group of cattle were given 10,000 irradiated schistosomula in a single percutaneous dose. Following percutaneous challenge with 10,000 cercariae, vaccinated calves showed higher growth rates and significantly lower faecal and tissue egg counts and lower adult worm counts.

The efficacy of this vaccine was evaluated in a field trial when 30 Zebu calves were immunised with a single intramuscular injection of 10,000 3Kr irradiated schistosomula [57]. At the end of a 10 month experiment faecal egg counts were 82% lower in the vaccinates than in the controls. On necropsy, there were significant reductions in tissue egg counts and on perfusion, adult worm burdens were about 69% lower in the vaccinates. These results suggested that with further study an effective vaccine against bovine schistosomiasis could be produced. It has also been shown that mice can be immunised against *S. bovis* by vaccination with irradiated cercariae [68]. A 60% reduction in worm count was observed following challenge in mice given 500 cercariae irradiated at 20Kr.

Following experiments with mice to find the optimal attenuating radiation dose for *S. japonicum*, monkeys were immunised with doses of between 6,000 and 9,000 cercariae irradiated at 0.5, 1.7 or 2Kr [39]. It was demonstrated that a significant level of immunity was conferred against challenge in these animals. Similar studies [38] were carried out later in chimpanzees using a human strain of *S. japonicum* cercariae but these failed to demonstrate any significant acquired resistance. More recently [40], it has been reported that cattle immunised intradermally or intramuscularly with 1 to 3 doses of 11,480 and 36,700 schistosomula irradiated at 24Kr, 36Kr or 48Kr at monthly intervals produced a significant reduction in worm burdens when challenged. Mean percentage worm reductions in the various challenge groups ranged from 55 to 87% and although the highest protection was shown in calves immunised with 36 Kr-irradiated larvae, double vaccination with larvae irradiated at 48Kr was recommended. It was thought that schistosomula irradiated at this dose would cause less lung damage than if exposed to smaller radiation doses.

S. incognitum is an important parasite of pigs and dogs in the Indian subcontinent. Reports of a possible irradiated vaccine against this parasite came from India when dogs were immunised with two doses of 3 or 4Kr irradiated *S. incognitum* cercariae given 21 days apart [84]. Worm recoveries 42 days after challenge showed 71% and 52% reductions, respectively, in the vaccinated groups suggesting that it might be possible to prepare an irradiated vaccine against this infection. In albino mice vaccinated with cercariae irradiated at 1-5Kr, it was found that radiation doses of 3Kr or above prevented worm maturation and significant resistance was induced by double vaccination of mice with doses of 100 and 200 cercariae irradiated at 2 and 3Kr [82]. The efficacy of irradiated *S. incognitum* vaccination was confirmed in mice immunised with one dose of 100 cercariae or two doses of 50 irradiated at 0-5Kr [73]. Following challenge, worms were reduced in all immunised groups and mice vaccinated with a single dose of 100 cercariae irradiated at 5Kr showed the highest resistance.

Fasciolidae

Fasciola spp. are liver flukes the adults of which inhabit the bile ducts, usually in sheep and cattle. *F. hepatica* is found in the temperate areas and at higher altitudes in the tropics and sub-tropics. Attempts to vaccinate animals against *F. hepatica* infection have, in the main, been disappointing. Hughes [41] showed that pathology and mortality was reduced in mice infected with 20 *F. hepatica* metacercariae irradiated at 2-3Kr. Serial killings of mice infected with 20 and 40 3Kr-irradiated metacercariae showed that flukes which had developed from irradiated metacercariae were smaller and weaker than normal flukes and their passage through the hepatic parenchyma was retarded [17]. Following challenge 22 days after vaccination, there was no evidence of protective immunity either in terms of reduced fluke burden or delayed maturation. These findings were confirmed in a more extensive study using mice and metacercariae irradiated at 3Kr [18]. Other studies [19] however, have pointed to reduced pathogenicity of irradiated metacercariae in rabbits, sheep and mice, both serologically and histologically, with fewer flukes recoverable from such metacercariae.

Boray [10] challenged sheep after immunisation with several doses of metacercariae irradiated at 20Kr. Except for a later development of anaemia in the vaccinated animals, there was no evidence of any protection against challenge. There was no significant reduction in fluke burdens or fluke size following challenge in similarly vaccinated calves. More encouraging results were later produced [69] when calves were vaccinated with three doses of 1,500 cercariae irradiated at 3Kr. After grazing over a full season, the vaccinated calves showed a 29% reduction in fluke burdens compared to the controls.

A study of the influence of ionising radiation on the antigenic potential of *F. hepatica* metacercariae in rabbits led to the conclusion that a full immunological response could not be expected at radiation doses greater than 3Kr [66].

Harness et al [34] demonstrated that immunisation of mice with one or two doses of 20 metacercariae irradiated at 3.8 Kr significantly reduced the number of flukes recovered two days after challenge. This led to the suggestion that some form of protective mechanism operated at the gut wall but this was latter disproved when the same workers [35] showed that immature flukes migrate more quickly from the peritoneal cavity in mice previously sensitised with irradiated larvae.

After many failures to stimulate immunity to *F. hepatica* in sheep by the use of a radiation attenuated vaccine, it was reported that 6 doses of 100 3Kr-irradiated metacercariae resulted in 66% reduction to challenge [16]. Later, other workers [13] immunised sheep with two doses of 100 or 1,000 metacercariae irradiated at 2Kr given six weeks apart and found no evidence of protection following challenge. Hughes et.al [42] studied the antibody pattern to the different antigens in cattle, sheep and rats following infection with 4Kr-irradiated metacercariae; the irradiated cercariae provided a normal antigenic stimulus for the development of circulating antibodies against cercariae, juvenile teguments and gut antigens in the host species. Titres against adult tegument appeared late in the vaccinated animals and were consistently lower than in the controls. These workers suggested that the latter stimulus came from flukes which 'broke through' the development barrier and which were found live at autopsy.

Fasciola gigantica is found mainly in tropical areas. In a study of the protective potential of irradiated *F. gigantica* metacercariae, 6 month-old calves were vaccinated with two doses of 1,000 cercariae irradiated at 3Kr on days 0 and 42 and challenged on day 140 [9]. Five challenged vaccinates had no worms or lesions while two had fluke burdens of 1–2%; worm recovery averaging 23% was recorded for the non-vaccinated challenge calves. Chandra [15] infected kids with 100 *F. gigantica* metacercariae irradiated at 1, 3, or 5Kr. Compared with fluke recovery from animals given a similar number of non-irradiated organisms, reductions of 91%, 84% and 97%, respectively were recorded. Flukes recovered from irradiated metacercariae were smaller than normal and were sexually retarded.

Paramphistomatidae

Paramphistomum spp. are usually found in the fore-stomachs of ruminants and sometimes cause serious economic losses from acute outbreaks in tropical and sub-tropical regions. The first report of successful immunisation of cattle with irradiated *P. microbothrium* came from South Africa [37]. It was later demonstrated that lambs vaccinated with 3000 *P. epiclitum* metacercariae irradiated at 2 or 3Kr showed significant resistance to challenge 40 days later, the latter group showing the greater resistance [33].

CESTODA

Taeniidae

The members of the family *Taeniidae* are tapeworms found in domestic carnivores and man. There are few reports of attempts to produce radiation-attenuated vaccines against cestodes. Movsesijan et al [65] compared the immunological properties of *Echinococcus granulosus* metacestodes irradiated at different dose rates, given in single or double doses. Pups vaccinated with larvae irradiated at 30Kr in doses of 500 followed by 1,000 two weeks later showed no evidence of resistance. In contrast, pups given one dose of 500 irradiated larvae at 20Kr showed 64% protection against challenge. Later, it was demonstrated that larvae irradiated at 25Kr conferred about 80% protection against challenge [36].

The intermediate stage of *Taenia saginata* is found in voluntary and involuntary muscles of cattle and buffalo and is a hazard to public health in many areas of the world (see Chapter 4). Lucker and Vegors [56] infected cattle with 220,000 *T. saginata* eggs irradiated at 50, 100, or 200 Kr. Five months later worm burdens were reduced by 80%, 91% and 99%, respectively, in the vaccinated groups. In a more recent study calves between 4 days and 2 months old were vaccinated with 16,000 eggs irradiated at 20–50Kr. On slaughter 12 to 13 weeks later, it was observed that metacestodes developing from 29 or 30Kr irradiated eggs were stunted and showed signs of degeneration and none were recovered from the eggs irradiated at 50Kr [70].

Taenia taeniaeformis is a tapeworm of cats. Its intermediate stage, *Cysticercus fasciolaris*, is found in the liver of mice and rats. Dow et al [27] vaccinated mice against *C. fasciolaris* with 200 oncospheres irradiated at 40–100 Kr. It was observed that metacestodes failed to establish from the irradiated oncospheres and the vaccinates were almost completely immune to challenge irrespective of the radiation dose used.

NEMATODA

Dictyocaulidae

Dictyocaulus spp.

In 1957 Jarrett & co-workers [45] showed that calves could be immunised against *D. viviparus* by dosing with irradiated larvae. To ascertain the level of radiation which would reduce the pathogenicity of *D. viviparus* larvae without significantly reducing their immonogenicity, groups of calves were dosed with 4,000 larvae irradiated at 0 – 60 Kr [49]. A high level of immunity, comparable to that resulting from natural infection, was stimulated by larvae irradiated at 20 or 40Kr. The numbers of worms found in the lungs of the vaccinates 35 days after vaccination were small and pathogenic effects of the 40Kr vaccine were correspondingly slight. Later, calves were vaccinated with 1,000 larvae irradiated at 40Kr to determine if fewer irradiated larvae would stimulate an acceptable level of immunity [46]. Faecal larval counts and respiratory rates were lower in the these animals compared with the controls given the same number of non-irradiated larvae.Following grazing on infected pastures, the mortality rate in the controls was 83% while it was only 29% in the vaccinates. In an attempt to produce a higher degree of immunity a double vaccination procedure was adopted [48]. Following vaccination of four groups of calves with 1,000 larvae irradiated at 40Kr on day 0, three groups were given a second dose of 1,000, 2,000, or 4,000 of the irradiated larvae on day 42. No worms were found in the lungs of the double vaccinated groups when challenged 50 days after the last vaccination and the single vaccinates had a mean burden of 820 worms. The mean adult worm burdens in the non-vaccinated controls challenged with 5,000 or 1,000 larvae were 1,020 and 879, respectively. This experiment demonstrated the superiority of double vaccination which became the basis for 'Dictol', the commercial vaccine against bovine *Dictyocaulus* introduced in 1960. The successful use of this vaccine in a number of countries, see review [4], led to may attempts to develop similar vaccines against a number of helminths.

The first attempt to produce a radiation attenuated vaccine against *Dictyocaulus filaria*, an important cause of bronchitis in sheep and goats, was made in Yugoslavia. It was demonstrated [52] that *D. filaria* larvae lost their pathogenicity yet retained antigenicity when irradiated at 40–60Kr, a finding later confirmed by other workers. In a trial involving almost 2,000 lambs, between 16 and 20 weeks old, about 50% were protected against challenge 60 days after vaccination with 2,000 larvae irradiated at 40Kr [76]. As with the cattle vaccine, enhanced protection was achieved by double vaccination. Vaccination with 40Kr-irradiated larvae, 1,000 larvae followed by 2,000 30 days later, gave 80% protection to a heavy challenge 15 days after final vaccination. In the field, vaccination reduced the number of infected animals and the severity of infection. The superiority of double vaccination was confirmed from Yogoslavia [53]. These experiments later led to the production and use in Yugoslavia of 'Difil', an anti-*D. filaria* irradiated vaccine.

Following a protocol similar to that used in Yugoslavia, the feasibility of developing a radiation attenuated vaccine against *D. filaria*, an important infection of sheep and goats in India (see Chapter 3), was investigated [83]. Larvae irradiated at 45 or 50Kr were given in doses of 1,000, followed by 2,000 one month later, to lambs 4 to 6 month-old. Following challenge 42 days later, the vaccinated lambs became worm-free, except one out of eight given 45Kr irradiated larvae which had 14 worms and one of the nine in the 50Kr vaccinates which harboured 11 worms. The remaining vaccinated lambs became worm-free. Four out of six non-vaccinated challenge controls died and the two survivors yielded 205 and 136 worms. In the endemic area of Kashmir, a pilot scheme was undertaken to compare the efficacy of the vaccine prepared in India with that produced in Yugoslavia [23]. About 3,000 lambs were vaccinated with either of the vaccines as previously described and the results indicated that both vaccines were equally effective. Later works confirmed the efficacy of the vaccine developed in India [24, 85].

Conflicting accounts of the efficacy of irradiated *D. filaria* came from Iraq. Failure of the irradiated vaccine to protect sheep and lambs was reported [54, 55]. On the other hand, there are reports of successful vaccination of lambs in Iraq with double doses of larvae irradiated at 40 or 50Kr [44]. Irradiated *D. filaria* larval vaccines have been used successfully in Bulgaria, Czechoslovakia, Ethiopia and Poland. It was also reported that guinea-pigs could be protected by two doses of irradiated vaccine [20, 86].

Whilst difil vaccine is no longer prepared and used comercially in Yugoslavia, the prospects are better in India for the use of an irradiated vaccine against *D. filaria* [24, 77].

Trichostrongylidae

Trichostrongylus spp.

T. colubriformis is an important contributor to parasitic enteritis in ruminants. The use of an irradiated larval vaccine against this parasite in lambs was demonstrated by Jarrett et al [50]. Fifteen young lambs were vaccinated with a single dose of 10,000 larvae irradiated at 40Kr. Five of which were killed 28 days later, harboured a mean worm burden of 3,900 and the maximum egg output was 420 epg compared with 5,900 worms and 2,200 epg in the controls. On re-vaccination and challenge 6 months later, the remaining ten lambs showed a 97% reduction in worm burdens. It was noted that a higher radiation dose would be required to produce a more effective vaccine. Gordon et al [30] succeeded in producing a marked resistance in guinea-pigs vaccinated with 5,000 *T. colubriformis* larvae irradiated at 40-100Kr against a challenge of 20,000 larvae given 30 days post-vaccination. As in studies on other species, these workers found male worms to be more radiosensitive than female worms. Mulligan et al [67] used doses of 80, 130 and 180Kr in a screening test in guinea-pigs and concluded that 80Kr was the optimal dose for attenuating *T. colubriformis* larvae. When ten 6 month-old lambs were immunised with two doses of 10,000 80Kr-irradiated larvae and challenged 34 days after the second vaccination, nine showed a high degree of resistance. Following a study using guinea-pigs and sheep [31], it was decided that a 50Kr radiation dose was suitable for the attenuation of *T. colubriformis* larvae prior to vaccination.

The response of vaccinated sheep to challenge given as a single dose or at regular and frequent intervals over a period following triple vaccination with 20,000 larvae irradiated at 50Kr was examined [32]. The vaccine was shown to be equally effective against single and sequential challenge and appeared more efficient after the removal of worms by anthelmintic treatment than in superimposed infections. The level of group protection calculated from worm burden or maximum egg excretion was about 98% following triple vaccination compared with 91% after double vaccination.

Immunological control of *T. colubriformis* infection by itself would perhaps have limited value since it normally occurs in the field along with other gastrointestinal helminths. This may be the reason why an anti-*T. colubriformis* vaccine has not been developed for routine use in the field.

Haemonchus spp.

Haemonchus contortus is a blood-sucking abomasal nematode of ruminants, found especially in tropical areas. Jarrett et al [47] vaccinated 7 month-old lambs, with 10,000 larvae irradiated at 10, 20, 40, 60 or 100Kr and challenged them with 8,000 larvae 116 days later. The mean number of worms recovered 47 days later, were 5, 0, 200, 0, 442 and 620 from the lambs given larvae irradiated at 10, 20, 40, 60, 100 and 0Kr, respectively. The challenge controls had a mean burden of 2,040 worms. Radiation doses of 40–60Kr were considered optimal. Later, these workers [51] vaccinated 8 month-old lambs with two doses of 10,000 40Kr irradiated larvae given 35 days apart. The lambs were challenged with either 10,000 or 50,000 larvae on day 64 and killed on day 113 of the experiment. Protection of approximately 99% was achieved in both challenge groups. The results suggested that an irradiated vaccine was effective in 7 to 8 month-old lambs.

Mulligan et al [67] concluded that 60Kr was the optimal irradiation dose and vaccinated 7 month-old lambs with 2,000 *H. contortus* larvae followed by 5,000, 35 days later, irradiated at this level. Following challenge 35 days later, only 60% of the lambs showed a strong resistance. Bitakaramire et al [8] succeeded in immunising adult sheep using a previously described protocol [51]. They observed protection ranging between 84% and 100% between days 9 and 16 after the challenge.

Attempts by Urquhart et al to produce an irradiated larval vaccine against *H. contortus* suffered a setback when either single or double doses of 10,000 larvae irradiated at 40Kr failed completely to protect 3 month-old lambs [88]. These workers [89] also failed to protect lambs 5, 13 and 21 weeks of age using the same vaccination protocol. A number of immunisation and challenge techniques were adopted but none could stimulate a response in lambs aged 5 or 13 weeks. Compared with animals 35 week-old, a poor response was detected in 23 week-old lambs. Several reasons were suggested for the failure of the young lambs to respond to the irradiated larvae (see Chapter 15). In an investigation of previous exposure to anthelmintic therapy and the response to gamma-irradiated *H. contortus* larvae in lambs of different ages, it was found that unresponsiveness of lambs could not be ameliorated by increasing the dose of irradiated larvae used [6]. Such a procedure led to severe haemonchosis following vaccination. Young lambs immunised at 10 and 14 weeks old failed to develop a significant resistance to challenge. Parasite naive lambs 9 and 10 month-old showed a highly significant immunity to challenge following double vaccination with larvae irradiated at 60Kr. This immunity was not significantly altered by dosing with spaced doses of normal larvae from 10 weeks of age until vaccination at 9 or 10 months of age. It was also shown that worm-free lambs vaccinated at 9 and 10 months of age showed no immunity when anthelmintic was given 3 weeks after each vaccination. It appeared that the presence of worms from the vaccination dose is important for the development of a significant degree of resistance.

Smith & Christie [75] demonstrated that the resistance to *H. contortus* challenge in 9 month-old lambs vaccinated with irradiated larvae was associated with increased IgG antibodies in serum and IgA and IgG antibodies in the abomasal mucosa. It was also shown that young lambs developed no immunity to the irradiated *H. contortus* vaccine and no significant increase in either serum IgG or mucosal IgA [28]. In spite of efforts by several workers the reason for unresponsiveness in young lambs is not fully understood. This is a major reason for the failure to produce an irradiated vaccine against *H. contortus* in sheep.

There is only one report of an attempt to immunise cattle with an irradiated *H. contortus* vaccine [71]. A 60% reduction in worm burdens was observed in calves vaccinated with 2 doses of 10,000 larvae irradiated at 60Kr and challenged 6 weeks after the second vaccination.

Ostertagia spp.

The use of a radiation attenuated larval vaccine against *O. circumcincta*, a major cause of parasitic gastritis in sheep and goats, was investigated by Smith et al [74]. The first experiment indicated that a radiation dose of 80Kr was optimal for this purpose. Subsequently, mature sheep were immunised with two doses of 10,000 larvae of *O. circumcincta* irradiated at 80Kr were found to be only slightly resistant to challenge 4 weeks later. In a concurrent experiment sheep immunised with two doses of *H. contortus* irradiated vaccine (10,000 60Kr-irradiated larvae) were found to be highly resistant to challenge with *H. contortus* larvae. Other sheep were immunised with irradiated *H. contortus* larvae and challenged with *O. circumcincta* larvae. Some evidence of cross protection was observed in the challenged animals, e.g. lower egg counts and more worms were retarded at the fourth larval stage, compared with the controls. These experiments showed that the immunity stimulated by the irradiated *O. circumcincta* larvae was much reduced compared to the degree of protection afforded by similar *T. colubriformis* or *H. contortus* vaccines.

O. ostertagi is one of the most common causes of gastritis in cattle and buffalo. The possibility of producing an irradiated larval vaccine against ostertagiasis in cattle was investigated by Armour [3]. Worm-free calves, 10 week-old, were vaccinated with two doses of 100,000 larvae irradiated at 60Kr or 80Kr, given one month apart. The calves failed to show any satisfactory degree of resistance to a challenge one month after vaccination. However, there was a decrease in faecal egg output, delayed onset of diarrhoea and an increase in the number of immature worms in the vaccinated calves following challenge, compared with the controls. Other workers [11] simultaneously immunised calves with 50,000 *O. ostertagi* and the same number of *Cooperia oncophora* larvae irradiated at 40Kr. This vaccination procedure was repeated 4 weeks later. Another group was given only the irradiated *C. oncophora* at the final immunisation followed by the irradiated *O. otertagi* one month later. Following a period of grazing vaccination did not have any obvious effect on the gastroenteritis in the vaccinates.

Trichonematidae

Oesophagostomum spp.
These nematodes are responsible for enteritis in ruminants and pigs, especially in the tropics and subtropics. In a study of the *in vitro* survival and the immunogenic effect of irradiated *O. columbianum* larvae 2 month-old kids were dosed with 1,000 larvae irradiated at 15-160Kr [22]. Doses of 15 to 40Kr resulted in progressively fewer worms reaching the adult stage and the formation of more gut wall nodules. Again, irradiation at 80Kr resulted in the failure of the larvae to develop and establish in the host and there was no nodular lesion in the wall. These results showed that the percentage take of worms was reduced with increasing larval irradiation with an increase in the number of nodules formed. Further, few larvae developed from eggs excreted in the faeces of animals infected with larvae irradiated at 30Kr or 40Kr. In a study of the immunogenicity of irradiated larvae, three kids were immunised with 2,000, 4,000 or 5,000 larvae irradiated at 40Kr. Following challenge with 1,000 larvae, a high degree of immunity was observed in the vaccinated kids. Unfortunately, there was more than a tenfold increase in the nodules of the vaccinates. In similar studies [72] lambs, 6 to 8 week-old, were infected with 2,000 infective *O. columbianum* larvae irradiated at 40 and 50Kr. When killed 35 days later, these animals had no worms but had a greater number of nodular lesions than lambs infected with 2,000 normal larvae. When lambs were immunised with 2,000 larvae irradiated either at 40 or 50Kr and then challenged virtually no protection was observed in the 50Kr group but partial (38.5%) protection was demonstrated in the 40Kr group. As in previous studies, more nodular lesions appeared in the vaccinates than in the controls.

Ancylostomatidae

Ancylostoma spp.
The first attempt to produce an irradiated larval vaccine against hookworm of dogs involved immunising 8–12 week-old pups with single dose of 1,000 *Uncinaria stenocephala* larvae irradiated at 40Kr [25]. The irradiated larvae conferred a higher degree of protection to challenge than an equal number of non-irradiated larvae. A further study [26] showed that double doses of vaccine were more effective than a single dose and that irradiation at 40Kr produced a more protective vaccine than 20Kr-irradiation. Although study of this species was not pursued, as it is not economically important, these experiments led to the search for an irradiated larval vaccine against *A. caninum*, the common hookworm of dogs.

In the first of a series of investigations using irradiated *A. caninum* larvae, Miller [58] observed that a 1,000 larval dose irradiated at 30Kr resulted in few worms capable of causing clinical disease. It was also shown in pups 3 to 5 month-old, that a single dose of *A. caninum* larvae irradiated at 40Kr reduced infectivity, measured by adult worm establishment and significantly reduced the male: female ratio [59]. The vaccinated pups also showed a marked resistance to challenge. Pups, 3 to 5 month-old, were subcutaneously vaccinated with double doses of 1,000 larvae irradiated at 40Kr [60]. Resistance to challenge in these animals was about 90%, more than twice that obtained by single vaccination.

In a series of experiments double subcutaneous vaccination was shown to be more effective than double oral vaccination with irradiated larvae [61]. However, both vaccination methods proved highly protective against a challenge sufficiently heavy to produce morbidity and mortality in unprotected animals and it was shown that a highly significant resistance to the vaccine persisted for at least 7 months following vaccination in 3 and 4-month old pups. It was further demonstrated that double vaccination against *A. caninum* in pups was effective as early as 72 hours after birth [62].

In other studies, Miller [63] used anthelmintic treatment before vaccinating pups and demonstrated that anthelmintic treatment could be used at any time without interfering with the efficacy of the vaccine. Double vaccination also gave strong protection against a heterologus challenge of *A. brasiliensis* and *U. stenocephala*. This gave rise to the idea that while the latter two nematodes are rare in dogs and cats, the interspecific immunity could contribute to a reduction in the incidence of 'creeping eruptions' in man sharing a common habitat with dogs.

Following successful field trials of the Miller vaccine in Scotland and in the U.S.A. when reductions in challenge worm burden in the order of 85–99% were observed, the vaccine was introduced commercially in the U.S. in 1973. Unfortunately, this vaccine was withdrawn in 1975 [64].

Gaigeria pachyscelis is a hookworm responsible for anaemia in sheep and goats in Africa, Asia and South America. Ansari and Singh [1] showed that infective *G. pachyscelis* larvae could withstand radiation doses of up to 160Kr without any significant effect on survival. When 4–6 month-old lambs were infected with 2,000 larvae irradiated at 0, 5, 10 and 20Kr mean worm burdens of 258, 166, 116 and 17, respectively were found. These workers [2] conducted a field trial involving immunisation of lambs with 2,000 larvae irradiated at 20, 40, 60 and 80Kr. Following challenge 95 days post-vaccination the lambs vaccinated with 40Kr attenuated larvae had the highest (76%) resistance to challenge.

Bunostomum trigonocephalum is also a hookworm of ruminants. Srivastava and Singh [78] infected lambs percutaneously with 6,000 larvae of this nematode irradiated at 20, 40, or 60Kr. The latter group failed to develop in the host. At 20Kr and 40Kr fewer worms, mostly small sterile females, developed compared with the controls. These lambs were not challenged to test their immune status.

Syngamidae

Syngamus spp.

S. trachea is a parasite of the upper respiratory tract of birds and can cause respiratory distress. Varga [91, 92, 93] studied the development, establishment and egg laying capacity of *S. trachea* larvae and decided to use 5Kr irradiated larvae in a series of vaccination experiments. Following immunisation of chickens, 1–3 week-old, with 800 and 1,200 irradiated larvae given one week apart, 76% of the birds were found to be fully immune to challenge, the remaining 24% showed a worm establishment two thirds less than in the controls. Day-old chicks immunised with 800 irradiated larvae on days 1 and 4 showed 94% protection to challenge and when the larval dose was reduced to 500, 98% protection was observed. A single dose of 150 irradiated larvae gave 47% protection to challenge in turkeys, double vaccination at 3 day intervals gave 93% protection. A high degree of protection from challenge (58%) was observed in pheasants in the field following single vaccination with 150 irradiated larvae and 80% after double vaccination. This vaccine conferred a high degree of protection as early as day 6 in chicks vaccinated when one day-old. Further evidence of the vaccine's efficacy was shown in a field trial [93] when 2 week-old chickens vaccinated with one dose of 150 irradiated larvae had a mortality rate of 2.8% compared with 74.5% in the control birds reared in the same enzootic area. Later, Feteanu et al [29] observed protection of about 90% against challenge in chickens immunised with 500–1,000 10Kr irradiated larvae. Although the efficacy of irradiated *S. trachea* irradiated larval vaccines have been proven in many trials it has not been developed commercially.

Heterakidae

Several workers have investigated the possibility of producing a radiation attenuated vaccine against *Ascardia galli* which is found in the small intestine of domestic and wild birds. It was observed that exposure of embryonated eggs of *A. galli* to between 20 and 40Kr prevented the eggs from developing and the larvae which did emerge were morphologically altered [5]. Varga [90], however was unable to demonstrate resistance to challenge in chickens infected with larvae irradiated at 10Kr to 20Kr although challenge worms were markedly retarded following double vaccination. Deo et al [21] immunised chickens, 20 and 50 day-old with 500 and 1,000 eggs, respectively, irradiated at 10Kr. A negligible number of worms were obtained on days 80 and 90 post-challenge and none on days 110 and 125. When the above procedure was repeated with a radiation dose of 15Kr complete immunity was observed. A comparative study of the worms recovered from the vaccinated birds and from vaccine controls indicated that the majority of the vaccine worms were eliminated after challenge and this increased as the birds became older, i.e. expulsion may be augmented by age resistance. Further evidence that expulsion of mature worms occurred after challenge in chickens vaccinated with irradiated eggs was reported by other workers [79].

Strongyloididae

There is little evidence of research into irradiated larval vaccines against parasites of this family. *Strongyloides* spp. are common parasites of the small intestine of young animals and of the caecum in poultry, particularly, in the tropics. Chaia [14] vaccinated chickens with 12,000 *S. avium* larvae irradiated at 140Kr. Following challenge 30 days later, 98% protection was observed compared with non-vaccinated controls.

Capillariidae

Capillaria spp. are found in the upper small intestine of birds and cause diarrhoea. Chickens vaccinated with 600. *C. obsignata* embryonated eggs irradiated at 20Kr gave 94% protection on challenge two weeks later; immunity was highest 31 days after vaccination [94].

CONCLUSIONS

The size of helminths, their complicated life-cycles and antigenic complexity presents a formidable challenge to the host's immune system, which the organisms are adept at evading. Many studies have indicated that live metabolising forms of helminths are necessary for protection and although they are antigenically superior, they can not be used for immunisation.

Studies on live radiation-attenuated vaccines, in the laboratory and in the field, have shown that controlled doses of radiation can reduce virulence without significantly affecting antigenicity. Such vaccines have been studied in many host-helminth systems. The most successful vaccines have been developed against helminths which evoke strong acquired immunity in natural infections. Some successful radio-attenuated vaccines developed in the laboratory have little practical application because the helminth involved has no economic importance.

Attempts to produce radio-attenuated vaccines against trematodes have been discouraging. Much effort has been applied to the development of an irradiated vaccine against the liver fluke *F. hepatica*, with little success. This is not surprising, since despite some evidence of acquired resistance in laboratory studies, it is known that this trematode invokes a poor immune response.

Except for *S. bovis*, trials of radiation-attenuated vaccines against *Schistosoma* spp. have not been promising. Further research into an anti-bovine schistomiasis vaccine could perhaps lead to the development of an irradiated vaccine which might have a wide application in the field.

Although there is some evidence of the effectiveness of radio-attenuated vaccines against *E. granulosus* and *T. saginata*, little effort has gone into developing radio-attenuated vaccines against cestodes.

Attempts to develop radiation-attenuated vaccines against nematodes have been more rewarding. The first commercially successful vaccine to be produced was 'Dictol', used in the control of the bovine lungworm, *D. viviparus*. This irradiated vaccine has been in production and use in the field for almost three decades.

Irradiated vaccines against the ovine lungworm *D. filaria* have been produced in Yugoslavia and India. The vaccine, 'Difil', was commercially produced in Yugoslavia but was later withdrawn. The advantages of this vaccine may not have been apparent to the livestock industry and production perhaps did not have the support it merited. The potential of this vaccine in endemic areas such as India and Iraq has not been realised.

The efficacy of irradiated vaccines against the fowl nematode *S. trachea* has been demonstrated, but a commercially available product has not been produced.

A successful vaccine against the canine hookworm *A. caninum* was produced in the U.S.A. The production of this vaccine was discontinued after a number of years. Miller, in a review [64], gives the reasons as unrealistic sales expectations, reluctance of veterinarians to use the product, high production costs and short shelf-life.

Radio-attenuated vaccines against other nematodes of major veterinary importance have been studied. Although it was shown that an effective vaccine against *T. colubriformis* could be produced, this was not pursued commercially. Perhaps such a vaccine would have limited value since *T. colubriformis* usually occurs in the field with other helminths. Unresponsiveness in young lambs has been suggested as the reason why attempts to vaccinate against *H. contortus* have been unsuccessful. Efforts to find an irradiated vaccine against *Ostertagia* spp. have also been unsuccessful.

Much activity followed the development of 'Dictol'. Unfortunately, although it was demonstrated that irradiated vaccines were effective against several helminths, commercial development did not follow due to economic reasons. Nevertheless, live irradiated vaccines may have a role to play in the future control of some of the economically important helminthic diseases.

REFERENCES

1. Ansari, M.Z. and Singh, K.S. Effects of gamma-irradiation on the survival and development of the infective larvae of the hookworm, *Gaigeria pachyscelis*. J. Helminthol. 52: 283–286 (1978).
2. Ansari, M.Z. and Singh, K.S.: The use of gamma-radiation-attenuated larvae of *Gaigeria pachyscelis*, Raillet and Henry, 1910, for vaccination in lambs. J. Helminthol. 52: 287–290 (1978).
3. Armour, J. *Ostertagia ostertagi* infections in the bovine: field and experimental studies. Ph. D. Thesis, University of Glasgow (1967).
4. Armour, J. Lungworms of cattle, sheep and pigs with special reference to their immunology, immunopathology and immunoprophylaxis. In: *Immune responses in parasitic infections: Immunology, immunopathology and immunoprophylaxis*. Vol. 1. (E.J.L. Soulsby ed), pp. 155–180. CRC Press Inc.: Boca Raton, Florida, USA (1987).

5. Babero, B.B. The effect of X-rays on the life cycle and morphology of *Ascaridia galli* (Schrank, 1788). Trans. Am. Microsc. Soc. **71**: 114–119 (1972).
6. Benitez-Usher, C., Armour, J., Duncan, J.L., Urquhart, G.M. and Gettinby, G. A study of some factors influencing the immunisation of sheep against *Haemonchus contortus* using attenuated larvae. Vet. Parasitol. **3**: 327–342 (1977).
7. Bickle, Q.D., Taylor, M.G., James, E.R., Nelson, G.S., Hussein, M.F., Andrews, B.J., Dobinson, A.R. and de C. Marshall, T.F. Further observations on immunisation of sheep against *Schistosoma mansoni* and *Schistosoma bovis* using irradiation attenuated schitosomula of homologous and heterologous species. Parasitology **78**: 185–193 (1979).
8. Bitakaramire, P.K. Studies on immunity to *Haemonchus contortus* infection; the elimination of challenge infection by immune sheep. Parasitology **56**: 619–622 (1966).
9. Bitakaramire, P.K. Preliminary studies on the immunisation of cattle against fascioliasis using gamma-irradiated metacercariae of *Fasciola gigantica*. In: *Isotopes and Radiation in Parasitology*. Vol. III, pp. 23–32. IAEA, Vienna (1973).
10. Boray, J.C. The effects of host reactions to experimental *Fasciola hepatica* infections in sheep and cattle. In: *The reaction of the host to parasitism* (E.J.L. Soulsby ed), pp. 84–96. Proc. Third Int. Conf. Wld Assoc. Adv. Vet. Parasitol., Lyon (1967).
11. Burger, H.J. and Pfeiffer, A. Experimental vaccination of calves with irradiated larvae of *Ostertagia ostertagi* and *Cooperia oncophora*. Zentrabl. Veterinaermed Reihe B **16**: 357–367 (1969).
12. Bushara, H.O., Hussein, M.F., Saad, A.M., Taylor, M.G., Dargie, J.D., de C. Marshall, T.F. and Nelson, G.S. Immunisation of calves against *Schistosoma bovis* using irradiated cercariae or schisotomula of *Schistosoma bovis*. Parasitology **77**: 303–311 (1978).
13. Campbell, N.J., Gregg, P., Kelly, J.D. and Dineen, J.K. Failure to induce homologous immunity to *Fasciola hepatica* in sheep vaccinated with irradiated metacercariae. Vet. Parasitol. **4**: 143–152 (1978).
14. Chaia, G. Immunisation of chickens with irradiated larvae of *Strongyloides avium*. Rev. Inst. Med. Trop. Sao Paulo **111**: 19–24 (1969).
15. Chandra, R. The effect of X-irradiation on the development of *Fasciola gigantica* (Cobbold, 1885), in goats. *Proceedings of the Symposium on Use of Radiation and Radioisotopes in Studies in Animal Production*. Izatnagar, pp. 312–317 (1975).
16. Dargie, J.D., Armour, J., Rushton, B. and Murray, M. Immune mechanisms and hepatic fibrosis in Fascioliasis. Proc. Sixth Int. Conf. Wld Assoc. Adv. Vet. Parasitol, pp. 249–271. Vienna (1974).
17. Dawes, B. Death of *Fasciola hepatica* L. weakened by X-irradiation. Nature (Lond.) **200**: 602–603 (1963).
18. Dawes, B. A preliminary study of the prospect of inducing immunity in fascioliasis by means of infection with X-irradiated metacercarial cysts and subsequent challenge with normal cysts of *Fasciola hepatica* L. Parasitology **54**: 369–389 (1964).
19. Dawes, B. and Hughes, D.L. Fascioliasis: The invasive stages of *Fasciola hepatica* in mammalian hosts. *Adv. Parasitol.* **2**: 97–168 (1964).
20. Denev, I. Trials of gamma-irradiated *Dictyocaulus filaria* larvae in the immunisation of guinea-pigs and lambs. Vet. Med. Nauki **11**: 7–12 (1974).
21. Deo, P.G., Sokolic, A. and Rao, Y.V.B.G. Further investigations on the use of irradiated eggs to develop immunity against *Ascaridia galli* in chickens. British Vet. J. **127**: 430–435 (1971).
22. Dhar, D.N. and Singh, K.S. Effect of irradiation on the infective stage of the nematode *Oesophagostomum columbianum* and their use as a vaccine. J. Helminthol. **44**: 11–26 (1970).
23. Dhar, D.N., Tewari, H.C. and Rao, Y.V.B.G. Studies on radiation attenuated vaccine. 2. Field trials with the radiation attenuated *Dictyocaulus filaria* in Kashmir. Indian J. Anim. Sci. **42**: 106–111 (1972).
24. Dhar, D.N. and Sharma, R.L. Immunisation with irradiated larvae against *Dictyocaulus filaria* in young lambs. Vet. Parasitol. **9**: 125–131 (1981).
25. Dow, C., Jarrett, W.F.H., Jennings, F.W., McIntyre, W.I.M. and Mulligan, W. The production of active immunity againt canine hookworm *Uncinaria stenocephala*. J. Am. Vet. Med. Assoc. **135**: 407–411 (1959).
26. Dow, C., Jarrett, W.F.H., Jennings, F.W. Studies on immunity to *Uncinaria stenocephala* infection in the dog-double vaccination with irradiated larvae. Am. J. Vet. Res. **22**: 352–354 (1961).

27. Dow, C., Jarrett, W.F.H., Jennings, F.W., McIntyre, W.I.M. and Mulligan, W. The production of immunity to *Cysticercus fasciolaris* using X-irradiated oncospheres. Am. J. Vet. Res. 23: 146–148 (1962).
28. Duncan, J.L., Smith, W.D. and Dargie, J.D. Possible relationship of levels of mucosal IgA and serum IgG to immune unresponsiveness of lambs to *Haemonchus contortus.* Vet. Parasitol. 4: 21–27 (1978).
29. Feteanu, A., Stoican, E., Megru, D. and Verdes, N. Experimental study on the protection of chickens and pheasants against syngamosis by use of irradiated larvae. I. Methods of immunisation, preparation of immunising material and identification of serum antibodies by immunofluorescence. In: *Isotopes and Radiation in Parasitology*. Vol. III, pp. 101–111. IAEA, Vienna (1973).
30. Gordon, H.McL., Mulligan, W. and Reineck, R.K. *Trichostrongylus colubriformis* in the guinea-pig. Aust. Vet. J. 36: 466–471 (1960).
31. Gregg, P., Dineen, J.K. and Griffiths, D.A. The effect of gamma-radiation in the development of infective larvae of *Trichostrogylus colubriformis* in guinea-pigs and sheep. Vet. Parasitol. 2: 363–375 (1976).
32. Gregg, P. and Dineen, J.K. The response of sheep vaccinated with irradiated *Trichostrongylus colubriformis* larvae to impulse and sequential challenge with normal larvae. Vet. Parasitol. 4: 49–53 (1978).
33. Hafeez, Md. and Rao, B.V. Studies on amphistomiasis in Andhra Pradesh (India) VI. Immunisation of lambs and kids with gamma-irradiated metacercariae of *Cercariae indicae XXVI*. J. Helminthol. 55: 29–32 (1981).
34. Harness, E., Hughes, D.L. and Doy, T.D. The demonstration of pre-hepatic immune response to *Fasciola hepatica* in the mouse. Int. J. Parasitol. 6: 15–17 (1976).
35. Harness, E., Doy, T.G. and Hughes, D.L. The early migratory behaviour of young *Fasciola hepatica* in sensitised mice. Int. J. Parasitol. 7: 51–54 (1977).
36. Herd, R.P., Chappel, R.J. and Biddell, D.G. Immunisation against canine *Echinococcus*. Proc. Third Int. Congr. Parasitol., Munich. 1: 570–571 (1974).
37. Horak, I.H. Host-parasite relationship of *Paramphistomum microbothrium* Fischoeder, 1901, in experimentally infested ruminants with particular reference to sheep. Onderstepoort J. Vet. Res. 34: 451–540 (1967).
38. Hsu, S.Y.L. Immunisation against *Schistosoma japonicum* in chimpanzees by administration of X-irradiated cercariae. Trans, R. Soc. Trop. Med. Hyg. 64: 597–600 (1970).
39. Hsu, H.F., Hsu, S.Y.L. and Osborne, J.W. Immunisation against *Schistosoma japonicum* in rhesus monkeys produced by irradiated cercariae. Nature (Lond.) 194: 98–99 (1962).
40. Hsu, W.Y.L., Hsu, H.F., Xu, S.T., Shi, F.H., He Y.X., Coarke, W.R. and Johnson, S.C. Vaccination against bovine schistosomiasis japonica with highly irradiated schistosomula. Am. J. Trop. Med. Hyg. 32: 367–370 (1983).
41. Hughes, D.L. Observations on the immunology of *Fasciola hepatica* infection in mice and rabbits. Parasitology 52: 4p (1962).
42. Hughes, D.L., Hanna, R.E.B. and Doy, T.G. Antibody response in cattle, sheep and rats to infection with gamma-irradiated metacercariae of *Fasciola hepatica*. Res. Vet. Sci. 32: 354–358 (1982).
43. Hussein, M.F. and Bushara, H.O. Investigations on the development of an irradiated vaccine for animal schistosomiasis. In: *Nuclear Techniques in Animal Production and Health*. pp. 421–431. IAEA, Vienna (1976).
44. Jabbir, M.H. and Singh, K.S. Successful trial of irradiated vaccine *Dictyocaulus filaria* in Iraq. Indian J. Anim. Sci. 51: 177–181 (1981).
45. Jarrett, W.F.H., McIntyre, W.I.M., Jennings, F.W. and Mulligan, W. The natural history of parasitic bronchitis with notes on prophylaxis and treatment. Vet. Rec. 69: 1329–1340 (1957).
46. Jarrett, W.F.H., Jennings, F.W., Martin, W.B., McIntyre, W.I.M., Sharp, N.C.C. and Urquhart, G.M. A field trial for a parasitic bronchitis vaccine. Vet. Rec. 70: 451–454 (1958).
47. Jarrett, W.F.H., Jennings, F.W., McIntyre, W.I.M., Mulligan, W. and Sharp. N.C.C. Studies on immunity to *Haemonchus contortus*—vaccination of sheep using a single dose of X-irradiated larvae. Am. J. Vet. Res. 20: 527–531 (1959).
48. Jarrett, W.F.H., Jennings, F.W., McIntyre, W.I.M., Mulligan, W., Sharp, N.C.C. and Urquhart, G.M. Immunological studies on *Dictyocaulus viviparus* infection in calves—double vaccination with irradiated larvae. Am. J. Vet. Res. 20: 522–526 (1959).
49. Jarrett, W.F.H., Jenning, F.W., McIntyre, W.I.M. and Sharp, N.C.C. Immunological studies on *Dictyocaulus viviparus* Infection. Immunity produced by the administration of irradiated larvae. Immunology 3: 141–151 (1960).

50. Jarrett, W.F.H., Jennings, F.W., McIntyre, W.I.M., and Sharp, N.C.C. Resistance to *Trichostrongylus colubriformis* produced by X-irradiated larvae. Vet. Rec. **72:** 884 (1960).
51. Jarrett, W.F.H., Jennings, F.W., McIntyre, W.I.M:, Mulligan, W. and Sharp, N.C.C. Studies on immunity to *Haemonchus contortus* infection—double vaccination of sheep with irradiated larvae. Am. J. Vet. Res. **22:** 186–188 (1961).
52. Jovanovic, M., Nevemic, V., Sokolic, A., Sofrenovic, D., Gligorijvic, J., Cuperlovic, K. and Movsesijan, M. Vaccination of sheep with irradiated *Dictyocaulus filaria* larvae, I. Influence of dose of irradiation on pathogenicity of the parasite. Vet. Glas. **15:** 445–464 (1961).
53. Jovanovic, M., Sokolic, A., Movsesijan, M. and Cuperlovic, K. Immunisation of sheep with irradiated larvae of *Dictyocaulus filaria*. British Vet. J. **121:** 119–131 (1965).
54. Kassai, T., Kadhim, J.L., Altaif, K. and Al-Jabir, M.H. Immunisation of sheep and goats with irradiated and non-irradiated larvae of *Dictyocaulus filaria*. Magy Allatrov Lapja **29:** 409–412 (1974).
55. Khadim, J.K., Kassai, T., Altaif, K.I. and Al-Jabir, M.H. Further studies on the efficacy of irradiated vaccine in the control of *Dictyocaulus viviparus* in Iraq. Res. Vet. Sci. **15:** 272–273 (1973).
56. Lucker, T.J. and Vegors. H.H. Vaccination against beef measles. J. Anim. Sci. **24:** 284 (1965).
57. Majid, A.A., de C. Marshall, T.F., Hussein, M.F., Bushara, H.O., Taylor, M.G., Nelson, G.S. and Dargie, J.D. Observations on cattle schistosomiasis in the Sudan. A study in comparative medicine. I. Epizootiological observations on *Schistosoma bovis* in the While Nile province. Am. J. Trop. Med. Hyg. **29:** 435–444 (1980).
58. Miller, T.A. The possibility of developing X-irradiated vaccines against hookworm disease. In: *Production and Utilization of Radiation Vaccines against Helminthic Diseases*. Technical Report Series, No. 3k, pp. 41–47. IAEA, Vienna (1964).
59. Miller, T.A. Effect of X-irradiation upon the infective larvae of *Ancylostoma caninum* and the immunogenic effects in dogs of a single infection with 40 Kr-irradiated larvae. J. Parasitol. **50:** 735–742 (1964).
60. Miller, T.A. Studies on canine Ancylostomiasis: Double vaccination with X-irradiated *Ancylostoma caninum* larvae. J. Am. Vet. Med. Assoc. **146:** 41–44 (1965).
61. Miller, T.A. Effect of route of administration of vaccine and challenge on the immunogenic efficiency of double vaccination with irradiated *Ancylostoma caninum* larvae. J. Parasitol. **51:** 200–206 (1965).
62. Miller, T.A. Effect of age of the dog on immunogenic efficiency of double vaccination with X-irradiated *Ancylostoma caninum* larvae. Am. J. Vet. Res. **26:** 1283–1390 (1965).
63. Miller, T.A. Pathogenesis and immunity in hookworm infection. Trans. R. Soc. Trop. Med. Hyg. **62:** 473–485 (1968).
64. Miller, T.A. Industrial development and field use of the canine hookworm vaccine. Adv. Parasitol. **16:** 333–342 (1978).
65. Movsesijan, M., Sokolic, A. and Meladenovic, Z. Studies on the immunological potential of irradiated *Echinococcus granulosus* forms: Immunisation experiments in dogs. British Vet. J. **124:** 425–432 (1968).
66. Movsesijan, M. and Cuperlovic, K. Pathophysiology and immunology of infections with non-irradiated and irradiated metacercariae of *Fasciola hepatica*. In: *Isotopes and Radiation in Parasitology*, Vol. II, pp. 23–24. IAEA, Vienna (1970).
67. Mulligan, W., Gordon, H. McL., Stewart, D.F. and Wagland, B.M. The use of irradiated larvae as immunising agents in *Haemonchus contortus* and *Trichostrongylus colubriformis* infections of sheep. Aust. J. Agric. Res. **12:** 1175–1187 (1961).
68. Murare, H.M. Studies on the host-parasitic relationship of *Schistosoma bovis* in the mouse. Ph.D. Thesis, University of London (1983).
69. Nansen, P. Resistance in cattle to *Fasciola hepatica* induced by gamma-ray attenuated larvae. Res. Vet. Sci. **19:** 278–283 (1975).
70. Onyango-Abuje, J.A. Preliminary report on the development of irradiated *Taenia saginata* oncospheres in calves. Bull. Anim. Health Prod. Afr. **26:** 17–22 (1987).
71. Ross, J.G., Armour, J., Hart, J. and Lee, R.P. *Haemonchus* spp. X-irradiated larval vaccine in Nigerian Zebu cattle, Vet. Rec. **71:** 751 (1959).
72. Sharma, R.L. and Dhar, D.N. Immunisation of lambs against *Oesophagostomum columbianum* using irradiated third stage larvae. Z. Parasitenkunde **61:** 53–61 (1979).

73. Shrivastava, H.O.P., Dutt, S.C., Shah, H.L. and Sahasrabudhe, V.K. Studies on immunity to *Schistosoma incognitum* in mice using irradiated cercariae. Indian vet. J. 58: 865–868 (1981).
74. Smith, W.D., Jackson, E. and Jackson, K. Attempts to immunise sheep against *Ostertagia circumcincta* with irradiated larvae. Res. Vet. Sci. 32: 101–105 (1982).
75. Smith, W.D. and Christie, M.G. *Haemonchus contortus:* Local and serum antibodies in sheep immunised with irradiated larvae. Int. J. Parasitol. 8: 219–223 (1978).
76. Sokolic, A., Jovanovic, M., Cuperlovic, K. and Movsesijan, M. Vaccination against *Dictyocaulus filaria* with irradiated larvae. British Vet. J. 121: 212–221 (1965).
77. Sokolic, A., Deorani, V.P.S., Tewari, H.C. and Rao, B.V. *Dictyocaulus filaria* in India. Present prospect for using radiation-attenuated vaccine in control. In: *Isotopes and Radiation in Parasitology.* Vol. II, pp. 125–128. IAEA, Vienna (1970).
78. Srivastava, V.K. and Singh, K.S. Effect of gamma-irradiation on the infective larvae of *Bunostomum trigonocephalum* (Rud. 1808) in lambs. J. Helminthol. 59: 43–45 (1985).
79. Subramanian, G. and Singh, K.S. Use of nuclear techniques in the control of *Ascaridia galli* in India. Investigations on immunisation of one-week-old chicks with irradiated eggs. In: *Isotopes and Radiation in Parasitology.* Vol. III, pp. 67–71. IAEA, Vienna (1973).
80. Taylor, M.G., James, E.R., Nelson, G.S., Bickle, Q., Dunne, D.W. and Webbe, G. Immunisation of sheep against *Schistosoma mattheei* using either irradiated cercariae or irradiated schistosomula. J. Helminthol. 50: 1–9 (1976).
81. Taylor, M.G., James, E.R., Bickle, Q., Hussein, M.F., Andrews, B.J., Dobinson, A.R. and Nelson, G.S. Immunisation of sheep against *Schistosoma bovis* using an irradiated schistomular vaccine. J. Helminthol. 53: 1–5 (1979).
82. Tewari, H.C. and Biswas, G. Experimental studies on the immunology of *Schistosoma incognitum* Chandler, 1962 by vaccination with gamma irradiated cercariae and passive transfer. Z. Parasitenkunde. 38: 48–53 (1972).
83. Tewari, H.C., Gangadhara Rao, Y.V.B. and Dhar, D.N. Studies on radiation-attenuated helminth vaccines: laboratory studies on the production and evaluation of gamma-irradiated vaccine against the sheep lungworm *Dictyocaulus filaria.* Indian J. Anim. Sci. 42: 32–37 (1972).
84. Tewari, H.C. and Singh, K.S. Acquired immunity in dogs against *Schistosoma incognitum.* J. Parasitol. 63: 945–946 (1977).
85. Tewari, H.C., Deo, P.G., Varshney, T.R. and Singh, Y.P. Note on the incidence of lungworm infection in Garhwal Hills, with observations on the efficacy of the radiation-attenuated *Dictyocaulus filaria* vaccine. Indian J. Anim. Sci. 7: 584–585 (1980).
86. Tomanek, J. and Sedelaček, M. Effects of the length of storage on the viability, infectivity and immunogenic potential of X-irradiated larvae of *Dictyocaulus filaria.* Acta Vet. Brno 44: 212–215 (1975).
87. Tyzzer, E.E. and Honeij, A.J. The effects of radiation on the development of *Trichinella spiralis.* J. Parasitol. 3: 43–56 (1916).
88. Urquhart. G.M., Jennings, F.W. and Mulligan, W. Immunity to *Haemonchus contortus* infection. Failure of X-irradiated larvae to immunise young lambs. Am. J. Vet. Res. 27: 1641–1643 (1966).
89. Urquhart, G.M., Jennings, F.W. and Mulligan, W. Immunity to *Haemonchus contortus* infection: Relationship between age and successful vaccination with irradiated larvae. Am. J. Vet. Res. 27: 1145–1148 (1966).
90. Varga, I. Immunisation experiments with irradiated larvae of *Ascaridia galli* II. The resistance inducing effect of treatment with irradiated larvae of *Ascaridia galli.* The effect of X-rays upon eggs and larvae of *Syngamus trachea.* Acta Vet. Hung. 14: 399–409 and 411–418 (1964).
91. Varga, I. Experimental immunisation of chicks with irradiated *Syngamus trachea* larvae. Magy. Allatorv. Lapja 20: 205–210 (1965).
92. Varga, I. Immunisation experiments with irradiated larvae of *Syngamus trachea* in chickens. In: *Istopes and Radiation in Parasitology.* Vol. I pp. 1–11. IAEA, Vienna (1968).
93. Varga, I. and Vetesi, F. immunisation of chickens with irradiated larvae of *Syngamus trachea.* In: *Isotopes and Radiation in Parasitology.* Vol. II, pp 52–62. IAEA, Vienna (1970).
94. Zeigler, K.: Vaccination of chickens with X-irradiated embryonated eggs of *Capillaria obsignata* (Madsen) 1945. Acta Vet. Brno 44: 115–122 (1975).

14. Chemotherapy of Helminths of Livestock and Man

A.J. Probert
School of Biological Sciences, University College of North Wales
Deiniol Road, Bangor, Gwynedd, UK

INTRODUCTION

Helminths of Livestock

Infection with helminths is an important factor in the health and productivity of livestock throughout the world. Intensive systems of management often lead to high levels of infestation particularly in grazing animals. Direct losses result either from mortalities or from reduced productivity [20, 26, 64, 65] caused by both clinical and subclinical parasitism. The major groups of parasites concerned are the gastrointestinal nematodes, *Fasciola* spp. and Paramphistomes of ruminants. A number of cestodes are also important either through their effects on the health and productivity of livestock or through their effects on human health.

Control can be achieved in a number of ways. Cultural methods include practices such as controlling stocking rates, rotational grazing, and the provision of clean grazing [64, 65]. Vaccination does not play a major role in helminth control with the exception of the use of Dictol for the control of the cattle lungworm *Dictyocaulus viviparus* throughout the world and the sheep lungworm *D. filaria* in northern India [76]. Control of intermediate hosts, with the exception of schistosomes, and/or reservoir hosts also plays a minor role in containing the parasite. However, chemotherapy remains the most widely used method of control where the economy will withstand the expense of the use of often costly drug treatment. Indeed, there is a bewildering choice of anthelmintics now available to the livestock producer.

Because animals are often infected with a wide range of helminths and the need for broad spectrum compounds active against trematodes, cestodes and nematodes and their larval stages is obvious. The appearance in 1961 of thiabendazole designed as a nematicide marked the beginning of the era of such compounds [28]. A whole series of benzimidazole analogues have since followed (parbendazole, 1967, cambendazole 1970, mebendazole 1971, oxibendazole 1973, fenbendazole 1974, oxfendazole 1975, albendazole 1976 and triclabendazole 1981 [79]) which have widened the spectrum of activity to include lungworms, flukes and tapeworms. Subsequently, the discovery of the avermectins in 1976 and with the commercial production of ivermectin has increased the spectrum of activity still further to include ectoparasitic arthropods.

The vast majority of anthelmintics are produced empirically with little attempt at design. This is because, our knowledge of the physiological and biochemical differences between parasites and their hosts is too meagre to exploit these differences. However, information continues to accumulate and the time when design will become a reality, is not too distant. Efforts to improve existing drugs by reducing their toxic side effects and increasing their spectrum of activity continue through the production of new analogues. The search for new drugs also continues although, in view of the increasing cost of development, now at a somewhat less frenetic pace. This is aided by the use of computerised modelling techniques to search for promising new lead compounds [39]. The current trend is to improve the existing drugs by novel methods of delivery such as in feed, slow release, pulse release and intraruminal delivery systems [88].

Helminths of Man

In many developing countries of the world particularly those of the tropics and sub tropics, human helminthiasis is very common. Schistosomes, hookworms, gastrointestinal worms, filarial nematodes and cestodes are important for health and productivity of man. Traditional methods of control include improved sanitation, the provision of clean water supplies and even changes in eating habits. While there is now a plethora of drugs available for the control of gastrointestinal nematodes, many of which are similar to those used against livestock parasites, relatively few drugs are available against filarial nematodes. Good drugs are also available against schistosomes (see later).

In the review that follows the range of drugs available for the control of human and livestock helminths, their efficacy, methods of delivery and the problems of resistance are briefly presented.

NEMATODES OF LIVESTOCK

Strategies Involved in the Chemical Control of Gastrointestinal Nematodes of Livestock

Two strategies are generally followed in the control of roundworms of ruminants: Suppressive and Non Suppressive treatment. In the former, the aim is to treat frequently to prevent the worms from becoming patent thereby preventing transmission and reducing mortality and morbidity. It is mostly used in animals in their first grazing season. However, the continuous use of drug therapy runs the risk of promoting the development of drug resistance. Whitelaw & Fawcett [87] have recommended a suppressive treatment using rafoxanide for the control of *Fasciola hepatica* in sheep.

Non suppressive treatment involves the strategic use of anthelmintics to reduce production losses and to treat animals showing clinical symptoms of parasitism.

In the control of gastrointestinal parasites of sheep the main aim is to protect the young animal since lambs usually acquire their infections from the ewe soon after birth. The ewe produces huge numbers of nematode eggs in the periparturient period, the so called 'spring rise' phenomenon. Protection of the lamb is, therefore, achieved by dosing the ewe from four weeks prior and eight weeks after lambing followed by movement of the lambs to safe pasture. If safe pasture is not available then regular dosing of young animals is needed.

Calves also need protection with anthelmintics during their first grazing season at the end of which they are likely to have acquired sufficient immunity to withstand attacks in their second and subsequent years. However, it may be necessary to dose mature animals at strategic intervals.

In case of both sheep and cattle scouring may occur due to the development of large numbers of arrested stomach worm larvae. In Britain this occurs in winter and early spring. Treatment with compounds with low efficacy against arrested larvae will only give temporary respite and the problem is likely to recur in which case dosing with a broad spectrum drug is recommended. A similar strategy to that of sheep is adopted for pigs with treatment of the sow prior to farrowing and again eight weeks later. This gives good protection to suckling piglets from attack by gastrointestinal nematodes and lungworms. Infection is usually highest in sows particularly when kept on grass or on solid floors and where animals are allowed to feed from the ground rather than from troughs [79].

Horses at grass are usually treated with a broad spectrum benzimidazole anthelmintic at 4 to 6 week intervals. Westcott [86] comments that the standard control programme in the U.S.A. involves dosing at

8 week intervals using drugs with good activity against *Parascaris equorum* in young horses and including one treatment against *Gastrophilus* spp. Using this strategy, problems are minimised but failures due to benzimidazole resistance [24] and because of the poor efficacy of most drugs against the cyathostomes are known. Herd et al [41] have suggested that better control can be achieved by shortening the interval of treatment during periods of the year when high risk of 'infestation' is known and lengthening the interval at others. In this way better control can be achieved using the same number of doses per year.

Routes of Administration

Most modern anthelmintics have high margins of safety with therapeutic indices greater than three. Therefore, the chances of overdosing are slight although care may be needed when treating pregnant or debilitated animals. The dose rate is calculated on the basis of mgs per kg body weight so the most likely problem is underdosing, if the estimated weight of the animal is too low. Anthelmintics used against gastrointestinal nematodes are given by parenteral, oral, topical or intraruminal routes. Oral formulation involves the use of tablets, gels, pastes, drenches or granules and powders for inclusion in-feed or water.

Accuracy of dose is best achieved by oral or parenteral administration because in-feed or in-water applications are notoriously inaccurate since the greedy animals consume more than the others. It is also important to realise that infection often suppresses appetite resulting in low intake of the drug by infected animals. Abbott et al [1] have shown that in moderate to heavy trichostrongylid 'infestations' of sheep food intake may be lowered by as much as twenty percent. Incorporation of drugs into feed blocks e.g. fenbendazole for sheep, is also practised but again dose control becomes impossible. It is important, therefore, to watch for reluctant feeders and to dose these individually. If in-feed medication is used thorough mixing with the food is necessary and powdered formulations should be avoided when pelletted feeds are used [79].

In-feed methods also run the risk of promoting the development of resistance since overdosing will lead to high selective pressure on resistance genes. In spite of these drawbacks, since handling and dosing are expensive on time and labour, in-feed preparations will continue to be attractive to the farmer.

Oral drenching of ruminants carries with it the risk that portions of the drench will bypass the rumen via the oesophageal reticular groove and hence pass directly to the omasum/abomasum. In some animals this oesophageal groove reflex persists after weaning and will occur in response to drenching.

To circumvent this problem a special gun is now available in some countries e.g. Australia [12] and Britain (Coopers Autoworm Injector) which delivers oxfendazole through the skin and directly into the rumen. This allows the treatment of large numbers of cattle over a short period of time.

Only levamisole and ivermectin are currently available as pour-on topical applications. Levamisole is used for the treatment of gastrointestinal nematodes of cattle. However, it does not give as good a protection as that of the subcutaneous or oral routes. Furthermore, the uptake and hence its efficacy varies considerably depending on the climatic conditions at the time of administration with lower absorption at low temperatures [31].

Slow and Pulse Release Methods

Because of the high labour costs involved in handling animals several slow release methods of delivering anthelmintics within the reticulo-rumen are in use in cattle while others are being developed. These are either based on the principle of a slow steady release of drug, which may not kill all the parasites, or the pulse release of a series of therapeutic doses which completely protect the animal. Bogan & Armour [12] and Zimmerman & Hoberg [87] have reviewed the technology of slow release and pulse release systems of delivering anthelmintics.

The slow release bolus (Paratect, Pfizer, [3]) delivers morantel tartrate when lodged in the reticulo rumen of cattle and gives protection for approximately 90 days. The drug is released by chemical diffusion through polyethylene discs. A similar device (Chronomintic) is also available for delivery of laevamisole in the reticulo-rumen of calves. Another continuous release bolus is the ivermectin slow release bolus (SRB, Merk & Co) which is based on the principle of an osmotic pump which gives a continuous release of ivermectin for 90–120 days in cattle.

A recent development in pulse release preparations is the oxfendazole pulse release bolus (Repidose, Autoworm, Coopers Animal Health) which is designed to release five 750 mg doses of oxfendazole individually at intervals of approximately three weeks in calves during their first grazing season. The bolus is 9 cm long and approximately 136 gm in weight with a corrodible core of magnesium alloy attached to a mild steel end weight of sufficient density to prevent regurgitation from the reticulum rumen. It is designed to be administered to calves 100–200 kg in weight at the time of spring turn out onto pasture. Herbert & Probert [40] demonstrated that it gives excellent protection against gastrointestinal nematodes resulting in improved live weight gains (see also Mitchell [60]) but for the advantage to be maintained in the second grazing year it is advisable in the second year to treat with conventional methods of drug application since treatment in the first year leaves the animals with little acquired immunity.

The most sophisticated pulse release bolue is the E bolus (Smith, Kline & French) which is an electronic device delivering three doses of albendazole at 31 day intervals to cattle. Contact with the rumenal fluid activates a timing device run by alkaline batteries which after the 31 day interval generates a gas (mainly Carbon Dioxide) which drives out the dose of anthelmintic [37].

Some metal boluses have, however, produced problems during meat processing by damaging the cutting and grinding equipment used in the preparation of offal. Attempts to find biodegradeable alternatives heavy enough to lodge the bolus in the rumen are being undertaken.

The use of slow release devices has the additional advantage of cleaning up the pasture since calves acquire their infection at spring turn out from larvae which have overwintered from the previous year. This advantage is only maintained if the animals continue to graze the same pasture. They will not be protected if they are moved to contaminated pasture in the second half of the grazing season. Prior to housing following the end of the grazing season, and to prevent possible outbreaks of ostertagiosis in late winter/early spring (in Britain), animals should be dosed with a compound effective against arrested larvae.

Future research may see the production of pulse release systems which can be externally started by radio transmitters. The technology is currently available to achieve this but whether they would be cost effective is doubtful.

The above discussion is directed towards slow and pulse release methods for the control of helminths of cattle. However, if costs could be kept low they would be particularly attractive in sheep. A continuous release device [50] delivering a continuous and steady release of oxfendazole within the rumen of sheep has been shown by Anderson et al [2] to give good protection against helminths in grazing animals.

The effect of slow release formulations on the development of resistance (see further) has yet to be assessed but it seems likely that it will be enhanced by their use. On the other hand, the use of Paratect for more than six years has yet to be associated with reports of resistance. Donald [22] suggested that these devices may not be so hazardous in this respect provided the total release time is less than the maximum life span of the parasite's free living stages, there is high kill rate and the release rate declines rapidly to zero at the end of its life. The device is to be infrequently used.

Drugs Currently Used Against Nematodes of Livestock

For a detailed list of the available drugs see Marriner and Armour [56], Blair & Klei [9], Taylor [79] and for their modes of action Rew & Fetterer [70]. The drugs available for the control of sheep and cattle nematodes have been summarised in Tables 1 and 2.

Table 1. Anthelmintics available for the control of nematodes in sheep

Chemical	Trade name	Gastrointestinal nematodes			Lung worm	Dose (mg/kg)
		Adults	Larvae	Inhib larvae		
Thiabendazole	Thibenzole	+	+	+/–	+/–	44
Parbendazole	Topclip: Helmatac	+	+	–	–	20
Mebendazole	Ovithelmin	+	+	–	+	15
Oxibendazole	Anthelworm	+	+	–	–	10
Cambendazole	Bonlam	+	+	–	–	20
Fenbendazole	Panacur	+	+	+	+	5
Oxfendazole	Systemex	+	+	+	+	5
Albendazole	Valbazan	+	+	+	+	5
Febantel	Bayverm	+	+	+	+	5
Thiophanate	Nemafax	+	+	+/–	+/–	50
Levamisole	Nilverm	+	+	+	+	7.5
Ivermectin	Ivomec	+	+	+	+	0.2
Morantel tartrate	Morantel	+	+	?	?	10

+ = effective; +/– = variable activity; – = ineffective

Mebendazole, Febantel, Fenbendazole, Oxfendazole and Albendazole are also active against tapeworms. Ivermectin is active against ectoparasites.

Based on Farmers Weekly Supplement Animal Health [4] and Marriner & Armour [56].

Table 2. Anthelmintics available for the control of nematodes in cattle

Chemical	Trade name	Gastrointestinal nematodes			Lung worm	Dose (mg/kg)
		Adults	Dev. larvae	Inhib. larvae		
Thiabendazole	Thibenzole	+	+	+/–	+/–	66–110
Parbendazole	Topclip	+	+	–	–	20–30
Oxibendazole	Anthelworm	+	+	–	–	10
Cambendazole	Novazole	+	+	–	–	25
Fenbendazole	Panacur	+	+	+	+	7.5
Oxfendazole	Systemex	+	+	+	+	4.5
Albendazole	Valbazan	+	+	+	+	7.5
Febantel	Bayverm	+	+	+/–	+	7.5
Thiophanate	Nemafaz	+	+	+/–	+/–	66–132
Levamisole	Nilverm	+	+	–	+	7.5
Ivermectin	Ivomec	+	+	+	+	0.2
Morantel*	Paratect	+	+	+	–	90/day for 60 days

Febantel, Fenbendazole, Oxfendazole and Albendazole are also active against tapeworms while Albendazole shows variable activity against liverfluke and Ivermectin shows activity against ectoparasites.

+ = effective; +/– = variable activity; – = inactive.

* Used as a slow release bolus.

Based on Farmers Weekly Supplement [4] and Marriner & Armour [56].

Broad Spectrum Anthelmintics

BENZIMIDAZOLES

A range of benzimidazoles is currently available for treatment of ruminants infected with gastrointestinal helminths of ruminants, pigs, and equines. These include thiabendazole, cambendazole, oxibendazole, mebendazole, flubendazole fenbendazole, albendazole and oxfendazole. Febantel and thiophanate are often included in this list since their activity resides in their metabolites which are benzimidazoles. They are, therefore, often called probenzimidazoles. The mammalian toxicity of the benzimidazoles is very low but cambendazole, oxfendazole, mebendazole, parbendazole and albendazole are teratogenic and are not recommended for use in pregnant animals. At the recommended dose rates oxfendazole, fenbendazole, albendazole and febantel are also effective against arrested larvae [55]. A wide range of formulations are available including slow release preparations.

IMIDAZOTHIAZOLES

This group contains tetramisole and levamisole. The former is a racemic mixture of the D and L forms but since the activity resides in the laevo isomer levamisole has replaced tetramisole thereby allowing half the dose rate to be used thus reducing its toxic side effects. It is restricted to use against gastro intestinal and lung nematodes of ruminants, pigs and poultry and is not recommended for use in dogs and cats because of its toxicity. It is also active against arrested nematode larvae in sheep. A variety of formulations are available including pour-on subcutaneous injections, drenches and in-feed preparations. It is safe for use in pregnant animals and is effective against nematodes of poultry and, with the exception of *Trichuris suis*, the nematode of pigs.

TETRAHYDROPYRIMIDINES

Pyrantel and its analogue morantel are included in this group. Both are active against adult nematodes but less so against the immature stages. Morantel is available as a slow release bolus for cattle and as a paste or in-feed preparation for horses.

ORGANOPHOSPHATES

This group contains some of the older anthelmintics but in view of their low therapeutic index and their toxic side effects on the nervous system of mammals are now largely obsolete. Haloxon, dichlorvos, coumaphos and trichlorphon are occasionally used against cattle and sheep nematodes and dichlorvos against porcine nematodes. They lack the broad spectrum activity of the benzimidazoles and are now almost restricted to use against ectoparasites. Trichlorphon at a concentration of 6–10% in petroleum jelly plus 4% sulphonamide is used for the control of *Stephanofilaria assamensis* which is the cause of 'hump sore' in cattle and buffaloes in India giving complete cure within 20–66 days [77].

AVERMECTINS

Ivermectin is the only member of this group so far developed as an anthelmintic. It shows broad spectrum activity against adult and larval nematodes at exceptionally low dose rates. Because of its long withdrawal time of 21–28 days, it is not used in lactating cows even though its mammalian toxicity is very low. It also shows activity against warbles, lice and mange mites. It is administered by subcutaneous injection to cattle and pigs and orally to sheep. Because of its insecticidal properties and persistence it slows down the biodegredation of faecal pats by insects.

NITROSCANATE
This compound is active against canine hookworms and ascarids but not *Trichuris vulpis* and is used in the treatment of gastrointestinal nematodes and tapeworms of dogs (see later).

Narrow Spectrum Anthelmintics

PIPERAZINE
This drug is available as the citrate, adipate, tartrate and calcium edetate. It is effective against ascarids of cattle, buffaloes, pigs, horses, dogs, cats and poultry. It is sometimes combined with thiabendazole to increase its spectrum of activity to include ascarids. Migratory stages, however, are not affected by it.

CLOSANTEL
This salicylanilide binds to plasma proteins and is, therefore, effective against blood feeding nematodes such as *Haemonchus contortus* in sheep and large strongyles including the migratory stages of *Strongylus vulgaris* of horses. It is also an effective flukicide (see later).

DISOPHENOL
This is an effective chemical against mature *H. contortus* and gives protection for thirty to sixty days when administered subcutaneously. It shows good activity against dog and cat hookworms.

Reasons for Anthelmintic Failure

The above account indicates that there is a wide choice of excellent compounds available for the treatment of helminths of livestock yet failures in control occur. The reasons for this are:

(a) The choice of the wrong drug, for example, the use of compounds with narrow specificity or poor activity against juvenile and arrested stages.
(b) Infrequency of dosing, particularly, where animals are kept on the same paddock for long periods.
(c) Incorrect calculation of dose or incorrect amounts delivered due to faulty equipment or uneven suspension of the active ingredient. In self-administered application in licks or feeds low dosage can also lead to failure.
(d) Reintroduction of dosed animals to contaminated pastures.
(e) Anthelmintic resistance.

Michel et al [59] commented that the value of most anthelmintic treatment of cattle in Britain was doubtful since often dosages, choice of drug and the timing of application were faulty.

NEMATODES OF MAN

Gastrointestinal Nematodes

Intestinal nematodes present high prevalence rate in man in many tropical and developing countries. Not only do they contribute to ill health but can also have a marked effect on the work efficiency of those people infected. Most infestations are associated with poor living standards, particularly, low standards of hygiene, sanitation and nutritional levels. Chemotherapy is the principal method of control.

The modern anthelmintics show good activity against the gastrointestinal nematodes, low toxic side effects and are efficient in a single oral dose. However, a single broad spectrum nematicide able to control all human gastrointestinal nematodes has yet to be discovered. Botero [13] has reviewed the drugs available for the treatment of the gastrointestinal nematodes of humans (see Table 3).

Table 3. Anthelmintics effective against nematodes of man

Chemical	Trade name	Gastrointestinal worms*	Filarial worms	Dose rate (mg/kg)
Albendazole	Valbazan	+	–	400–800
Mebendazole	Vermox	+	+/–	100/day for 3 days
Pyrantel pamoate	Antiminth Helmex	+	–	10/day for 3 days
Diethylcarbamizine	Filazine Franocide	–	+	50–72
Levamisole/ Mebendazole	?	+	+	(needs long treatment)
Ivermectin	Mectizan	?	+	?

* *Ascaris lumbricoides, Enterobius vermicularis, Necator americanus, Ancylostoma duodenale.*
The drug of choice against Strongyloidiasis is Thiabendazole at a dose of 25–50 mg/kg/d for 5–8 days.
Various side effects are seen with the use of DEC against different species of filarial nematodes.
\+ = effective; +/– = variable activity; – = ineffective.

Pyrantel pamoate, albendazole and mebendazole are all effective and the drugs of choice against *A. lumbricoides* and *E. vermicularis.* Community treatment is recommended since reinfection is likely to occur if the whole community is not treated. Piperazine is an alternative drug since it is inexpensive. However, toxic side effects are seen and it is not recommended for patients with a history of epilepsy, or renal or hepatic malfunction.

Iron deficiency anaemia is associated with the hookworms *N. americanus* and *A. duodenale* and even light infections lead to lowered levels of haemoglobin in ill nourished patients. Pyrantel pamoate, albendazole and mebendazole are all efficient drugs.

For *Strongyloides stercoralis* the drug of choice is thiabendazole but for disseminated infections no reliable chemical is available at present. Ivermectin has been found to be effective against migratory larval stages in mice [35].

Trichuris trichuria is an important parasite of the colon of undernourished children but it responds to treatment with mebendazole or oxantel pamoate (an analogue of pyrantel). The most important extraintestinal nematodes of man are the filarial worms. Diethylcarbamazine (DEC) has been widely used against *Wuchereria bancrofti* and *Brugia* spp. [21]. It is microfilaricidal and produces few side-effects against *W. bancrofti* but severe side effects against *Brugia* spp. [67]. When given in a full course, DEC is also effective against adult *W. bancrofti.* It is also recommended for the treatment of *Loa loa* but care is needed when treating patients with heavy microfilarial counts which may cause meningoencephalitis [21]. Few other drugs have been found to be effective against lymphatic filariasis. Mebendazole used in conjunction with levamisole appears to be active but needs to be used at high concentrations over a prolonged period to be effective [66].

Onchocerca volvulus in a particularly debilitating parasite of humans which causes appalling dermatitis and blindness yet no satisfactory drug is available. DEC is highly microfilaricidal but it induces skin lesions when the larvae are destroyed. Ivermectin has been found to be microfilaricidal and extensive mass trials involving 140,000 people in 14 African and Central American countries are planned by WHO [5]. It looks to have great promise. Merck, Sharp & Dohme have decided to donate ivermectin (Mectizan) to countries where onchocerciasis is endemic [14].

TREMATODES

The chemotherapy of trematode infections is largely directed towards the treatment and control of schistosomiasis, fascioliasis and paramphistomiasis. Since the schistosomes and species of *Fasciola* occur outside the digestive tract, the drugs active against them reach the parasites via the blood even though they may be administered by the oral route. Consequently, side effects on the host are a serious problem with many of the anthelmintics currently available. *Fasciola* species reside in the liver, which is the major detoxifying organ of mammalian hosts. Despite the close biological relationship between these two groups of trematodes few drugs, however, are effective against both genera. Furthermore, few drugs are efficient against all species of human and livestock schistosomes. For this reason the two groups are considered separately. It is also important to note that with the exception of albendazole (and then only at increased dose rates) anthelmintics effective against liver flukes are not always active against gastrointestinal worms and vice versa.

Antischistosomal Compounds

Schistosoma bovis, S. mattheei, S. nasalis, S. indicum, S. spindale and *S. japonicum* infect sheep, goats, cattle and buffaloes in India, Africa and China. While there is an extensive literature on the drugs available for treatment of human schistosomes [44, 62] relatively little information is available on the treatment of ovine and bovine schistosomiasis. Boray [11] has summarised the efficiency of antischistosomal drugs so far tried against livestock schistosomes. However, caution is necessary in their use since in heavy infections the death of parasites within the blood vascular system may lead to the formation of emboli, occlusion of the portal vessels and liver infarction. The following compounds include those used against human schistosomes and where applicable against livestock schistosomes (see also Table 4).

Table 4. Anthelmintics used against *Schistosoma* species

Chemical	Trade name	*S. mansoni*	*S. haematobium*	*S. japonicum*	Dose (mg/kg)
Oxamniquine	Mansil	+	NR	NR	15 × 2/day 3 day
Metrifonate	Bilarcil	NR	+	NR	7.5–10 × 3 over 3 weeks
Hycanthone	Etrenol	+	+	–	toxic and little used
Praziquantel	Biltricide	+	+	+	40–50 the choice drug
Niridazole	Ambilhar	+/–	+	+	25/day for 5–10 days
Amoscanate	Nithiocyanamide	+	+	?	20–40 but toxic

+ = effective; +/– = variable activity; – = ineffective
The most promising drug for livestock schistosomes is Praziquantel.
Based on Cambell & Garcia [17] and Jordan & Webbe [44].

Oxamniquine

This drug is active against *S. mansoni* and is given in capsules or as a syrup. It is inactive against the remaining species. No marked side effects are seen but the urine is coloured red or orange after use. It appears not to have been used against livestock schistosomes.

METRIFONATE
This compound is active against *S. haematobium* and is the drug of first choice against this species. Some activity is seen against *S. mansoni*. It is given in tablet from and inhibits cholinesterases. Slight toxic side effects include muscle tremor, weakness, bronchiospasms but the frequency of these effects is low. Again it is not used in animals.

HYCANTHONE
This is active only against *S. mansoni* and *S. haematobium*. Parentral dosing under medical control is needed because of the side effects, which include impaired liver function. This drug is now going out of favour because of these effects. It has been tried against *S. mattheei* in sheep where it was found to be 98–100% efficient [51].

PRAZIQUANTEL
This is a heterocyclic pyrazino-isoquinoline compound which is highly effective against all stages and also most trematodes and cestodes (see Harnett [38] for mode of action). It causes immobilization of schistosomes and structural damage to the tegument possibly leading to exposure of normally disguised antigen to the host's immune system [52]. No side effects have been reported at clinical dose rates. It is given in tablet form and is likely to become the drug of choice for treatment of all human and livestock species. Bushara et al [15] found that it was well toleratd by Sudanese cattle and that it removed 99% of *S. bovis*.

NIRIDAZOLE
Niridazole is administered in tablet form but does cause toxic side effects more so in adults. In view of the wide range of these side effects its importance is now fading. Muraleedhran et al [63] showed that it effectively removed *S. nasalis* from Indian cattle but Hurter & Potgeiter [43] doubted its value in the treatment of sheep and goats and also showed that it was toxic.

AMOSCANATE
This compound is active against all species and also hookworms. It is used extensively in China but it induces jaundice. Leitch & Probert [53] showed that it causes tegumental damage in *S. haematobium* and effects on the vitelline cells resulting in the ceasation of egg laying. It binds irreversibly to amino acid groups of proteins. This drug has been found to be effective as a 2% intravenous injection for the control of *S. japonicum* of cattle and buffaloes in China [69].

OLTIPRAZ
Oltipraz has been seen to be active in trials against *S. mansoni* and *S. haematobium* giving good cure rates. It appears to decrease glutathione levels in schistosomes. Toxic side effects have been reported and it does not look to be a promising drug in humans. Trials on livestock have not yet been reported. The older drugs such as the antimonials have been investigated for use against ruminant schistosomes but they are toxic at effective dose rates [34]. Variable results including toxicity are also seen with haloxon and trichlorophon [11].

Fasciolicides
Most of the available anthelmintics used against *Fasciola* spp. are administered orally although parenteral application is often more efficient. However for practical reasons the oral route is preferred for sheep and goats and the subcutaneous route for cattle [11]. The ideal compound against *Fasciola* spp. should show

good efficiency against both larval and adult stages. It should have a high therapeutic index and leave no residues in the tissues or milk. The drugs currently available against *Fasciola* spp. (Table 5) are described by Boray [11] and are:

OXYCLOZANIDE
It is orally administered to sheep and cattle but shows poor efficacy against flukes younger than 12–14 weeks of age.

NICLOFOLAN
This compound is orally administered to both sheep and cattle but shows poor efficacy against flukes younger than 12 weeks of age.

NITROXYNIL
When subcutaneously administered to sheep and cattle adult worms are killed but it is not effective in removing flukes below 8–10 weeks of age. It has been used for the successful treatment of the cattle filarial nematode *Parafilaria bovicola* in South Africa [85].

BROTIANIDE
Oral administration to sheep is effective against adults but poor efficacy is seen against flukes below 12 weeks of age.

RAFOXANIDE
This is orally administered to both sheep and cattle and also subcutaneously to cattle. It is effective down to flukes 6 weeks of age in sheep and 12 weeks of age in cattle.

DIAMPHENETIDE
It is effective against flukes 1 day to 7 weeks of age when orally administered to sheep. Since acute fascioliasis is not a problem in cattle, it is not used.

CLOSANTEL
When orally administered to sheep and subcutaneously to cattle it is not effective against flukes younger than 6–8 weeks in sheep and 12 weeks in cattle.

ALBENDAZOLE
This can be administered orally to both sheep and cattle but is effective only against flukes older than 12–17 weeks of age. Its advantage lies in the fact that it is also a broad spectrum nematicide.

TRICLABENDAZOLE
This is a very efficient drug when orally administered to sheep and cattle and is effective against flukes 1 week and older. It is thus the drug of choice against fascioliasis.

CLORSULON
This compound shows poor efficacy against flukes below 6–8 weeks of age when orally administered.

Anthelmintic treatment is the chief method of controlling *F. gigantica* in most endemic areas in the tropics [26] and it is advisable to treat cattle and sheep at the start of the dry season and at the beginning of the wet season [27]. The drugs available for treatment of *F. hepatica* given above are equally of use in the control of *F. gigantica*. Early immature *F. gigantica* are less susceptible than those of *F. hepatica* to treatment at comparable dose rates but adults are more susceptible [42]. Bithionol given on alternate days for 2–3 weeks is believed to be effective against *Fasciola* spp. in man [17].

Drugs Against Paramphistomes

Paramphistomiasis is a serious problem in many tropical and subtropical regions with substantial losses seen in cattle, sheep and goats [8, 75]. It is difficult to diagnose since clinical disease precedes the appearance of eggs in the faeces although immature flukes may be seen in the faeces in acute infections. Boray [11] has listed the drugs currently available for treatment together with their efficiencies (Table 5).

Table 5. Anthelmintics used against *Fasciola* spp. and Paramphistomes of sheep and cattle

Chemical	Trade name	*Fasciola* spp.			Dose	Paramphistomes		Dose
		Adults 10–12 w	Imm 6–10 w	Imm > 6 w	(mg/kg)	Adults	Immatures	(mg/kg)
Oxyclozanide	Zanil	+	–	–	15 s	+	+	15
					13–15 c			
Nitroxynil	Trodax	+	+/–	–	10	NR	NR	
Rafoxanide	Flukanide	+	+/–	–	7.5	–	+*	15
Albendazole	Valbazan	+	–	–	4.75 s	NR	NR	
					10 c			
Diamphenetide	Coriban*	+/–	+	+	80–120	NR	NR	
Brotianide	Dirian	+	–	–	5.6*	+	+	15
Triclabendazole	Fasinex	+	+	+	5–10 s	NR	NR	
						12 c		
Clorsulon	—	+	+/–	–	20 s	NR	NR	
					7 c			
Closantel	Flukiver	+	+/–	–	7.5–10*	–	+	10*
Niclosamide	Mansonil	NR	NR	NR	—	+/–	+	50
Bithionol	Bitin S	+	–	–	75 s	+	+	100–200
					30 c			
Resorantel	Terenol	NR	NR	NR	—	+	+	65

Bithionol is toxic and only recommended where losses are likely and when no other drug is available.
NR = not recommended; * = Not used for cattle; s = sheep, c = cattle, Imm=Immature flukes.
+ = effective; +/ – = variable activity; – = ineffective
Based on Marriner & Armour [56] and Boray [11].

Bithionol, although effective against immature and mature flukes, is toxic at the effective dose rate. It is only recommended for use where no other drugs are available and where losses are likely to occur. Niclosamide is the safest and most effective drug against acute paramphistomiasis of sheep but unfortunately is rather erratic in its efficiency against immature flukes of cattle. It shows little effect against adult worms of both hosts. Niclofolan is effective only against immature paramphistomes of sheep. Oxyclozanide either alone or with levamisole shows reasonable efficiency (60–90%) against immature flukes of sheep and cattle and 73–90% efficacy against adults of sheep and cattle. Its use lies in the fact that it is also an efficient compound against *Fasciola gigantica*. Following treatment with any of these anthelmintics against paramphistomes, animals should be moved to safe pasture if they are not to become reinfected. Young animals in particular should be kept away from habitats where the snail hosts are prevalent.

Drugs Against Other Trematodes

Several additional species of trematodes cause problems in tropical and subtropical regions of the world. *Fasciolopsis buski* occurs in the small intestine of man and pigs in Asia, particularly, India and China. Hexylresorcinol is effective while tetrachlorethylene reduces egg output by 97% and produces cure rates of 77% in humans [77]. Paragonimiasis is also an important disease of man, pig, dog, cat, goat and cattle in parts of Asia and Africa. Albendazole (50–100 mg/kg/d) for 14–21 days reduced the egg output in cats while Bithionol (100 mg/kg/d) for 7 days is also effective as is Niclofolan (1 mg/kg/d) for 3 days [77].

Flukes of the family Opisthorchiidae, such as *Clonorchis sinensis* and *Opisthorchis viverrini* are pathogenic flukes of man, cats and dogs in South East Asia. No entirely satisfactory drugs are available. In experimental tests hexachlorophene (15 doses of 20 mg/kg) caused 100% removal of *C. sinensis* from the livers of rabbits while disophenol, clioxanide and dithiazanine were also 99% effective. However, high doses of these drugs in man may lead to toxic side effects. Niclofolan (1–2 mg/kg) for 2–3 days and praziquantel (1 dose of 50 mg/kg) both look promising [77]. Indeed, Kumar [48] considers that praziquantel is a 'panacea' for the treatment of zoonotic trematode infections (such as schistosomiasis, clonorchiasis, paragonimiasis, fasciolopsiasis and metagonimiasis) in South East and Far East Asian countries where cure rates of 81–100% have been demonstrated in man.

The use of human faeces in pisciculture increases the transmission of these species and ammonium sulphate treatment of the faeces is recommended, if it is to be used as a fertilizer in fish ponds.

CESTODES

Drugs Against Cestodes of Mammals

Adult tapeworms are found in sheep and to a lesser extent in cattle, throughout the world; the most widespread species being *Moniezia expansa*, *M. benedeni* and *Thysaniezia giardi*. None of these species are particularly pathogenic in the adult stage but various clinical symptoms have been reported. They are most common in lambs and kids which become infected by ingesting the cysticercoid stage in oribatid mites. Since the segments are easily observed in the faeces, they are often associated with unthriftiness in these animals when it is much more likely that this is the result of nematodes [7]. Larval cestode infections of ruminants are more important either as a human health hazard, when infected meat is consumed, or as a direct cause of ill health as in the case of coenuriasis caused by *Taenia multiceps*, hydatidosis by *Echinococcus* spp. or cysticercosis by a wide range of taeniid tapeworms.

Albendazole, bunamidine hydroxynaphthoate, fenbendazole, niclosamide, oxfendazole and praziquantel are all used against the adult worms in ruminants. For details of the chemistry of anticestode agents refer to Andrews and Bonse [6] and for their modes of action to Van den Bossche [80]. The control of the larval stages is best approached by worming dogs where they act as final hosts, or by sanitation which prevents access to the eggs of the tapeworms.

Some of the more common compounds used against cestodes of cats and dogs (Table 6) are:

Arecoline Hydrobromide

This alkaloid has been in use for more than one hundred years in both human and veterinary medicine and was for many years the only drug effective against *E. granulosus* [71]. However, in view of its low efficacy and its side effects it is now seldom used. It has a paralysing effect on tapeworms and a powerful purgative action. It is not recommended for cats and pregnant bitches or pups.

Table 6. Anthelmintics used against cestodes of dogs, cats and sheep

Chemical	Trade name	Dose (mg/kg)	*Taenia* spp.	*D. caninum*	*Spirometra* spp.	*E. granulosus*	*Mesocestoides* spp.	*Moniezia / Thysaniezia* spp.	Dose (mg/kg) sheep
Bunamidine	Scoloban	50	+	+	+	+	+	+	50
Fenbendazole	Panacur	50/3 d	+	+	–	–	–	+	5
Mebendazole	Cvithelmin	10/3 d	+/–	+/–	–	–	–	+/–	
Niclosamide	Yomesan	100	+	+/–	–	–	–	+	50
Nitroscanate	Lopatol	50	+	+	–	–	–	NR	
Praziquantel	Droncit	5	+	+	–	+	+	+	2.5
Humans		Dose (mg/kg)	*T. solium*	*T. saginata*	*D. latum*	*H. nana*			
Dichlorophen	Dicestal	60–100	+	+	+	+/–			
Niclosamide	Yomesan	2 × 0.5 g	+	+	+	+			
Praziquantel	Cesticide	10–25	+	+	+	+			

Albendazole, Mebendazole, Flubendazole and Fenbendazole show activity against hydatidosis of humans.
+ = effective; +/– = variable activity; – = ineffective.
Based on Arundel [7] and Webbe [84].

BUNAMIDINE HYDROCHLORIDE

It is effective against *Taenia* spp. *Dipylidium caninum* and *Echinococcus* spp. when given as a single dose. Little toxicity is seen except occasional diarrhoea and vomiting although in field use 1 in 2,000 dogs collapsed and died [33]. It is irritating to the mucous membrane and is thus coated in lactose. It was for many years the drug of choice against *Echinococcus* spp. until the advent of praziquantel.

NICLOSAMIDE

This is an effective and safe drug against *Taenia* spp. of dogs and cats and is also safe in pregnant and lactating animals. It is not effective against *D. caninum* [73] or *E. granulosus* [32]. Although still in use in some countries but being replaced by praziquantel.

BENZIMIDAZOLES

Although developed as nematicides, mebendazole and fenbendazole show activity against cestodes and in view of their low toxicity, are widely used against tapeworms of cats and dogs. Albendazole is also used for the treatment of hydatid cysts in man (Nelson 1987 pers com.) and [84].

NITROSCANATE

This is a broad spectrum compound effective against canine roundworms (including hookworms), *Taenia* spp. and *D. caninum.* It is not totally effective against *Echinococcus* spp. Vomiting is seen but this can be minimised by administration in an empty stomach.

PRAZIQUANTEL

This drug was introduced in 1975 and is safe, highly specific compound active in a single dose against a wide range of cestodes including *Echinococcus* spp. It is also used for the treatment of human cysticercosis although expensive ($65/dose) [29, 49].

Three species of cestodes commonly found in horses throughout the world are: *Anoplocephala perfoliata, A. magna* and *Paranoplocephala mamillana.* Clinical signs of disease are uncommon but unthriftiness and diarrhoea are seen when oribatid mites which act as intermediate hosts, are abundant on the herbage. Niclosamide shows high efficacy and is safe. Pyrantel pamoate and its analogue morantel are also effective.

RESISTANCE TO ANTHELMINTICS

Drug resistance occurs when "there is a greater frequency of individuals within a population able to tolerate doses of a compound than in a normal population of the same species. Resistance is a heritable character" (Prichard et al [68]; see Chapters 5 and 15). Anthelmintic resistance is largely restricted to animal parasitic nematodes although it has been reported from *Fasciola hepatica* by Dorsman [23] and from *Schistosoma mansoni* by Jordan & Webbe [44]. The subject of resistance to chemical treatment by nematodes has been extensively reviewed by Waller [82] and Waller & Prichard [83].

Genes for resistance are known to occur in majority of the important nematode species of sheep and goats [82] and also the cyathostome or small strongyles of horses [24]. The prevalence and geographical distribution of this resistance can be correlated with those areas of the world where frequency of application of anthelmintics is high. It seems likely that in view of the genetic capacity of nematodes to counteract the effects of drug treatment, resistance will become more widespread.

At present anthelmintic resistance is largely confined to the gastrointestinal nematodes of sheep, goats and horses. Unlike insecticidal resistance, anthelmintic resistance has been slow to develop. This is probably because nematodes are less mobile than insects. Furthermore, resistance genes will only spread when animals containing resistant populations are moved to new areas. Most animals are likely to be slaughtered before resistance has had the opportunity to pass to new areas.

In Australia, where resistance is most evident, there is a direct correlation between the presence of resistance and the areas where there is a high frequency of drenching against nematodes with the same drug or at least the same group of drugs. The development of resistance will be slowed down by the small percentage of the population of nematodes which are subjected to drug pressure since a large proportion of the population will be free living.

Le Jambre [4] estimated that only three percent of the total population of *Haemonchus contortus* are present in sheep at any one time, the remainder being in refugia (i.e. free living). In addition only a proportion of this three percent will be under drug pressure since all the animals are not likely to be treated by the same chemical. Furthermore, since most of the broad spectrum anthelmintics lack persistence thereby again few of the parasitic members of the population will be under selective pressure.

The detection of resistance is, however, difficult since it is often impossible to detect the failure of drug treatment unless acute disease is a feature of the infestation. Most resistance is therefore, likely to go unrecognised. In this context Sutherland et al [78] described a method for detecting benzimidazole resistant trichostrongyle nematodes based on the staining of acetylcholinesterase.

Resistance has, however, been detected in gastrointestinal nematodes of sheep and possibly cattle first with phenothiazine then with benzimidazoles and most recently with levamisole/morantel treatment. Resistance by the cyathostome nematodes of horses to phenothiazine and the benzimidazoles has also been reported. Clear cut examples of resistant nematodes of cattle are lacking although low efficacy of levamisole and thiabendazole against *Ostertagia ostertagi* have been reported.

The problem of resistance is the greatest in those countries where *H. contortus* is endemic and causes major losses in productivity. Anthelmintic resistance is most commonly found in Australia, the resistance being usually to the benzimidazoles by *H. contortus* (see Chapter 5), *Ostertagia* spp. and *Trichostrongylus* spp. Resistance to levamisole/morantel has been slower to develop but appears to be on the increase as this drug combination has been used to replace the benzimidazoles where resistance is seen to this group of anthelmintics [82].

Surveys in New Zealand have also demonstrated low levels of benzimidazole resistance by *H. contortus* and other Trichostrongyle nematodes with about twenty percent of sheep and seventy five percent of farms showing problems [46, 47].

In Britain low levels of benzimidazole resistance have been detected in *O. circumcincta* and *H. contortus* of sheep [19, 18] and the small strongyles of horses [72].

In the rest of Europe benzimidazole resistant *H. contortus* has been reported in Switzerland [45], the Netherlands [10, 25] and France [36] but it has not reached the levels seen in Australia. It is difficult to assess the importance of resistance in North America since few widespread surveys have been conducted. However, cases of phenothiazine, and benzimidazole resistance have been reported in sheep and goats [22] and horses [86]. Resistance to benzimidazoles has also been reported from Brazil [74] and South Africa [81].

To consider resistance as a problem only when control fails is clearly foolhardy [82] since once this stage is reached the resistance genes will be already at a high frequency in the nematode population. Martin [57, 58] has suggested that the control of resistance can be achieved by moderation, saturation or multiple attack. Moderation involves low dose rates and infrequent dosing therby ensuring that susceptible genes are kept in the population. Saturation involves high dose rates lethal to heterozygotes, a strategy which is thought to delay the onset of resistance. However, where resistance has already appeared and the genes are in the homozygous state, saturation will speed up the development of resistance. Multiple attack consists of the utilisation of either mixtures of anthelmintics such as levamisole/morantel or rotation of anthelmintics with different modes of action. The expectation is that any parasites surviving one drug are unlikely to be also fit to survive the other provided, of course, that the drugs are structurally unrelated or have entirely different modes of action.

The comments above are clearly directed towards nematodes and other helminths of livestock and the situation in plant parasitic nematodes is quite different. Plant parasitic nematodes are largely controlled by soil fumigants which act rapidly and are unspecific cell poisons. This might explain why resistance has been rarely encountered in the field or in the laboratory [16]. Foot [30], however, has reported resistance to nematicides by plant nematodes in New Zealand.

CONCLUSIONS

The control of helminth parasites is central to the improved health of man and his livestock and the increased productivity of the latter. At present, chemotherapy is the dominant means of control and this is likely to continue into the foreseeable future. However, in view of the difficulties and high production cost, it seems unlikely that new compounds will appear at the same rate as they have been in the past. Consequently, the current trend is to investigate novel methods of delivery of existing anthelmintics. Slow and pulse release methods, which involve considerable savings in time and labour, are currently in use against gastrointestinal nematodes of cattle and these methods are likely to be extended for use in other animals including sheep [2] and perhaps man.

In order to conserve the armoury of available drugs attempts to control or at least slow down the development of resistance are necessary. At present drug resistance is largely confined to the gastrointestinal nematodes of sheep, goats, cattle and horses. This is a reflection of the intensity of selective pressure which the high frequency of drug treatment exerts on nematodes of these animals. Indeed, it is paradoxical that the greater the efficiency of anthelminitic treatment the faster is the onset of resistance. Strategic dosing, at intervals combined with pasture management and relaxation of drug treatment can be effective in slowing down the appearance of resistance. The influence of slow and pulse release methods of drug delivery on drug resistance has yet to be determined since this approach is relatively new.

At present we rely on the availability of a wide spectrum of anthelmintics although many are structurally related. This means that where different compounds share similar modes of action cross resistance is a distinct likelihood. This danger has given impetus to the search for new drugs, with attention focussing in some cases on the search for plant products (see Mott [61] and the application of plant products as molluscicides for schistosome control) and alternative methods of control. These methods include breeding for host resistance and the exploitation of the host's immune system (Chapter 15). The genetic engineering techniques may provide some hope in the development of vaccines against helminths (see also Chapters 10, 12 and 13).

REFERENCES

1. Abbott, E.M., Parkins, J.J. and Holmes, P.H. Influence of dietary protein on parasite establishment and pathogenesis in Finn Dorset and Scottish Blackface lambs given a single moderate infection of *Haemonchus contortus*. Res. Vet. Sci. **38**: 16–13 (1985).
2. Anderson, N, Laby, R.M., Prichard, R.K. and Henessy, D. Controlled release of anthelmintic drugs a new concept for prevention of helminthiasis in sheep. Res. Vet. Sci. **29**: 333–341 (1980).
3. Anon. Special issue on Morantel sustained release bolus. Vet. Parasitol. **12**: 215–363 (1985).
4. Anon. Animal health. Farmers Weekly Suppl., January **6**: 1–23 (1984).
5. Anon. Ivermectin freely donated for onchocerciasis control. Parasitol. Today 3: 359 (1987).
6. Andrews, P. and Bonse, G. Chemistry of anticestodal agents. In: *Chemotherapy of Parasitic Diseases* (W.C. Campbell and R.S. Rew eds), pp. 447–456. Plenum Press: New York (1986).
7. Arundal, J.H. Cestode infections of domestic animals. In: *Chemotherapy of Parasitic Diseases*. (W.C. Campbell and R.S. Rew eds), pp. 479–494. Plenum Press: New York (1986).
8. Bagatko, M. Mass mixed infection with *Paramphistomum cervi* and *Fasciola gigantica* in cattle in Northern Nigeria. Medycyna Wet. **31**: 469–470 (1975).
9. Blair, L.S. and Klei, T.R. Nematode infections of domestic animals: extraintestinal infections. In: *Chemotherapy of Parasitic Diseases* (W.C. Campbell and R.S. Rew eds), pp. 307–320. Plenum Press: New York (1986).
10. Boersema, J.H., Lewing-van der Weil, P.J., and Borgsteede, F.H.M. Benzimidazole resistance in a field strain of *Haemonchus contortus* in the Netherlands. Vet. Rec. **110**: 203–204 (1982).
11. Boray, J.C. Trematode infections of domestic animals. In: *Chemotherapy of Parasitic Disease* (W.C. Campbell and R.S. Rew eds), pp. 401–426. Plenum Press: New York (1986).
12. Bogan, J. and Armour, J. Anthelmintics for ruminants. Proc. Sixth Int. Congr. Parasitol. pp. 483–491 (1986).
13. Botero, D. Nematode infections of man. Intestinal infections. In: *Chemotherapy of Parasitic Diseases* (W.C. Campbell and R.S. Rew eds), pp 267–276. Plenum Press: New York (1986).
14. Bradshaw, H. Onchocerciasis and the Mectizan donation programme. Parasitol. Today **5**: 63: (1989).
15. Bushara, H.C., Hussein, M.F., Majid, M.A. and Taylor, M.G. Effects of praziquantel and metrifonate on *Schistosoma bovis* infections in Sudanese cattle. Res. Vet. Sci. **33**: 125–126 (1982).
16. Busvine, J.R. Pest resistance to pesticides. In: *Pesticides and Human Welfare* (D.L. Gunn and J.G.R. Stevens eds), pp. 193–205. Oxford University Press: Oxford, England (1976).
17. Campbell, W.C. and Garcia, E.G. Trematode Infections of Man. In: *Chemotherapy of Parasitic Disease* (W.C. Campbell and R.S. Rew eds), pp. 385–399. Plenum Press: New York (1986).
18. Cawthorne, R.J.G. and Cheong, F.H. Prevalence of anthelmintic resistant nematodes in south east England. Vet. Rec. **114**: 562–564 (1984).

19. Cawthorne, R.J.G. and Whitehead, J.D. Isolation of benzimidazole resistant strains of *Ostertagia circumcincta* from British sheep. Vet. Rec. **112**: 274–277 (1983).
20. Dargie, J.D. The Impact on Production and Mechanisms of Pathogenesis of Trematode Infections in Cattle and Sheep. Proc. Sixth Int. Congr. Parasitol. pp. 453–463 (1986).
21. Denham, D.A. Nematode infections of man: Extraintestinal infections. In: *Chemotherapy of Parasitic Diseases* (W.C. Campbell and R.S. Rew eds), pp. 277–286. Plenum Press: New York (1986).
22. Donald, A.D. The development of anthelmintic resistance in nematodes of grazing animals. In: *Facts and Reflections IV. Resistance of Parasites to Anthelmintics* (F.H.M. Borgsteede, S.A. Henriksen and H.J. Over eds), pp. 15–28. Central Veterinary Institute: Lelystad, The Netherlands (1983).
23. Dorsman, W. Resistance of *Fasciola hepatica* towards hexachlorophene. Wiad. Parazyt. **14**: 631–638 (1983).
24. Drudge, J.H., Lyons, E.T. and Tolliver, S.C. Parasite control in horses: A summary of contemporary Drugs. Vet. Med. Small Anim. Clin. **77**: 1479–1489 (1981).
25. Eysker, M., Jansen, J., Boersema, J.H. and Lewing Van der Wiel, P.J. Development of benzimidazole resistance in *Haemonchus contortus* strain in the Netherlands following fenbendazole treatment of ewes at parturition. Res. Vet. Sci. **34**: 46–49 (1983).
26. Fabiyi, J.P. Production losses and control of helminths in ruminants of tropical regions. Proc. Sixth Int. Congr. Parasitol. pp. 435–442 (1986).
27. Fabiyi, J.P. and Adeleye, G.A. Bovine fascioliasis on the Jos Plateau, Northern Nigeria, with particular reference to economic importance. Bull Anim. Hlth. Prod. Afr. **30**: 41–43 (1982).
28. Fisher, M.H. Chemistry of antinematodal agents. In: *Chemotherapy of Parasitic Diseases* (W.C. Campbell and R.S. Rew eds), pp. 239–266. Plenum Press: New York (1986).
29. Flisser, A. Neurocysticercosis in Mexico. Parasitol. Today **4**: 131–137 (1988).
30. Foot, M.A. Nematicide resistance in plant parasitic nematodes. New Zeal. J. Zoo. **7**: 599 (1980).
31. Forsyth, B.A., Gibbon, A.J. and Pryor, D.E. Seasonal variation in anthelmintic response by cattle to dermally applied levamisole. Aust. Vet. J. **60**: 141–146 (1983).
32. Gemmell, M.A., Johnstone, P.D. and Oudemans, G. The effect of niclosamide on *Echinococcus granulosus, Taenia hydatigena* and *Taenia ovis* infection in dogs. Res. Vet. Sci., **22**: 389–391 (1977).
33. Gemmell, M.A., and Shearer, G.C. Bunamidine hydrochloride: Its efficiency against *Echinococcus granulosus.* Vet. Rec. **82**: 252–256 (1968).
34. Gibson, T.E. *Veterinary Anthelmintic Medication.* pp. 172. Commonwealth Agricultural Bureau: Slough, England (1975).
35. Grove, D.I. The effects of 22, 23-dihydroavermectin Bl on *Strongyloides ratti* and *S. stercoralis* infections in mice. Ann. Trop. Med. Parasitol. **77**: 401–410 (1983).
36. Gruner, L., Kerboeuf, D., Beaumont, C. and Hubert, J. Resistance to benzimidazole of *Haemonchus contortus utkalensis* in sheep in Martinique. Vet. Rec. **118**: 276: (1986).
37. Gyurik, R.J. and Bagnall, B.G. The electronic bolus: A novel intermittent-release anthelmintic delivery system for cattle. Proc. Am. Assoc. Vet. Parasitol. **31**: 46: (1986).
38. Harnett, W. The anthelmintic action of praziquantel. Parasitol. Today 4: 144–146 (1988).
39. Hart, D., Langridge, A. Barlow, D. and Sutton B. Antiparasitic drug design. Parasitol. Today **5**: 117–120 (1989).
40. Herbert, I.V. and Probert, A.J. Use of an oxfendazole pulse release bolus in calves exposed to natural subclinical infection with gastrointestinal nematodes. Vet. Rec. **121**: 536–540 (1987).
41. Herd, R.P., Miller, T.B, and Gabel, A.A. A field evaluation of pro-bendimidazoles, benzimidazoles and non benzimidazole anthelmintics in horses. J. Am. Vet. Med. Assoc. **179**: 686–691 (1981).
42. Horak, I.G., Sdijders, A.J. and Louw, J.P. Trials with rafoxanide 5. Efficacy studies against *Fasciola hepatica, Fasciola gigantica, Paramphistomum microbothrium* and various nematodes in sheep. J.S. Afr. Vet. Med Assoc. **43**: 397–403 (1972).
43. Hurter, L.R. and Potgeiter, L.N.D. Schistosomiasis in small stock in the Potgeitersrus veterinary area. J.S. Afr. Vet. Med. Assoc. 38: 444–446 (1967).
44. Jordan, P. and Webbe, G. *Schistosomiasis Epidemiology and Control.* p. 361. William Heinemann Medical books: London (1982).

45. Jordi, R. Study of resistance to anthelmintics in sheep trichostrongyles. Schweiz. Arch. Tierrheilk. **122**: 679–694 (1980).
46. Kemp, G.K. and Smith, C.F. Anthelmintic resistance survey in New Zealand. N. Z. Vet. J. **30**: 141–144 (1982).
47. Kettle, P.R., Vlassoff, A., Reid, T.C. and Horton, C.T. A survey of nematode control measures used by milking goat farmers and of anthelmintic resistance on their farms. N. Z. Vet. J. **31**: 139–143 (1983).
48. Kumar, V. Zoonotic trematodes in South East and Far East Asian countries. In: *Helminth Zoonoses* (S. Geerts, V. Kumar and J. Brandt eds), J. Martinus Nijhoff Publishers: The Netherlands (1987).
49. Kumar, V., Geerts, S. and Brandt, J. Praziquantel. In: *Helminth Zoonoses* (S. Geerts, V. Kumar and J. Brandt eds), J. Martinus Nijhoff Publishers: The Netherlands (1987).
50. Laby, R.H. Australian Patent Application No 35908/78 (1978).
51. Lawrence, J.A. and McKenzie, R.L. Treatment of *Schistosoma mattheei* infestation in sheep: further observations. J.S. Afr. Vet. Med. Assoc. **41**: 298–306 (1970).
52. Leitch, B. and Probert, A.J. The effect of praziquantel on the ultrastructure of *Schistosoma haematobium.* Parasitology **87**: LViii (1983).
53. Leitch, B. and Probert, A.J. *Schistosoma haematobium*: Amoscanate and adult worm ultrastructure. Exp. Parasitol. **58**: 278–289 (1984).
54. Le Jambre, L.F. Anthelmintic resistance in gastrointestinal nematodes of sheep. In: *The Epidemiology and Control of Gastrointestinal Parasites of Sheep in Australia* (A.D. Donald, W.H. Sothcott, and J.K. Dineen eds), C.S.I.R.O., Division of Animal Health, Australia pp. 109–120 (1978).
55. M.A.F.F. Anthelmintics for cattle and sheep. Ministry of Agriculture Fisheries and Food. pp. 1–26. Weybridge, Surrey (1982).
56. Marriner, S. and Armour, J. Nematode infections of domestic animals: Gastrointestinal infections. In: *Chemotherapy of Parasitic Diseases* (W.C. Campbell and R.S. Rew eds), pp. 287–305. Plenum Press: New York (1986).
57. Martin, P.J. Nematode control schemes and anthelmintic resistance. In: *Resistance of Nematodes to Anthelmintic Drugs* (N. Anderson and P.J. Waller eds), C.S.I.R.O., Division of Animal Health, Australia. pp. 29–43 (1985).
58. Martin, P.J.: Development and control of resistance to anthelmintics. Proc. Sixth Int. Cong. Parasitol. pp. 493–501 (1986).
59. Michel, J.F., Latham, J.C., Church, B.M. and Leech, P.K. Use of anthelmintics for cattle in England and Wales during 1978. Vet. Rec. **108**: 252–258 (1981).
60. Mitchell, G.B.B. Assessment of an oxfendazole pulse release bolus for control of parasitic gastroenteritis in calves in a rotational grazing system. Vet. Rec. **121**: 377–378 (1987).
61. Mott, K.E. *Plant Molluscicides.* John Wiley & Sons: London (1987).
62. Mrozik, H. Chemistry of antitrematodal agents. In: *Chemotherapy of Parasitic Diseases* (W.C. Campbell and R.S. Rew eds), pp. 365–383. Plenum Press: New York (1986).
63. Muraleedharan, K., Kumar, S.P., Hegde, K.S. and Alwar, V.S. Comparative efficiency of neguvon, ambilhar and Sodium antimony tartrate against nasal schistosomiasis in cattle. Indian vet. J. **54**: 703–708 (1977).
64. Nansen, P. Production losses and control of helminths in ruminants of temperate regions. Proc. Sixth Int. Congr. Parasitol. pp. 425–433 (1986).
65. Nansen, P. Production losses and control of helminths in ruminants of temperate regions. Int. J. Parasitol. **17**: 425–433 (1986).
66. Narasimham, M.V.V.L., Roychowdhury, S.P., Das, M. and Rao, C.K. Levamisole and Mebendazole in the treatment of Bancroftian infection. Southeast Asian J. Trop. Med. Pub. Hlth. **9**: 571–575 (1978).
67. Portono, F., Purnomo and Soewarta, A. A simple method to control *Brugia timori* by diethylcarbamazine administration. Trans. R. Soc. Trop. Med. Hyg. **73**: 536–542 (1979).
68. Prichard, R.K., Hall, C.A., Kelly, J.D., Martin, I.C.A. and Donald, A.D. The problem of anthelmintic resistance in nematodes. Aust. Vet. J. **56**: 239–250 (1980).
69. Quin, L.R., Liu, Z.L., Chen, Y.Z., Ma, L.H., Zeng, X.G. and Xu, M.G. Treatment of *Schistosoma japonicum* infestation in farm cattle with nithiocyanamin. Chinese J. Vet. Med. **8**: 2–6 (1982).
70. Rew, R.S. and Fetterer, R.H. Mode of action of antinematodal drugs. In: *Chemotherapy of Parasitic Diseases* (W.C. Campbell and R.S. Rew eds), pp. 321–338. Plenum Press: New York (1986).

71. Rickard, M.A. and Arundel, J.H. The chemotherapy of tapeworm infections: veterinary medicine. In: *Chemotherapy of Intestinal Helminths, Handbook of Experimental Pharmacology* **77**: 557–611 (1985).
72. Round, M.C., Simpson, D.J., Haselden, C.S., Glendinning, E.S.A. and Baskerville, R.E. Horse strongyles' tolerence to anthelmintics. Vet. Rec. **95**: 517–518 (1974).
73. Roberson, E.L. Comparative effects of uredofos, niclosamide and bunamidine hydrochloride against tapeworm infections in dogs. Am. J. Vet. Res. **37**: 1483–1484 (1976).
74. Santiago, M.A.M., da Costa, U.C. and Benevenga, S.F. Anthelmintic activity of tetramiscle and thiabendazole against a strain of *Trichostrongylus colubriformis* resistant to levamisole. Revta Cent. Cienc. Rurais **8**: 257–261 (1978).
75. Schillhorn Van Veen, T. and Bida, S.A. Acute paramphistomiasis in sheep in Zaria. Proc. Twentieth Wld Vet. Congr. Vol. **1**: 202–203 (1975).
76. Sharma, R.L., Bhat, T.K. and Dhar, D.D. Control of lungworm in India. Parasitol. Today **4**: 33–36 (1988).
77. Soulsby, E.J.L. *Helminths, Arthropods and Protozoa of Domesticated Animals*. Bailliere and Tindall: London (1982).
78. Sutherland, I.A., Lee, D.L. and Lewis, D. Detection of benzimidazole resistance in trichostrongylid nematodes Parasitol. Today **4**: 22–24 (1988).
79. Taylor, M. The ADAS guide to anthelmintics. Ways with worms. Farmers Weekly suppl. May. pp. 3–14 (1987).
80. Van den Bossche, H. Mode of Action of Anticestodal Agents. In: *Chemotherapy of Parasitic Diseases* (W.C. Campbell and R.S. Rew eds), pp. 495–506. Plenum Press: New York (1986).
81. Van Wyk, J.A. and Gerber, H.M. A field strain of *Haemonchus contortus* showing slight resistance to rafoxanide. Onderstepoort J. Vet. Res. **47**: 137–142 (1980).
82. Waller, P.J. Resistance to anthelmintics and the implications for animal production. In: *Resistance in Nematodes to Anthelmintic Drugs* (N. Anderson and P.J. Waller eds), pp. 1–11. C.S.I.R.O., Australia (1985).
83. Waller, P.J. and Prichard, R.K. Drug Resistance in Nematodes. In: *Chemotherapy of Parasitic Diseases* (W.C. Campbell and R.S. Rew eds), pp. 339–364. Plenum Press: New York (1986).
84. Webbe, G. Cestode Infections of Man. In: *Chemotherapy of Parasitic Diseases* (W.C. Campbell and R.S. Rew eds), pp. 457–477 (1986).
85. Wellington, A.C. The effect of nitroxynil in *Parafilaria bovicola* infestations in cattle. J.S. Afric. Vet. Med. Assoc. **49**: 131–132 (1978).
86. Westcott, R.B. Anthelmintics for Horses. Proc. Sixth Int. Congr. Parasitol. pp. 503–510 (1986).
87. Whitelaw, A. and Fawcett, A.R. Further studies on the control of ovine fascioliasis by strategic dosing. Vet. Rec. **109**: 118–119 (1981).
88. Zimmerman, G.L. and Hoberg, E.P. Controlled release devices for the delivery of anthelmintics in cattle. Parasitol. Today **4**: 55–56 (1988).

15. Genetic Control of Immunity to Helminths

D. Wakelin
Department of Life Science, University of Nottingham
University Park, Nottingham, UK

INTRODUCTION

Variation in the degree of resistance or susceptibility of individual humans or individual animals to infectious disease has been known empirically for many years. Such knowledge predates both the evidence that resistance can have an immunological basis, and that variation in resistance can be genetically determined. In the 1920's and 30's empirical observations of variation in resistance to helminth infections were reinforced by careful quantitative studies in man [103], sheep [51] and chickens [2]. These firmly established the phenomenon as an important aspect of human and veterinary parasitology. Despite the progress that was being made elsewhere at this time in the genetic analysis of variation in resistance to viral and bacterial infections [49, 136], there were no comparable studies dealing with worms until the seminal studies of Whitlock [137–139] with trichostrongyles of sheep. During the last 20 years, this situation has changed dramatically. It is now accepted as self-evident that genetically heterogeneous populations of hosts, from mice to men, will show extensive variation in ability to resist helminth infections, and that a major component of this variation can be attributed to variation in ability to generate and express protective immune responses. The evidence that this is so has come from both field and laboratory studies, involving humans, domestic animals, laboratory rodents and feral populations of small mammals, reviewed by Wakelin [117, 123]. The interactions between genetics and immunology that underlie variations in resistance can now be analysed and described in detail, because we now have a greater understanding of immunity to parasites *per se*, as well as of the wider disciplines of immunology and immunogenetics. Before discussing specific examples taken from studies dealing with the major helminth infections, it will be helpful to review briefly the progress in 3 areas, namely epidemiology, helminth immunity and immunogenetics, which has helped to establish present concepts of variation in resistance to helminth infection.

Epidemiology

Detailed surveys of host populations infected with helminths have demonstrated, almost universally, that the worm population is not normally or randomly distributed but is aggregated, such that most hosts have few or no parasites, and most parasites are concentrated into a few hosts [5]. Although many factors, operating at different points in the host-parasite relationship could generate such aggregation, e.g. differential exposure, different nutritional or physiological states [120], it has become clear that an important factor may be the ability of the host to develop protective immunity, and thus regulate the level of infection. Several lines of evidence can be cited to support this view. Thus:

(a) Aggregation occurs even when the host population itself and the degree of exposure to infection are relatively uniform, implying that variations in worm burdens arise after infection and are independent of nutritional, physiological and age variation.

(b) Mathematical modelling of the infection process, based upon simple assumptions regarding the existence of genetically-determined resistance controlled by a single dominant allele influencing immune responses, shows that aggregated distribution will be generated as time-related development of resistance restricts the level of infection in homozygous and heterozygous individuals (Fig. 1).

Mathematical modelling of a specific host-parasite infection, that of the tapeworm *Hymenolepis citelli* in deer mice (*Peromyscus maniculatus*), and based upon direct evidence of genetically determined resistance to infection, demonstrates the same conclusion [132].

(c) Laboratory studies, where all host variables can be rigorously controlled, and specified changes in genotype or in immune response manipulated selectively, show that the level of worm burden from uniform single or multiple infections is genetically determined through immune-dependent resistance mechanisms [117, 123].

Helminth Immunity

Although progress in the analysis of immunity to helminths has been slower than that to protozoa, we now have a clear idea of the components of the immune and inflammatory responses that are necessary for the expression of resistance [119, 122]. Earlier and simpler concepts of immunity being either humoral or cellular have been replaced by the view that, for any given worm, resistance is multifactorial, involving interactions between several components. Although in particular circumstances particular responses may make a major contribution to resistance, the absence of such responses does not necessarily inactivate resistance, as other components are brought into play. There are very few cases where antibodies alone can effect resistance, and none known where T cells alone can do so. Resistance requires interaction between antibodies and complement, or between antibodies and cells (antibody-dependent cellular cytotoxicity or ADCC) (see Chapter 9); it requires T cell-mediated inflammatory reactions, involving macrophages, eosinophils, platelets, mast cells and basophils; it requires hypersensitivity responses involving reaginic antibodies and amine-containing cells. Each of these components is known to be under genetic control, and so individuals in genetically heterogeneous populations (or different inbred strains) are likely to vary in their ability to develop and express these component functions (Table 1).

Immunogenetics

Analysis and understanding of the genes concerned with the control of the immune and inflammatory responses are now well advanced. It is clear that every phase of these responses is genetically regulated, and that marked variations in immune response capacity can, therefore, be expected to occur in genetically diverse populations. Control operates both quantitatively, influencing the level of the response expressed, and qualitatively, determining whether responses are expressed or not (Table 2). Attention has been focussed recently upon the role of major histocompatibility complex-linked (MHC-linked) genes in controlling the process of antigen recognition by T cells and in determining the consequences of recognition. Much of this work has used simple defined peptide antigens with few epitopes available, and in these circumstances MHC genes can determine responsiveness in a qualitative (all-or-nothing) manner, i.e. the antigen is either recognized or not recognized. It had been thought for many years that the molecular complexity of helminths would mean that MHC-linked restriction of recognition and of responsiveness to helminth antigens was unlikely. Not only is this now known not to be the case (see below) but it is accepted that for most worms there are relatively few antigens that are critical for the stimulation of protective responses [61].

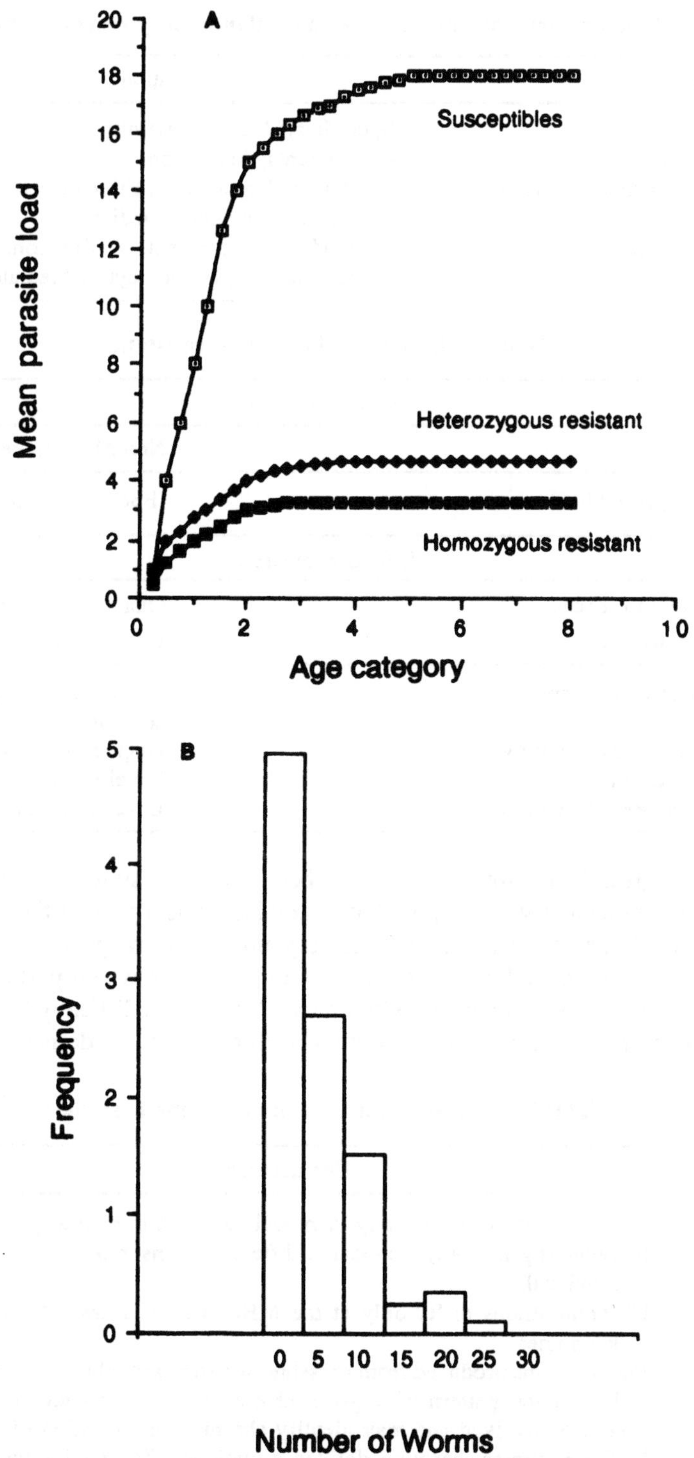

Fig. 1. Mathematical modelling of changes in mean parasite load with time (A) in a population where resistance to infection is determined in homozygotes and heterozygotes by a single dominant allele. The frequency distribution of worm burdens in the total population (homozygous and heterozygous resistant, and homozygous susceptible) is shown in (B) (Redrawn from Anderson [5]).

Table 1. Genetic variations in immune and inflammatory responses to helminths

Response	Variation
Antibody	Specificity; Isotype; Affinity; Level
Complement	Components; Activation
Antibody-dependent cellular cytotoxicity (ADCC)	Antibody; Complement; Effector cells—numbers, receptors, enzymes, mediators
Hypersensitivity	Antibody; Complement; Amine-containing cells; Macrophages; Granulocytes; Mediators

Table 2. Genetic influences on resistance

Genes involved	
MHC-linked	Non-MHC-linked
(HLA-Man; H-2-Mice)	(Background genes)
Effects expressed	
Qualitative differences between individuals	Quantitative differences between individuals
(All-or-nothing Responses) seen in: Resistance or Susceptibility Antigen Recognition Lymphocyte Responsiveness	(Level of Response) seen in: Degree of Resistance Level of Antibody Response Level of Inflammation

The rapidity of progress in immunogenetics has been made possible by technical developments in molecular biology and molecular genetics, but has depended critically upon the availability of immunologically- and genetically-defined strains of laboratory mice. The range of strains available makes it possible to identify particular gene loci and so to define the ways in which particular immune responses are controlled. Careful use of these strains provides data which can realistically be extrapolated to provide an understanding of the genetic control of immunity in humans and in domestic animals (Table 3).

Table 3. Mouse strains for immunogenetic studies

Strains	Characteristics
Random-Bred	Individuals differ genetically ∴ model variation in natural populations.
Inbred	Individuals genetically identical ∴ different strains model variation seen between different individuals
MHC Congenic	Different strains differ only at the MHC loci ∴ allow influence of specific loci to be determined
Recombinant Inbred	Inbred strains produced from crossing two standard inbred strains. Determination of strain distribution pattern of a given characteristic allows identification of nature of gene control involved and may identify chromosomal location of genes
Mutant	Different strains, carrying defined mutations affecting immunological or inflammatory responses.

It will not be possible in a short chapter to cover the very extensive literature that now exists on the genetic control of immunity to helminths. A number of comprehensive reviews of the topic are available, and the reader is referred to there for detailed descriptions [63, 72, 95, 102, 117, 123, 133]. The approach of this review will be, therefore, to introduce the subject area, to identify key concepts and to provide an up-to-date view of the field.

TREMATODE INFECTIONS

Experimental studies in rodent hosts have demonstrated both individual and interstrain variation in resistance to a number of flukes, but the only substantial body of work concerns variation in resistance to schistosomes. Most of this work has been carried out in mice, primarily using *S. mansoni.*

Early workers [104] showed clearly that strain differences in levels of adult worm burden after a standard infection reflected an important component of innate resistance. This influenced the proportion of parasites that successfully penetrated the skin and migrated to the liver. Variation in ability to develop immunologically-mediated resistance has been most easily studied in mice challenged after vaccination; but components of the immune response to primary infections have also received attention. Differences in levels of primary infections have been recorded in Biozzi high- and low-responder mice, in which differences in degree of innate resistance are likely to be small [14]. High-responder mice, which produce high antibody levels, were more susceptible than low-antibody responders, perhaps reflecting their reduced macrophage activity. It is also possible that their higher antibody responsiveness leads to higher titres of potential blocking antibody that can interfere with protective ADCC mechanisms. No evidence for MHC-linked control of resistance to primary infections has been obtained, although there is a clear influence of MHC-linked genes on serological recognition of specific worm antigens [60]. Differences in recognition of a major protective antigen (26 kDa) were also reported within the 129/J inbred strain when infected with *S. japonicum.* About half of these mice fail to develop patent infections, and these "resistant" mice respond to the 26 kDa antigen [73]. Differential resistance is also seen with *S. mansoni* in this strain. However, recent work has shown that "resistant" mice have inherent vascular abnormalities in their livers [28, 36, 71], which result in diversion of developing schistosomes into other organs where they die (and presumably elicit high antibody responses to the antigens they release).

Genetic variation in resistance to challenge infections with *S. mansoni* in mice primed by an active immunizing infection has been demonstrated by several workers, but the interpretation of this variation is complicated by the fact that immunizing infections induce profound liver pathology through hypersensitivity responses to eggs, and this contributes directly to resistance in a non-immunological manner. Dean et al [32] showed a clear correlation between level of resistance (reduction in number of challenge worms) and degree of portal hypertension. The most resistant of a panel of 10 strains (CF1 and NMR1—< 90% reduction) had the highest portal pressure values implying the most severe pathology. When mice are immunized by exposure to radiation-attenuated cercariae or schistosomula resistance is generated without immunopathology: variations in resistance therefore reflect differences in anti-parasite immune and inflammatory responses. Considerable inter-strain variation has been recorded, influenced by both background and MHC-linked genes [78, 79]. Comparison of immunological parameters between resistant and susceptible mice failed to identify significant difference in either T cell or antibody function, suggesting that perhaps some non-immunological (? inflammatory) component was involved [57]. Analysis of responses in P strain mice, which are virtually unresponsive to irradiated larval vaccines, has identified a failure in macrophage function as a key defect [27, 56]. Macrophages of these mice have a poor response to T cell cytokines, the mice also produce less cytokine, so that antiparasite (ADCC) responses are impaired. The macrophage defect is controlled by recessive alleles at a signal major locus *Rsm-1* [26]. Although it is clear that macrophages play an important role in vaccine-induced immunity in mice, other inflammatory cells may be more important in other host species. Evidence from other parasite systems (see below) show clearly that the production and activity of inflammatory cells such as eosinophils and mast cells are also under genetic control, and thus variable, and it can be concluded that anti-schistosome responses involving these cells are also likely to show genetically-determined variation. Although this has not been investigated directly, it is interesting to note that anti-schistosome ADCC mechanisms involving eosinophils in the rat are IgE-dependent, and IgE-responses are also strain variable in this system [97].

A current focus of immunological research in schistosomiasis is the development of a vaccine using genetically-engineered or chemically-synthesized antigens. Trials with a major candidate antigen (P28) in baboons gave variable results between the individuals tested [8], which may reflect genetically determined variations in responsiveness. Certainly when tested in inbred mouse strains there were very clear genetic differences in immunological response to this antigen [69].

Schistosomiasis in humans is a long-term chronic infection and development of effective resistance has been debatable. Recently, detailed epidemiological studies which have monitored prevalence and intensity of infection (in both *S. haematobium* and *S. mansoni*) before and after chemotherapy, and related these parameters to degree of water contact, have shown convincing evidence that resistance does develop and does control the level of infection [see 20]. A striking aspect of these studies was the fact that individuals appeared to be predisposed to heavy or to light infection independently of water contact; this predisposition was maintained when infections were reacquired after chemotherapy, suggesting that a strong genetic component was influencing the degree of resistance expressed [21].

Genetic factors also influence the degree of immunopathology associated with schistosomiasis. Mouse studies have shown this very clearly [45] but there is also evidence relating to genetic influences on pathology in humans for example HLA-linked genes (HLA A1 and B5) have been associated with hepato-splenomegaly in *schistosomiasis mansoni* [1] and with post-schistosomal liver cirrhosis attributable to *S. japonicum* [82]. Detailed *in vitro* studies with *S. japonicum* have shown that this MHC (HLA)-linked control is probably expressed through regulation of T cell responsiveness to worm antigens. Individuals carrying DDW1 and DR2 (HLA class II genes) respectively have low or high responsiveness and show mild/severe infection-induced pathology [54]. Susasuki and colleagues have referred to the DQ gene involved as an immune-suppression gene [81]. Mitchinson [77] has proposed the intriguing hypothesis that there is a selective pressure of the evolution of such genes in animals exposed to chronic, pathology-inducing, chronic parasitic infections, the pressure arising from the decreased fitness associated with high immune responsiveness and greater pathology (see Chapter 10). If this hypothesis is correct it is clearly advantageous to parasites to produce molecules that are strongly immunogenic (= pathogenic) so that hosts evolve immune suppression genes in order to survive. These genes, in turn, not only promote host survival but, by reducing host responsiveness, promote parasite survival (Fig. 2).

INFECTION	Acute ↓	Chronic ↓
CONSEQUENCE OF IMMUNE RESPONSE TO PARASITE ANTIGENS	Resistance ↓	Pathology ↓
SELECTION PRESSURE ON HOSTS	For Immune response genes (Ir) ↓	For immune suppression genes (Is) ↓
SELECTION PRESSURE ON PARASITE	Make antigens less immunogenic— Survival prolonged	Keep antigens that induce suppression— Survival prolonged

Fig. 2. Selection pressures for the evolution of immunosuppression genes in hosts and maintenance of immunopathogenic antigens in parasites (Adapted from Mitchison & Oliveira [77]).

CESTODE INFECTIONS

Genetic variation in resistance to tapeworms has been recorded for both adult and larval stages, but only in the latter have the genetic and immunological components been analysed in any detail. One important exception to this is the work of Wassom on variation in resistance of deer mice (*Peromyscus maniculatus*) to the adult worms of *Hymenolepis citelli* [131, 132]. The importance of this study is that it defined genetic variation in a natural host-parasite relationship and showed how this variation produced a characteristic aggregated parasite distribution. It also identified the genetic control as apparently due to a single dominant gene, possibly acting via some T cell-dependent mechanism.

The best investigated tapeworm systems are those using larval stages (metacestodes) in mice. (It is pointed out that evidence exists for aggregated distributions of both the adult and larval stages of one of the most serious metacestode infections of man (*Echinococcus granulosus*—Fig. 3) implying that such model systems have an important part to play in elucidating genetic aspects of clinically significant host-parasite relationships.) Perhaps the most useful of these systems has been *Taenia taeniaeformis*. Here the

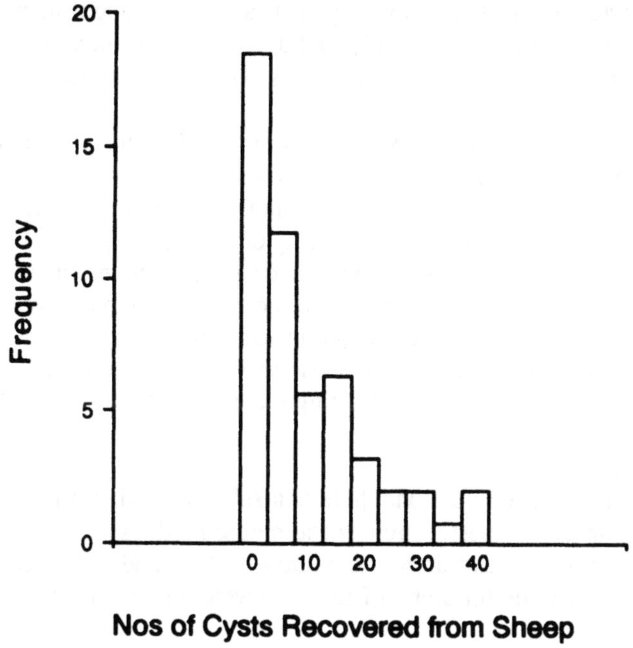

Fig. 3. Frequency distribution of viable hydatid cysts recovered from sheep infected with 2500 eggs of *Echinococcus granulosus* (Data from Gemmell et al [47]).

metacestodes develop in the liver of the rodent host, after oral infection of eggs passed by the definitive host—cat. Mouse strains differ quite dramatically in their ability to support the development of the liver stages. When infected with the same number of eggs, resistant strains allow the development of only a few, small larvae, whereas susceptible strains have numerous, large fully-cystic stages [23]. MHC-linked genes do not play an important role in this variation [74], which appears to result from background gene influences on the rate of anti-larval antibody production [76]. Both the onchosphere stage in the intestine and the early metacestode in the liver are susceptible to IgA-and IgG-mediated responses, the larvae in the liver being easily damaged by complement in the early stage of development. As they mature these larvae develop anti-complement activity which protects them against further damage. In order for hosts to express immunity to an initial infection they must be able to produce specific antibodies, before the larvae become insusceptible. This ability is genetically-determined. Because antibody production is always rapid on challenge, even susceptible hosts can express immunity when reinfected, or when they are challenged after vaccination [87]. It is likely that genetically-determined susceptibility to *T. taeniaeformis* reflects not a general defect in rate of antibody production (as in Biozzi low-line mice) but a defect in production of specific isotypes against specific target antigens. Gibbens et al [48] have in fact shown that susceptible C3H/He mice produce a wide range of isotypes, recognizing a broader spectrum of *T. taeniaeformis* antigens than do the more resistant C57BL/6.

Although attention has been focussed upon genetic control of antibody as an explanation of variation in immunity in this system, it is likely that variation in ability to mount inflammatory cell responses may also be relevant. Development of metacestodes in the liver elicits a strong inflammatory response in which eosinophils and neutrophils are prominent. In resistant strains these cells appear to come into contact with the larvae and may contribute to parasite damage, in susceptible strains this does not occur and the larvae become surrounded by a thick (? protective) mucopolysaccharide layer [68]. Variation in ability to mount granulocyte-mediated responses against larvae is thought to play an important role in limiting resistance to *T. taeniaeformis* in the rat [7], and detailed studies in another system, *Mesocestoides corti* in the mouse, have also shown that marked variations occur in ability to amount eosinophil responses to infection.

M. corti larvae develop in the liver and body cavity of mice, and can stimulate an intense eosinophilia. Striking differences in the level of eosinophilia occur when infections are followed in different strains of mice, and these are primarily determined by background, rather than by MHC-linked genes [65]. The expression of these genes appears to determine the ability of mice to respond, at a stem-cell or precursor-cell level, to the T cell lymphokine IL5 that controls eosinophil development [64]. No evidence for genetically-determined differences in T cell activity have been detected in this system. Although resistance to *M. corti* is transferable adoptively with T cells, and is accompanied by increased eosinophilia [66], the results of a detailed study of responsiveness to infection show that there is, in fact, an inverse correlation between resistance and eosinophilia, perhaps reflecting the extent of parasite invasion and development.

NEMATODE INFECTIONS

In contrast to trematode and cestode infections there is a substantial literature on variation in resistance to nematode infections in humans, domestic animals and mouse model systems. This section will highlight important concepts from the human and domestic animal studies, and use data from the mouse models to provide mechanistic explanations for some of the observed genetically-determined differences.

Humans

Almost all large-scale studies on nematode parasites in humans have shown individual and racial differences in prevalence of infection, implying strong genetic influence. The likelihood that these influences may operate at the level of resistance to infection is particularly strong where the parasites are soil-transmitted (*Ascaris*, hookworms, *Trichuris*) or vector-borne (filariasis). Under these condition exposure to infection can be assumed to be relatively uniform within host populations, yet it is a universal observation that, where worm burdens can be quantified, parasite distribution is aggregated [19], and where intensity of infection can be determined only indirectly, the population includes uninfected, lightly-infected and heavily infected individuals [121]. Although it has been concluded that humans develop little or no immunity to nematode infections, because of the ease with which reinfection can occur, it is hard to reconcile this with the evidence from recent chemotherapy intervention-reinfection studies with intestinal nematodes which clearly show predisposition to heavy or to light infections (Fig. 4), even though differential exposure to infection cannot be completely excluded. Similarly, the clear evidence that individuals in

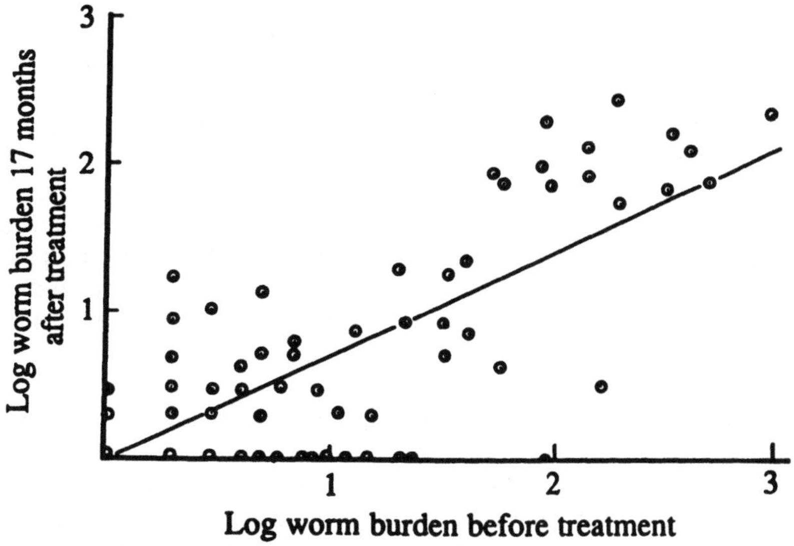

Fig. 4. Predisposition to infection with *Trichuris trichiura.* Numbers of worms recovered from individuals given anthelmintic compared with numbers present after 17 months of reinfection (Redrawn from Bundy [18], figure taken from Wakelin [123] with permission of the publishers: Taylor & Francis).

endemic areas differ markedly in immunopathological responses to filarial infections (both lymphatic and onchocercal) strongly suggests genetically-determined differences in immunological responsiveness. Descriptive rather than analytic experimental data to support these interpretations are now accumulating. Examples include:

1. comparisons showing a relationship between antibody and cellular responses in individuals with different pathological status arising from infection with lymphatic filariae [79, 85] or with *Onchocerca volvulus* [46].
2. differences between individuals in their serological recognition of filarial antigens [22, 55], and of gastro-intestinal nematodes [62, 93].
3. correlations between HLA-linked genes and resistance to *Trichuris trichiura* [19].

Domestic Animals

The evidence that individuals and breeds of all major domestic species differ in their resistance to nematode infections is well-established and of long-standing [117]. The most detailed studies have been concerned with differential resistance in sheep to infection with the gastro-intestinal trichostrongyles. Although this phenomenon has been identified with all the major ovine parasites (*Dictyocaulus, Haemonchus, Nematodirus, Ostertagia, Trichostrongylus*) most attention has been focussed on two species *Haemonchus contortus* and *Trichostrongylus colubriformis* [112]. Resistance to these nematodes is thought to involve immunologically-mediated inflammatory mechanisms acting on the intestine, as well as direct antibody-mediated effects against the worms themselves but the precise mechanisms are not clearly defined. In the case of *H. contortus* there is, in addition, a physiological component to resistance (resilience) which enables some strains better to withstand the pathological effects of the worms' blood-feeding habits.

Haemonchus

Individual variation in resistance to *H. contortus* was described by Whitlock, who also showed that this variation had a heritable basis [137–139]. Differences in resistance between breeds have been described by many workers [30] and there is an association between resistance and haemoglobin B. In Australia Merino sheep are selectively bred for wool production and are highly susceptible to haemonchosis. Its control in endemic areas relies heavily on anthelmintic treatment. The development of drug resistance in *H. contortus* has prompted a search for additional means of control (see Chapter 14). Vaccination has still to prove effective in the field [112] and the possibility of selectively breeding for resistance is an attractive option. Genes conferring resistance to *H. contortus* can be found in Merinos, and rams selected for this characteristic produce progeny with enhanced resistance [50]. This resistance has a clear immunological basis [86] and detailed studies of the responses of resistant progeny are now being undertaken. It is clear that both immune responses *per se* (antibody levels) and immune-mediated inflammatory mechanisms (e.g. intestinal mast cell levels) are higher in these animals than in non-selected controls [Gill—personal communication]. No negative association has been found between parasite resistance and wool or meat production

Selective breeding for resistance would be much enhanced if predictive tests could be carried out before infection. Attempts to find pre-infection correlates of resistance have so far been unsuccessful; for example, no linkage to the ovine MHC has been demonstrated [25]. An alternative approach to improving livestock performance in areas endemic for *H. contortus*, in the face of heavy parasite challenge, may be to introduce resistance genes by crossing more productive breeds with indigenous breeds that have been naturally selected for resistance. Red Masai sheep in Africa are one such breed, and crosses with European breed have produced progeny with enhanced resistance [30].

Trichostrongylus

Interest in genetic variation in resistance to *T. colubriformis* has come from attempts to vaccinate against infection using irradiated third-stage larvae. When a population of Merino lambs was vaccinated in this way and challenged with normal larvae, there was no significant difference in mean values between the vaccinated and control groups [33]. However, analysis of the data from the vaccinated animals showed that lambs fell into two distinct groups, "non-responders", with worm counts similar to controls, and "responders", with significantly lower (10x lower) worm counts. Subsequent breeding experiments showed that responder parents could pass on substantial levels of enhanced resistance to their progeny [reviewed in [34].

Recent studies with this system have concentrated upon two areas: (a) the detection of genetic or immunological markers by which animals with the responder phenotype might be identified and thus used for selective breeding, and (b) analysis of the mechanisms through which resistance is expressed.

Conflicting results have been obtained when lymphocyte proliferative responses to *T. columbriformis* antigen were used as a predictive marker of resistance. Originally no correlation was observed when uninfected animals were tested, but a positive correlation was found when cells were taken from naturally infected sheep [91]. More recently it has been shown that SY1, an ovine lymphocyte antigen can be used to type for resistance [83].

The mechanisms of resistance to *T. columbriformis* have been studied in sheep, and also in the guinea-pig host, making use of genetically-defined resistant and susceptible strains [96]. Resistance in this host shows MHC-linkage, is T cell-dependent and correlates positively with inflammatory responses in the infected intestine. Numbers of mucosal mast cells and of tissue eosinophils are markedly increased in resistant animals, but it is not clear whether they reflect causal or contingent correlates of resistance. Lambs selected for resistance to *T. colubriformis* similarly show a positive correlation between mucosal mastocytosis [35], eosinophilia [31] and greater responsiveness to challenge. No differences could be found between the antigen-recognition profiles of post-infection sera taken from resistant and susceptible lines of either guinea-pigs or sheep [80], although this observation may have no bearing on differential recognition of those antigens which trigger intestinal inflammatory responses.

Mouse Models

A wide variety of mouse models is available for genetic studies (Table 4), covering the groups that are of major clinical interest in humans and domestic stock. In the majority, differences between resistant and susceptible strains have been studied not only at the level of differential ability to express protective immunity but also in terms of the immunological and inflammatory responses involved. Of the species listed in Table 4 five will be used to illustrate specific aspects of genetically-determined variation.

Table 4. Mouse models for studying genetic variation in resistance to nematodes

Species	Infection modelled
Ascaris suum	Ascariasis—man
Brugia malayi	Lymphatic filariasis—man
Brugia pahangi	Lymphatic filariasis—man
Heligmosomoides polygyrus	Chronic trichostrongyle infections—domestic stock + general intestinal infections.
Necator americanus	Hookworm—man
Nippostrongylus brasiliensis	Trichostrongyle infections—domestic stock + general intestinal infections.
Onchocerca lienalis	Onchocerciasis—man
Trichinella spiralis	Trichinellosis—man + general intestinal infections
Trichuris muris	Trichuriasis—man

Ascaris suum

Although development in the mouse does not proceed much beyond the lung stages, infections can be used to study both resistance to infection and the range of immune responses made. Considerable strain variation in ability to control primary and challenge infections has been described [75], greater resistance being regulated by background rather than by MHC-linked genes and inherited as a dominant characteristic. MHC-restriction does occur at the level of antibody recognition of larval ES and adult internal antigens [110], there being quite clear-cut differences between mice of different H-2 haplotype. Of particular interest is an MHC-linked recognition of a 14kDa major allergenic molecule. Not only do MHC-linked gene products control recognition of this allergen, there is also background genetic control of the level of IgE antibody made, such that certain $H\text{-}2^s$ strains produce enough anti-14 kDa IgE to generate PCA responses in recipient rats, whereas others, with different backgrounds, do not. A similar allergen occurs in *A. lumbricoides* and infected individuals vary in their ability to recognize this serologically, raising the possibility of a genetic component in the allergic conditions associated with ascariasis [62].

Acanthocheilonema (= Dipetalonema) viteae

Mouse models of filarial infections are rather limited in their closeness to human filariases, but *A. viteae* and *Brugia malayi* provide useful species with which to study factors regulating the level of infection, especially the extent and duration of microfilaraemia. Although mice cannot be infected directly with *A. viteae*, microfilaraemia can be established by subcutaneous implantation of fertile female worms. Striking differences are seen when mice of different strains are used as recipients [53, 107], BALB-background and B10-background mice providing the extremes of susceptibility and resistance (Fig. 5). The genes controlling this resistance are inherited as a dominant characteristic, BALB/c x B10 F1 progeny behaving as B10 mice; no MHC-linked influence is evident. Resistance is expressed through a cell population derived from bone marrow stem cells, as shown by studies using radiation chimaeras. The response phenotype of such chimaeras is determined by that of the bone marrow donor.

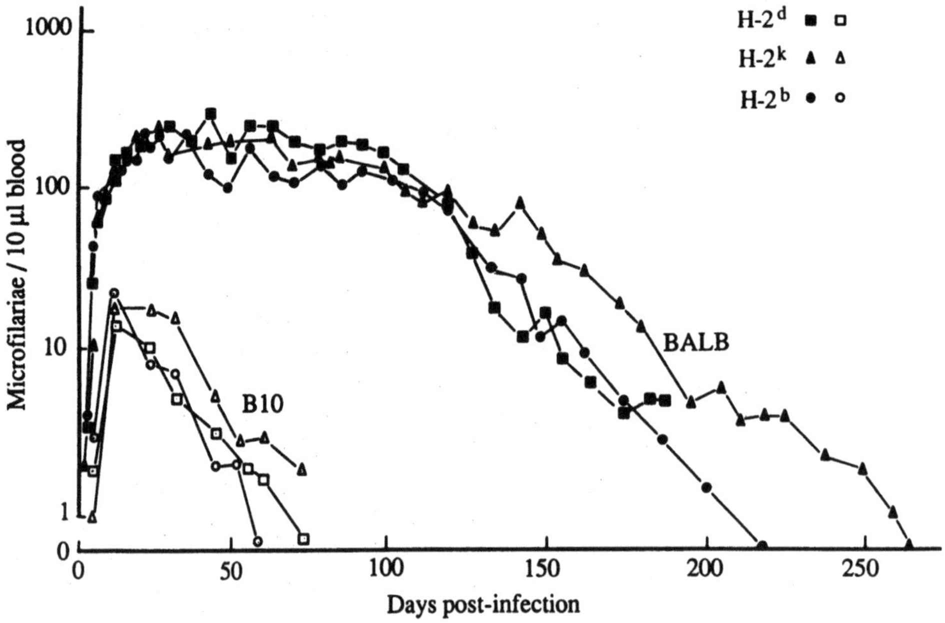

Fig. 5. Time course of microfilaraemia after implanting adult female *Acanthocheilonema (Dipetalonema) viteae* into H-2 congenic strains of BALB- and B10-background mice (Data from Storey et al [107], figure taken from Wakelin [123] with permission of the publishers: Taylor & Francis).

It seems most likely that the ability to control microfilaraemia reflects an antibody-mediated (ADCC) response to cuticular antigens [84]. The most obvious difference between resistant and susceptible strains concerns the level of cuticle-specific IgM antibodies [84, 105]. Susceptible strains have very low, or no IgM against these antigens; it is also thought that low antibody affinity may also contribute to an ineffective, protective response. Mutant CBA/N mice, which are genetically-deficient in IgM production are likewise susceptible to *A. viteae* [109] as they are also to *B. malayi* [108]. Genetic variation in the cellular component of ADCC may also be involved in the *A. viteae* system, as transfer of immune serum from resistant mice brings about only a temporary decline in microfilaraemia in susceptible recipients [106].

Helmigmosomoides polygyrus (= Nematospiroides dubius)

It is characteristic of nematode infections in humans and in domestic animals that they are chronic, the worms surviving for prolonged periods. *H. polygyrus* provides a good mouse model for this situation, infections persisting for several months in the most susceptible strains [94]. Both background and MHC-linked genes regulate the degree of susceptibility to infection; the precise interplay between the two depending on the level of infection used and whether susceptibility is assessed in terms of primary or challenge infections [10, 43, 113]. During primary infections the strongest influence is exerted by background genes, SJL mice, for example, being much more resistant than mice carrying the same MHC haplotype [$H\text{-}2^s$] on the B10 background. In turn B10. S mice are more resistant than other B10 background mice such as B10.BR ($H\text{-}2^k$). or B10 ($H\text{-}2^b$), showing that MHC-linked genes do influence responsiveness (Fig. 6). When strain differences are assessed in mice challenged after a drug-abbreviated primary infection MHC-linked genes appear to exercise the major influence on resistance, $H\text{-}2^q$ haplotype mice, for example being uniformly more resistant than $H\text{-}2^k$ mice, regardless of background. Background genes are responsible for a minor modulating effect. On the basis of this work Enriquez et al [43] concluded

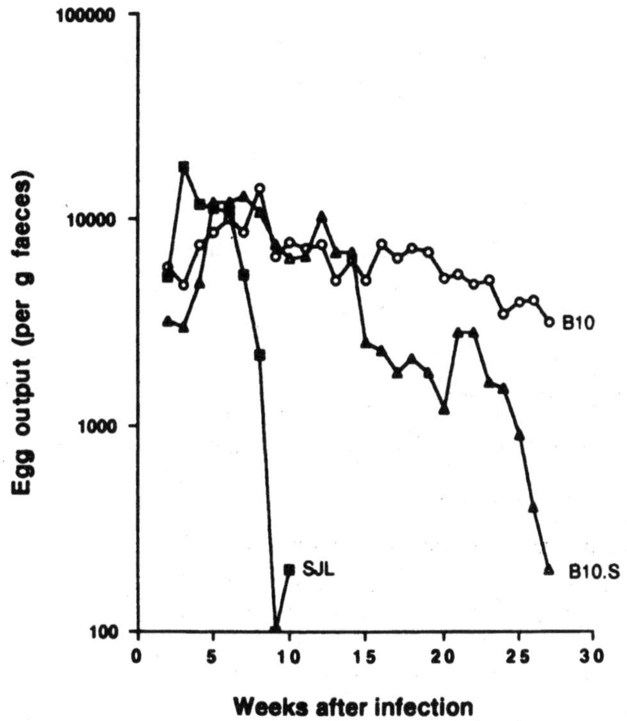

Fig. 6. Faecal egg output of B10 [H-2^b], B10.S [H-2^s] and SJL [H-2^s] mice infected with *Heligmosomoides polygyrus (Nematospiroides dubius)* (Figure redrawn from Behnke & Wahid [11]).

that two MHC-linked genes (class II gene and D end gene) control the expression of resistance. The susceptibility of *H*-2^k mice, which are Class II I-E + ve, and the resistance of *H*-2^q mice, which are I-E -ve, led to the view that presentation of worm antigen via I-E class II molecules might lead to suppression of immunity (as has been proposed for *Trichinella spiralis*—see below). However this interpretation does not accord with the data of Behnke & Wahid [11], who found that although *H*-2^k mice were susceptible to primary infections, *H*-2^b [I-E-ve] mice were more susceptible still. The answer to these inconsistencies may be in the relative degrees of immunosuppression exerted by *H. polygyrus* at different infection levels in primary or challenge situations [44, 94]. The evidence from these and other studies implies that the degree of suppression induced is itself a genetically variable host characteristic. As parasite-induced suppression may well be involved in maintaining the chronicity of intestinal nematode infections in humans and domestic stock [9], understanding of the genetic component involved is clearly of considerable importance.

In addition to its usefulness in the study of the genetics and immunology of chronic infections, *H. polygyrus* has provided a useful model for long-term selection experiments designed to analyse changes in host phenotype, at the level of innate and acquired resistance to infection, and changes in parasite phenotype. Both are crucial aspects of any programme designed to improve genetic resistance in domestic animals. Dobson and colleagues have carried out many experiments in this field, using two different starting populations of outbred mice. Two-way selection using primary infections increased levels of both resistance and susceptibility, influencing innate and acquired immunity separately [15–17]. Selection for acquired immunity *per se*, using challenge infections, established distinct high and low responder lines after five generations [99]. After 10 generations the difference between the lines, in terms of faecal egg output from a standard infection, was 10 fold [101]. Interestingly, when exposed to drug-abbreviated, graded immunizing infections, the two lines showed markedly different levels of resistance to challenge depending upon the size of the immunizing dose. High responders were more efficiently immunized at all doses, but low-responders showed progressively less resistance at the higher doses, implying an increasing degree of immuno-suppression from the immunizing infection in this line [100].

Trichinella spiralis

This nematode is the most intensively studied in terms of helminth immunogenetics, and data from the mouse system have defined a number of important concepts related to the genetic control of resistance to intestinal and parenteral infections. Almost all parameters of infection, parasitological as well as immunological, are genetically variable, the variation in the latter contributing to the marked differences in resistance that exist between strains (Table 5). Only four aspects of this variation can be considered here: (1) the roles of MHC-linked and background genes in controlling resistance, (2) control of antibody responses, (3) control of immune-mediated inflammatory responses, (4) response to vaccination.

Table 5. Genetic variation in parameters of the response to infection with *Trichinella spiralis*

Parasitological	Immunological
Site of infection in gut	Antibody response—specificity, isotype and level
Survival of adult worms	Proliferative response of T cells to antigen
Growth of adult worms	Cytokine production
Fecundity of female worms	Inflammatory response
Number of muscle larvae	Degree of immune depression
Response to challenge infection	Response to vaccination

A very large number of inbred mouse strains has been typed for response to *T. spiralis*, response being determined by speed of expulsion of intestinal worms in primary and challenge infections, and by the numbers of muscle larvae produced [13, 118, 129]. Comparisons between strains and within congenic lines show that the major influence on resistance appears to be expressed through background genes, although MHC-linked genes can exert a marked effect when the background is held constant. Two background genes have been identified to date, one *Ihe-1* (chromosomal location unknown) regulating the rapid expulsion of challenge infections [12], the other, *Ts-3*, closely linked to *Fv*-1^a or chromosome 4, influencing resistance to a primary infection [134]. Two MHC-linked genes have also been identified, *Ts-1* and *Ts-2;* both regulate resistance to primary infection [128, 130], *Ts-2* modulating the influence of *Ts-1* in a complex fashion. Attention has been focussed on *Ts-1*, resistance originally being linked with the expression of this gene at the Class II I-A locus. However, it is possible that resistance reflects less the presence of a resistance gene at I-A than the absence of alleles at I-E, I-E + ve mice in general being more susceptible to *T. spiralis* than I-E-ve mice. The original suggestion that presentation of parasite antigen to T cells via I-E molecules led to suppressor function, whilst presentation via I-A led to enhanced T helper function [134] has now been modified [133]. Mice expressing I-E are known to lose some T cell clones at the stage of intrathymic development because of the presence of certain variable-region amino acid sequences in the T cell receptor [59]. If these clones were to include some with *T. spiralis* specificity there could be a defective T cell response to infection. Certainly I–E+ve B10.BR mice show a slower T cell response to infection than do IA+ve mice, a response reflected in reduced cytokine production and responsiveness [141], but this could equally well reflect some suppressive influence.

The interplay between MHC-linked and background genes influences all components of protective immune responses to *T. spiralis* and a model has been proposed to interpret this interaction [135]. MHC-linked genes are thought to regulate antigen-presentation to T cells and control the amplification or down-regulation of T cell activity; background genes then regulate development of the intestinal inflammatory responses which bring about worm expulsion (see Chapters 1 and 9). To this model can now be added the genetic control of antibody responses, although the role of antibody may be limited to effects on growth and reproduction of adult worms and to the destruction of released new-born larvae [Fig. 7]. MHC-restriction of *T. spiralis* antigen recognition has been described very recently [Kennedy—personal communication] the antibodies of strains carrying different MHC haplotypes recognizing quite distinct antigen profiles. Strain-dependent differences in functional and isotype-specific responses to surface and released antigens have also been described [4, 58]. Rapid-responder strains produce parasite-specific antibodies earlier in infection, there being a positive correlation between production of adult-worm-specific IgA and earlier worm expulsion, and of ADCC-promoting antibodies and lower muscle larval burdens.

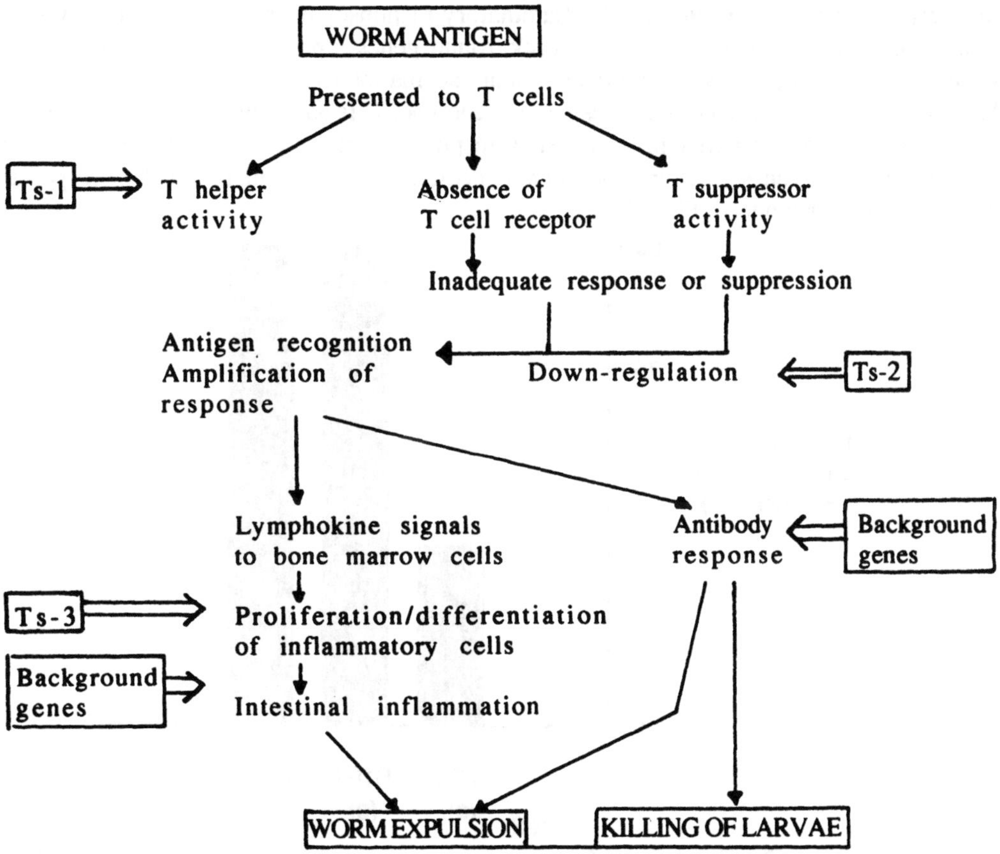

Fig. 7. Roles of MHC-linked [*Ts-1* and *Ts-2*] and other genes [Ts-3, background] in controlling resistance to infection with *Trichinella spiralis* in the mouse.

A great deal of experimental data points to the conclusion that the rate of response to the intestinal stages in a primary infection is determined by background genes that regulate the capacity to translate cytokine signals from T cells into an effective intestinal inflammatory response (see reviews in [124, 126]). These data have come from *in vivo* experiments involving adoptive transfer of effector T cells (known to be CD4$^+$) into intact and bone-marrow reconstituted mice, using histocompatible recipients of similar or contrasting response phenotypes. Two prominent components of the intestinal response are mucosal mast cells and eosinophils, both of which are likely to play important functional roles in worm expulsion [70, 96]. Positive correlations exist between response phenotype in terms of speed of worm expulsion and response phenotype in terms of mastocytosis and eosinophilia [3, 125] and this correlation has now been extended to include mast cell functional activity during infection [111].

Both mastocytosis and eosinophilia can be accelerated, as can worm expulsion, by transfer of parasite-specific T cells [90], but *in vitro* studies with both cell types show that the limiting factor in responsiveness is the ability of precursor populations to respond to T cell-derived cytokines-cytokines themselves are not rate-limiting in the intact animal [64, 89]. Support for the involvement of mucosal mast cells in the expression of resistance comes also from studies with *T. spiralis* infections in mutant mast cell-deficient W/W^v mice [88].

The important role played by intestinal inflammatory responses in resistance to the intestinal stages of *T. spiralis*, and the dominant influence of background genes, upon precursor cell responsiveness to lymphocyte-derived signals, carry the implication that vaccination to raise immunological responsiveness should have relatively little influence upon response phenotype. This has been demonstrated experimentally, using a muscle larval antigen in rapid- and slow-responder strains [127]. Despite clear evidence of enhanced T cell responsiveness, vaccination did not accelerate worm expulsion in slow-responders to the degree seen in rapid-responders (Fig. 8).

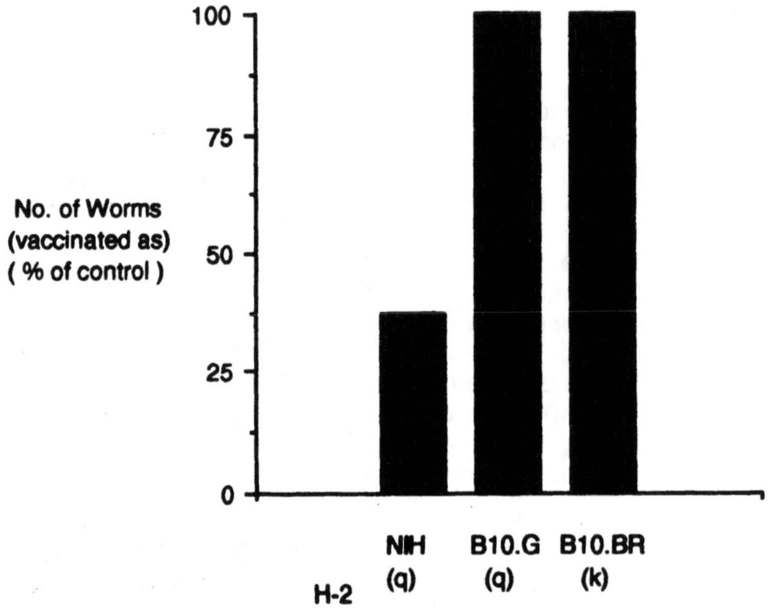

	NIH	B10.G	B10.BR
Proliferation to Ag	++	+	++
IL-2 production	++	++	++
IL-3 production	++	++	++

Fig. 8. Response to vaccination with larval antigens of *Trichinella spiralis* in MHC-compatible (NIH and B10.G, both *H-2q*) , and MHC-congenic mice [B10.G and B10.BR, *H-2*^{q} and *H-2*^{k}]. Numbers of worms recovered after challenge and responses of lymphocytes to the immunizing antigen.

Trichuris muris

The current recognition of human trichuriasis as a major helminthiasis [24] has renewed interest in the mouse model, both for immunological and for immunogenetic studies. As with the other species discussed in this section there is marked strain variation in resistance to infection, most mouse strains expelling primary infections well before the worms become mature [115]. Analysis of those strains in which worms do mature has revealed some interesting aspects of the host-parasite relationship. In one outbred strain a constant proportion of animals fails to develop resistance to primary or challenge infections and allows adult worms to persist for many weeks. Relatively few genes control this susceptibility, and selective breeding rapidly separated lines with contrasting response phenotypes [116]. The available evidence suggests that susceptibility results from a genetically determined influence upon cellular elements of resistance, susceptible mice being able to produce antibodies that could transfer immunity into competent recipients [114]. Surprisingly, differential responsiveness has also been described in certain inbred (and presumptively genetically homogeneous) strains [67]. Some progress in explaining this anomaly has come from analysis and experimental manipulation of infections in a series of H-2 congenic strains [37, 40]. When BALB-background and B10 background congenic strains are compared it is clear that background genes provide the dominant influence upon resistance, all BALB mice losing worms before any B10 strain (Fig. 9). However MHC-linked genes modulate the background influence in a consistent way, $H\text{-}2^k$ haplotype mice being the least resistant on either background. On the B10 background different haplotypes not only reduced the degree of resistance (in the order $k > d > q > b$) but also influenced the proportion of mice within the strain that failed to control infection at all, resulting in the development of a patent infection. B10.BR [$H\text{-}2^k$] mice fail completely to develop resistance, and all individuals harbour mature worms.

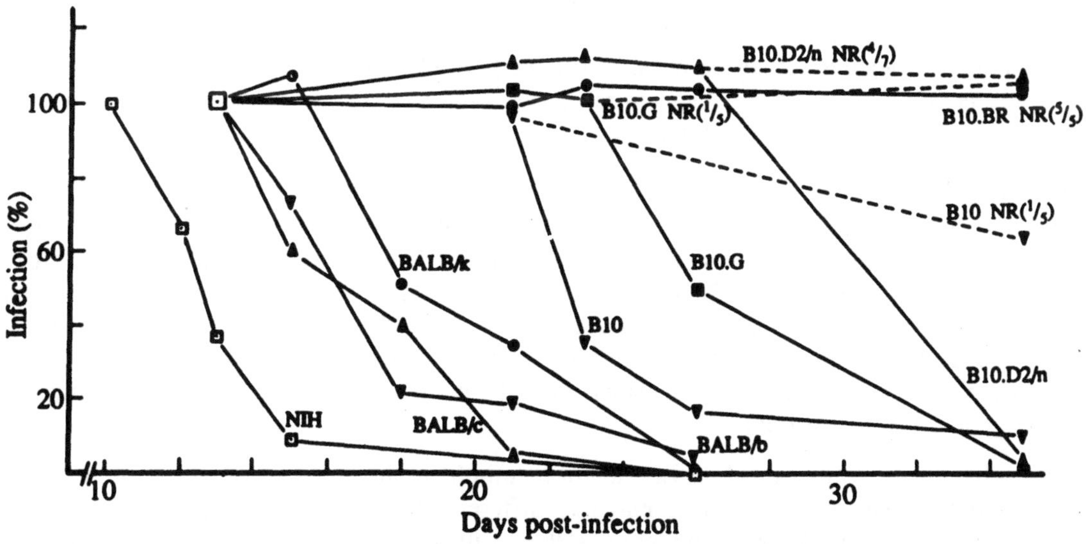

Fig. 9. Time course of infection with *Trichuris muris* in inbred strains of mice (high-responder NIH, intermediate BALB congenic strains and low-responder B10 congenic strains). In the B10 strains the proportion of non-responder (NR) mice is shown (Redrawn from Else & Wakelin [37]; figure taken from Wakelin [123] by permission of the publishers: Taylor & Francis].

The explanation of this situation appears to lie in a critical phase of parasite development at about the 3rd-4th week. If worms persist at this time they appear to exert an irreversible suppressive influence on the host, preventing not only the development of resistance to a primary infection, but also immunity to subsequent challenge. The mechanisms of this suppression are not understood. They do not involve an interference with inflammatory responses [67] but may involve other aspects of T cell-mediated resistance. In general, resistant strains produce lower antibody titres and have a more restricted antigen-recognition profile than susceptible strains [38], suggesting that the greater antibody response of the latter may reflect some blocking activity. However, other data showed that within congenic strains protective immunity can be positively correlated with genetic influences upon isotype and antigen-specific responses [38, 41]. When B10 background mice are compared, the more resistant $H\text{-}2^q$ made greater IgG_1 responses (as do other $H\text{-}2^q$ mice) than susceptible $H\text{-}2^k$. $H\text{-}2^q$ mice also recognize antigens not seen by more susceptible mice. Significantly, vaccination with worm antigen can induce protection, greater IgG_1 responses and specific antigen recognition in susceptible $H\text{-}2^k$ mice [39].

Detailed immunogenetic studies have pin-pointed class II MHC alleles in the control of resistance [42]. As with other systems [133] I-A alleles appear to be critical, *q* and *b* alleles being associated with resistance, *k* and *d* with susceptibility. The influence of both resistance and susceptibility I-A alleles was modulated by alleles at the D end of the MHC. $I\text{-}A^q$ alleles also appear to restrict antibody recognition of particular high MW antigens, whose recognition correlates positively with greater resistance; they also regulate the level of specific IgG_1 responses [41].

The extensive similarities between immune responsiveness to *T. muris* in the mouse and *T. trichiura* in humans implies that data from the model system can be extrapolated to explain the dynamics of infection in man. The clear demonstration of genetic control of isotype- and antigen-specificity in antibody response in mice suggests that similar genetic factors may well influence the development of protective resistance in man. Any delay in the development of resistance in humans may lead to responsiveness being down-regulated by immunomodulatory factors from the worm, with the establishemt of chronic infection and impaired ability to respond to re-infection. IgA-specific responses have been implicated in the expression of protective resistance in both systems [19, 92], but the mouse data suggest that other isotypes, and possibly other T cell-dependent responses, may be critically genotype-dependent and thus also determine the balance between resistance and susceptibility.

CONCLUSIONS

It will be clear from this short review that resistance to helminth infection is not a fixed host characteristic, but is genetically variable, such that some individuals or breeds are markedly less able to control their levels of infection than are others. The detailed analyses that have been undertaken to define both the genetic and immunological elements of this variation have revealed a great deal of basic information that has direct relevance to the understanding and eventual control of these ubiquitous infections. Four aspects relevant to human helminthology can be considered briefly, these are (a) epidemiology, (b) chemotherapy; (c) pathology and (d) vaccination.

(a) The characteristic aggregated distribution of helminths in the host can now be seen to reflect not only differential exposure to infection, but also differential individual susceptibility. Analysis of epidemiological data must take this into account if the results are to explain fully the processes by which populations acquire and regulate their worm infections [5].

(b) The continuing transmission of schistosomes and of gastro-intestinal helminths depends upon release of infective stages into the environment. The aggregated distributions of adult worms in the host population means that certain individuals ("wormy persons" [29]) will contribute disproportionately to the reservoir of infective stages and thus to the maintenance of transmission. One effective strategy to break the transmission cycle is to use chemotherapy selectively against the most heavily infected individuals [52]. Given the evidence that these individuals may be genetically susceptible to more than one of the major helminths [6] then targeted chemotherapy should have a widespread beneficial effect.

(c) A significant component of the pathology associated with helminth infection is of immunopathological origin. The immune responses that underlie this pathology, being under genetic control, will vary between individuals, some individuals will therefore suffer more pathological consequences than others. Differential pathology occurs in schistosomiasis and filariasis, and some of the likely mechanisms have already been discussed [54, 81, 82]. To these can be added known genetic variation in responses likely to contribute to hypersensitivity and its pathological consequences (IgE [62]; mast cells [88]; eosinophils [64]). It is possible that knowledge of the control of these responses will lead ultimately to some form of rational therapy, e.g. anti-pathology vaccines capable of preventing pathological changes and designed to increase beneficial host responses and decrease harmful responses

(d) Vaccination as a means of preventing helminth infection is a desirable aim, because it removes many of the difficulties associated with chemotherapeutic or environmental control. However, the existence of genetic variation in the development and expression of resistance, at all levels from antigen recognition to ultimate effector mechanisms, creates particular problems for the use of vaccines in host populations. The development of defined peptides as vaccine antigens, by chemical synthesis or DNA recombinant technology, may solve the difficulties of obtaining adequate amounts of material, but increases the chances that some individuals will fail to respond because they lack the appropriate MHC molecules for presentation of such restricted epitopes to T cells [140]. The extent to which vaccines will potentiate protection in all individuals will also vary even if antigen recognition is achieved. Resistance dependent directly upon antibody-mediated responses can be enhanced effectively [87] but that dependent upon inflammatory responses may not be if the genetic defect occurs at that level [127]. In addition the well known ability of helminths to induce suppression in the host, the degree of which is itself genetically variable, will reduce the effectiveness of vaccination in certain individuals.

The role of immunogenetic influences in the complex interplay between host and parasite is, therefore, a challenging and critically important field of helminthology. Progress is at present rapid and has as its goal the explanation of immunogenetic control in molecular terms, i.e. the interaction between parasite- and host-derived molecules in initiating, controlling and expressing the mechanisms of resistance by which helminth infections are successfully regulated.

REFERENCES

1. Abdel Salem, E., Ishaac, S. and Mahmoud, A.A.F. Histocompatibility-linked susceptibility for hepatosplenomegaly in human *Schistosomiasis mansoni*. J. Immunol. **123**: 1829–1832 (1979).
2. Ackert, J.E., Eisenbrandt, L.L., Wilmoth, J.H., Glading, B. and Pratt, I. Comparative resistance of five breeds of chickens to the nematode *Ascaridia lineata* (Schneider). J. Agric. Res. **50**: 607–624 (1935).
3. Alizadeh, H. and Wakelin, D. Genetic factors controlling the intestinal mast cell response in mice infected with *Trichinella spiralis*. Clin. Exp. Immunol. **49**: 331–337 (1982).
4. Almond, N.R. and Parkhouse, R.M.E. Immunoglobulin class specific responses to biochemically active defined antigens of *Trichinella spiralis*. Parasite Immunol. **8**: 391–406 (1986).
5. Anderson, R.M. The population biology and genetics of resistance to infection. In: *Genetics of Resistance to Bacterial and Parasitic Infection*. (D. Wakelin and J.M. Blackwell eds) pp. 223–263. Taylor & Francis: London (1988).
6. Anderson, R.M. and Medley, G.F. Community control of helminth infections in man by mass and selective chemotherapy. Parasitology **90**: 629–660 (1985).
7. Ansari, A. and Williams, J.F. The eosinophilic response of the rat to infection with *Taenia taeniaeformis*. J. Parasitol. **62**: 728–736 (1976).

8. Balloul, J.M., Boulanger, D., Sondermeyer, P., Dreyer, D., Capron, M., Grzych, J.M., Pierce, R.J., Carvallo, D., LeCocq, J.P. and Capron, A. Vaccination of baboons with a P28 antigen of *Schistosoma mansoni* expressed in *E. coli*. In: *Molecular Paradigms for Eradicating Helminthic Parasites* (A.J. MacInnis ed), pp. 77–79. Alan R. Liss Inc.: New York (1987).
9. Behnke, J.M.: Evasion of immunity by nematode parasites causing chronic infections. Adv. Parasitol. **26**: 1–71 (1987).
10. Behnke, J.M. and Robinson, M. Genetic control of immunity to *Nematospiroides dubius*: a 9-day anthelmintic abbreviated immunizing regime which separates weak and strong responder strains of mice. Parasite Immunol. **7**: 235–253 (1985).
11. Behnke, J.M. and Wahid, F. Immunological relationships during primary infections with *Heligosomoites polygyrus (Nematospiroides dubius)*: H-2 genes determine worm survival. Parasitology 103: 157–164 (1991).
12. Bell, R.G., Adams, L.S. and Ogden, R.W. A single gene determines rapid expulsion of *Trichinella spiralis* in mice. Infect. Immun. **45**: 273–275 (1984).
13. Bell, R.G., Adams, L.S. and Ogden, R.W. *Trichinella spiralis*: genetics of worm expulsion in inbred and F1 mice infected with different worm doses. Exp. Parasitol. **58**: 345–355 (1984).
14. Blum, K. and Cioli, D. Behaviour of Biozzi high and low responder mice upon infection with *Schistosoma mansoni*. Eur. J. Immunol. **8**: 52–56 (1978).
15. Brindley, P.J. and Dobson, C. Multiple infections with *Nematospiroides dubius* in mice selected for liability to a single infection. Aust. J. Exp. Biol. Med. Sci. **60**: 319–327 (1982).
16. Brindley, P.J. and Dobson, C. Genetic control of liability to infection with *Nematospiroides dubius* in mice: direct and correlated responses to selection of mice for faecal parasite egg count. Parasitology **87**: 113–127 (1983).
17. Brindley, P.J. and Dobson, C. Partitioning innate and acquired immunity in mice after infection with *Nematospiroides dubius*. Int. J. Parasitol. **13**: 503–507 (1983).
18. Bundy, D.A.P. Epidemiological aspects of *Trichuris* and trichuriasis in Caribbean communities. Trans. R. Soc. Trop. Med. Hyg. **80**: 706–718 (1986).
19. Bundy, D.A.P. Population ecology of intestinal helminth infections in human communities, Phil. Trans. R. Soc. Lond. *B*., **321**: 405–420 (1988).
20. Butterworth, A.E. Immunity and morbidity in human schistosomiasis. In: *New Strategies in Parasitology* (K.P.W. McAdam ed), pp. 193–210. Churchill Livingstone: Edinburgh (1989).
21. Butterworth, A.E. and Hagan, P. Immunity in human schistosomiasis. Parasitol. Today. **3**: 11–16 (1987).
22. Cabrera, Z., Buttner, D.W. and Parkhouse, R.M.E. Unique recognition of a low molecular weight *Onchocera volvulus* antigen by IgG3 antibodies in chronic hyperactive oncho-dermatitis (Sowda). Clin. Exp. Immunol. **74**: 223–229 (1988).
23. Conchedda, M. and Ferretti, G. Susceptibility of different strains of mice to various levels of infection with the eggs of *Taenia taeniaeformis*. Int. J. Parasitol. **14**: 541–546 (1984).
24. Cooper, E.S. and Bundy, D.A.P. *Trichuris* is not trivial. Parasitol. Today. **4**: 301–306 (1988).
25. Cooper, D.W., Van Oorschot, R.A.H., Piper, L.R and Le Jambre, L.F. No association between the ovine leucocyte antigen (OLA) system in the Australian Merino and susceptibility to *Haemonchus contortus* infection. Int. J. Parasitol. **19**: 695–697 (1989).
26. Correa-Oliveira, R., James, S.L., McCall, D. and Sher, A. Identification of a genetic locus, *Rsm-1*, controlling protective immunity against *Schistosoma mansoni*. J. Immunol. **137**: 2013–2019 (1986).
27. Correa-Oliviera, R., Sher, A. and James, S.L. Defective vaccine-induced immunity to *Schistosoma mansoni* in P strain mice. J. Immunol. **133**: 1581–1586 (1984).
28. Coulson, P.S. and Wilson, R.A. Portal shunting and resistance to *Schistosoma mansoni* in 129 strain mice. Parasitology **99**: 383–389 (1989).
29. Croll, N.A. and Ghadirian, E. Wormy persons: contributions to the nature and patterns of over-dispersion with *Ascaris lumbricoides, Ancylostoma duodenale, Necator americanus* and *Trichuris trichiura*. Trop. Geogr. Med. **33**: 241–248 (1981).
30. Dargie, J.D.: The influence of genetic factors on the resistance of ruminants to gastointestinal nematode and trypanosome infections. In: *Animal Models in Parasitology* (D.G. Owen ed), pp. 17–51. McMillan: London (1982).

31. Dawkins, H.J.S., Windon, R.G. and Eagleson, G.K. Eosinophil responses in sheep selected for high and low responsiveness to *Trichostrongylus colubriformis*. Int. J. Parasitol. **19**: 199–207 (1989).
32. Dean, D.A., Bukowski, M.A. and Cheever, A.W. Relationship between acquired resistance, portal hypertension and lung granulomas in ten strains of mice infected with *Schistosoma mansoni*. Am. J. Trop. Med. Hyg. **30**: 806–814 (1987).
33. Dineen, J.K., Gregg, P. and Lascelles, A.K. The response of lambs to vaccination at weaning with irradiated *Trichostrongylus colubriformis* larvae: segregation into 'responders' and 'non-responders'. Int. J. Parasitol. **8**: 59–63 (1978).
34. Dineen, J.K. and Outteridge, P.M. (eds.). *Immunogenetic Approaches to the Control of Endoparasites with Particular Reference to Parasites of Sheep*. C.S.I.R.O.: Australia (1984).
35. Dineen, J.K. and Windon, R.G. The effect of sire selection on the response of lambs to vaccination with irradiated *Trichostrongylus colubriformis* larvae. Int. J. Parasitol. **10**: 189–196 (1980).
36. Elsaghier, A.A.F. and McLaren, D.J. *Schistosoma mansoni*: evidence that vascular abnormalities correlate with the "non-permissive" trait in 129/01a mice. Parasitology **99**: 377–381 (1989).
37. Else, K. and Wakelin, D. The effects of H-2 and non-H-2 genes on the expulsion of the nematode *Trichuris muris* from inbred and congenic mice. Parasitology **96**: 543–550 (1988).
38. Else, K.J. and Wakelin, D. Genetic variation in the humoral immune response of mice to the nematode *Trichuris muris*. Parasite Immunol. **11**: 77–90 (1989).
39. Else, K.J. and Wakelin, D. Genetically determined influences on the ability of poor responder mice to respond to vaccination against *Trichuris muris*. Parasitology **100**: 479–489 (1990).
40. Else, K.J., Wakelin, D. and Roach, T.I.A. Host predisposition to trichuriasis: the mouse—*T. muris* model. Parasitology **98**: 275–282 (1989).
41. Else, K.J., Wakelin, D., Wassom, D.L. and Hauda, K.M. MHC-restricted antibody responses to *Trichuris muris* excretory-secretory (E-S) antigens. Parasite Immunol. **12**: 509–527 (1990).
42. Else, K.J., Wakelin, D., Wassom, D.L. and Hauda, K.M. The influence of genes mapping within the major histocompatibility complex on resistance to *Trichuris muris* infections in mice. Parasitology (in press).
43. Enriquez, F.J., Brooks, B.O., Cypess, R.H., David, C.S. and Wassom, D.L. *Nematospiroides dubius*: two H-2-linked genes influence levels of resistance to infection in mice. Exp. Parasitol. **67**: 221–226 (1988).
44. Enriquez, F.J., Cypess, R.H. and Wassom, D.L. Influence of immunizing dose and presence or absence of adult worms on the development of resistance to *Nematospiroides dubius* challenge infections of mice. J. Parasitol. **74**: 409–414 (1988).
45. Fanning. M.M. and Kazura, J.W. Further studies on genetic variation of hepatosplenic disease and modulation in murine *Schistosomiasis mansoni*. Parasite Immunol. **7**: 213–222 (1985).
46. Gallin, M., Edmonds, K., Ellner, J.J., Erttmann, K.D., White, A.T., Newland, H.S., Taylor, H.R. and Green, B.M. Cell-mediated immune responses in human infection with *Onchocerca volvulus*. J. Immunol. **140**: 1999–2007 (1988).
47. Gemmell, M.A., Lawson, J.R. and Roberts, M.G. Population dynamics in echinococcosis and cysticercosis: biological parameters of *Echinococcus granulosus* in dogs and sheep. J. Parasitol. **92**: 599–620 (1986).
48. Gibbens, J.C., Harrison, L.J.S. and Parkhouse, R.M.E. Immunoglobin class responses to *Taenia taeniaeformis* in susceptible and resistant mice. Parasite Immunol. **8**: 491–502 (1986).
49. Gowen, J.W. Genetics and disease resistance. In: *Genetics in the 20th Century* (L.C. Dunn ed), pp. 401–429. McMillan: New York (1951).
50. Gray, G.D. Genetic resistance to haemonchosis in sheep. Parasitol. Today **3**: 253–255 (1987).
51. Gregory, P.W. The possibility of establishing within breeds lines of sheep that are genetically resistant to stomach worms. Proc. Am. Soc. Anim. Prod. **10**: 316–324 (1937).
52. Haswell-Elkins, M.R., Elkins, D.B. and Anderson, R.M. Evidence for predisposition in humans to infection with *Ascaris*, hookworm, *Enterobius* and *Trichuris* in a South Indian fishing community. Parasitology **95**: 323–337 (1987).
53. Haque, A., Worms, M.J., Ogilvie, B.M. and Capron, A. *Dipetalonema viteae*: microfilariae production in various mouse strains and in nude mice. Exp. Parasitol. **49**: 398–407 (1980).
54. Hirayama, K., Matsushita, S., Kikuchi, I., Iuchi, M., Ohta, N. and Sasazuki, T. HLA-DQ is epistatic to HLA-

DR in controlling in the immune response to schistosoma antigen in humans. Nature (Lond.) **327**: 426–430 (1987).

55. Hussain, R., Grogl, M. and Ottesen, E.A. IgG antibody subclasses in human filariasis: differential subclass recognition of parasite antigens correlates with different clinical manifestation of infection. J. Immunol. **139**: 2794–2798 (1987).
56. James, S.L., Correa-Oliviera, R. and Leonard, E.J. Defective vaccine-induced immunity to *Schistosoma mansoni* in P strain mice. J. Immunol. **133**: 1587–1593 (1984).
57. James, S.L. Labine, M. and Sher, A. Mechanisms of protective immunity against *Schistosoma mansoni* infection in mice vaccinated with irradiated cercariae. 1. Analysis of antibody and T lymphocyte responses in mouse strains developing different levels of immunity. Cell Immunol. **65**: 75–83 (1981).
58. Jungery, M. and Ogilvie, B.M. Antibody response to stage-specific *Trichinella spiralis* surface antigens in strong and weak responder mouse strains. J. Immunol. **129**: 839–843 (1982).
59. Kappler, J.W., Roehm, N. and Marrack, P.: T cell tolerance by clonal elimination in the thymus. Cell **49**: 263–271 (1987).
60. Kee, K.C., Taylor, D.W., Cordingley, J.S., Butterworth, A.E. and Munro, A.J. Genetic influence on the antibody response to antigens of *Schistosoma mansoni* in chronically infected mice. Parasite Immunol. **8**: 565–574 (1986).
61. Kennedy, M.W. Genetic control of the immune repertoire in nematode infections. Parasitol Today **5**: 316–324 (1989).
62. Kennedy, M.W., Tomlinson, L.A., Fraser, E.M. and Christie, J.F. The specificity of the antibody response to internal antigens of *Ascaris*: heterogeneity in infected humans, and MHC (H-2) control of the repertoire in mice. Clin. Exp. Immunol. **80**: 219–224 (1990).
63. Krco, C.J. and David, C.S. Genetics of immune response: a survey. C.R.C. Crit. Rev. Immunol. 211–257 (1981).
64. Lammas, D.A., Mitchell, L.A. and Wakelin, D. Genetic control of eosinophilia. Analysis of production and response to eosinophil differentiating factor in strains of mice infected with *Trichinella spiralis*. Clin. Exp. Immunol. **77**: 137–143 (1989).
65. Lammas, D.A., Mitchell, L.A. and Wakelin, D. Genetic influences upon eosinophilia and resistance in mice infected with *Mesocestoides corti*. Parasitology **101**: 291–299 (1990).
66. Lammas, D.A., Wakelin, D. and Mitchell, L.A. Adoptive transfer of enhanced eosinophilia and resistance to infection in mice by an *in vitro* generated T-cell line specific for *Mesocestoides corti* larval antigens. Parasite Immunol. **9**: 591–601 (1987).
67. Lee, T.D.G. and Wakelin, D. The use of host strain variation to assess the significance of mucosal mast cells in the spontaneous cure response of mice to the nematode *Trichuris muris*. Int. Archs Allergy Appl. Immunol. **67**: 302–305 (1982).
68. Letonja, T., Rikihisa, Y. and Hammerberg, C. Differential cellular response of resistant and susceptible rodents to the early stages of infection of *Taenia taeniaeformis*. Int. J. Parasitol. **14**: 551–558 (1984).
69. Mendlovic, F., Tarrab-Hazdai, R., Puri, J. and Arnon, R. Genetic control of immune response to a purified *Schistosoma mansoni* antigen I. Effect of MHC class II antigens on the cellular, humoral and protective responses. Parasite Immunol. **11**: 667–682 (1989).
70. Miller, H.R.P. The protective mucosal response against gastrointestinal nematodes in ruminants and laboratory animals. Vet. immunol. Immunopathol. **6**: 167–259 (1984).
71. Mitchell, G.F. Portal system peculiarities may contribute to resistance against schistosomes in 129/J mice. Parasite Immunol. **11**: 713–717 (1989).
72. Mitchell, G.F., Anders, R.F., Brown, G.V., Handman, E., Roberts-Thomson, I.C., Chapman, C.B., Forsyth, K.P., Kahl, L.P. and Cruise, K.M. Analysis of infection characteristics and antiparasite immune responses in resistant compared with susceptible hosts. Immunol. Rev. **61**: 137–188 (1982).
73. Mitchell, G.F., Beall, J.A., Cruise, K.M., Tiu, W.V. and Garcia, E.G. Antibody responses to the antigen Sj26 of *Schistosoma japonicum* worms that is recognized by genetically resistant 129/J mice, Parasite Immunol. **7**: 165–178 (1985).

74. Mitchell, G.F., Goding, J.W. and Richard, M.S. Studies on immune responses to larval cestodes in mice: increased susceptibility of certain mouse strains and hypothymic mice to *Taenia taeniaeformis* and analysis of passive transfer with serum. Aust. J. Exp. Biol. Med. Sci. **55**: 165–186 (1977).
75. Mitchell, G.F., Hogarth-Scott, R.S., Edwards, R.D., Lewers, H.M., Cousins, G. and Moore, T. Studies on immune responses to parasite antigens in mice. I. *Ascaris suum* larvae numbers and antiphosphorylcholine responses in infected mice of various strains and in hypothymic nu/nu mice. Int. Archs Allergy Appl. Immunol. **52**: 64–78 (1976).
76. Mitchell, G.F., Rajasekariah, G.R. and Rickard, M.D. A mechanism to account for mouse strain variation in resistance to the larval cestode, *Taenia taeniaeformis*. Immunology **39**: 481–489 (1980).
77. Mitchison, N.A. and Oliveira, D.B.G. Chronic infection as a major force in the evolution of the suppressor T-cell system. Parasitol. Today **2**: 312–313 (1986).
78. Murrell, K.D., Clark, D.S., Dean, D.A. and Vannier, W.E. Influence of mouse strain on induction of resistance with irradiated *Schistosoma mansoni* cercariae. J. Parasitol. **65**: 829–834 (1979).
79. Nutman, T.B., Kumaraswami, V. and Ottesen, E.A. Parasite-specific anergy in human filariasis: insights after anaysis of parasite-antigen-driven lymphokine producton. J. Clin. Invest. **79**: 1516–1523 (1987).
80. O'Donnell, I.J., Dineen, J.K., Rothwell, T.L.W. and Marshall, R.C. Attempts to probe the antigens and protective immunogens of *Trichistrongylus colubriformis* in immunoblots with sera from infected and hyperimmune sheep and high- and low-responder guinea pigs. Int. J. Parasitol. **15**: 129–135 (1985).
81. Ohta, N., Minai, M. and Sasazuki, T. Antigen-specific suppressor T lymphocytes (Leu-2a$^+$3a$^-$) in human Schistosomiasis japonica. J. Immunol. **131**: 2524–2528 (1983).
82. Ohta, N., Nishimura, Y.K., Iuchi, M. and Sasazuki, T. Immunogenetic analysis of patients with post-schistosoma liver cirrhosis in man. Clin. Exp. Immunol. **49**: 493–499 (1982).
83. Outteridge, P.M., Windon, R.G. and Dineen, J.K. An ovine lymphocyte antigen marker for acquired resistance to *Trichostrongylus colubriformis*. Int. J. Parasitol. **18**: 853–858 (1988).
84. Philipp, M., Worms, M.J., Maizels, R.M. and Ogilvie, B. Rodent models of filariasis. In: *Contemporary Topics of Immunobiology* Vol. 12 (J.J. Marchalonis ed), pp. 275–321. Plenum Publishing Crop.: New York (1984).
85. Piessens, W.F., McGreevy, P.B., Ratiwayanto, S., McGreevy M., Piessens, P.W., Koiman, I., Saroso, J.S. and Dennis, D.T. Immune responses in human infections with *Brugia malayi*: correlation of cellular and humoral reactions to microfilarial antigens with clinical status. Am. J. Trop. Med. **29**: 563–570 (1980).
86. Presson, B.L., Gray, G.D. and Burgess, S.K. The effect of immunosuppression with dexamethasone on *Haemonchus contortus* infections in genetically resistant Merino sheep. Parasite Immunol. **10**: 675–680 (1988).
87. Rajasekariah, G.R., Mitchell, G.F. and Rickard, M.D. *Taenia taeniaeformis* in mice: protective immunization with oncospheres and their products. Int. J. Parasitol. **10**: 155–160 (1980).
88. Reed, N.D. Function and regulation of mast cells in parasite infections. In: *Mast Cell and Basophil Differentiation and Function in Health and Disease*. (S.J. Galli and K.F. Ansten eds), Raven Press: New York (1989).
89. Reed, N.D., Wakelin, D., Lammas, D.A. and Grencis, R.K. Genetic control of mast cell development in bone marrow cultures. Strain-dependent variation in cultures from inbred mice. Clin. Exp. Immunol. **73**: 510–515 (1988).
90. Riedlinger, J., Grencis, R.K. and Wakelin, D. Antigen specific T-cell lines transfer protective immunity against *Trichinella spiralis in vivo*. Immunology **58**: 57–61 (1986).
91. Riffkin, G.C. and Yong, W.K. Recognition of sheep which have innate resistance to trichostrongylid nematode parasites. In: *Immunogenetic Approaches to the Control of Endoparasites* (J.K. Dineen and J.K. Outteridge eds), pp. 30–40. C.S.I.R.O.: Australia (1984).
92. Roach, T.I.A., Else, K.J., Wakelin, D., McLaren, D.J. and Grencis, R.K. *Trichuris muris*: antigen recognition and transfer of immunity in mice by IgA monoclonal antibodies. Parasitology **101**: 61–67 (1990).
93. Roach, T.I.A., Wakelin, D., Else, K.J. and Bundy, D.A.P. Antigenic cross-reactivity between the human whipworm, *Trichuris trichiura* and the mouse trichuroids *Trichuris muris* and *Trichinella spiralis*. Parasite Immunol. **10**: 279–291 (1988).
94. Robinson, M., Wahid, F., Behnke, J.M. and Gilbert, F.S. Immunological relationships during primary infection with *Heligmosomoides polygyrus* (*Nematospiroides dubius*): dose-dependent expulsion of adult worms. Parasitology **98**: 115–124 (1989).

95. Rosenstreich, D.L., Weinblatt, A.C. and O' Brien, A.D. Genetic control of resistance to infection in mice. C.R.C. Crit. Rev. Immunol. **3**: 263–330 (1982).
96. Rothwell, T.L.W. Immune expulsion of parasitic nematodes from the alimentary tract. Int. J. Parasitol. **19**: 139–168 (1989).
97. Rousseaux-Prevost, R., Capron, M., Basin, H. and Capron, A. IgE in experimental schistosomiasis. II Quantitative determination of specific IgE antibodies against *S. mansoni*: a follow-up study of two strains of infected rats. Correlation with protective immunity. J. Immunol. **35**: 33–39 (1978).
98. Sher, A., Hieny, S. and James, S. Mechanisms of protective immunity against *S. mansoni* infection in mice vaccinated with irradiated cercariae. VI. Influence of the major histocompatibility complex. Parasite Immunol. **6**: 319–328 (1984).
99. Sitepu, P. and Dobson, C. Genetic control of resistance to infection with *Nematospiroides dubius* in mice: selection of high and low immune responder populations of mice. Parasitology **85**: 73–84 (1982).
100. Sitepu, P., Dobson, C. and Brindley, P.J. Immunization and immunosuppression in mice reared for high or low immune responsiveness against *Nematospiroides dubius*. Int. J. Parasitol. **15**: 277–281 (1985).
101. Sitepu, P., Dobson. C. and Brindley, P.J. *Nematospiroides dubius*: multiple infections in mice bred for immune responsiveness. J. Parasitol. **72**: 559–564 (1986).
102. Skamene, E. (ed). *Genetic Control of Host Resistance to Infection and Malignancy* (Progress in Leukocyte Biology Vol. 3). Alan R. Liss: New York (1985).
103. Smillie, W.G. and Augustine, D.L.: Intensity of hookworm infestation in Alabama. Its relationship to residence, occupation, age, sex and race. J. Am. Med. Assoc. **85**: 1958–1963 (1925).
104. Stirewalt, M.A., Shepperson, J.R. and Lincicome, D.R. Comparison of penetration and maturation of *Schistosoma mansoni* in four strains of mice. Parasitology **55**: 227–235 (1965).
105. Storey, N., Behnke, J.M. and Wakelin, D. Immunity to *Dipetalonema viteae* (Filarioidea) infections in resistant and susceptible mice. Acta Trop. **44**: 43–54 (1987).
106. Storey, N., Behnke, J.M. and Wakelin, D. *Acanthocheilonema viteae (Dipetalonema viteae)* in mice: attempts to correct the low-responder phenotype of the BALB/c host. Int. J. Parasitol. **19**: 723–727 (1989).
107. Storey, N., Wakelin, D. and Behnke, J.M. The genetic control of host responses to *Dipetalonema viteae* (Filarioidea) infections in mice. Parasite Immunol. **7**: 349–358 (1985).
108. Thompson, J.P., Crandall, R.B. and Crandall, C.A. Microfilaraemia and antibody responses in CBA/H and CBA/N mice following injection of microfilaraemia of *Brugia malayi*. J. Parasitol. **67**: 728 (1981).
109. Thompson, J.P., Crandall, R.B., Crandall, C.A. and Neilson, J.T. Clearance of microfilariae of *Dipetalonema viteae* in CBA/n and CBA/H mice. J. Parasitol. **65**: 966–969 (1974).
110. Tomlinson, L.A.T., Christie, J.F., Fraser, E.M., McLaughlin, D., McIntosh, A.E. and Kennedy, M.W. MHC restriction of the antibody repertoire to secretory antigens, and a major allergen, of the nematode parasite *Ascaris*. J. Immunol. **143**: 2349–2356 (1989).
111. Touhy, M., Lammas, D.A., Wakelin, D., Huntley, J.F., Newlands, G.F.J. and Miller, H.R.P. Functional correlations between mucosal mast cell activity and immunity to *Trichinella spiralis* in high and low responder mice. Parasite immunol. **12**: 675–685 (1990).
112. Urquhart, G.M. Application of immunity in the control of parasite disease. Vet. Parasitol. **6**: 217–239 (1980).
113. Wahid, F.N., Robinson, M. and Behnke, J.M. Immunological relationships with *Heligmosomoides polygyrus (Nematospiroides dubius)*: expulsion of adult worms from fast responder syngeneic and hybrid strains of mice. Parasitology **98**: 459–469 (1989).
114. Wakelin, D. Genetic control of immunity to parasites. *Parasitology* **71**: xxv (1975).
115. Wakelin, D. Genetic control of immune responses to *Trichuris muris* in inbred and random-bred strains of mice. Parasitology **71**: 51–60 (1975).
116. Wakelin, D. Genetic control of immune responses to parasites: selection for responsiveness and non-responsiveness to *Trichuris muris* in random-bred mice. Parasitology **71**: 377–384 (1975).
117. Wakelin, D. Genetic control of susceptibility and resistance to parasitic infection. Adv. Parasitol, **16**: 219–308 (1978).
118. Wakelin, D. Genetic control of immunity to parasites. Infection with *Trichinella spiralis* in inbred and congenic mice showing rapid and slow responses to infection. Parasite Immunol. **2**: 85–98 (1980).

119. Wakelin, D. *Immunity to Parasites. How Animals Control Parasite Infections*. Edward Arnold: London (1984).
120. Wakelin, D. Genetics, immunity and parasite survival. In: *Ecology and Genetics of Host-Parasite Interactions* (D. Rollinson and R.M. Anderson eds), pp. 39–54. Academic Press: London & New York (1985).
121. Wakelin, D. Genetically determined variation in immune response to filarial infections. J. Commun. Dis. **18**: 234–242 (1986).
122. Wakelin, D. Immunity to helminths. EOS. Riv. Immunol. Immunofarmacol. **6**: 35–46 (1986).
123. Wakelin, D. Helminth infections. In: *Genetics of Resistance to Bacterial and Parasitic Infection* (D. Wakelin and J.M. Blackwell eds), pp. 153–224. Taylor & Francis: London (1988).
124. Wakelin, D. and Denham, D.A. The immune response. In: *Trichinella and Trichinosis* (W.C. Campbell ed), pp. 265–308. Plenum Press: New York (1983).
125. Wakelin, D. and Donachie, A.M. Genetic control of eosinophilia: mouse strain variation in response to antigens of parasitic origin. Clin. Exp. Immunol. **51**: 239–246 (1983).
126. Wakelin, D. and Grencis, R.K. Immunological responses to intestinal parasite infection. In: *Immunology of the Gastrointestinal Tract*. (K. Miller and S. Nicklin eds), pp. 1–23. C.R.C. Press Inc.: Boca Raton, Florida (1987).
127. Wakelin, D., Mitchell, L.A., Donachie, A.M. and Grencis, R.K.: Genetic control of immunity to *Trichinella spiralis* in mice. Response of rapid- and slow-responder strains to immunization with parasite antigens. Parasite Immunol. **8**: 159–170 (1986).
128. Wassom, D.L., Brooks,, B.O., Badisch, J.G. and David, C.S. A gene mapping between the S and D regions of the H-2 complex influences resistance to *Trichinella spiralis* infections of mice. J. Immunogenet. **10**: 371–378 (1983).
129. Wassom, D.L., Brooks, B.O., Cypess, R.H.P. and David, C.S. A survey of susceptibility to infection with *Trichinella spiralis* of inbred mouse strains sharing common H-2 alleles but different genetic backgrounds. J. Parasitol. **69**: 1033–1037 (1983).
130. Wassom, D.L., David, C.S. and Gleich, G.J. Genes within the major histocompatibility complex influence susceptibility to *Trichinella spiralis* in the mouse. Immunogenetics **9**: 491–496 (1979).
131. Wassom, D.L., DeWitt, C.W. and Grundmann, A.W. immunity to *Hymenolepis citelli* by *Peromyscus maniculatus*: genetic control and ecological implications. J. Parasitol. **60**: 47–52 (1974).
132. Wassom, D.L., Dick, T.A., Arnason, N., Strickland, D. and Grundmann, A.W. Host genetics; a key factor in regulating the disribution of parasites in natural host populations. J. Parasitol. **72**: 334–337 (1986).
133. Wassom, D.L. and Kelly, E.A. The role of the major histocompatibility complex in resistance to parasite infection. C.R.C. Crit. Rev. Immunol. **10**: 31–52 (1990).
134. Wassom, D.L., Krco, C.J. and David, C.S. I-E expression and susceptibility to parasite infection. Inmmunol. Today. **8**: 39–43 (1987).
135. Wassom, D.L., Wakelin, D., Brooks, B.O., Krco, C.J. and David. C.S. Genetic control of immunity to *Trichinella spiralis* infections of mice. Hypothesis to explain the role of H-2 genes in primary and challenge infections. Immunology **51**: 625–631 (1984).
136. Webster, L.T. Inheritance of resistance of mice to enteric, bacterial and neurotropic virus infections. J. Exp. Med. **65**: 261–280 (1937).
137. Whitlock, J.H. A study of inheritance of resistance to trichostrongylidosis in sheep. Cornell Vet. **45**: 422–439 (1955).
138. Wnitlock, J.H. The inheritance of resistance to trichostrongylidosis in sheep. I. Demonstration of the validty of the phenomena. Cornell Vet. **48**. 127–133 (1958).
139. Whitlock, J.H. and Madsen, H. The inheritance of resistance to trichostrongylidosis in sheep. II Observations on the genetic mechanism in trichostrongylidosis. Cornell Vet. **48**: 134–145 (1958).
140. Wolowczuk, I., Auriault, C., Gras-Masse, H., Vendeville, C., Balloul, J.M., Tartar, A. and Capron, A. Protective immunity in mice vaccinated with the *Schistosoma mansoni* P-28–1 antigen. J. Immunol. **142**: 1342–1350 (1989).
141. Zhu, D. and Bell, R.G. IL-2 production, IL-2 receptor expression and IL-2 responsiveness of spleen and mesenteric lymph node cells from inbred mice infected with *Trichinella spiralis*. J. Immunol. **142**: 3262–3267 (1989).

INDEX

Bold type indicates pages with illustrations or tables.
Note that 'antibodies', 'antigens', 'diagnosis', 'treatment' etc. are to be found umbrella entries under 'immunology', 'immunodiagnosis' and 'chemotherapy'.

MIX
Papier aus verantwortungsvollen Quellen
Paper from responsible sources
FSC® C105338

If you have any concerns about our products,
you can contact us on
ProductSafety@springernature.com

In case Publisher is established outside the EU,
the EU authorized representative is:
Springer Nature Customer Service Center GmbH
Europaplatz 3, 69115 Heidelberg, Germany

Printed by Libri Plureos GmbH
in Hamburg, Germany